PLANAR MULTIBODY DYNAMICS

Formulation, Programming, and Applications

PLANAR MULTIBODY DYNAMICS

Formulation, Programming, and Applications

Parviz E. Nikravesh

CRC Press is an imprint of the
Taylor & Francis Group, an **informa** business

CRC Press
Taylor & Francis Group
6000 Broken Sound Parkway NW, Suite 300
Boca Raton, FL 33487-2742

CRC Press is an imprint of Taylor & Francis Group, an Informa business

Printed in the United States of America on acid-free paper
10 9 8 7 6 5 4 3 2 1

International Standard Book Number-13: 978-1-4200-4572-7 (Hardcover)

Library of Congress Cataloging-in-Publication Data

Nikravesh, Parviz E., 1946-
Planar multibody dynamics : formulation, programming, and applications / Parviz E. Nikravesh.
p. cm.
Includes bibliographical references and index.
ISBN 978-1-4200-4572-7 (alk. paper)
1. Machinery, Kinematics of. 2. Machinery, Dynamics of. I. Title.

TJ175.N527 2008
621.8'11--dc22 2007022018

Visit the Taylor & Francis Web site at
http://www.taylorandfrancis.com

and the CRC Press Web site at
http://www.crcpress.com

Dedication

This book is dedicated to the memory of a remarkable man,
a true engineer, and a great humanitarian,
Mohandes Saleh Khorramabadi,
my mentor

Table of Contents

Chapter 15

Appendix

Preface

The basic premise of this textbook is to introduce fundamental theories, computational methods, and program development for analyzing simple to complex planar mechanical systems. Such a combination of theory, computational methods, and programming did not exist in any mechanical engineering curricula three decades ago, but since then it has become a standard course in most programs. Graduate-level courses on these combined subjects began to be offered in different institutions in the late 1970s. Several textbooks and monograms were published soon after, all at the graduate level. Eventually, selected chapters of these books were used for courses that were offered as technical electives at the senior undergraduate level. More recently, these topics are being introduced at the junior level, either replacing or complementing some of the more traditional courses.

It was not the author's intention for this book to become a simpler version of a graduate-level research monogram; this book was written specifically for the undergraduate students and practicing engineers. Therefore, the prerequisite for this book is the first course on dynamics, and the corequisites are courses on numerical methods and programming with MATLAB®.

The book covers a variety of methods for formulating the kinematic and dynamic equations of motion. The methodologies for deriving the equations of motion, even for complex systems, are all based on Newton's laws of motion. As much as possible, the use of graduate-level principles of mechanics has been avoided.

Due to its popularity and ease of use, MATLAB was chosen as the programming language for this textbook. Because the reader may not be a skilled programmer, the examples and exercises provide a tutorial for learning MATLAB. The examples begin with basic commands before introducing the reader to more advanced programming techniques. The routines that are developed in each chapter are eventually put together to form complete programs for different types of analyses.

The analytical and computational subjects are organized around practical application examples. Several examples are repeatedly used in various chapters to provide the reader with a basis for comparison between different formulations. Other application examples are used in problem assignments and in the more extensive projects.

This textbook is organized for both classroom teaching and self-study. Following the introductory chapter, Chapter 2 provides an overview of the notation associated with the fundamentals of matrix algebra. The reader is strongly encouraged to become familiar with the notation and to comprehend the concepts thoroughly. Chapters 3 and 4 provide the necessary fundamentals of particle and rigid-body kinematics and dynamics. An understanding of these fundamentals is essential to learning the remainder of the textbook. Chapters 5 and 6 discuss the point-coordinate formulation, Chapters 7 and 8 concentrate on the body-coordinate formulation, and Chapters 9 and 10 cover the joint-coordinate formulation. Different types of analyses are discussed in Chapters 11 to14. These include kinematic, inverse dynamic, forward dynamic, static, and static equilibrium analyses. Finally, Chapter 15 presents several multibody systems with practical application to be considered as projects.

An instructor, or a reader, may choose to cover the book in its entirety or to consider only a selected number of chapters. For instance, if the interest is in kinematic analysis and only one method of formulation such as the point-coordinates, Chapters 1, 2, 3, 5, and 11 suffice. With these few chapters, a wide range of planar mechanical systems can be analyzed. Addition of Chapters 4, 6, 12, and 13 to the previous list extends the capabilities to dynamic analysis and, therefore, more variety of problems can be studied. Another possibility would be to substitute Chapters 7 and 8 on the body-coordinate formulation for Chapters 5 and 6. However, if one is interested in the joint-coordinate formulation, Chapters 7 and 8 must be studied before Chapters 9 and 10. This textbook can also be used as a graduate-level introductory course on multibody dynamics. For such a course, all of the chapters could be covered during one semester.

Acknowledgment: Since my first book on the subject of multibody dynamics was published almost two decades ago, I have received many compliments and acknowledgments from colleagues and readers, many of whom I have not had the privilege of ever meeting. I am humbled by their kind words, and because of their encouraging remarks I decided to write another textbook.

Over the years, I have received many useful suggestions and ideas from students who have taken my courses. It is their appreciation that makes writing a textbook worthwhile. I am grateful to all of them. I am also thankful to my graduate research assistants over the years, in particular to Yi-shih Lin, who has assisted me in many ways during the course of this project.

Website: A dedicated website for this textbook provides the computer programs (MATLAB M-file) that are listed in the book for download. Since the listed programs may contain programming, logical, or typesetting errors, the posted programs will be revised as the errors are found. Additional complimentary programs and other relevant materials will be posted on the website as they become available. The reader is encouraged to visit the website on a regular basis. The materials can be accessed at http://www.crcpress.com/e_products/downloads/default.asp.

Author

Parviz Nikravesh has been a researcher and an educator for more than thirty years. He is currently a professor in the Department of Aerospace and Mechanical Engineering at the University of Arizona. He is the author of a large number of journal publications in theoretical and computational dynamics. His first book, titled *Computer-Aided Analysis of Mechanical Systems*, has been translated from English to several other languages and is considered to be the first textbook on the subject of multibody dynamics.

Professor Nikravesh is a member of ASME (American Society of Mechanical Engineers) and SAE (Society of Automotive Engineers). He has served on the editorial board of the journal *Multibody System Dynamics* since its conception. He has received many awards for his contributions to the field of computational dynamics including an honorary doctorate degree.

1 Introduction

The major goal of the engineering profession is to design and manufacture marketable products of high quality. Today's industries are utilizing computers in every phase of the design and manufacture of their products. The process of design and manufacture, beginning with an idea and ending with a final product, is a closed-loop process. The design process requires a thorough understanding and ability to analyze the product. Computer-aided analysis allows an engineer to simulate the behavior of a product. Based on the analysis results, the product design can be optimized prior to actual production.

To simulate the behavior of a product, we must know the individual components that make up the product. A product may contain mechanical, electrical, or other components. If the mechanical components are allowed to move relative to one another, the product is called a multibody system. The interconnection between various components, or bodies, can be through kinematic joints, springs, dampers, simple contact, or other elements. Bodies of a mechanical system are in general deformable. But in most cases, they can be assumed rigid (nondeformable) due to their negligible deformation. The behavior (e.g., the motion) of a multibody system can be analyzed by pencil-and-paper (classical methods) only if the system is extremely simple and simplifying assumptions are made. Even for simple systems it may not be feasible to perform an analysis without a computer. This is definitely true for realistically complex multibody systems. Therefore, it is the objective of this textbook to present computational analysis techniques that can be applied systematically to systems composed of nondeformable bodies undergoing large planar motion regardless of their complexity.

1.1 MULTIBODY MECHANICAL SYSTEMS

Multibody mechanical systems range from the very simple to the very complex. A pendulum and a slider-crank mechanism are examples of simple systems, whereas the suspension and steering systems of an automobile, a bicycle, and an exercise machine are examples of more complex systems. The bodies of some multibody systems, such as a pendulum, a slider-crank, or a stationary exercise bicycle, can only move in parallel planes. Such systems are called *planar* or two-dimensional. Systems whose bodies do not move in parallel planes are called *spatial* or three-dimensional. For example, a robotic device that is capable of operating in all three directions. The motion of some spatial systems, if projected onto a plane, could be approximated as planar.

Most common components in multibody systems are bodies (also referred to as links), kinematic joints, and compliance elements. Examples of kinematic joints are pins, sliders, gears, and cam-followers. Typical compliance elements are springs and dampers. Kinematic joints and compliance elements provide the connectivity between the bodies of a multibody system. A multibody system may contain non-mechanical components such as an electronic controller. Analyzing the dynamics of such a system must include an analytical model of the controller as well. The bodies of most multibody systems can be considered as nondeformable. However in some applications, deformation of a link may not be negligible and should be considered in the analysis.

Studying a multibody system involves two fundamental steps: *modeling* and *analysis*. Modeling or *formulation* is the process of constructing the necessary equations that, if solved, would reveal the behavior of a system. In this textbook we will consider several methods of formulation, each having its own advantages and disadvantages. Depending on the application of a multibody system, different types of analyses can be considered.

1.2 TYPES OF ANALYSES

There are two different aspects to the study of a mechanical system: *analysis* and *design*. When a mechanical system is acted upon by a given excitation, for example an external force, the system exhibits a certain response. The process that allows an engineer to study the response of an already existing system to a known excitation is called *analysis*. This requires a complete knowledge of the physical characteristics of the mechanical system, such as material composition, shape, and arrangement of parts. Conversely, the process of determining what physical characteristics are necessary for a mechanical system to perform a prescribed task is called *design* or *synthesis*. The design process requires the application of scientific techniques along with the engineering judgment. The scientific techniques in the design process, such as *analysis* and *optimization*, are merely tools to be used by the engineer. Although these methodologies can be applied in a systematic manner, the overall design process hinges on the judgment of the designer. Because the scientific aspect of the design process requires analysis techniques as tools, it is important to learn about methods of analysis prior to design.

The branch of analysis that studies motion, time, and force is called *mechanics*. It consists of two parts—*statics* and *dynamics*. In statics we analyze stationary systems—systems in which time is not a factor. Dynamics, on the other hand, deals with systems that are nonstationary—systems that change their positions with respect to time. Dynamics is divided into two disciplines—*kinematics* and *kinetics*. Kinematics is the study of motion regardless of forces that produce the motion. More explicitly, kinematics is the study of displacement, velocity, and acceleration. Kinetics, on the other hand, is the study of motion and its relationship with the forces that produce that motion. It is, however, very common to refer to *kinetic analysis* as *dynamic analysis*, because kinetic analysis must be based on the knowledge of the kinematics of a system as well. Therefore, in this textbook we will use the term *dynamic* instead of *kinetic*. We will discuss several computational methods of analyses—*kinematic analysis*, *inverse dynamics analysis*, *forward dynamic analysis*, *static* and *static-equilibrium analysis*.

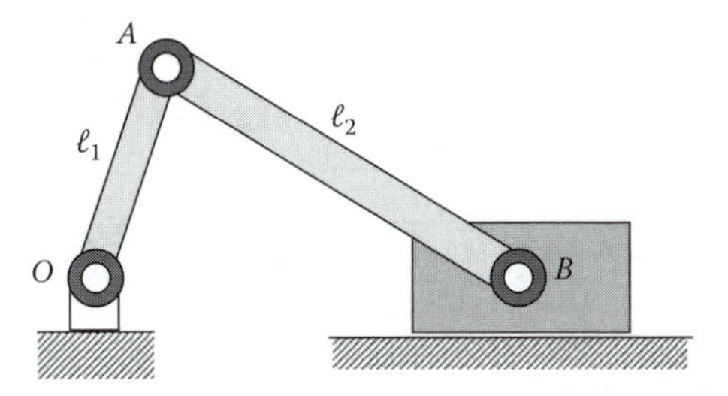

FIGURE 1.1 A slider-crank mechanism.

1.3 METHODS OF FORMULATION

Classical methods of analysis in mechanics have relied upon graphical and often quite complex techniques. These techniques are mostly based on geometrical interpretations of the system under consideration. Furthermore, these techniques have been developed for hand derivation and solution of the equations. Some of these techniques can be implemented in computer programs. However, new solution techniques have been developed to take full advantage of the capabilities of computational methods. In this section, we provide an overview of some of the formulation methods that are discussed in this textbook through several simple examples.

We consider the slider-crank mechanism shown in Figure 1.1. The lengths of the crank and connecting rod are given as $\ell_1 = 2.0$ and $\ell_2 = 4.0$ (any unit system). The crank, link *OA*, rotates with a constant angular velocity of $\omega = 1.5$ rad/sec in the counterclockwise direction. The objective is to determine the position and the velocity of the slider at the configuration where the crank makes a 30° angle with a horizontal axis. Because we are not interested in the forces that cause or are the result of this motion, the process is purely a kinematic analysis.

The first method of analysis that we consider is the classical graphical technique. For position analysis, the triangle *OAB* is drawn as accurately as possible because the lengths *OA* and *AB* and the angle of the crank are given. As shown in Figure 1.2(a) we draw a horizontal line first to establish the axis of the slider. Then we construct a line with a length of 2.0 units, making a 30° angle with the horizontal line. This establishes the position of the crank and point *A*. We then draw a circle with a radius of 4.0 units centered at *A*. This circle intersects the horizontal line at two points, *B* and *B'*, which indicates that there are two possible solutions—the center of the slider block could be either at *B* or at *B'*. Although both solutions are feasible, based on the original diagram, we choose point *B* which represents the position of the slider. We directly measure the length of *OB* on the diagram to be 5.6 units. If necessary, we can also measure the angle of *AB*. This completes the graphical position analysis for this mechanism in the specified configuration.

To perform velocity analysis, for the given orientation of the slider-crank, a *velocity vector polygon* can be constructed as shown in Figure 1.2(b). We first establish a point of zero absolute velocities, such as point *O*. Because we know the length and the angular velocity of link *OA*, we can determine the magnitude ($2.0 \times 1.5 = 3.0$ units) and the direction of the velocity of *A*, $\mathbf{v}^A$. We draw this

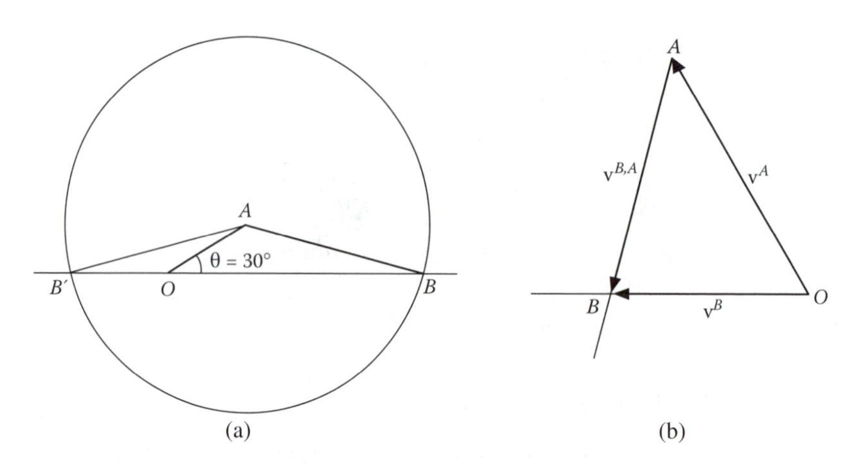

FIGURE 1.2 Graphical methods for position and velocity analyses.

vector perpendicular to link *OA* as the first vector of the velocity polygon. We also know that the velocity of the slider, $\mathbf{v}^B$, must be along the axis of the slider. We draw a line parallel to the axis of the sliding joint starting at *O*. Point *B* must be on this line. We further know that the velocity of point *B* relative to point *A*, $\mathbf{v}^{B,A}$, must be perpendicular to the link *AB*. Therefore we draw a line perpendicular to link *AB* on the polygon starting at *A*. This line intersects the horizontal line at point *B*. This completes the velocity polygon, which provides the magnitude and the direction of the velocity of the slider, $\mathbf{v}^B$. We measure its magnitude to be 2.2 units and the direction to the left. If necessary, we can determine the angular velocity of the connecting rod based on the measured magnitude of $\mathbf{v}^{B,A}$ and the length of the link *AB*.

The graphical process provides a visual understanding of the kinematics of a system. However, the process is not accurate and it could become impractical if we are required to repeat the process for many different configurations of a system. Extending the graphical approach from kinematics to dynamic analysis may only be possible for very simple systems. But even then, the process is inaccurate and time consuming. The use of graphical methods for analysis of multibody systems is not a feasible one—we need to rely on analytical methods.

A classical analytical formulation for kinematics of planar mechanisms is known as the vector-loop method. For the slider-crank mechanism, this method requires defining variables that describe the position of each moving link, either relative to another moving link or relative to the ground. As shown in Figure 1.3, the variables are two angles, θ_1 and θ_2, and a length *d*. The angles describe the orientation of the two links, and *d* defines the distance of point *B* on the slider from the pin at *O*. A vector-loop for the closed triangle *OABO* yields the following equations:

$$\begin{aligned} 2\cos\theta_1 - 4\cos\theta_2 - d &= 0 \\ 2\sin\theta_1 - 4\sin\theta_2 &= 0 \end{aligned} \tag{1.1}$$

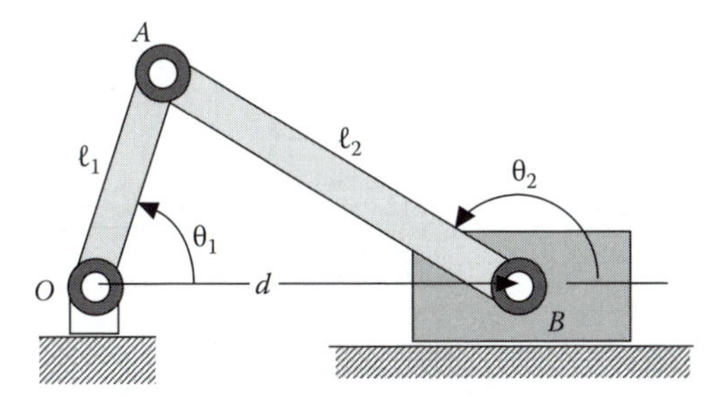

FIGURE 1.3 Variables describing the orientation of the links of a slider-crank.

For 30° crank angle, we rewrite the equations as

$$2\cos 30^\circ - 4\cos\theta_2 - d = 0$$

$$2\sin 30^\circ - 4\sin\theta_2 = 0$$

These two equations can be solved for the two unknown variables, θ_2 and d. The method of solution for such nonlinear algebraic equations will be discussed later in this textbook. At this point we are only interested in the concept and not in the details. Such a solution provides $\theta_2 = 165.5247^\circ$ and $d = 5.6057$. These values are much more accurate than those obtained graphically.

For velocity analysis, the time derivative of Eq. (1.1) provides the velocity equations as:

$$\begin{aligned} -\ell_1 \sin\theta_1 \dot{\theta}_1 + \ell_2 \sin\theta_2 \dot{\theta}_2 - \dot{d} &= 0 \\ \ell_1 \cos\theta_1 \dot{\theta}_1 - \ell_2 \cos\theta_2 \dot{\theta}_2 &= 0 \end{aligned} \tag{1.2}$$

Because in this configuration we already know the values of θ_1 and θ_2, for the given angular velocity of the crank, $\dot{\theta}_1 = \omega = 1.5$, the velocity equations are rewritten as

$$-2\times 1.5\sin 30^\circ + 4\sin 165.5247^\circ\,\dot{\theta}_2 - \dot{d} = 0$$

$$2\times 1.5\cos 30^\circ - 4\cos 165.5247^\circ\,\dot{\theta}_2 = 0$$

The solution to these equations yields $\dot{d} = -2.1707$ (compare this value against the value we obtained graphically) and $\dot{\theta}_2 = -0.6708$ rad/sec. The negative value of $\dot{d}$ indicates that the slider is moving to the left. It should be obvious that the computational process can be repeated easily for different values of the crank angle.

The vector-loop formulation is a simple process for kinematic analysis of planar systems. However, extending this formulation to dynamic analysis, using classical methods, requires knowledge of more advanced principles. In this textbook, in addition to the vector-loop method, several other methods of formulation are discussed. To form a

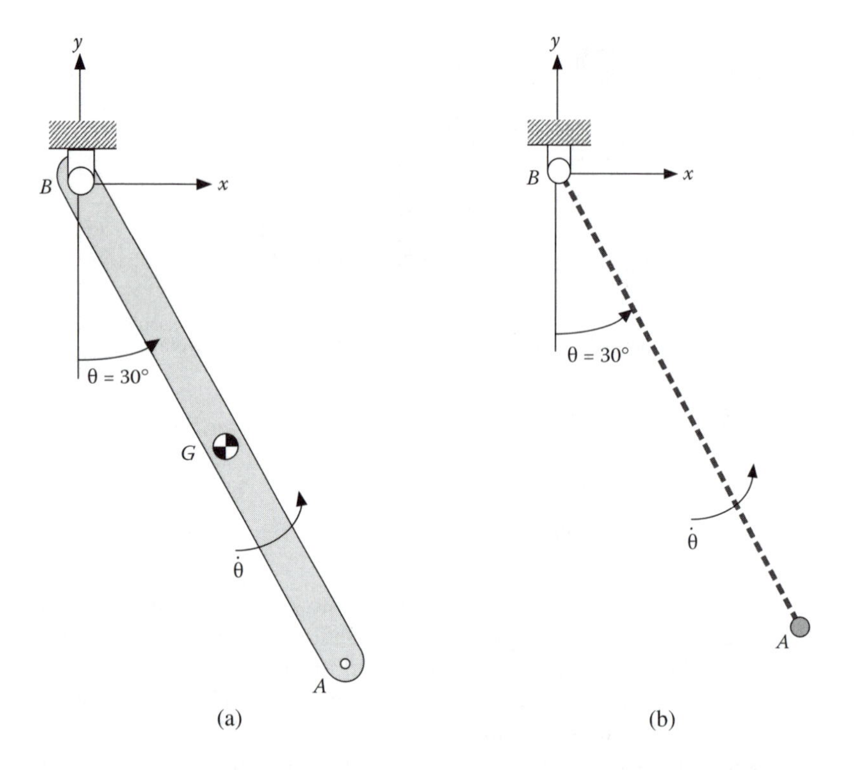

FIGURE 1.4 A single pendulum.

general idea about these formulations, in particular when the formulation is extended to dynamic analysis, we use an example which is much simpler than the slider-crank.

The simplest multibody system to consider for our discussion is the single pendulum shown in Figure 1.4(a). The pendulum is composed of a slender rod *AB*, pinned to the ground at *B*. The objective is to derive the dynamic equations of motion for this pendulum and then solve the equations for the accelerations, assuming that the position and velocities are known. We consider the following data for the pendulum (SI units): The length of the rod is $\ell = 4$, the mass is $m = 6$ and uniformly distributed (the mass center is at point *G*), and the moment of inertia about an axis through the mass center and perpendicular to the rod is $J = \frac{1}{12} m\ell^2 = 8$. Because gravity is the only external force that acts on the system, we consider the gravitational constant $g = 9.81$. Assume that we want to determine the acceleration of the pendulum when it makes an angle of $\theta = 30°$ with the vertical axis and its angular velocity is $\dot{\theta} = 1.0$ rad/sec. Attaching a Cartesian reference frame *x-y* at *O*, the given values of θ and $\dot{\theta}$ can be used to compute the coordinates and velocity components of the mass center, point *G*, and the end point *A*:

$$x^G = 1.0,\ y^G = -\sqrt{3},\ \dot{x}^G = \sqrt{3},\ \dot{y}^G = 1.0,\ x^A = 2.0,\ y^A = -2\sqrt{3},\ \dot{x}^A = 2\sqrt{3},\ \dot{y}^A = 2.0$$

Some of these values will be used in the following formulations.

The first formulation that we discuss here is known as the *joint-coordinate* method. This method can yield the same kinematic formulation as the classical vector-loop method when applied to mechanisms. For the dynamics of the single pendulum, this formulation yields a single equation for the rotational acceleration of the rod as

$$(\tfrac{1}{3}m\ell^2)\ddot{\theta} + \tfrac{1}{2}(m)g\ell\sin\theta = 0 \tag{1.3}$$

This equation represents the rotation of the rod about point B. The coefficient of the rotational acceleration (shown within parentheses) is the moment of inertia of the rod about an axis through B and perpendicular to the rod. We substitute the known values in Eq. (1.3) to find the rotational acceleration:

$$\ddot{\theta} = -1.8394$$

This value can be used to compute the Cartesian acceleration of G and A if needed.

The second formulation is the *body-coordinate* method. This method, unlike the joint-coordinate method, yields a large number of equations for the single pendulum:

$$\begin{aligned}
&(m)\ddot{x}^G - \lambda_1 = 0\\
&(m)\ddot{y}^G - \lambda_2 = -(m)g\\
&(\tfrac{1}{12}m\ell^2)\ddot{\theta} + \tfrac{1}{2}\ell\cos\theta\,\lambda_1 + \tfrac{1}{2}\ell\sin\theta\,\lambda_2 = 0\\
&\ddot{x}^G - \tfrac{1}{2}\ell\cos\theta\,\ddot{\theta} = -\tfrac{1}{2}\ell\sin\theta\,\dot{\theta}^2\\
&\ddot{y}^G - \tfrac{1}{2}\ell\sin\theta\,\ddot{\theta} = \tfrac{1}{2}\ell\cos\theta\,\dot{\theta}^2
\end{aligned} \tag{1.4}$$

These equations contain the Cartesian acceleration of the mass center, $\ddot{x}^G$ and $\ddot{y}^G$, and the rotational acceleration, $\ddot{\theta}$. The coefficients of these acceleration components (all shown within parentheses) are exactly the mass and the moment of inertia about the mass center. In addition, two other unknown quantities, λ_1 and λ_2, appear in these equations. These quantities, as we will learn in the upcoming chapters, are directly related to the reaction force that acts at the pin joint B. Substituting the known values in Eq. (1.4) and solving the equations for $\ddot{x}^G$, $\ddot{y}^G$, $\ddot{\theta}$, λ_1, and λ_2 yield:

$$\ddot{x}^G = -4.1859,\ \ddot{y}^G = -0.1073,\ \ddot{\theta} = -1.8394,\ \lambda_1 = -25.1153,\ \lambda_2 = 58.2161$$

The computed value for the angular acceleration is exactly the same as that computed from Eq. (1.3).

The third formulation, called the *point-coordinate* method, is based on a classical idea of representing a mechanical system as a collection of particles. For such representation, the mass of the bodies and the forces that act on them must be distributed among the defined particles. The mass distribution can be performed either as an approximate or as an exact representation. We show this formulation for both approximate and exact representations.

In the *approximated* point-coordinate formulation, the rod is replaced by two particles, one at A and the other at B as shown in Figure 1.4(b). Half of the mass of the rod, $\frac{1}{2}m$, is given to particle A and the other half to particle B. The two particles must keep a constant distant ℓ from each other. The gravity acts on each particle. Because particle B does not move, the equations of motion contain only contributions from particle A:

$$\begin{aligned} (\tfrac{1}{2}m)\ddot{x}^A - x^A\lambda &= 0 \\ (\tfrac{1}{2}m)\ddot{y}^A - y^A\lambda &= -(\tfrac{1}{2}m)g \\ x^A\ddot{x}^A + y^A\ddot{y}^A &= -(\dot{x}^A)^2 - (\dot{y}^A)^2 \end{aligned} \tag{1.5}$$

The mass of particle A, as the coefficient of $\ddot{x}^A$ and $\ddot{y}^A$, appears in the first two equations and it is exactly half the mass of the rod. The unknown quantity λ that appears in these equations is related to the reaction force. Substituting the known values in Eq. (1.5) and solving the equations yield:

$$\ddot{x}^A = -6.2479,\ \ddot{y}^A = 1.0116,\ \lambda = -9.3718$$

These values should be exactly twice the values that we obtained in the previous formulation for the acceleration of the mass center G, but they are not! In this example, in particular, the error is large but, in general, the error will not be significant in every example. The approximation in this formulation comes from the fact that we did not consider the moment of inertia of the rod when we distributed the mass to the two particles.

In the *exact* point-coordinate formulation, the mass is distributed in such a way that the complete inertial characteristics of the rod are preserved. The proper mass distribution yields the following equations:

$$\begin{aligned} (\tfrac{1}{3}m)\ddot{x}^A - x^A\lambda &= 0 \\ (\tfrac{1}{3}m)\ddot{y}^A - y^A\lambda &= -(\tfrac{1}{2}m)g \\ x^A\ddot{x}^A + y^A\ddot{y}^A &= -(\dot{x}^A)^2 - (\dot{y}^A)^2 \end{aligned} \tag{1.6}$$

Here the mass of the particle at A is reduced to $\frac{1}{3}m$ instead of $\frac{1}{2}m$. We may wonder how the inertial characteristics of the rod are preserved! We will discuss the reason in Chapter 6. Substituting the known quantities in Eq. (1.6) and solving for the unknowns yield:

$$\ddot{x}^A = -8.3718, \ \ddot{y}^A = -0.2146, \ \lambda = -8.3718$$

These values are exactly two times the acceleration of point G as expected.

The single pendulum example demonstrates clearly that a multibody system can be formulated in different ways. If the principles of mechanics are followed and no approximations are considered, all of the formulations should lead to the same results. In the upcoming chapters we will discuss the different analytical methods that were applied to the pendulum example.

Analytical methods are more systematic in implementation when compared to the graphical methods. A problem formulated analytically can be solved repeatedly for different values of input, especially when a computer program is developed to perform the computation. The results can be obtained accurately and the process can be performed efficiently. The usefulness of using a computer program becomes even more apparent when the mechanical system under consideration is complex. Regardless of the complexity of the equations of motion, it is always possible to find a numerical solution describing the response.

1.4 COMPUTER PROGRAMMING

Analytical methods for formulating equations of motion and numerical methods for solving them are the basis for developing computer programs to determine the response of a multibody system. This requires systematic techniques for formulating and generating the kinematic and dynamic equations of motion, numerical methods for solving them, and preferably a front-end graphical interface for communicating the input and output to the user. Such a computer program can be developed either as a *special-purpose* or as a *general-purpose* program.

A special-purpose program is a computer code that deals with only a particular multibody system, for example, a slider-crank mechanism. The equations of motion for that system are derived a priori and coded in the program. As input to the program, the user provides the data on the physical characteristics of the system. The numerical algorithms for solving the equations of motion can be fine-tuned for that particular form of equations to gain computational efficiency. The main disadvantage of a special-purpose program is its lack of flexibility for handling other types of systems.

In contrast to a special-purpose program, a general-purpose program can analyze different multibody systems without making any changes in the program. For example, the motion of a slider-crank mechanism under a given set of applied loads, the response of an automobile and its suspension system driven over a rough terrain, and the opening of the panels of a satellite antenna can all be simulated with the same general-purpose program. The input data to such a program must completely describe the mechanical system under consideration. The input must contain all the necessary information (e.g., number of bodies, connectivity between the bodies, joint types, applied forces, and geometric and inertial characteristics). The program generates the equations of motion for the requested type of analysis and solves the equations numerically. A general-purpose program may not be computationally as efficient as a special-purpose program, but it provides flexibility in its use.

The formulations and methodologies that are discussed in this textbook can be used to develop either special- or general-purpose programs for multibody systems. The choice of the programming language is up to us. We can develop programs in Fortran, C, C++, or any other languages. We can use high-performance languages that integrate programming, computation, and visualization such as Mathematica, Maple, or MATLAB®.

1.4.1 MATLAB®

In this textbook we exclusively use MATLAB due to its popularity and ease of use. It is assumed that the reader has some fundamental knowledge of MATLAB but may not be an expert programmer. Although, the MATLAB examples in the book start with some fundamental commands and exercises, the burden of learning to program with MATLAB is upon the reader—the main objective of this textbook is about learning analytical and computational methods in kinematics and dynamics and not programming.

Wherever possible, brief explanation is provided for a fundamental command or an operation that is used for the first time in this textbook. Because each MATLAB statement in a program refers to an already discussed formulation, it should be easy for the inexperienced programmer to figure out what a command means and how it operates. If the use of a command or a function is not clear, the reader should consult the `help` command in MATLAB to obtain the necessary explanation.

A MATLAB program is saved in one or more so-called *M-files*. Some of the M-files that are developed in this textbook are used in more than one program, in different chapters. Therefore it is important to establish a file organization scheme that would allow us to track a file conveniently and to make it easy to define the so-called *path*. Because three different formulations are discussed in this textbook, the corresponding M-files are organized in three different main folders as well. At the beginning of some of the chapters, the file organization related to that particular formulation is provided in a table. A typical file organization is shown in Table 1.1, where the icon represents a *folder* or a *directory*. Within a folder we may have other folders and/or M-files. The pointed arrow in front of a folder indicates whether the folder is open or closed—that is, whether we can see its contents or not. Overall, there are four main folders: one for the point-coordinate formulation, `Point_Coord`; one for the body-coordinate formulation, `Body_Coord`; one for the joint-coordinate formulation, `Joint_Coord`; and one for arbitrary exercises, `Exercises`. A number within parentheses follows the name of a folder or an M-file. This number denotes the chapter in which the folder or the M-file has been discussed. When we execute a program, the folder containing that program must be the *Current Directory* of MATLAB.

The M-files that are developed in some of the exercises, especially those in Chapters 2 to 4, can be kept in the folder `Exercises`. The organization of the files in this folder will not be discussed—it is up to the reader how to organize them. With a few exceptions, the M-files from these exercises will not be used in any of the programs in the other chapters. The M-files that are developed in Chapters 5 to14 must be organized in the other three main directories as recommended. An experienced

TABLE 1.1
A Typical File Organization for MATLAB®

Folder / File	Chapter
Textbook	
Exercises	
Point_Coord	(5) (6)
Body_Coord	
BC_formulation.m	(7)
BC_global.m	(7)
Basics	(3) (4)
BC_Basics	
BC_Phi_rev.m	(7)
BC_gamma_tran.m	(7)
...	
Joint_Coord	(9) (10)

MATLAB user should be able to reorganize the directory structure as he/she wishes. The present structure may not be the best way to organize our M-files. This organization is adopted here because of its simplicity and ease of establishing the `path` to access different M-files.

The programs that are presented in Chapters 5 to 14 are developed with two main objectives in mind: (1) to assist us in understanding a formulation or a methodology; and (2) to be flexible for easy adaptation from one formulation or analysis to another. The programs are not developed with computational efficiency as an objective. Although many issues related to computational efficiency will be pointed out throughout the book. When a subject is well understood, these programs could be revised to improve their efficiency.

1.5 APPLICATION EXAMPLES

Throughout this textbook several examples are used repeatedly to clarify concepts and formulations. While the examples reflect actual multibody systems with application, they are simple enough not to distract us from understanding a concept. Through these examples we will learn how to construct analytical models, formulate multibody systems using different formulations, and observe major and subtle differences between formulations.

In the following subsections five simple examples of planar multibody systems, with the necessary data, are presented. A short description is provided for each example. Some of the listed data are realistic but some are not. Some data have been scaled or changed completely to simplify the numerical values that appear in the formulations, or to avoid encountering numerical problems too early in our discussions. For example, if in a model a body with a small mass is connected to a stiff spring, the response may contain high frequencies. Numerical integration of the equations of motion for such a system may require employment of special algorithms. Because such issues will not be discussed until in the final chapters, the masses and stiffnesses are scaled in some cases to avoid such problems.

1.5.1 Double A-Arm Suspension

One of the most commonly used independent front and rear suspension systems in automobiles is the double A-arm, also known as double wishbone or short-long arm (SLA). As a rear suspension, it provides an up-down motion of the wheel-assembly with respect to the frame. As a front suspension, in addition to the up-down motion, it allows a steering degree of freedom. When this system is viewed from the front or the back, it exhibits planar motion as a four-bar mechanism, as depicted in Figure 1.5. The steering or tie rod does not have any effect on the planar motion of this system. The multibody system shown in Figure 1.5 is referred to as a quarter-car representation, which could be either the front or the rear view of a car. We assume that this is the frontal view of the left-front suspension.

For modeling purposes, the joints, the attachment points, and some of the mass centers are labeled in the schematic view shown. In this presentation, body (0) is the frame with its mass center shown as point G_0 (this is actually the mass center of the entire frame and not just the quarter of the frame). Bodies (1) and (3) are the lower and upper arms respectively, and body (2) is the wheel assembly with its mass center shown as G_2. The wheel/tire is also shown in scaled size in the background for reference. The coil-over-shock is the suspension spring-damper.

The individual multibody components of this system and the corresponding dimensions are shown in Figure 1.6. Approximated coordinates of the moving points

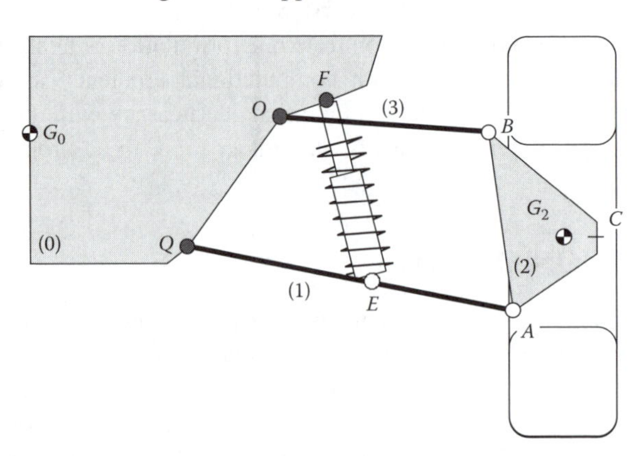

FIGURE 1.5 Planar frontal view of a double A-arm suspension.

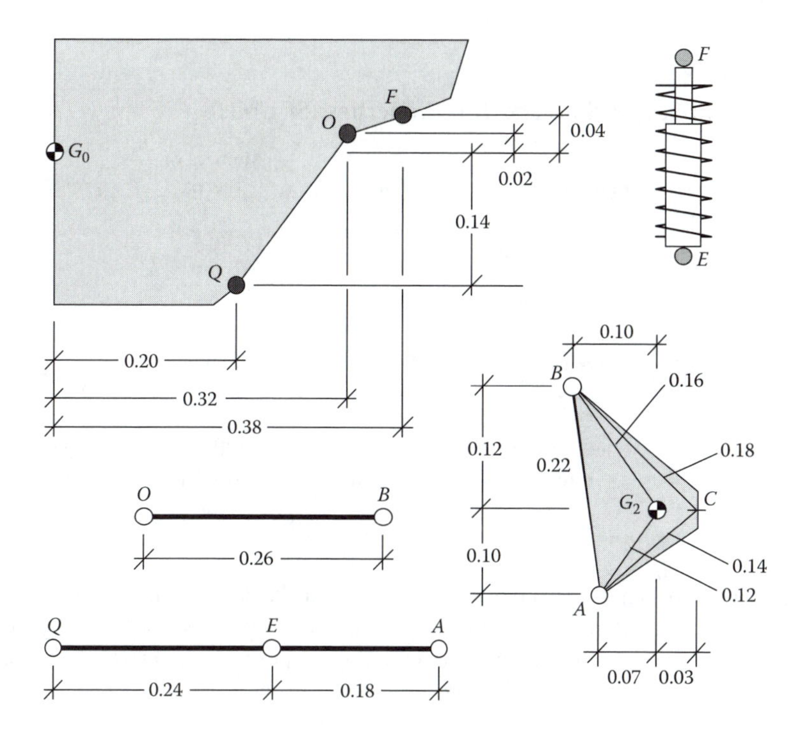

FIGURE 1.6 Geometrical dimensions in meters.

A, *B*, *C*, and *E* with respect to a nonmoving *x*-*y* frame attached to G_0 are listed in Table 1.2. The mass and roll moment of inertias, the characteristics of the spring-damper for the suspension, and tire spring-damper data in the radial direction (vertical) are provided in Table 1.3.

1.5.2 MacPherson Strut Suspension

Another common type of front suspension system in private cars is the MacPherson strut and link. The fundamental principles of this system are very similar to those of the double A-arm suspension. Variations of this basic strut and link can be found in different automobiles.

TABLE 1.2
Estimated Initial Positions with Respect to G_0 (SI Units)

Point	*x*	*y*
A	0.61	–0.22
B	0.58	0.0
C	0.72	–0.13
E	0.44	–0.19
G_2	0.68	–0.13

TABLE 1.3
Inertia, Suspension, and Tire Properties (SI Units)

Inertia	Body (Index)	Mass	Moment of Inertia	
	1/4 chassis (0)	340	325	
	(1)	2	0.5	
	(2)	30	2.5	
	(3)	1	0.5	
Suspension	**Free length**	**Stiffness**	**Damping**	
	0.23	91,600	1,433	
Tire	**Undeformed radius**	**Deformed radius at rest**	**Stiffness (radial)**	**Damping (radial)**
	0.35	0.34	58,770	2,500

When this system is viewed from the front (or the back), it exhibits planar motion as an inverted slider-crank mechanism. The steering rod does not have any effect on the planar motion of this system. For modeling purposes, the joints, the attachment points, and some of the mass centers are labeled in the schematic presentation of Figure 1.7. In this presentation, body (0) is the frame with its mass center shown

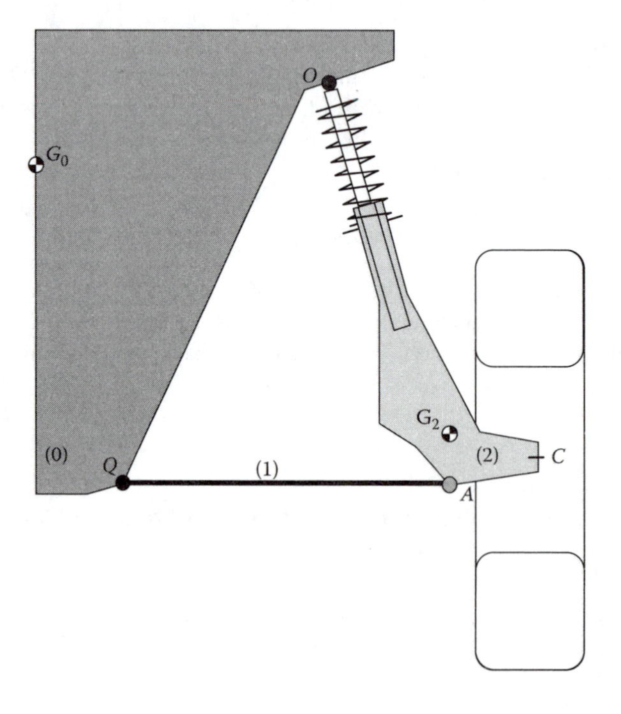

FIGURE 1.7 Planar view of the MacPherson suspension.

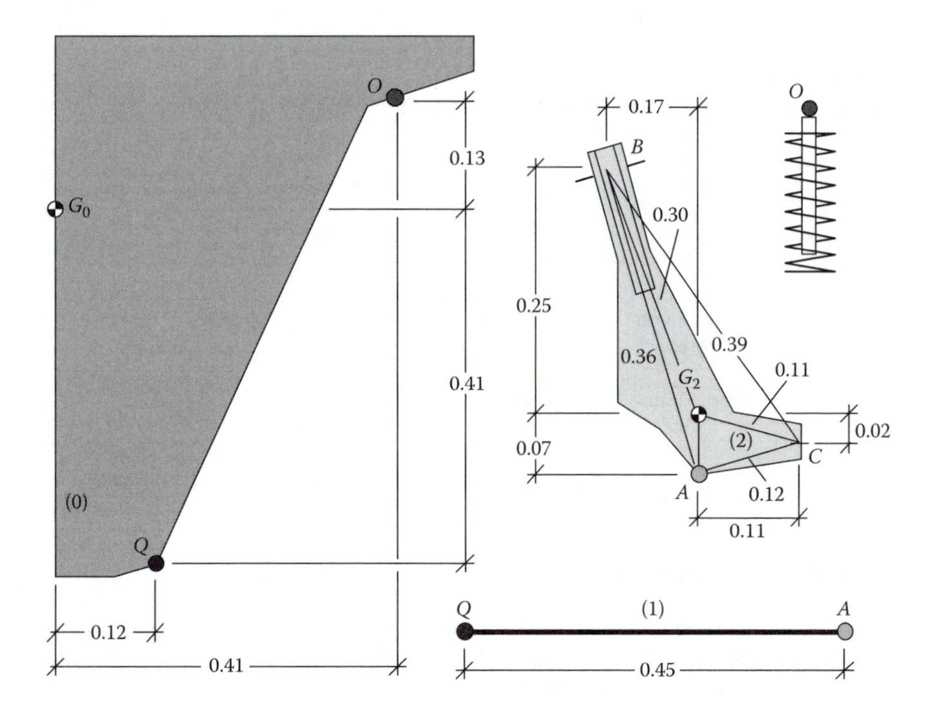

FIGURE 1.8 Geometric dimensions in meters.

as point G_0. Body (1) is the lower arm and body (2) is the wheel assembly with its mass center shown as G_2. The wheel/tire is also shown for reference. The coil-over-shock absorber also acts as a sliding joint (a cylindrical joint in 3D).

The dimensions of the links and the positions of the attachment points are given in Figure 1.8. Link numbers and the joint/point indices are also provided. For link (2) point *B* is defined on the strut axis at a convenient location as a reference point for some of the dimensions. Approximated values of the coordinates of the moving points *A*, *B*, and *C* with respect to a nonmoving *x*-*y* frame attached to G_0 are listed in Table 1.4. The mass and moment of inertia of the bodies, the characteristics of the suspension spring-damper, and the tire spring-damper data in the radial direction are listed in Table 1.5.

TABLE 1.4
Estimated Initial Positions with Respect to G_0 (SI Units)

Point	x	y
A	0.57	–0.41
B	0.45	–0.05
C	0.68	–0.38
G_2	0.55	–0.33

TABLE 1.5
Inertia, Suspension, and Tire Properties (SI Units)

Inertia	Body (Index)	Mass	Moment of Inertia	
	1/4 chassis (0)	300	100	
	(1)	2	0.5	
	(2)	20	2.5	
Suspension	**Free Length**	**Stiffness**	**Damping**	
	0.34	22,600	1,270	
Tire	**Undeformed Radius**	**Deformed Radius at Rest**	**Stiffness (Radial)**	**Damping (Radial)**
	0.30	0.29	200,000	2,000

1.5.3 Film-Advancer Mechanism

This simple mechanism can be found in most textbooks on analysis and design of mechanisms. The mechanism provides an intermittent motion to advance a filmstrip in a primitive type movie projector. A schematic presentation of this four-bar mechanism is shown in Figure 1.9 where the coupler point *C* periodically engages and disengages from the sprocket holes in the filmstrip as the crank rotates clockwise. If the crank is rotated at 1800 rpm (60 π rad/sec), the film is advanced 30 frames a second.

This linkage has three moving links and four pin joints. The dimensions of the links are given in Figure 1.10. This example will be used for kinematic analysis only; therefore there is no need to list its inertial data.

1.5.4 Web-Cutter

A crank-rocker four-bar mechanism has been adopted in the design of a web-cutter as shown in Figure 1.11. As the crank rotates, the cutting blades at *C* and *D* cut through a sheet moving from left to right. The velocity of the web at the time of cutting must be adjusted to match the average velocity of the two cutting points. It is assumed that the crank rotates with a constant angular velocity of 60 rpm CCW. The dimensions for the system are shown in Figure 1.12. Assume that the mass centers are at *Q* for link (1) and at *B* for links (2) and (3).

1.5.5 Variable-Length Pendulum

This is an open-chain system and is the simplest example we have presented so far. This example will be used to demonstrate a transformation process from the body-coordinate formulation to the joint-coordinate formulation. The pendulum is a two-degrees-of-freedom system consisting of two bodies. As shown

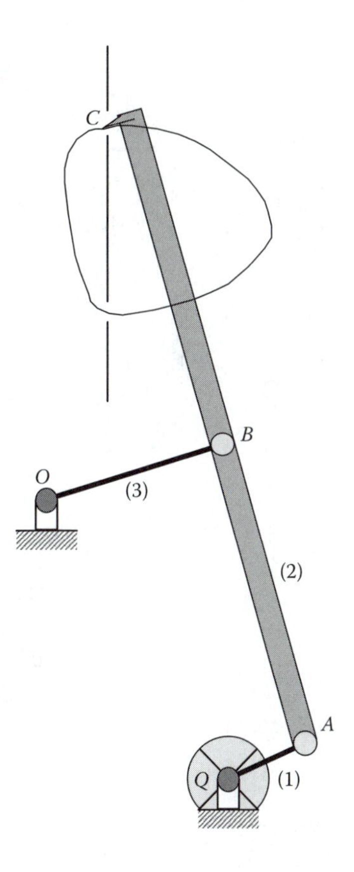

FIGURE 1.9 A film-advancer mechanism.

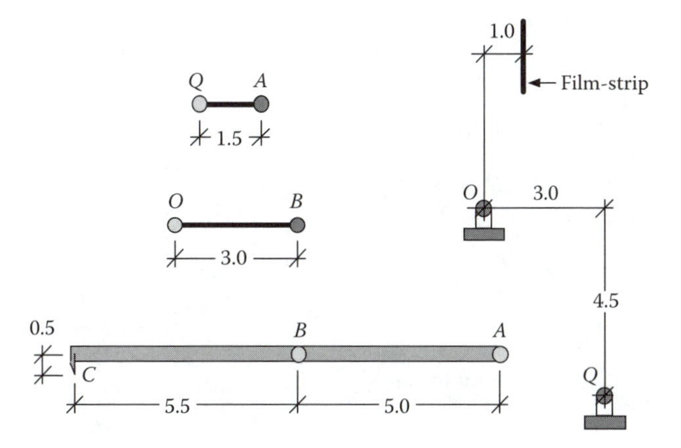

FIGURE 1.10 Geometrical dimensions of the film-advancer mechanism in centimeters.

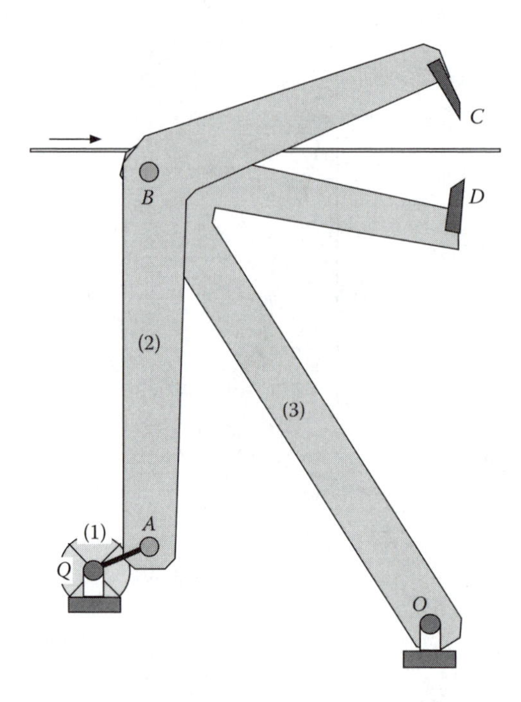

FIGURE 1.11 A web-cutter mechanism.

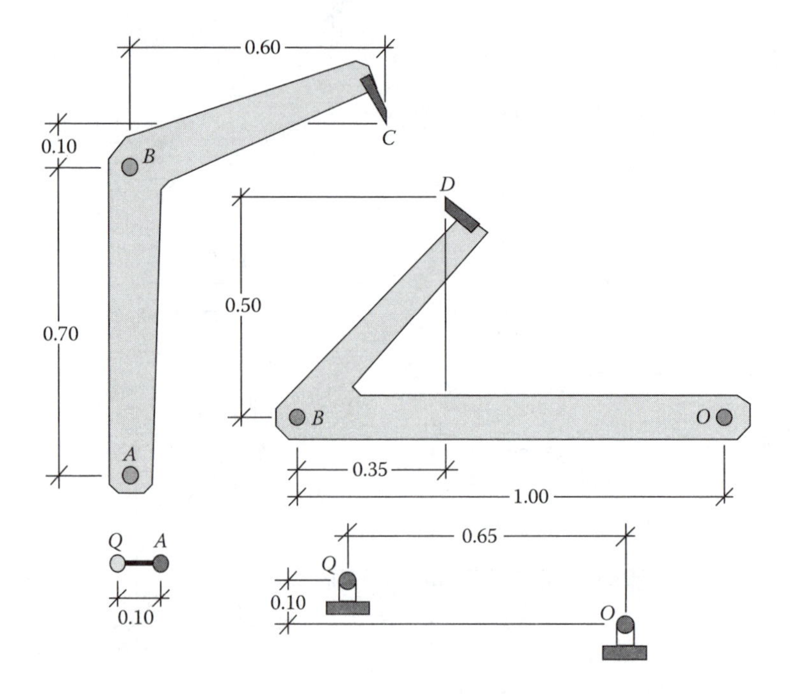

FIGURE 1.12 Geometrical dimensions of the web-cutter mechanism in meters.

TABLE 1.6
Inertia Properties (SI Units)

Body	Mass	Moment of Inertia
(1)	1	0.1
(2)	10	5.0
(3)	10	4.0

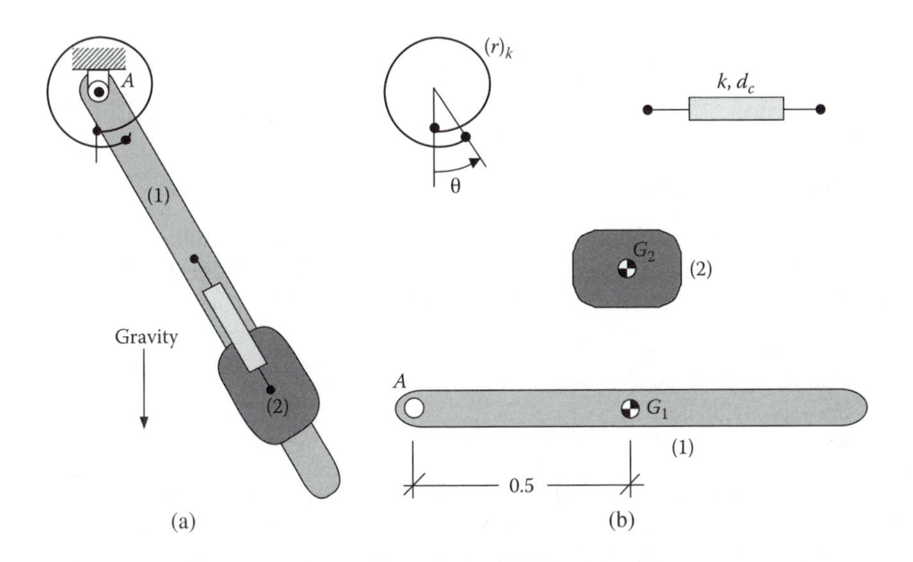

FIGURE 1.13 (a) A variable-length pendulum and (b) its corresponding geometric data.

in Figure 1.13(a), body (1) is connected to the ground by a pin joint at A and body (2) is connected to body (1) by a sliding joint. A rotational spring acts about the pin joint axis and a point-to-point spring-damper element is connected between the two mass centers. The geometric data are provided in Figure 1.13(b). The inertial and spring-damper data are listed in Table 1.7.

1.6 UNIT SYSTEM

The system of units adopted in this book is, unless otherwise stated, the international system of units (SI). In most examples and problems, the constants and variables are organized as the elements of vectors and arrays suitable for computer programming. These elements represent different entities and therefore have different units. If the unit of each element of an array were to be stated, it would cause notational chaos and confusion. Therefore, to eliminate this problem, the units of the constants and variables are not stated in most parts. The reader should assign correct units to all entities. For entities representing angular quantities, we may use the unit of *degree* or *radian*, and therefore, the unit for angular quantities is stated clearly.

TABLE 1.7
Inertia and Spring-Damper Properties (SI Units)

Inertia	Body (Index)	Mass	Moment of Inertia
	(1)	2	0.04
	(2)	1	0.1
Rotational Spring	**Free Angle**	**Stiffness**	**Damping**
	0	60	0
Point-to-Point Spring-Damper	**Free Length**	**Stiffness**	**Damping**
	0.2	20	5

1.7 REMARKS

This textbook explores several methods of formulation for planar multibody systems and discusses various analysis techniques suitable for computational procedures. The subjects that are covered in the book can be divided into four categories:

Fundamentals—Chapters 2 to 4
Formulations—Chapters 5 to 10
Analyses—Chapters 11 to 14
Projects—Chapter 15

The reader is encouraged to study the fundamental chapters before proceeding to other chapters. The formulation chapters introduce three methods for constructing the kinematic and dynamic equations of motion. These chapters could be studied in sequence or the reader could study one formulation and then move to the analyses chapters. For example, one could study the kinematic formulation with point coordinates in Chapter 5, or with body coordinates in Chapter 7, then learn about kinematic analysis in Chapter 11. We should point out that for the dynamic formulations, one must study the corresponding chapter on the kinematics prior to the chapter on dynamics.

2 Preliminaries

Geometry of motion is at the heart of both the kinematics and the dynamics of mechanical systems. Vector analysis is the time-honored tool for describing geometry; hence, vector algebra forms the mathematical foundation for kinematics and dynamics. However, vector algebra, in its *geometric form* of presentation, is not well suited to computer implementation.

In this chapter, a systematic matrix formulation of vector algebra, referred to as *algebraic vector* representation, is discussed for use throughout the text. This form of vector representation, in contrast to the more traditional geometric vector representation, is easier to use for formula manipulation and computer implementation. Elementary properties of vector and matrix algebra are reviewed in this chapter without proof. It is assumed that the reader has the fundamental knowledge of vector and matrix algebra. Some of the fundamental formulas involving vectors are shown in both classical vector and the algebraic representations to assist readers in relating the two representations. In all other chapters only the algebraic representation is used.

2.1 REFERENCE AXES

To specify the configuration of a vector, we need to define one or more *Cartesian reference axes*, also called *reference frames*. Two typical reference frames, the x-y and ξ-η frames, are shown in Figure 2.1. We normally define a frame at a convenient location relative to the vectors (or the mechanical system). Reference frames are further discussed in Chapter 3.

2.2 SCALARS AND VECTORS

A quantity possessing only a magnitude is called a *scalar*. Length, mass, and speed are examples of scalar quantities. In this book, scalars are shown in *lightface italic* characters such as ℓ, m, J, and v.

A vector is represented by a *lowercase boldface* character. Examples of vectors in their geometric forms are shown in Figure 2.2: a position vector $\mathbf{r}$; a velocity vector $\mathbf{v}$; an acceleration vector $\mathbf{a}$; and a force vector $\mathbf{f}$. All of these vectors possess a line-of-action with a direction, and a magnitude.

The magnitude of a typical vector $\mathbf{a}$ is denoted by the scalar a.* The direction of a vector is determined by the angle it makes with respect to a known reference axis. The reference axis may be a stationary or a moving axis; it may be an axis of a reference frame or another vector. Unless stated otherwise, in this text we define

* Traditionally, the magnitude of a vector, such as $\mathbf{a}$, is described as $|\mathbf{a}|$ or $||\mathbf{a}||$.

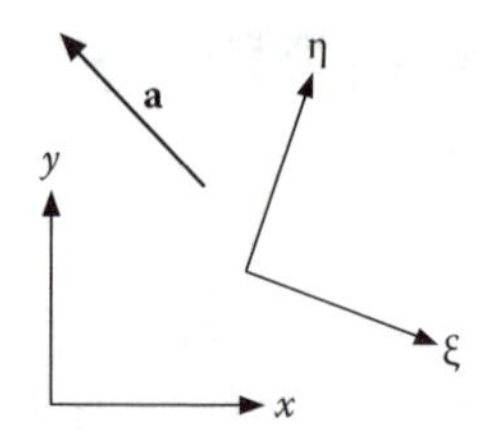

FIGURE 2.1 Two Cartesian reference frames and a vector.

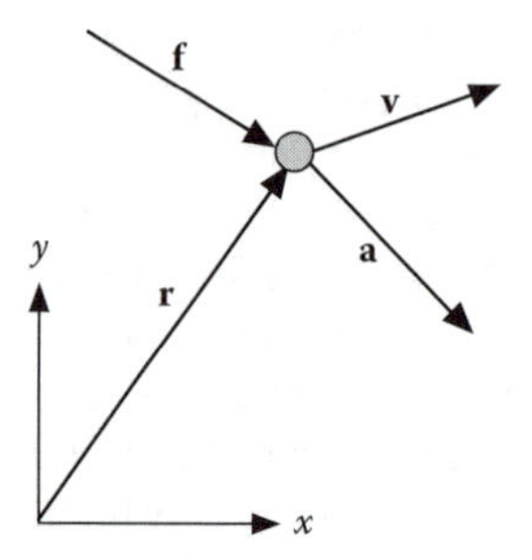

FIGURE 2.2 Typical vectors.

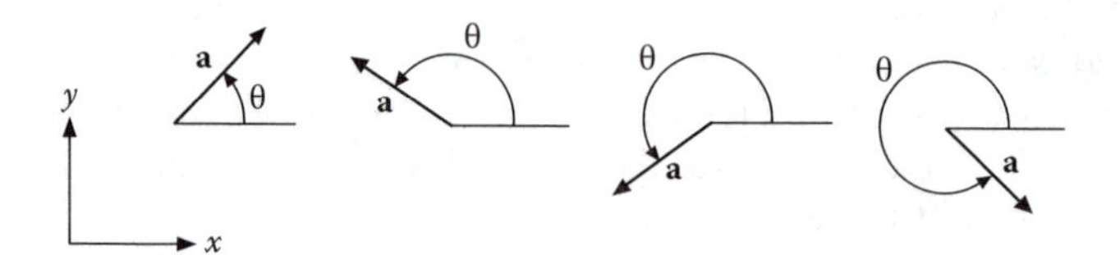

FIGURE 2.3 Angle of a vector.

the angle of a vector as the angle that it makes with the x-axis of a nonmoving reference frame. We adopt the *convention* to measure the angle in a counterclockwise (CCW) direction, known as the positive direction, between the x-axis and the axis of the vector as shown for four cases in Figure 2.3. If the angle is measured in a clockwise (CW) direction, it must be assigned a negative sign.

Example 2.1

Vector **a** makes a 30° angle with the y-axis as shown in Figure 2.4(a). Determine the angle of this vector according to our convention.

Solution:

The angle of this vector is $\theta = 240°$ or $\theta = -120°$ as shown in Figure 2.4(b). The preference is to use $\theta = 240°$.

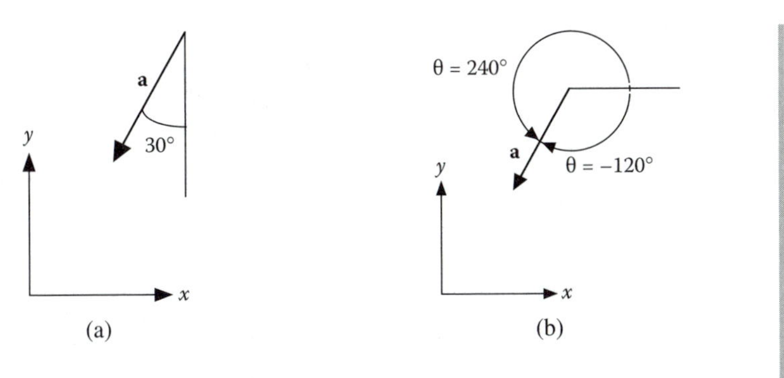

FIGURE 2.4 Angle of a vector can be described with respect to different axes.

A typical vector **a** can be resolved into its *Cartesian components* $a_{(x)}$ and $a_{(y)}$ along the x- and y-axes. If the angle of **a** with respect to the x-axis is θ , the components of the vector are computed as

$$
\begin{aligned}
a_{(x)} &= a \cos\theta \\
a_{(y)} &= a \sin\theta
\end{aligned}
\tag{2.1}
$$

The *algebraic* representation of a vector describes the vector in terms of its Cartesian components. For vector **a** with Cartesian components as described by Eq. (2.1), the algebraic representation is

$$
\mathbf{a} \equiv \begin{Bmatrix} a_{(x)} \\ a_{(y)} \end{Bmatrix} = \begin{Bmatrix} a\cos\theta \\ a\sin\theta \end{Bmatrix} \tag{2.2}
$$

By definition, the algebraic representation of a vector is a *column* array. The *transpose* operation, denoted by a right *apostrophe*, transforms an algebraic *column* vector to a *row* vector (or vise versa) as*

$$
\mathbf{a}' \equiv \begin{Bmatrix} a_{(x)} & a_{(y)} \end{Bmatrix}
$$

EXAMPLE 2.2

A typical vector with its x- and y-components is shown in Figure 2.5. The angle of the vector is $\theta = 120°$ and its magnitude is 2 units. For this vector, because $90° < \theta < 180°$ (i.e., $\cos\theta < 0$ and $\sin\theta > 0$), $a_{(x)}$ is in the direction of negative x-axis and $a_{(y)}$ is in the direction of positive y-axis. Therefore the x-component of

* In MATLAB® there are two ways to transpose an array or a matrix. If matrix (array) X is followed by an apostrophe, such as X', it is called *conjugate transpose*. If it is followed by a period and an apostrophe, such as X.', it is called a *nonconjugate transpose*. Because in this textbook we deal with matrices (arrays) that do not have imaginary parts, X' and X.' yield the same result. Therefore, we will use the simpler of the two (i.e., X').

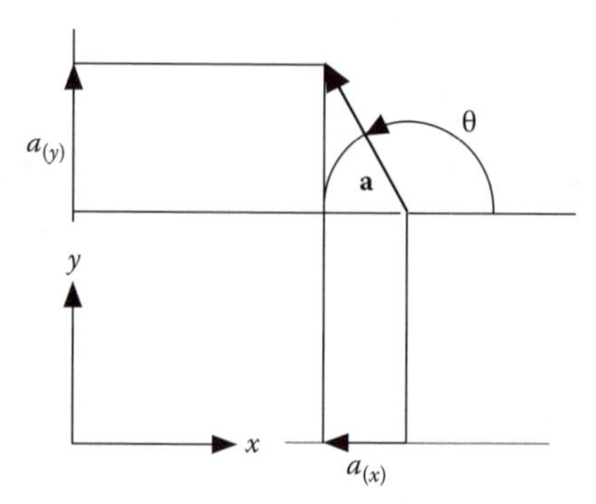

FIGURE 2.5 Components of vector **a**.

this vector is negative and its y-component is positive. This shows that due to our convention for measuring angles, regardless of the value of θ, the components of a vector always end up with correct signs (directions).

When we program in MATLAB, we must be clear whether we define a vector as a row or a column vector. In the following MATLAB program we compute the components of vector **a** and report it in different forms.

```
% Define magnitude and angle for vector
    a_mag = 2.0;
    teta_deg = 120; % degrees
% Convert degrees to radians
    teta = teta_deg*pi/180;
% Compute components of the vector
    a_x = a_mag*cos(teta);
    a_y = a_mag*sin(teta);
% Construct a column vector
    a = [a_x; a_y]
% Construct an equivalent row vector
    b = a'
```

Executing these statements yields vectors **a** and **b** as:

```
a =
    -1.0000
     1.7321

b =
    -1.0000      1.7321
```

When we use the trigonometric functions `sin` and `cos` in MATLAB, we must have the angles in radians. If we wish to leave the angles in degrees, we must then use functions `sind` and `cosd`.* The above example can be rewritten as:

* In the Command Window of MATLAB, type `help elfun` to view a list of all the available elementary math functions.

```
% Define magnitude and angle for vector a
    a_mag = 2.0;
    theta_deg = 120; % degrees
% Compute components of the vector
    a_x = a_mag*cosd(theta_deg);
    a_y = a_mag*sind(theta_deg);
% Construct a column vector
    a = [a_x; a_y]
% Construct an equivalent row vector
    b = a'
```

Because a geometric vector may be resolved into its components in any Cartesian reference frame and not necessarily in the familiar *x-y* reference frame, to generalize the following observations, in this section we denote the components with subscripts *1* and *2*, instead of *x* and *y*. Assume that two vectors **a** and **b** are described in algebraic form as

$$\mathbf{a} = \begin{Bmatrix} a_1 \\ a_2 \end{Bmatrix}, \quad \mathbf{b} = \begin{Bmatrix} b_1 \\ b_2 \end{Bmatrix}$$

These are 2-dimensional vectors, or 2-vectors (or 2-arrays), because each has only two components.

Multiplication of a vector **a** by a scalar α is defined as a vector in the same direction as **a** that has a magnitude $\alpha\, a$. This operation in algebraic form occurs component by component; that is,

$$\alpha\,\mathbf{a} = \begin{Bmatrix} \alpha\, a_1 \\ \alpha\, a_2 \end{Bmatrix} \tag{2.3}$$

The negative of a vector is obtained by multiplying the vector by –1; it changes the direction of the vector.

The vector sum of two vectors **a** and **b** is written as (the geometric form appears on the left and the algebraic form on the right)

$$\mathbf{c} = \mathbf{a} + \mathbf{b} \qquad \begin{Bmatrix} c_1 \\ c_2 \end{Bmatrix} = \begin{Bmatrix} a_1 + b_1 \\ a_2 + b_2 \end{Bmatrix} \tag{2.4}$$

It is also true that $\mathbf{a} = \mathbf{b}$ if $a_1 = b_1$ and $a_2 = b_2$.

A vector with a unit magnitude is called a *unit vector* and it is denoted as **u**. If a unit vector is defined along the axis of vector **a**, the unit vector is written as $\mathbf{u}_{(a)}$.

Any vector, such as **a**, can be described as the product of its magnitude and a unit vector defined along its axis as

$$\mathbf{a} = a\,\mathbf{u}_{(a)} \qquad \begin{Bmatrix} a_1 \\ a_2 \end{Bmatrix} = a \begin{Bmatrix} u_1 \\ u_2 \end{Bmatrix} = \begin{Bmatrix} a\,u_1 \\ a\,u_2 \end{Bmatrix} \tag{2.5}$$

In situations when it is clear that vector **u** is along the axis of vector **a**, for notational simplification we may not use the subscript *(a)*; we simply write $\mathbf{a} = a\,\mathbf{u}$.

Vectors having the same or parallel lines of action are called *parallel vectors*, as those shown in Figure 2.6(a). *Collinear vectors* have the same direction and the same line of action, as those shown in Figure 2.6(b). A *zero* or *null vector* has zero magnitude and therefore no specified direction. A zero vector, denoted by **0**, has its every component equal to zero.

The *scalar* or *dot product* of two vectors **a** and **b** is defined as the product of the magnitudes of the two vectors and the cosine of the angle between them—that is, $\theta_{a,b}$. The angle could be measured either from **a** to **b** or from **b** to **a**. In algebraic form the scalar product of two vectors is determined in component form. The geometric and algebraic presentations are

$$\mathbf{a}\,.\,\mathbf{b} = a\,b\cos\theta_{a,b} \qquad \mathbf{a}'\mathbf{b} = a_1\,b_1 + a_2\,b_2 \tag{2.6}$$

If the two vectors are nonzero (i.e., if $a \neq 0$ and $b \neq 0$), then their scalar product is zero only if $\cos\theta = 0$. Two nonzero vectors are thus said to be *orthogonal* (perpendicular) if their scalar product is zero; that is,

$$\mathbf{a}\,.\,\mathbf{b} = 0 \qquad \mathbf{a}'\mathbf{b} = 0 \tag{2.7}$$

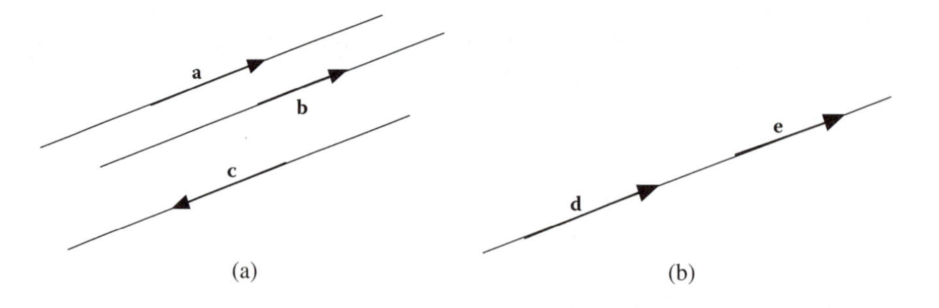

FIGURE 2.6 (a) Parallel and (b) collinear vectors.

We also note that the order of the two vectors in a scalar product operation could be reversed; that is,

$$\mathbf{a} \cdot \mathbf{b} = \mathbf{b} \cdot \mathbf{a} \qquad \mathbf{a}'\mathbf{b} = \mathbf{b}'\mathbf{a} \tag{2.8}$$

It should be obvious that the scalar product of a vector by itself yields the square of the magnitude of the vector:

$$\mathbf{a} \cdot \mathbf{a} = a^2 \qquad \mathbf{a}'\mathbf{a} = a_1^2 + a_2^2 = a^2 \tag{2.9}$$

EXAMPLE 2.3

The following MATLAB program is an exercise for some of the equations from this section.

```
% Define vectors a and b and scalar alpha
     a = [1 2]';      b = [2 -4]';
     alpha = -1;
% Compute the sum of vectors a and b
     c = a + b
% Compute a unit vector along vector a
     a_mag = sqrt(a'*a); % compute magnitude of a
     u = a/a_mag
% Compute the dot product of a and b
     d = a'*b
```

Executing these statements yields:

```
c =
      3
     -2

u =
     0.4472
     0.8944

d =
     -6
```

Vector **a** rotated 90° in the positive sense (CCW) is denoted by $\breve{\mathbf{a}}$ as shown in Figure 2.7. The components of $\breve{\mathbf{a}}$ in terms of the components of **a** are*

$$\breve{\mathbf{a}} = \begin{Bmatrix} -a_2 \\ a_1 \end{Bmatrix} \tag{2.10}$$

* The semicircle overscore is called a *breve*. We could read $\breve{\mathbf{x}}$ as x-rotated.

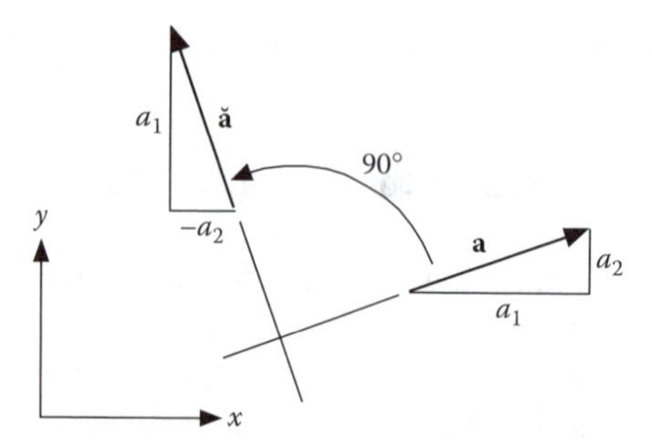

FIGURE 2.7 Rotating a vector 90° in positive sense.

It is obvious that

$$\breve{\mathbf{a}} \cdot \mathbf{a} = 0 \qquad \breve{\mathbf{a}}'\mathbf{a} = 0 \tag{2.11}$$

The *vector* or *cross product* of two vectors **a** and **b** is defined as vector **c**:

$$\mathbf{c} = \mathbf{a} \times \mathbf{b} \tag{2.12}$$

where **c** is perpendicular to the plane of **a** and **b**. In planar kinematics and dynamics, because **a** and **b** are in the x-y plane, vector **c** will be directed out of (or into) the plane. The magnitude of vector **c** can be computed as

$$c = ab\sin\theta_{a,b} \qquad c = \breve{\mathbf{a}}'\mathbf{b} \tag{2.13}$$

where $\theta_{a,b}$ is the angle between **a** and **b** measured by rotating **a** toward **b** in a positive sense. If the computed value of c is positive, vector **c** is coming out of the plane; if the value is negative, the vector is going into the plane.

Proof 2.1

The following is the proof for $c = ab\sin\theta_{a,b} = \breve{\mathbf{a}}'\mathbf{b}$. We denote the angles of **a** and **b** with respect to the x-axis as θ_a and θ_b. Therefore the angle between the two vectors is $\theta_{a,b} = \theta_b - \theta_a$. We then write:

$$\begin{aligned} c &= ab\sin\theta_{a,b} \\ &= ab\sin(\theta_b - \theta_a) \\ &= ab(\sin\theta_b\cos\theta_a - \cos\theta_b\sin\theta_a) \\ &= (b\sin\theta_b)(a\cos\theta_a) + (b\cos\theta_b)(-a\sin\theta_a) \end{aligned}$$

$$= (b_2)(a_1) + (b_1)(-a_2)$$

$$= \{-a_2 \quad a_1\}\begin{Bmatrix} b_1 \\ b_2 \end{Bmatrix}$$

$$= \breve{\mathbf{a}}'\mathbf{b}$$

Because reversal of the order of the two vectors in Eq. (2.13) yields a vector in the opposite direction, it is clear that

$$\mathbf{a} \times \mathbf{b} = -\,\mathbf{b} \times \mathbf{a} \qquad \breve{\mathbf{a}}'\mathbf{b} = -\,\breve{\mathbf{b}}'\mathbf{a} \tag{2.14}$$

For two vectors **a** and **b** to be parallel, we must have

$$\mathbf{a} \times \mathbf{b} = \mathbf{0} \qquad \breve{\mathbf{a}}'\mathbf{b} = 0 \tag{2.15}$$

This equation states that the condition for vector **a** to be parallel to vector **b** is equivalent to the condition for vector $\breve{\mathbf{a}}$ to be perpendicular to vector **b**.

EXAMPLE 2.4

The components of two planar vectors **a** and **b** are given as $\mathbf{a} = \{2 \quad 3\}'$ and $\mathbf{b} = \{4 \quad 5\}'$. Compute $c = \breve{\mathbf{a}}'\mathbf{b}$ and $d = \breve{\mathbf{b}}'\mathbf{a}$.

Solution:

Equation (2.13) yields $c = \breve{\mathbf{a}}'\mathbf{b} = (-3)(4) + (2)(5) = -2$ and $d = \breve{\mathbf{b}}'\mathbf{a} = (-5)(2) + (4)(3) = 2$. The results can be interpreted that **c** is a vector with a magnitude of 2 going into the plane and **d** is a vector with a magnitude of 2 coming out of the plane.

The following MATLAB program performs the above computations.

```
% Define vectors a and b
    a = [2 3]'; b = [4 5]';
% Rotate vectors a and b positively by 90 degrees
    a_rot = [-a(2) a(1)]';
    b_rot = [-b(2) b(1)]';
% Compute the magnitude of cross products a x b and b x a
    c = a_rot'*b
    d = b_rot'*a
```

Executing these statements yields:

```
    c =

         -2

    d =

          2
```

If two vectors, **a** and **b**, are both rotated, then their scalar product can be expressed as

$$\breve{\mathbf{a}}'\breve{\mathbf{b}} = \mathbf{a}'\mathbf{b} \tag{2.16}$$

We can also show that the rotation of the sum of two (or more) vectors is distributive; that is,

$$\mathbf{a} \breve{+} \mathbf{b} = \breve{\mathbf{a}} + \breve{\mathbf{b}} \tag{2.17}$$

where $\breve{+}$ denotes adding the vectors before rotating.

2.2.1 Arrays

A set of elements or functions stacked and arranged in *curly brackets* is called an *array*. For example, for vector **a**, its two Cartesian components, $a_{(x)}$ and $a_{(y)}$, and its angle, θ, are arranged in a 3-array as

$$\begin{Bmatrix} a_{(x)} \\ a_{(y)} \\ \theta \end{Bmatrix}$$

An array can have a dimension of one or greater. An algebraic vector as given by Eq. (2.2) is a special case of a 2-array containing only the Cartesian components of a vector. By our definition, an array is a *column stack* of elements. An array in compact form will be represented by *lowercase boldface* letters. Sometimes for convenience, we may show an array as a *row stack* with a *transpose* superscript. For example

$$\mathbf{q} = \{q_1 \quad q_2 \quad \cdots \quad q_n\}'$$

is an n-array. Stacking up arrays or algebraic vectors forms a new array. For example, if $\mathbf{s}_i$; $i = 1, 2, \cdots, n$, are 2-vectors, then

$$\mathbf{s} = \begin{Bmatrix} \mathbf{s}_1 \\ \mathbf{s}_2 \\ \vdots \\ \mathbf{s}_n \end{Bmatrix} \tag{2.18}$$

is a 2n-array. Note that if we need to write Eq. (2.18) as the transpose of a row array, we must write it as

$$\mathbf{s} = \{\mathbf{s}_1' \quad \mathbf{s}_2' \quad \cdots \quad \mathbf{s}_n'\}'$$

For two arrays **a** and **b** with dimensions n, the array summation is defined as

$$\mathbf{c} = \mathbf{a} + \mathbf{b} \tag{2.19}$$

The scalar product of these two arrays is defined as

$$\mathbf{a}'\mathbf{b} = \mathbf{b}'\mathbf{a} = a_1 b_1 + a_2 b_2 + \cdots + a_n b_n \tag{2.20}$$

These two arrays are said to be orthogonal if

$$\mathbf{a}'\mathbf{b} = 0 \tag{2.21}$$

EXAMPLE 2.5

The following MATLAB program is an exercise showing how vectors or arrays can be put together to form another array or a matrix.

```
% Define vectors a and b
    a = [2 3]'; b = [4 5]';
% Construct a 4 x 1 array c by stacking a and b
    c = [a; b]
% Construct a 2 x 2 matrix by placing b next to a
    D = [a  b]
```

Executing these statements yields:

```
c =
       2
       3
       4
       5

D =
       2     4
       3     5
```

We define the square root of the sum of squares of the elements of an array **a** as the *norm* of **a**:

$$norm(\mathbf{a}) = \sqrt{\mathbf{a}'\,\mathbf{a}} \tag{2.22}$$

There is a more general definition for the norm of an array or a matrix. However, in this textbook we use the special case of the definition for an array as given in Eq. (2.22).

2.3 MATRICES

Compact matrix notation often allows one to concentrate on the form of a system of equations and what it means, rather than on the minute details of its construction. Matrix manipulation also allows for the organized development and simplification of systems of equations. In this section several useful definitions and identities are stated that are used extensively throughout this text.

A matrix with m rows and n columns is said to be of *dimension* $m \times n$ and is denoted by an *uppercase, boldface* letter; it is written in the form

$$\mathbf{A} \equiv \left[a_{ij}\right] \equiv \begin{bmatrix} a_{11} & a_{12} & \cdots & a_{1n} \\ a_{21} & a_{22} & \cdots & a_{2n} \\ \vdots & \vdots & \ddots & \vdots \\ a_{m1} & a_{m2} & \cdots & a_{mn} \end{bmatrix}$$

where a typical element a_{ij} is located at the intersection of the i-th row and j-th column. The *transpose* of a matrix is formed by interchanging rows and columns and is designated by the right *apostrophe*. Thus, if a_{ij} is the ij element of matrix **A**, a_{ji} is the ij element of its transpose $\mathbf{A}'$.

A matrix with only one column is called a *column* matrix or an array. A matrix with only one row is called a *row* matrix. Therefore, the transpose of a column matrix is a row matrix and vice versa.

A *square* matrix has an equal number of rows and columns. A *diagonal* matrix is a square matrix with $a_{ij} = 0$ for $i \neq j$ and at least one nonzero diagonal term. An $n \times n$ diagonal matrix may be denoted by

$$\mathbf{A} \equiv diag\,[a_{11} \quad a_{22} \quad \cdots \quad a_{nn}] \tag{2.23}$$

If square matrices $\mathbf{B}_i$, $i = 1, \cdots, k$, are arranged along the diagonal of a matrix **D**, it is called a *block-diagonal* matrix and may be denoted as

$$\mathbf{D} = blkdiag\,[\mathbf{B}_1 \quad \mathbf{B}_2 \quad \cdots \quad \mathbf{B}_k] \tag{2.24}$$

An $n \times n$ *unit* or *identity* matrix, denoted by **I**, is a diagonal matrix with $a_{ii} = 1$, $i = 1, \cdots, n$. A null or zero matrix, designated by **0**, has $a_{ij} = 0$ for all i and j.

If $a_{ij} = a_{ji}$ for all i and j, the matrix **A** is called *symmetric*; that is, $\mathbf{A} = \mathbf{A}'$.

2.3.1 Matrix Operations

If two matrices **A** and **B** are of the same dimension, they are defined to be *equal* matrices if $a_{ij} = b_{ij}$ for all i and j. The sum or the difference of two *equidimensional* matrices **A** and **B** is a matrix with the same dimension, defined as

$$\mathbf{C} = \mathbf{A} \pm \mathbf{B} \tag{2.25}$$

where $c_{ij} = a_{ij} \pm b_{ij}$ for all i and j.

Multiplication of a matrix by a scalar is defined as

$$\mathbf{C} = \alpha \mathbf{A} \tag{2.26}$$

where $c_{ij} = \alpha a_{ij}$ for all i and j.

Let **A** be an $m \times p$ matrix and let **B** be an $p \times n$ matrix, written in the form

$$\mathbf{A} \equiv \begin{bmatrix} \mathbf{a}_1' \\ \mathbf{a}_2' \\ \vdots \\ \mathbf{a}_m' \end{bmatrix}, \qquad \mathbf{B} \equiv [\mathbf{b}_1 \quad \mathbf{b}_2 \quad \cdots \quad \mathbf{b}_n] \tag{2.27}$$

where the $\mathbf{a}_i'$; $i, \cdots, m$, are row p-arrays and $\mathbf{b}_i$; $i, \cdots, n$, are column p-arrays. Then the matrix product of **A** and **B** is defined as the $m \times n$ matrix

$$\mathbf{C} = \mathbf{A}\mathbf{B} = \begin{bmatrix} \mathbf{a}_1'\mathbf{b}_1 & \mathbf{a}_1'\mathbf{b}_2 & \cdots & \mathbf{a}_1'\mathbf{b}_n \\ \mathbf{a}_2'\mathbf{b}_1 & \mathbf{a}_2'\mathbf{b}_2 & \cdots & \mathbf{a}_2'\mathbf{b}_n \\ \vdots & \vdots & \ddots & \vdots \\ \mathbf{a}_m'\mathbf{b}_1 & \mathbf{a}_m'\mathbf{b}_2 & \cdots & \mathbf{a}_m'\mathbf{b}_n \end{bmatrix} \tag{2.28}$$

where $c_{ij} = \mathbf{a}_i'\mathbf{b}_j$. It is important to note that the product of two matrices is defined only if the number of columns in the first matrix equals the number of rows in the second matrix. It is clear from the definition that, in general,

$$\mathbf{AB} \neq \mathbf{BA} \tag{2.29}$$

In fact, the products **AB** and **BA** are defined only if both **A** and **B** are square and of equal dimensions.

Example 2.6

This exercise shows some of the basic matrix operations in MATLAB.

```
% Define matrices A and B
     A = [1 2 3; 4 5 6]
     B = [7 8 9; -1 -2 -3]
% Compute the sum of A and B
     C = A + B
% Construct the transpose of B
     B_t = B'
% Compute the product A*B'
     D = A*B_t
```

Executing these statements yields:

```
A =
     1     2     3
     4     5     6

B =
     7     8     9
    -1    -2    -3

C =
     8    10    12
     3     3     3

B_t =
     7    -1
     8    -2
     9    -3

D =
    50   -14
   122   -32
```

The following identities are valid, assuming that the matrices have proper dimensions:

$$(\mathbf{A}+\mathbf{B})\mathbf{C} = \mathbf{AC}+\mathbf{BC} \tag{2.30}$$

$$(\mathbf{AB})\mathbf{C} = \mathbf{A}(\mathbf{BC}) = \mathbf{ABC} \tag{2.31}$$

$$(\mathbf{A}+\mathbf{B})' = \mathbf{A}'+\mathbf{B}' \tag{2.32}$$

$$(\mathbf{AB})' = \mathbf{B}'\mathbf{A}' \tag{2.33}$$

Consider an $m \times p$ matrix **A**. If linear combinations of the rows of **A** are nonzero; that is, if

$$\mathbf{A}'\boldsymbol{\alpha} \neq \mathbf{0} \tag{2.34}$$

for all $\boldsymbol{\alpha} = \{\alpha_1 \quad \alpha_2 \quad \cdots \quad \alpha_m\}' \neq \mathbf{0}$, the rows of **A** are said to be *linearly independent*. Otherwise, if

$$\mathbf{A}'\boldsymbol{\alpha} = \mathbf{0} \tag{2.35}$$

for at least one $\boldsymbol{\alpha} \neq \mathbf{0}$, then the rows of **A** are said to be *linearly dependent* and at least one row can be written as a linear combination of the other rows.

The *row rank* (*column rank*) of a matrix **A** is defined as the largest number of linearly independent rows (columns) in the matrix. The row and the column ranks of any matrix are equal. Each of them can thus be called the *rank* of the matrix. A square matrix with linearly independent rows (columns) is said to have *full rank*. When a square matrix does not have full rank, it is called *singular*. For a nonsingular matrix there is an *inverse*, denoted by $\mathbf{A}^{-1}$, such that

$$\mathbf{A}\mathbf{A}^{-1} = \mathbf{A}^{-1}\mathbf{A} = \mathbf{I} \tag{2.36}$$

The following identities are valid:

$$(\mathbf{A}^{-1})' = (\mathbf{A}')^{-1} \tag{2.37}$$

$$(\mathbf{AB})^{-1} = \mathbf{B}^{-1}\mathbf{A}^{-1} \tag{2.38}$$

Example 2.7

This exercise shows how to compute the inverse of a square matrix.

```
% Define square matrix A
     A = [1 2; 3 4]
% Compute inverse of A
     A_inv = inv(A)
% Test that A*A_inverse is identity
     B = A*A_inv
```

Executing these statements yields:

```
       A =
             1        2
             3        4

     A_inv =
          -2.0000      1.0000
           1.5000     -0.5000

     B =
           1.0000           0
           0.0000      1.0000
```

A special nonsingular matrix that arises often in kinematics and dynamics is called an *orthonormal** matrix, with the property that

$$\mathbf{A}' = \mathbf{A}^{-1} \tag{2.39}$$

It is obvious that for an orthonormal matrix

$$\mathbf{A}'\mathbf{A} = \mathbf{I} \tag{2.40}$$

Because constructing the inverse of a nonsingular matrix is generally time-consuming, it is important to know when a matrix is orthonormal. In this special case, the inverse is trivially constructed by transposing the matrix.

2.4 VECTOR, ARRAY, AND MATRIX DIFFERENTIATION

In the kinematics and dynamics of mechanical systems, vectors representing the position of points or bodies, or quantities describing the geometry of the dynamics of the motion, are often functions of time or some other variables. In analyzing these quantities, time derivatives or partial derivatives with respect to some variables are needed. In this section, these derivatives are defined and the notation used in the text is explained.

2.4.1 TIME DERIVATIVES

In analyzing velocities and accelerations, time derivatives of vectors and arrays that describe the position of points and bodies or the geometry of motion must often be calculated. The following identities are expressed for arrays, but they are also applicable to 2-vectors. Consider an array

$$\mathbf{a} \equiv \mathbf{a}(t) = \begin{Bmatrix} a_1(t) \\ \vdots \\ a_n(t) \end{Bmatrix} \tag{2.41}$$

where t is the independent variable (or parameter) *time*. The time derivative of $\mathbf{a}$ is denoted by

$$\dot{\mathbf{a}} \equiv \frac{d}{dt}\mathbf{a} = \begin{Bmatrix} \frac{d}{dt}a_1(t) \\ \vdots \\ \frac{d}{dt}a_n(t) \end{Bmatrix} \tag{2.42}$$

* In some textbooks, the term *orthogonal* is used for *orthonormal*.

The derivative of the sum of two arrays gives

$$\frac{d}{dt}(\mathbf{a}+\mathbf{b}) = \dot{\mathbf{a}} + \dot{\mathbf{b}} \tag{2.43}$$

which is completely analogous to the ordinary differentiation rule that the derivative of the sum is the sum of the derivatives. If $\alpha = \alpha(t)$ is a scalar function of time, the following vector/array forms of the product rule of differentiation can also be verified:

$$\frac{d}{dt}(\alpha\,\mathbf{a}) = \dot{\alpha}\,\mathbf{a} + \alpha\,\dot{\mathbf{a}} \tag{2.44}$$

$$\frac{d}{dt}(\mathbf{a}'\,\mathbf{b}) = \dot{\mathbf{a}}'\,\mathbf{b} + \mathbf{a}'\,\dot{\mathbf{b}} \tag{2.45}$$

$$\frac{d}{dt}\breve{\mathbf{a}} = \dot{\breve{\mathbf{a}}} = \breve{\dot{\mathbf{a}}} \tag{2.46}$$

In Eq. (2.46) $\mathbf{a}$ is assumed to be a 2-vector.

The second time derivative of a vector/array $\mathbf{a} \equiv \mathbf{a}(t)$ is denoted as

$$\ddot{\mathbf{a}} \equiv \frac{d}{dt}\dot{\mathbf{a}} = \frac{d}{dt}\left(\frac{d}{dt}\mathbf{a}\right) \tag{2.47}$$

Thus the differentiation is performed on every component of the array.

Note that for a vector the *time derivative of the magnitude* and the *magnitude of the time derivative* may or may not be the same. To clarify this, assume that vector $\mathbf{a}$ described by Eq. (2.5) has a variable magnitude. The time derivative of Eq. (2.5) is written as

$$\dot{\mathbf{a}} = \dot{a}\,\mathbf{u}_{(a)} + a\,\dot{\mathbf{u}}_{(a)} \tag{2.48}$$

where we have denoted the *time derivative of the magnitude* of vector $\mathbf{a}$ as $\dot{a}$. The *magnitude of the time derivative* of vector $\mathbf{a}$ will be denoted as $|\dot{\mathbf{a}}|$ where $|\dot{\mathbf{a}}| \neq \dot{a}$.

The second time derivative of Eq. (2.48) is expressed as

$$\ddot{\mathbf{a}} = \ddot{a}\,\mathbf{u}_{(a)} + a\,\ddot{\mathbf{u}}_{(a)} + 2\dot{a}\,\dot{\mathbf{u}}_{(a)} \tag{2.49}$$

where we have denoted the second time derivative of the magnitude of the vector as $\ddot{a}$.

Just as in the differentiation of a vector or an array function, the derivative of a matrix whose components depend on t may be defined. The derivative of $\mathbf{A}(t) = [a_{ij}(t)]$ is defined as

$$\dot{\mathbf{A}} \equiv \frac{d}{dt}\mathbf{A} = \left[\frac{d}{dt}a_{ij}(t)\right] \tag{2.50}$$

With this definition, and assuming that $\mathbf{B} = \mathbf{B}(t)$, $\alpha = \alpha(t)$, and $\mathbf{a} = \mathbf{a}(t)$, it can be verified that

$$\frac{d}{dt}(\mathbf{A} + \mathbf{B}) = \dot{\mathbf{A}} + \dot{\mathbf{B}} \tag{2.51}$$

$$\frac{d}{dt}(\mathbf{A}\,\mathbf{B}) = \dot{\mathbf{A}}\mathbf{B} + \mathbf{A}\dot{\mathbf{B}} \tag{2.52}$$

$$\frac{d}{dt}(\alpha\,\mathbf{B}) = \dot{\alpha}\,\mathbf{B} + \alpha\,\dot{\mathbf{B}} \tag{2.53}$$

$$\frac{d}{dt}(\mathbf{A}\,\mathbf{a}) = \dot{\mathbf{A}}\mathbf{a} + \mathbf{A}\dot{\mathbf{a}} \tag{2.54}$$

2.4.2 Partial Derivatives

Assume that $\mathbf{q}$ is an n-array, where every element of $\mathbf{q}$ is a variable, and Φ is a function of $\mathbf{q}$; that is, $\Phi = \Phi(\mathbf{q})$. The partial derivative of Φ with respect to $\mathbf{q}$, using j as the column index, is defined as

$$\frac{\partial \Phi}{\partial \mathbf{q}} \equiv \left[\frac{\partial \Phi}{\partial q_j}\right] \equiv \left[\frac{\partial \Phi}{\partial q_1} \quad \frac{\partial \Phi}{\partial q_2} \quad \cdots \quad \frac{\partial \Phi}{\partial q_n}\right] \tag{2.55}$$

This expression indicates that the partial derivative of a single function with respect to an n-array is a row matrix with n columns.

Example 2.8

Array $\mathbf{q}$ designates four variables as $\mathbf{q} = \{x_1 \quad x_2 \quad x_3 \quad x_4\}'$. Find the partial derivative of a scalar function Φ with respect to $\mathbf{q}$ where $\Phi = -x_1 + 3x_2x_4^2$.

Solution:

Because $\partial\Phi/\partial x_1 = -1$, $\partial\Phi/\partial x_2 = 3x_4^2$, $\partial\Phi/\partial x_3 = 0$, and $\partial\Phi/\partial x_4 = 6x_2x_4$, then

$$\frac{\partial \Phi}{\partial \mathbf{q}} = \begin{bmatrix} -1 & 3x_4^2 & 0 & 6x_2x_4 \end{bmatrix}$$

If $\mathbf{\Phi} \equiv \mathbf{\Phi}(\mathbf{q}) \equiv \{\Phi_1(\mathbf{q}) \quad \Phi_2(\mathbf{q}) \quad \cdots \quad \Phi_m(\mathbf{q})\}'$ is an m-array of differentiable functions of $\mathbf{q}$, using i as row index and j as column index, the partial derivative of $\mathbf{\Phi}$ with respect to $\mathbf{q}$ is defined as

$$\frac{\partial \mathbf{\Phi}}{\partial \mathbf{q}} \equiv \left[\frac{\partial \Phi_i}{\partial q_j}\right] \equiv \begin{bmatrix} \frac{\partial \Phi_1}{\partial q_1} & \frac{\partial \Phi_1}{\partial q_2} & \cdots & \frac{\partial \Phi_1}{\partial q_n} \\ \frac{\partial \Phi_2}{\partial q_1} & \frac{\partial \Phi_2}{\partial q_2} & \cdots & \frac{\partial \Phi_2}{\partial q_n} \\ \vdots & \vdots & \ddots & \vdots \\ \frac{\partial \Phi_m}{\partial q_1} & \frac{\partial \Phi_m}{\partial q_2} & \cdots & \frac{\partial \Phi_m}{\partial q_n} \end{bmatrix} \tag{2.56}$$

where the result is an $m \times n$ matrix.

EXAMPLE 2.9

Array $\mathbf{q}$ designates six variables as $\mathbf{q} = \{x_1 \quad y_1 \quad x_2 \quad y_2 \quad x_3 \quad y_3\}'$. Determine the partial derivative of two functions $\mathbf{\Phi} = \{\Phi_1 \quad \Phi_2\}'$ with respect to $\mathbf{q}$ where

$$\Phi_1 = x_1 + 3y_1 - x_2 + 2x_3 - y_3$$

$$\Phi_1 = x_1 + 3y_1 - x_2 + 2x_3 - y_3$$

Solution:

The partial derivative of $\mathbf{\Phi}$ with respect to $\mathbf{q}$ is a 2×6 matrix:

$$\frac{\partial \mathbf{\Phi}}{\partial \mathbf{q}} = \begin{bmatrix} 1 & 3 & -1 & 0 & 2 & -1 \\ y_1 & x_1 & 0 & 1 & 0 & 2 \end{bmatrix}$$

Note that the rows of this matrix follow the order of the functions and the columns follow the order of variables in $\mathbf{q}$.

The partial derivative of the scalar product of two n-array functions $\mathbf{a}(\mathbf{q})$ and $\mathbf{b}(\mathbf{q})$ is a row matrix as

$$\frac{\partial (\mathbf{a}'\mathbf{b})}{\partial \mathbf{q}} = \mathbf{b}'\frac{\partial \mathbf{a}}{\partial \mathbf{q}} + \mathbf{a}'\frac{\partial \mathbf{b}}{\partial \mathbf{q}} \tag{2.57}$$

where the dimension of the resultant row matrix is the same as the dimension of array $\mathbf{q}$.

Example 2.10

Arrays **a** and **b** are functions of a single variable x. Determine the partial derivative of $\mathbf{a}'\mathbf{b}$ with respect to x if

$$\mathbf{a} = \begin{Bmatrix} 2x \\ -1 \\ x \end{Bmatrix} \quad \mathbf{b} = \begin{Bmatrix} -3 \\ x \\ -x \end{Bmatrix}$$

Solution:

The derivatives of **a** and **b** with respect to x are

$$\frac{\partial \mathbf{a}}{\partial x} = \begin{bmatrix} 2 \\ 0 \\ 1 \end{bmatrix} \quad \frac{\partial \mathbf{b}}{\partial x} = \begin{bmatrix} 0 \\ 1 \\ -1 \end{bmatrix}$$

Using Eq. (2.57), it is found that

$$\frac{\partial(\mathbf{a}'\mathbf{b})}{\partial x} = \{-3 \quad x \quad -x\} \begin{bmatrix} 2 \\ 0 \\ 1 \end{bmatrix} + \{2x \quad -1 \quad x\} \begin{bmatrix} 0 \\ 1 \\ -1 \end{bmatrix} = [-2x - 7]$$

where the result is a 1×1 matrix. This result can also be obtained directly, to verify Eq. (2.57), by determining the scalar product $\mathbf{a}'\mathbf{b} = -x^2 - 7x$ then taking its partial derivative with respect to x to get $-2x - 7$.

Example 2.11

Array **q** contains two variables as $\mathbf{q} = \{x \quad y\}'$. Arrays **a** and **b** are functions of **q** as:

$$\mathbf{a} = \begin{Bmatrix} x - y \\ x + y \\ y - 1 \end{Bmatrix} \quad \mathbf{b} = \begin{Bmatrix} -x + y \\ x + 2 \\ -x - y \end{Bmatrix}$$

Determine the partial derivative of $\mathbf{a}'\mathbf{b}$ with respect to **q**.

Solution:

The derivatives of **a** and **b** with respect to **q** are

$$\frac{\partial \mathbf{a}}{\partial \mathbf{q}} = \begin{bmatrix} 1 & -1 \\ 1 & 1 \\ 0 & 1 \end{bmatrix} \quad \frac{\partial \mathbf{b}}{\partial \mathbf{q}} = \begin{bmatrix} -1 & 1 \\ 1 & 0 \\ -1 & -1 \end{bmatrix}$$

We can now compute the following terms:

$$\mathbf{b}'\frac{\partial \mathbf{a}}{\partial \mathbf{q}} = \{-x+y \quad x+2 \quad -x-y\}\begin{bmatrix} 1 & -1 \\ 1 & 1 \\ 0 & 1 \end{bmatrix} = [y+2 \quad x-2y+2]$$

$$\mathbf{a}'\frac{\partial \mathbf{b}}{\partial \mathbf{q}} = \{x-y \quad x+y \quad y-1\}\begin{bmatrix} -1 & 1 \\ 1 & 0 \\ -1 & -1 \end{bmatrix} = [y+1 \quad x-2y+1]$$

Using Eq. (2.57), it is found that

$$\frac{\partial(\mathbf{a}'\mathbf{b})}{\partial \mathbf{q}} = [2y+3 \quad 2x-4y+3]$$

This result can also be verified by computing the scalar product $\mathbf{a}'\mathbf{b}$ first and then taking its partial derivative with respect to x and y.

2.5 EQUATIONS AND EXPRESSIONS

In this text, we use the terms *equations* and *expressions* to denote different forms of identities. The term "equation" or "a set of equations" will refer to one or more analytical identities that need analytical or numerical solution to determine the unknown(s). For example, a set of two algebraic equations

$$\begin{aligned} (x^B-2)^2+(y^B-3)^2-9&=0 \\ (x^B-4)^2+(y^B-1)^2-4&=0 \end{aligned} \tag{2.58}$$

need be solved simultaneously to determine the unknowns x^B and y^B. Where as, in the two identities

$$\begin{aligned} x^A &= \ell_1 \cos\theta \\ y^A &= \ell_1 \sin\theta \end{aligned} \tag{2.59}$$

If ℓ_1 and θ are known quantities, a simple substitution yields the values for the unknowns x^A and y^A. The identities that require solution of the equations, such as Eq. (2.58), are referred to as *equations*, whereas the identities that require only substitutions, such as Eq. (2.59), will be referred to as *expressions*.

2.5.1 Compact and Expanded Forms

In this textbook, an equation or expression may appear in compact form, or in matrix form, or in expanded form. As an example consider the following equation:

$$\mathbf{r}^P = \mathbf{r} + \mathbf{A}\,\mathbf{s}^P_{(\xi-\eta)} \tag{2.60}$$

This is the *compact form* of presenting an equation. The vectors in component form are

$$\mathbf{r}^P \equiv \begin{Bmatrix} x^P \\ y^P \end{Bmatrix}, \quad \mathbf{r} \equiv \begin{Bmatrix} x \\ y \end{Bmatrix}, \quad \mathbf{s}^P_{(\xi-\eta)} \equiv \begin{Bmatrix} \xi^P \\ \eta^P \end{Bmatrix}$$

and the 2×2 matrix is

$$\mathbf{A} = \begin{bmatrix} \cos\phi & -\sin\phi \\ \sin\phi & \cos\phi \end{bmatrix}$$

Equation (2.60) can also be presented in *matrix form* as

$$\begin{Bmatrix} x^P \\ y^P \end{Bmatrix} = \begin{Bmatrix} x \\ y \end{Bmatrix} + \begin{bmatrix} \cos\phi & -\sin\phi \\ \sin\phi & \cos\phi \end{bmatrix} \begin{Bmatrix} \xi^P \\ \eta^P \end{Bmatrix} \tag{2.61}$$

The *expanded form* of Eq. (2.60) or (2.61) is given as

$$\begin{aligned} x^P &= x + \xi^P \cos\phi - \eta^P \sin\phi \\ y^P &= y + \xi^P \sin\phi + \eta^P \cos\phi \end{aligned} \tag{2.62}$$

All these representations express the same relationship. The compact form provides an overall view of a relationship between several entities; the matrix form shows the details within an equation in an organized vector/matrix form; and the expanded form also provides the details without the use of vector/matrix brackets.

In the following chapters we will use all the three representations. An equation number may refer to more than one form of representing an equation. For example we may see the compact and matrix representations on one line with a single equation number such as

$$\mathbf{r}^P = \mathbf{r} + \mathbf{A}\,\mathbf{s}^P_{(\xi-\eta)} \qquad \begin{Bmatrix} x^P \\ y^P \end{Bmatrix} = \begin{Bmatrix} x \\ y \end{Bmatrix} + \begin{bmatrix} \cos\phi & -\sin\phi \\ \sin\phi & \cos\phi \end{bmatrix} \begin{Bmatrix} \xi^P \\ \eta^P \end{Bmatrix} \tag{2.63}$$

However, when we refer to an equation by an equation number for which there is more than one representation, we must use the form that is most convenient for our purpose.

2.6 REMARKS

The compact notation for performing matrix algebra will be used extensively throughout this textbook. The notation allows a simple transition from the formula derivation to MATLAB programming. The reader is highly encouraged to become comfortable with the notation—this will lead to taking full advantage of MATLAB's capabilities. Although we use compact notation in most examples and discussions, in certain cases we revert to the expanded notation to clarify certain points.

2.7 PROBLEMS

Solve the following problems on paper—do not use MATLAB:

2.1 Verify that for any unit vector $\mathbf{u}$ the Cartesian components are those expressed by Eq. (2.2). Assume an arbitrary value for the angle in either the first, second, third, or fourth quadrant. Draw the vector in an x-y reference frame and compare its measured components against the computed values. Repeat the problem for angles in other quadrants.

2.2 Vectors $\mathbf{a}$ and $\mathbf{b}$, and a constant α have been defined as:

$$\mathbf{a} = \begin{Bmatrix} 1.5 \\ 0.4 \end{Bmatrix}, \quad \mathbf{b} = \begin{Bmatrix} -0.6 \\ 1.2 \end{Bmatrix}, \quad \alpha = -0.3$$

Determine the following vectors or scalars:

(a) $\mathbf{c} = \alpha\,\mathbf{b}$
(b) $\mathbf{d} = \mathbf{a} + \alpha\,\mathbf{b}$
(c) $\mathbf{u}_{(a)}$ and $\mathbf{u}_{(b)}$
(d) $\mathbf{a}'\mathbf{b}$
(e) $\mathbf{b}'\mathbf{a}$
(f) the angle between $\mathbf{a}$ and $\mathbf{b}$
(g) magnitudes of $\mathbf{a}$ and $\mathbf{b}$
(h) $\breve{\mathbf{a}}$ and $\breve{\mathbf{b}}$
(i) the angle between $\breve{\mathbf{a}}$ and $\mathbf{a}$
(j) $\breve{\mathbf{a}}'\mathbf{b}$
(k) $\breve{\mathbf{b}}'\mathbf{a}$

2.3 Based on the scalar product expressions of Eq. (2.6), prove that $a\,b\cos\theta_{a,b} = a_1\,b_1 + a_2\,b_2$.

2.4 Show that if vector $\mathbf{a}$ is rotated by 90°, the components of $\breve{\mathbf{a}}$ are those expressed by Eq. (2.10) regardless of the angle of $\mathbf{a}$ being in any of the four quadrants.

2.5 Vector $\mathbf{a}$ is rotated 90° in a negative sense (CW direction) to obtain vector $\hat{\mathbf{a}}$. What are the components of $\hat{\mathbf{a}}$?

2.6 Show that for any arbitrary angle ϕ, matrix $\mathbf{A} = \begin{bmatrix} \cos\phi & -\sin\phi \\ \sin\phi & \cos\phi \end{bmatrix}$ is orthonormal.

2.7 The following matrices and arrays are defined as:

$$\mathbf{A} = \begin{bmatrix} 2 & 1 & -1 \\ -1 & 3 & 0 \\ 0 & -2 & 1 \end{bmatrix}, \mathbf{B} = \begin{bmatrix} -1 & 0 & 2 & 1 \\ 1 & 2 & -1 & 3 \\ 3 & 1 & 0 & -2 \end{bmatrix}, \mathbf{c} = \begin{Bmatrix} 2 \\ 1 \\ -1 \end{Bmatrix}, \mathbf{d} = \begin{Bmatrix} 2 \\ 0 \\ -1 \\ 1 \end{Bmatrix}$$

Determine:

(a) $\mathbf{Ac}$
(b) $\mathbf{Bd}$
(c) $\mathbf{AB}$
(d) $\mathbf{BA}$
(e) $\mathbf{B'A}$
(f) $\mathbf{c'd}$
(g) $\mathbf{c'B}$

2.8 Show that Eq. (2.16) is valid.

2.9 Consider matrix **B** from Problem 2.7.
What are the column arrays that make up this matrix?
What are the row arrays that make up this matrix?

2.10 Consider the following matrices:

$$\mathbf{A} = \begin{bmatrix} 2 & 1 & -1 \\ -1 & 3 & 0 \\ 0 & -2 & 1 \end{bmatrix}, \mathbf{B} = \begin{bmatrix} 2 & 4 & -2 \\ -1 & -3 & 2 \\ 0 & -2 & 2 \end{bmatrix}$$

Is **A** singular? Why?
Is **B** singular? Why?

2.11 Arrays $\mathbf{\Phi}$ and $\mathbf{q}$ are defined as:

$$\dot{\mathbf{\Phi}} = \begin{Bmatrix} 2x - 3xy + y^2 - xz + yz^2 - 4xyz \\ -x^2 + xy^2 - 2y + 5yz - xz^2 \end{Bmatrix}, \mathbf{q} = \{x \quad y \quad z\}^{\mathrm{T}}$$

where x, y, and z are function of time.

(a) Construct $\dfrac{\partial \mathbf{\Phi}}{\partial \mathbf{q}}$ by taking the partial derivative of the function.
(b) Determine $\dot{\mathbf{\Phi}}$ by taking the time derivative of the function.
(c) Show that $\dot{\mathbf{\Phi}} = \dfrac{\partial \dot{\mathbf{\Phi}}}{\partial \mathbf{q}} \dot{\mathbf{q}}$.

2.12 If vector **a** has constant length, analytically show that $\dot{\mathbf{a}}'\mathbf{a} = 0$. Describe the components of $\dot{\mathbf{a}}$ in terms of the components of **a**. Explain the results graphically.

2.13 Vector **a** is a 2-vector and its elements are functions of time. Verify the validity of Eq. (2.46).

Solve the following problems with MATLAB:

2.14 Repeat Problem 2.2.

2.15 Arrays $\mathbf{a}$ and $\mathbf{b}$ are defined as

$$\mathbf{a} = \begin{Bmatrix} 1 \\ 2 \\ 3 \end{Bmatrix}, \quad \mathbf{b} = \begin{Bmatrix} 4 \\ 5 \\ 6 \end{Bmatrix}$$

Construct the following entities:

(a) $\mathbf{c} = \begin{Bmatrix} \mathbf{a} \\ \mathbf{b} \end{Bmatrix}$

(b) $\mathbf{d} = \{\mathbf{a}' \quad \mathbf{b}'\}$

(c) $\mathbf{E} = [\mathbf{a} \quad \mathbf{b}]$

2.16 Verify Problem 2.6 in MATLAB. Assign any desired value to ϕ.

2.17 Given $\mathbf{a} = \begin{Bmatrix} 4 \\ 2 \end{Bmatrix}$ and $\mathbf{b} = \begin{Bmatrix} 2 \\ 3 \end{Bmatrix}$:

(a) Compute a unit vector, $\mathbf{u}_{(\mathbf{a})}$, in the $\mathbf{a}$ direction.

(b) Project $\mathbf{b}$ onto $\mathbf{u}_{(\mathbf{a})}$ to obtain $\mathbf{b}_{(\mathbf{a})}$.

(c) Compute $\mathbf{c} = \mathbf{b} - \mathbf{b}_{(\mathbf{a})}$.

(d) Show that $\mathbf{c}$ is perpendicular to $\mathbf{a}$.

(e) Show that $\breve{\mathbf{a}}'\mathbf{c} = \breve{\mathbf{a}}'\mathbf{b}$ (i.e., $\mathbf{a} \times \mathbf{c} = \mathbf{a} \times \mathbf{b}$).

2.18 Repeat Problem 2.7.

2.19 Try to invert matrices $\mathbf{A}$ and $\mathbf{B}$ from Problem 2.10. What do you get?

2.20 Try the following:

(a) Construct a 3 × 3 diagonal matrix where every diagonal element is equal to "5".

(b) Try the statement `diag([5 5 5])`.

(c) Try the statement `5*eye(3)`.

2.21 Answer the questions in Problem 2.10 by determining the rank of $\mathbf{A}$ and $\mathbf{B}$; that is, use the command `rank(A)` and `rank(B)` in MATLAB.

2.22 Vectors $\mathbf{a}$ and $\mathbf{b}$ are defined as $\mathbf{a} = \mathbf{A}_1\,\mathbf{c}_1$ and $\mathbf{b} = \mathbf{A}_2\,\mathbf{c}_2$, where

$$\mathbf{A}_i = \begin{bmatrix} \cos\phi_i & -\sin\phi_i \\ \sin\phi_i & \cos\phi_i \end{bmatrix} i = 1, 2 \quad \mathbf{c}_1 = \begin{Bmatrix} 1.2 \\ -0.5 \end{Bmatrix} \quad \mathbf{c}_2 = \begin{Bmatrix} -0.3 \\ 0.8 \end{Bmatrix}$$

(a) Let $\Phi = \mathbf{a}'\,\mathbf{b}$ and $\mathbf{q} = \{x_1 \quad y_1 \quad \phi_1 \quad x_2 \quad y_2$. Evaluate $\frac{\partial \Phi}{\partial \mathbf{q}}$ for $\phi_1 = 30°$ and $\phi_2 = 45°$.

(b) Let $\mathbf{d} = \begin{Bmatrix} x_2 - x_1 \\ y_2 - y_1 \end{Bmatrix}$ and $\Phi = \breve{\mathbf{a}}'\,\mathbf{d}$. Evaluate $\frac{\partial \Phi}{\partial \mathbf{q}}$ for $\phi_1 = 30°$, $\phi_2 = 45°$, $x_1 = 6.2$, $y_1 = 1.0$, $x_2 = -1.9$, and $y_2 = 2.3$.

2.23 Consider the following matrix:

$$\mathbf{B} = \begin{bmatrix} -1 & 0 & 2 & 1 \\ 1 & 2 & -1 & 3 \\ 3 & 1 & 0 & -2 \end{bmatrix}$$

Learn how to use the colon operator in MATLAB to extract specific rows or columns from **B** to construct new arrays or matrices:

(a) The second column of **B**
(b) The third row of **B**
(c) The third and fourth columns of **B**

2.24 After you have constructed arrays **c**, **d**, and **E** in Problem 2.15, perform the following tasks:

(a) Extract **a** from **c**.
(b) Extract **b** from **d**.
(c) Calculate the scalar product of **a** and **b**, exclusively using matrix **E**.
(d) Transform (do not redefine) matrix **E**, using only one MATLAB command, to

$$\mathbf{B} = \begin{bmatrix} \mathbf{a} & \begin{matrix} a_1 b_1 \\ a_2 b_2 \\ a_3 b_3 \end{matrix} \end{bmatrix}$$

2.25 A vector is given as $\mathbf{a} = \begin{Bmatrix} 1.2 \\ 0.4 \end{Bmatrix}$:

(a) Calculate the length of **a**.
(b) Calculate the matrix $\mathbf{A} = [\mathbf{u}_{(\mathbf{a})} \quad \mathbf{u}_{(\breve{\mathbf{a}})}]$.
(c) Calculate the angle $\phi_{\mathbf{u}_{(\mathbf{a})}}$.
(d) Construct the matrix $\mathbf{B} = \begin{bmatrix} \cos\phi_{\mathbf{u}_{(\mathbf{a})}} & -\sin\phi_{\mathbf{u}_{(\mathbf{a})}} \\ \sin\phi_{\mathbf{u}_{(\mathbf{a})}} & \cos\phi_{\mathbf{u}_{(\mathbf{a})}} \end{bmatrix}$ and compare it against **A**.
(e) Calculate vectors $\mathbf{b} = \mathbf{A}'\mathbf{a}$ and $\mathbf{c} = \mathbf{A}'\breve{\mathbf{a}}$. Explain the results.

2.26 For vector $\mathbf{a} = \begin{Bmatrix} 1.2 \\ -0.4 \end{Bmatrix}$:

(a) Calculate the angle $\phi_{\mathbf{a}}$.
(b) Calculate the angle $\phi_{-\mathbf{a}}$.
(c) Draw **a** and –**a** on paper. Verify your results from (a) and (b). What problems occur if you use the functions `asin`, `acos`, or `atan` to calculate the angles? Illustrate your answers graphically.
(d) Learn about the MATLAB function `atan2` using the `help` command. Use this function to redo (a) and (b).

3 Fundamentals of Kinematics

This chapter presents a summary of some of the fundamental concepts in planar kinematics. We first review the kinematics of a single particle, and then we state the fundamental formulas for the kinematics of a single planar rigid body. Definitions for arrays of coordinates, constraint equations, and degrees of freedom are also discussed.

The kinematic formulas that are presented in this chapter will be used intensively throughout the book. Therefore, it is important to understand these simple but fundamental formulas at this point. Exercises with MATLAB® are provided to carry out most of the computations, even though most of the examples are very simple and the calculations may be performed on paper. It is our intent to use MATLAB as much as possible and be prepared to solve more complex and realistic problems whenever needed.

3.1 A PARTICLE

The simplest body arising in the study of motion is a *particle,* or a *point mass*. Analysis of the behavior of a point mass can lead to the kinematic and dynamic analysis of a body, and eventually to the analysis of a system of bodies. Therefore it is essential to study the kinematics of particles first. In this textbook, we use the terms *particles* and *points* interchangeably.

3.1.1 KINEMATICS OF A PARTICLE

The most fundamental step in planar kinematic is to describe the position of a particle or a point in a plane. A plane is described by a Cartesian nonmoving reference frame x-y. The position of a typical particle A in the plane is described by vector $\mathbf{r}^A$ as shown in Figure 3.1. The components of the position vector $\mathbf{r}^A$, or the x-y coordinates of particle A, represent the algebraic vector of coordinates for this particle. This vector and its components can be described in any of the following forms:

$$\mathbf{r}^A = \begin{Bmatrix} r^A_{(x)} \\ r^A_{(y)} \end{Bmatrix} = \begin{Bmatrix} x^A \\ y^A \end{Bmatrix} \tag{3.1}$$

The velocity of particle A is described by the time derivative of its position vector or of its coordinates as:

$$\mathbf{v}^A = \dot{\mathbf{r}}^A = \begin{Bmatrix} \dot{x}^A \\ \dot{y}^A \end{Bmatrix} \tag{3.2}$$

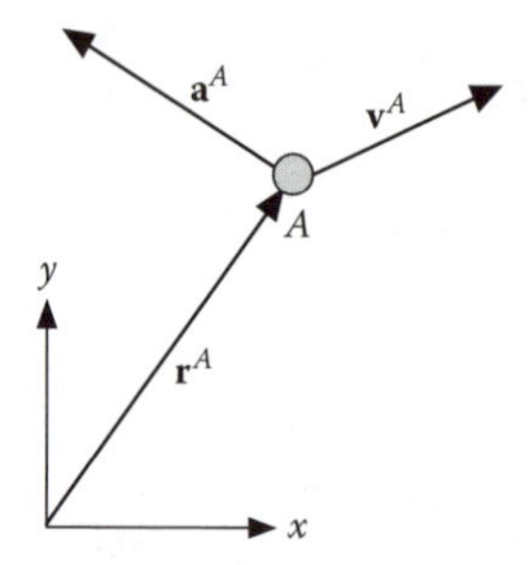

FIGURE 3.1 Position, velocity, and acceleration vectors for a particle.

The acceleration of A is described by the second time derivative of the position vector, or the first time derivative of the velocity vector, as:

$$\mathbf{a}^A = \dot{\mathbf{v}}^A = \ddot{\mathbf{r}}^A = \begin{Bmatrix} \ddot{x}^A \\ \ddot{y}^A \end{Bmatrix} \tag{3.3}$$

Typical velocity and acceleration vectors, $\mathbf{v}^A$ and $\mathbf{a}^A$, for particle A are shown in Figure 3.1.

3.2 KINEMATICS OF A RIGID BODY

A *rigid body* is defined as a system of particles for which distances between particles remain unchanged. If one of these particles is positioned by a vector that is fixed to the body, the vector never changes its position relative to the body, even when the body is in motion. In reality, all solid materials change shape to some extent when forces are applied to them. Nevertheless, if movement associated with the changes in shape is small compared with the overall movement of the body, then the concept of rigidity is applicable. For example, displacements due to elastic vibration of the connecting rod of an engine may be of no consequence in the description of engine dynamics as a whole, so in this case the rigid-body assumption is clearly in order. On the other hand, if the problem is one of describing stress in the connecting rod due to vibration, then the deformation of the rod becomes of prime importance and cannot be neglected. In this text, essentially all analysis is based on the rigid body assumption unless specified otherwise. Therefore the term *body* refers to a *rigid body*.

To study the kinematics of a body, we need to define a nonmoving reference frame. In this text we denote the *nonmoving* frame as x-y. This is also called a *global* or an *inertial* frame. We reserve ξ-η to denote a *moving* frame, also called a *local* or a *body-fixed* frame. Such two frames are shown in Figure 3.2. While in a system we can only have one x-y frame, we can define as many ξ-η frames as the number of moving links or vectors in the system.

A free body in a plane can move along the x- and y-axes (*translational* motion) and it can also rotate about an axis perpendicular to the plane (*rotational* motion). Defining a set of coordinates to monitor the position and orientation of the body is essential in kinematic analysis. The first and second time derivatives of these coordinates are used to define the velocity and acceleration of the body in the plane.

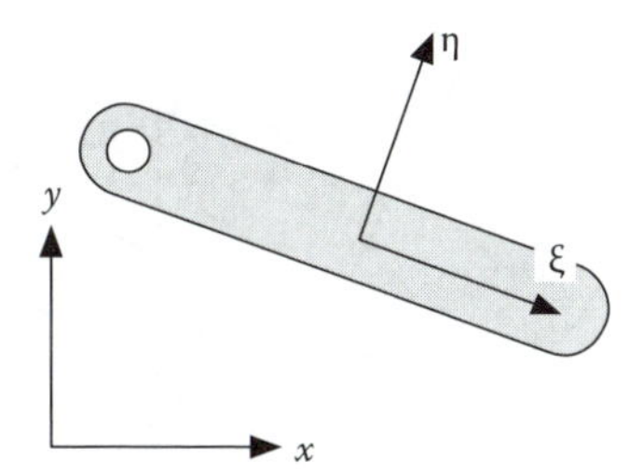

FIGURE 3.2 A nonmoving and a moving reference frames.

3.2.1 Coordinates of a Body

To specify the configuration of a planar body, we need to define three coordinates: position along the x-axis; position along the y-axis; and rotational orientation about the z-axis—an axis perpendicular to the x-y plane. As shown in Figure 3.3(a), we place a reference point O on the body. The position vector **r** is used to locate point O from the origin of the global reference frame x-y. The x- and y-components of this vector are used as the *translational coordinates* of the body; that is,

$$\mathbf{r} = \begin{Bmatrix} r_{(x)} \\ r_{(y)} \end{Bmatrix} = \begin{Bmatrix} x \\ y \end{Bmatrix}$$

The angle of *any* vector attached to the body can be used as the rotational coordinate of that body. However, we attach a body-fixed ξ-η frame to the body with its origin at point O. The angle that the ξ-axis makes with the x-axis, measured in the positive sense (counterclockwise), is used as the *rotational coordinate* of the body and is denoted by ϕ. The *array of body-coordinates* (or *array of coordinates*) for the body is now defined as

$$\mathbf{c} = \begin{Bmatrix} x \\ y \\ \phi \end{Bmatrix} = \{x \quad y \quad \phi\}'$$

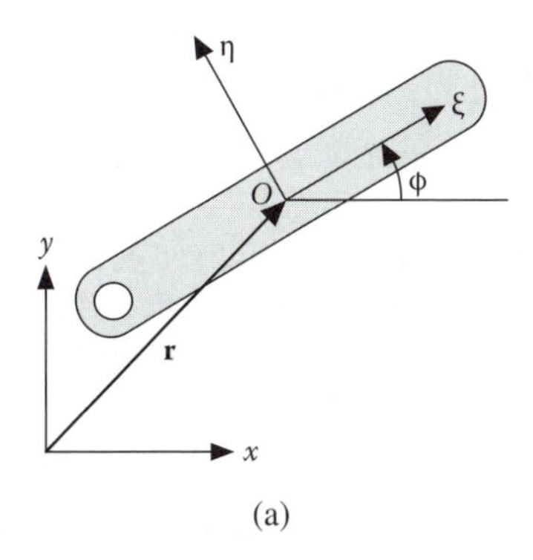

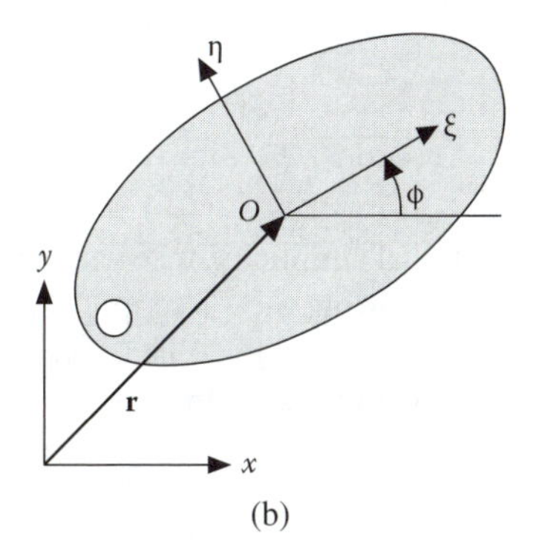

FIGURE 3.3 Coordinates of a body.

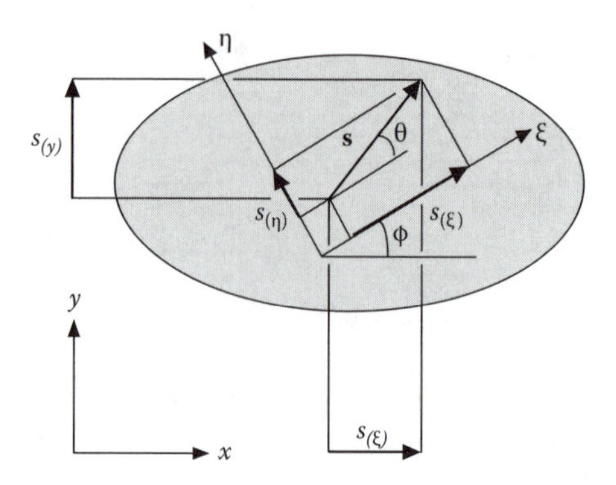

FIGURE 3.4 Components of a vector attached to a body.

The position and orientation of a rigid body is completely described by vector **c**. The shape of a rigid body is not of any importance in most of the discussions we have in this textbook. Therefore, in some of the Figures that deal with general discussions of bodies, the shape is represented as an oval, instead of the actual shape, as depicted in Figure 3.3(b).

As it will be seen in the upcoming sections, a vector can be defined on a body to describe the axis of a sliding joint. Assume that vector **s**, as shown in Figure 3.4, is defined on a body. Although this vector rotates and translates with the body, it has a *fixed* length. The projection of this vector onto the ξ-η frame results into two components represented as

$$\mathbf{s}_{(\xi-\eta)} \equiv \not{\mathbf{s}} = \begin{Bmatrix} s_{(\xi)} \\ s_{(\eta)} \end{Bmatrix}$$

These components will be referred to as the *local* or *body-fixed* components. The same vector can also be projected onto the global *x*-*y* axes to yield

$$\mathbf{s}_{(x-y)} \equiv \mathbf{s} = \begin{Bmatrix} s_{(x)} \\ s_{(y)} \end{Bmatrix}$$

For notational simplicity, when an algebraic vector contains body-fixed components, instead of showing the subscript (ξ-η), cross the vector with a line, such as $\not{\mathbf{s}}$ —we will read it as s-local*. When a vector contains the *x*-*y* components, we represent the vector without showing the subscript (*x*-*y*).

* In some literature and textbooks, the body-fixed components of a vector are denoted by an apostrophe, such as $\mathbf{s}'$. In this textbook we have reserved the apostrophe for vector and matrix transpose to be consistent with the MATLAB notation.

Because $\mathbf{s}$ and $\mathbf{s}'$ represent components of the same vector in two different reference frames, their components are related as (shown in both compact and expanded forms)

$$\mathbf{s} = \mathbf{s}_{(\xi)} + \mathbf{s}_{(\eta)} = \mathbf{A}\,\mathbf{s}' \qquad \begin{aligned} s_{(x)} &= s_{(\xi)} \cos\phi - s_{(\eta)} \sin\phi \\ s_{(y)} &= s_{(\xi)} \sin\phi + s_{(\eta)} \cos\phi \end{aligned} \tag{3.4}$$

where $\mathbf{A}$ is a 2×2 matrix as

$$\mathbf{A} = \begin{bmatrix} \cos\phi & -\sin\phi \\ \sin\phi & \cos\phi \end{bmatrix} \tag{3.5}$$

The columns of $\mathbf{A}$ are unit vectors along the ξ- and η-axes expressed in the x-y frame. Matrix $\mathbf{A}$ is called the *rotational transformation matrix* which is an *orthonormal* matrix; that is, its transpose is equal to its inverse. Because $\mathbf{A}^{-1} = \mathbf{A}'$, the inverse transformation of Eq. (3.4) is found easily as

$$\mathbf{s}' = \mathbf{A}'\,\mathbf{s} \qquad \begin{aligned} s_{(\xi)} &= \;\; s_{(x)} \cos\phi + s_{(y)} \sin\phi \\ s_{(\eta)} &= -s_{(x)} \sin\phi + s_{(y)} \cos\phi \end{aligned} \tag{3.6}$$

PROOF 3.1

To prove that the elements of matrix $\mathbf{A}$ are those stated in Eq. (3.5), we assume that vector $\mathbf{s}$ makes an angle θ with the ξ-axis, and the ξ-axis itself makes an angle ϕ with the x-axis. Therefore the local and global components of vector $\mathbf{s}$ can be expressed as

$$\begin{aligned} s_{(\xi)} &= s \cos\theta \qquad & s_{(x)} &= s \cos(\phi + \theta) \\ s_{(\eta)} &= s \sin\theta \qquad & s_{(y)} &= s \sin(\phi + \theta) \end{aligned}$$

The global components of vector $\mathbf{s}$ are expanded as

$$\begin{aligned} s_{(x)} &= s(\cos\phi\cos\theta - \sin\phi\sin\theta) \\ s_{(y)} &= s(\sin\phi\cos\theta + \cos\phi\sin\theta) \end{aligned}$$

We can write these equations in matrix form as

$$\begin{Bmatrix} s_{(x)} \\ s_{(y)} \end{Bmatrix} = \begin{bmatrix} \cos\phi & -\sin\phi \\ \sin\phi & \cos\phi \end{bmatrix} \begin{Bmatrix} s\cos\theta \\ s\sin\theta \end{Bmatrix} = \begin{bmatrix} \cos\phi & -\sin\phi \\ \sin\phi & \cos\phi \end{bmatrix} \begin{Bmatrix} s_{(\xi)} \\ s_{(\eta)} \end{Bmatrix}$$

The coefficient matrix is the transformation matrix $\mathbf{A}$.

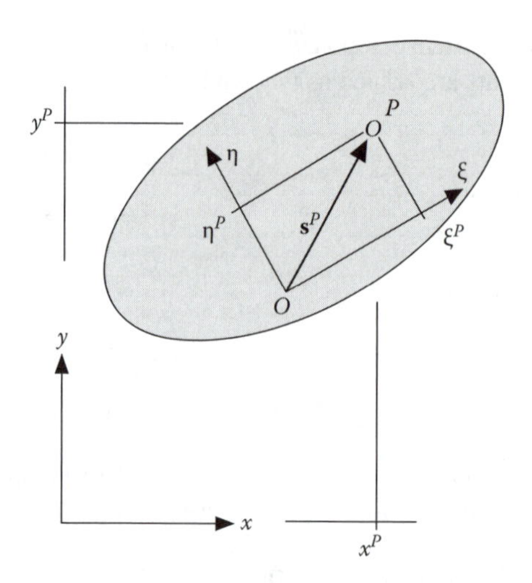

FIGURE 3.5 A point attached to a body.

The magnitude of a vector that is defined on a body can be obtained from either its local or global components:

$$s^2 = \mathbf{s}'' \mathbf{s}' = \mathbf{s}' \mathbf{s} \tag{3.7}$$

Vectors are also used to position body-attached points with respect to the origin of the body. As shown in Figure 3.5, vector $\mathbf{s}^P$ is defined on a body to locate point P from its origin. The components of $\mathbf{s}^P$ in the ξ-η and x-y frames are denoted as

$$\mathbf{s}'^P \equiv \begin{Bmatrix} s^P_{(\xi)} \\ s^P_{(\eta)} \end{Bmatrix} \equiv \begin{Bmatrix} \xi^P \\ \eta^P \end{Bmatrix}, \quad \mathbf{s}^P \equiv \begin{Bmatrix} s^P_{(x)} \\ s^P_{(y)} \end{Bmatrix}$$

The ξ-η and x-y components are related through the transformation matrix $\mathbf{A}$ as

$$\mathbf{s}^P = \mathbf{A}\,\mathbf{s}'^P \tag{3.8}$$

The x-y coordinates of point P are found from the vector relation

$$\mathbf{r}^P = \mathbf{r} + \mathbf{s}^P \tag{3.9}$$

or

$$\mathbf{r}^P = \mathbf{r} + \mathbf{A}\,\mathbf{s}'^P \qquad \begin{Bmatrix} x^P \\ y^P \end{Bmatrix} = \begin{Bmatrix} x \\ y \end{Bmatrix} + \begin{bmatrix} \cos\phi & -\sin\phi \\ \sin\phi & \cos\phi \end{bmatrix} \begin{Bmatrix} \xi^P \\ \eta^P \end{Bmatrix} \tag{3.10}$$

In expanded form we have

$$x^P = x + \xi^P \cos\phi - \eta^P \sin\phi$$
$$y^P = y + \xi^P \sin\phi + \eta^P \cos\phi$$

Equations (3.4) through (3.10) form some of the fundamental formulas in planar rigid-body kinematics.

Example 3.1

A body-fixed frame is defined on a link of a mechanism as shown in Figure 3.6. In a nonmoving frame, the link has translational and rotational coordinates

$$\mathbf{r} = \begin{Bmatrix} 2.5 \\ 1.2 \end{Bmatrix}, \quad \phi = 5.6723 \text{ rad (325 degrees)}$$

Points A and B on the link have local coordinates

$$\mathbf{s}'^{A} = \begin{Bmatrix} 2.18 \\ 0 \end{Bmatrix}, \quad \mathbf{s}'^{B} = \begin{Bmatrix} -1.8 \\ 1.3 \end{Bmatrix}$$

Find (a) the global components of vectors $\mathbf{s}^A$ and $\mathbf{s}^B$, (b) the global coordinates of points A and B, and (c) the global components of vector $\mathbf{s}^{B,A} = \mathbf{s}^B - \mathbf{s}^A$.

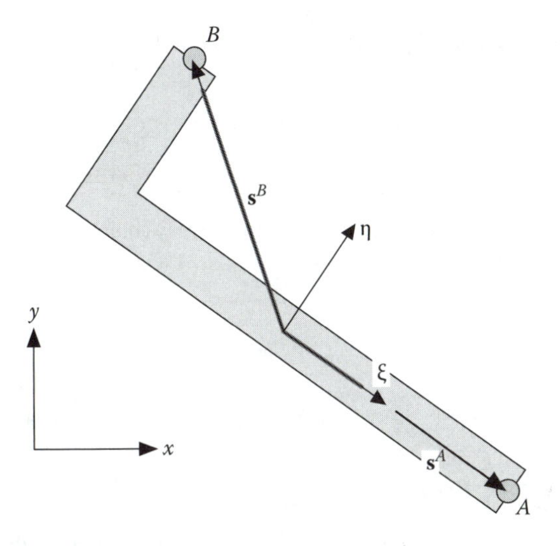

FIGURE 3.6 Points A and B are defined on a link.

Solution

The following MATLAB program is used to perform the computation.

```
% Define coordinates of the link
    r = [2.5; 1.2]; phi = 5.6723;
% Define constant coordinates (local)
    s_A_p = [2.18; 0]; s_B_p = [-1.8; 1.3];
% Compute matrix A
    A = [cos(phi) -sin(phi); sin(phi) cos(phi)]
% Compute global components of s_A and s_B
    s_A = A*s_A_p
    s_B = A*s_B_p
% Compute global coordinates of A and B
    r_A = r + s_A
    r_B = r + s_B
% Compute s_BA
    s_BA_1 = A*(s_B_p - s_A_p)
    s_BA_2 = r_B - r_A
```

(a) The program first computes the transformation matrix from Eq. (3.5), then it uses Eq. (3.8) to find:

```
A =
      0.8191     0.5736
     -0.5736     0.8191

  s_A =                        s_B =
      1.7857                       -0.7288
     -1.2504                        2.0973
```

(b) The program applies Eq. (3.9) to compute the following:

```
r_A =                          r_B =
      4.2857                        1.7712
     -0.0504                        3.2973
```

(c) Vector $\mathbf{s}^{B,A}$ can be determined in two ways. We may compute it as $\mathbf{s}^{B,A} = \mathbf{r}^B - \mathbf{r}^A$ or as $\mathbf{s}^{B,A} = \mathbf{A}(\not{s}^B - \not{s}^A)$. Either way (s_BA_1 or s_BA_2) the program finds:

```
  s_BA =
     -2.5145
      3.3478
```

3.2.2 Velocity of a Body

The translational and rotational velocities of a body are described by the time derivative of the coordinates as $\dot{\mathbf{r}} = \{\dot{x} \quad \dot{y}\}'$ and $\dot{\phi}_i$ respectively. The *3-array of velocities* for body i is defined as $\mathbf{v} \equiv \dot{\mathbf{c}} = \{\dot{x} \quad \dot{y} \quad \dot{\phi}\}'$.

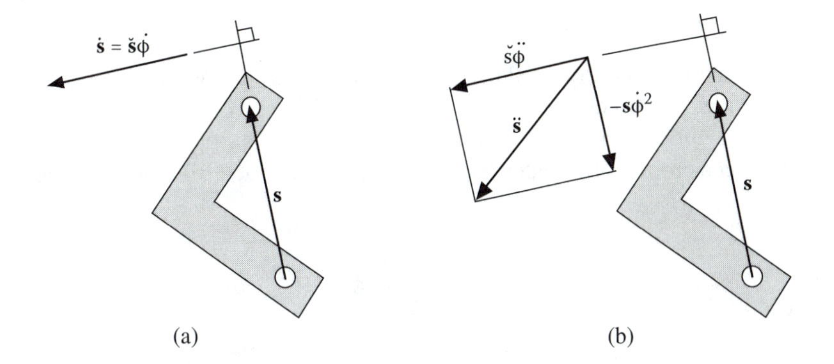

FIGURE 3.7 (a) Velocity and (b) acceleration vectors corresponding to vector **s** that has a constant magnitude.

The velocity equation for a vector fixed to body i is obtained from the time derivative of Eq. (3.4),

$$\dot{\mathbf{s}} = \dot{\mathbf{A}}\,\mathbf{s}' \qquad \begin{aligned} \dot{s}_{(x)} &= -(s_{(\xi)} \sin\phi + s_{(\eta)} \cos\phi)\,\dot{\phi} \\ \dot{s}_{(y)} &= \;\;\,(s_{(\xi)} \cos\phi - s_{(\eta)} \sin\phi)\,\dot{\phi} \end{aligned} \tag{3.11}$$

Note that $\dot{\mathbf{s}}' = \mathbf{0}$ because $\mathbf{s}'$ contains constant components. Equation (3.11) can also be written as

$$\begin{Bmatrix} \dot{s}_{(x)} \\ \dot{s}_{(y)} \end{Bmatrix} = \begin{bmatrix} -\sin\phi & -\cos\phi \\ \cos\phi & -\sin\phi \end{bmatrix} \begin{Bmatrix} s_{(\xi)} \\ s_{(\eta)} \end{Bmatrix} \dot{\phi} = \begin{Bmatrix} -s_{(y)} \\ s_{(x)} \end{Bmatrix} \dot{\phi} \qquad \dot{\mathbf{s}} = \breve{\mathbf{s}}\,\dot{\phi} \tag{3.12}$$

where the definition for vector $\breve{\mathbf{s}}$ is given in Eq. (2.10). Figure 3.7(a) illustrates how vector $\dot{\mathbf{s}}$ is constructed on an axis perpendicular to vector **s**. It is assumed that $\dot{\phi}$ is positive (CCW), otherwise $\dot{\mathbf{s}}$ would be in the opposite direction.

The first time derivative of Eq. (3.10) yields the velocity equations for point P as

$$\dot{\mathbf{r}}^P = \dot{\mathbf{r}} + \breve{\mathbf{s}}^P\,\dot{\phi} \qquad \begin{aligned} \dot{x}^P &= \dot{x} - s^P_{(y)}\dot{\phi} \\ \dot{y}^P &= \dot{y} + s^P_{(x)}\dot{\phi} \end{aligned} \tag{3.13}$$

or,

$$\begin{aligned} \dot{x}^P &= \dot{x} - (\xi^P \sin\phi + \eta^P \cos\phi)\dot{\phi} \\ \dot{y}^P &= \dot{y} + (\xi^P \cos\phi - \eta^P \sin\phi)\dot{\phi} \end{aligned} \tag{3.14}$$

3.2.3 Acceleration of a Body

The translational and rotational accelerations of a body are described by the second time derivative of its coordinates as $\ddot{\mathbf{r}} = \{\ddot{x} \quad \ddot{y}\}'$ and $\ddot{\phi}$ respectively. Hence, the 3-*array of accelerations* for the body is defined as $\dot{\mathbf{v}} \equiv \ddot{\mathbf{c}} = \{\ddot{x} \quad \ddot{y} \quad \ddot{\phi}\}'$.

The acceleration of a vector fixed to a body is obtained from the time derivative of Eq. (3.11) as

$$\ddot{\mathbf{s}} = \ddot{\mathbf{A}}\, \mathbf{s}' \qquad \begin{aligned} \ddot{s}_{(x)} &= -(s_{(\xi)} \sin\phi + s_{(\eta)} \cos\phi)\ddot{\phi} - (s_{(\xi)} \cos\phi - s_{(\eta)} \sin\phi)\dot{\phi}^2 \\ \ddot{s}_{(y)} &= \;\;(s_{(\xi)} \cos\phi - s_{(\eta)} \sin\phi)\ddot{\phi} - (s_{(\xi)} \sin\phi + s_{(\eta)} \cos\phi)\dot{\phi}^2 \end{aligned} \tag{3.15}$$

or,

$$\ddot{\mathbf{s}} = \breve{\mathbf{s}}\,\ddot{\phi} - \mathbf{s}\,\dot{\phi}^2 \tag{3.16}$$

The two components of the acceleration vector are illustrated in Figure 3.7(b) where $\breve{\mathbf{s}}\,\ddot{\phi}$ is known as the *tangential* component and $-\mathbf{s}\,\dot{\phi}^2$ as the *normal* component. In this illustration it is assumed that $\ddot{\phi}$ is positive otherwise the tangential component would be in the opposite direction of that shown. Note that the normal component is always in the opposite direction of $\mathbf{s}$ because $\dot{\phi}^2$ is always positive.

The time derivative of Eq. (3.13) yields the acceleration equations for point P as

$$\ddot{\mathbf{r}}^P = \ddot{\mathbf{r}} + \breve{\mathbf{s}}^P\,\ddot{\phi} - \mathbf{s}^P\,\dot{\phi}^2 \qquad \begin{aligned} \ddot{x}^P &= \ddot{x} - s^P_{(y)}\ddot{\phi} - s^P_{(x)}\dot{\phi}^2 \\ \ddot{y}^P &= \ddot{y} + s^P_{(x)}\ddot{\phi} - s^P_{(y)}\dot{\phi}^2 \end{aligned} \tag{3.17}$$

or,

$$\begin{aligned} \ddot{x}^P &= \ddot{x} - (\xi^P \sin\phi + \eta^P \cos\phi)\ddot{\phi} - (\xi^P \cos\phi - \eta^P \sin\phi)\,\dot{\phi}^2 \\ \ddot{y}^P &= \ddot{y} + (\xi^P \cos\phi - \eta^P \sin\phi)\,\ddot{\phi} - (\xi^P \sin\phi + \eta^P \cos\phi)\,\dot{\phi}^2 \end{aligned} \tag{3.18}$$

Example 3.2

This is an extension of Example 3.1. In addition to the coordinate data given in Example 3.1, assume that the link has the following velocity and acceleration:

$$\dot{\mathbf{r}} = \begin{Bmatrix} 1 \\ -2 \end{Bmatrix},\ \dot{\phi} = 1,\ \ddot{\mathbf{r}} = \begin{Bmatrix} -0.4 \\ 0.4 \end{Bmatrix},\ \ddot{\phi} = 4.65$$

In addition to the coordinates, compute the velocity and acceleration of points *A* and *B*.

Solution

Equations (3.9), (3.13), and (3.17) are implemented in the following program to perform the computation.

```
% Define coordinates
    r = [2.5; 1.2]; phi = 5.6723;
% Define constant coordinates (local)
    s_A_p = [2.18; 0]; s_B_p = [-1.8; 1.3];
% Define velocities and accelerations
    r_d = [1; -2]; phi_d = 1;
    r_dd = [-0.4; 0.4]; phi_dd = 4.65;
% Compute matrix A
    A = A_matrix(phi)
% Compute global components of s_A and s_B
    s_A = A*s_A_p
    s_B = A*s_B_p
% Compute global coordinates of A and B
    r_A = r + s_A
    r_B = r + s_B
% Compute veloccity of A and B
    r_A_d = r_d + s_rot(s_A)*phi_d
    r_B_d = r_d + s_rot(s_B)*phi_d
% Compute acceleration of A and B
    r_A_dd = r_dd + s_rot(s_A)*phi_dd - s_A*phi_d^2
    r_B_dd = r_dd + s_rot(s_B)*phi_dd - s_B*phi_d^2
```

```
function A = A_matrix(phi)
% This function computes matrix A
    cp = cos(phi); sp = sin(phi);
    A = [cp -sp; sp cp];
```

```
function s_r = s_rot(s)
% This function rotates a vector 90 degrees positively
    s_r = [-s(2); s(1)];
```

Executing the program yields the following results:

```
r_A =                    r_B =
     4.2857                   1.7712
    -0.0504                   3.2973

r_A_d =                  r_B_d =
     2.2504                  -1.0973
    -0.2143                  -2.7288

r_A_dd =                 r_B_dd =
     3.6288                  -9.4239
     9.9541                  -5.0862
```

In this MATLAB program, for the first time we have used function M-files. Here we have written two functions: one to compute the rotational transformation matrix, and one to rotate a vector 90° in the positive sense. Such functions will simplify our programming efforts and will be reused in some of the upcoming programs.

Note: These function M-files are listed following the main program. However, we must save them in separate files.

EXAMPLE 3.3

In this exercise three function M-files are developed to compute the coordinates, velocity and acceleration of a point attached to a body according to Eqs. (3.9), (3.13), and (3.17) respectively.

```
function r_Point = r_Point(r, s_P)
% This function computes the x-y coordinates of a point P
     r_Point = r + s_P;
```

```
function r_Point_d = r_Point_d(r_d, s_P, phi_d)
% This function computes the velocity of a point P
     r_Point_d = r_d + s_rot(s_P)*phi_d;
```

```
function r_Point_dd = r_Point_dd(r_dd, s_P, phi_d, phi_dd)
% This function computes the acceleration of a point P
r_Point_dd = r_dd + s_rot(s_P)*phi_dd - s_P*phi_d^2;
```

3.3 DEFINITIONS

A *mechanism* is a set of rigid elements that are arranged to produce a specified motion. This definition of mechanism includes classical linkages, as well as interconnected bodies that make up a vehicle, a vending machine, aircraft landing gear, an engine, and many other mechanical systems.

The individual bodies that collectively form a mechanism are called *links*. The combination of two links in contact constitutes a *kinematic pair,* or *joint*. An assemblage of interconnected links is called a *kinematic chain*. A mechanism is formed when at least one of the links of the kinematic chain is held fixed and any of its other links can move. The fixed link is called the *ground, frame,* or *chassis*. In this text, the terms *links* and *bodies* are used interchangeably.

If all the links in a multibody system move in a plane or in parallel planes, it is called a *planar system*. If some links undergo motion in three-dimensional space, it is called a *spatial system*.

A system that is formed from a collection of links or bodies that are kinematically connected to one another, but for which it is not possible to move to successive links

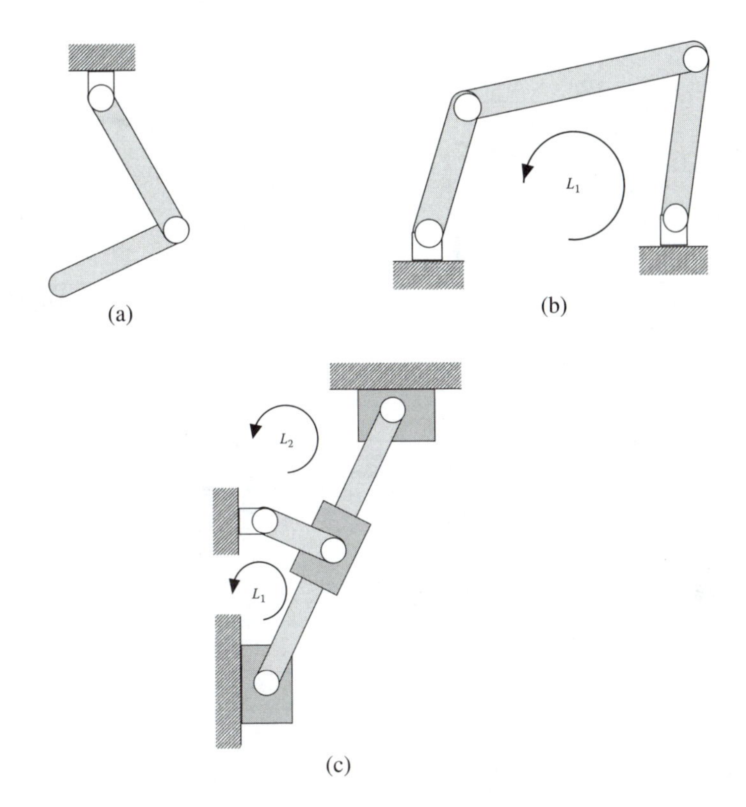

FIGURE 3.8 (a) An open-chain system, (b) a single-loop closed-chain system, and (c) a multiloop closed-chain system.

across kinematic joints and return to the starting link, is called an *open-chain system*. An example is the double pendulum shown in Figure 3.8(a). A *closed-chain system* is formed where each link is connected to at least two other links of the mechanism and it is possible to transverse a loop. Figure 3.8(b) shows a four-bar linkage, which is a closed-chain mechanism.

A closed-chain system may contain one or more loops in its kinematic structure. If the number of loops in a closed-chain mechanism is one, then the mechanism is called *a single-loop system*. If the closed-chain mechanism contains more than one loop, then the mechanism is called a *multiloop system*. Figure 3.8(b) is an example of a single-loop system, and Figure 3.8(c) shows a double-loop system.

Mechanisms and kinematic pairs can be classified generally into two groups. If we consider joints with their three-dimensional geometry, the joints with surface contact are referred to as *lower-pairs*, and those with line or point contact as *higher-pairs*. The derivation of kinematic formulation for most lower-pair joints is generally simpler than that of higher-pair joints. *Revolute* (pin) joints and *translational* (sliding) joints are examples of lower-pairs. *Gears* and *cam-followers* are examples of higher-pairs.

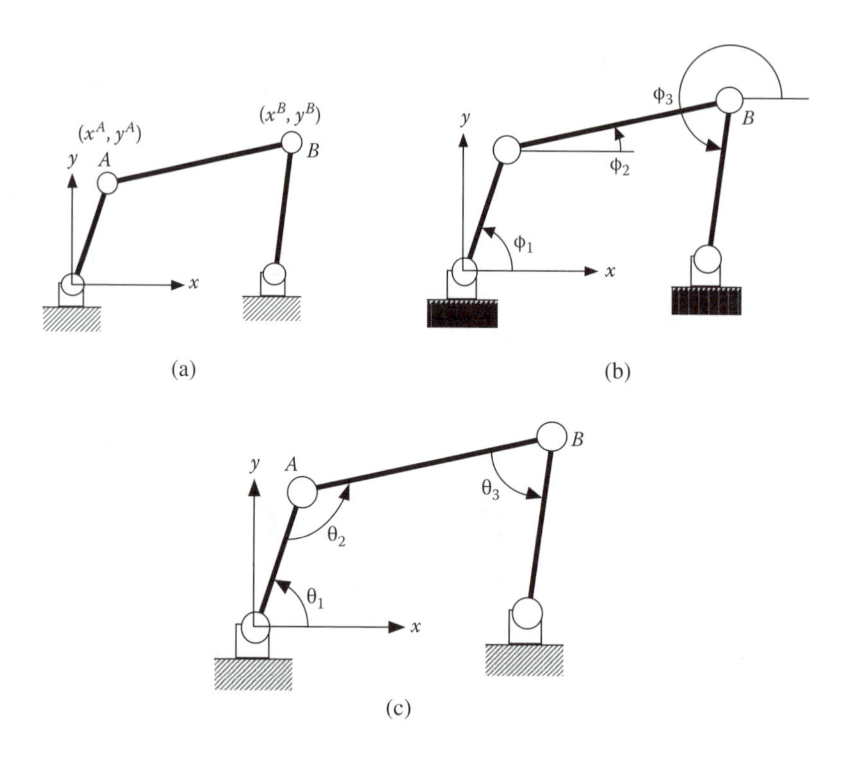

FIGURE 3.9 Different sets of coordinates for a four-bar mechanism.

Springs, dampers, and actuators are considered as force elements. These and other types of force elements are discussed in the next chapter.

3.3.1 Array of Coordinates

Any set of parameters that can specify the position (configuration) of all the bodies in a system is called a set of *coordinates*. For systems in motion, these parameters vary with time, therefore their first and second time derivatives could specify the *velocity* and *acceleration* of the bodies. Depending on the formulation used to perform an analysis, the defined set of coordinates could be different. For example, for the four-bar mechanism shown in Figure 3.9(a), we can define the Cartesian coordinates of points A and B as a set of position coordinates. For this set, an *array of coordinates* is defined as:

$$\mathbf{q} = \{x^A \quad y^A \quad x^B \quad y^B\}' \tag{3.19}$$

We can also perform kinematic analysis on the same mechanism by defining three angles as the position parameters, also shown in Figure 3.9(b). These three angles can be formed into an array of coordinates as:

$$\mathbf{q} = \{\phi_1 \quad \phi_2 \quad \phi_3\}' \tag{3.20}$$

Position parameters can be measured with respect to the global reference frame, therefore called *absolute* coordinates, such as the coordinates used in Eqs. (3.19) and (3.20). If the position parameters are defined between two moving bodies, they are called *relative* coordinates. Such relative coordinates are shown in Figure 3.9(c) which yield an array of coordinates as

$$\mathbf{q} = \{\theta_1 \quad \theta_2 \quad \theta_3\}' \tag{3.21}$$

An array of coordinates may contain a mix of absolute and relative coordinates. For example, for the four-bar mechanism, the array of coordinates may be defined as:

$$\mathbf{q} = \{x^A \quad y^A \quad \theta_2\}' \tag{3.22}$$

In the upcoming chapters, different formulations are presented that use different sets of coordinates. Therefore, the array of coordinates for one formulation may be denoted differently than the array of coordinates for another formulation. For example, in the point-coordinate formulation, the array of coordinate is denoted as **r**, and in the body-coordinate formulation it is denoted as **c**. However, in general, when we are not emphasizing any particular formulation, the array of coordinates is denoted as **q**.

The total number of coordinates that are defined for a system is denoted by n_v (number of variables). Because the first and second time derivatives of the coordinates are used to denote *velocities* and *accelerations,* n_v also represents the number of velocities and the number of accelerations for the system. For example, the arrays of velocities and accelerations for the coordinates described by Eq. (3.22) are expressed as:

$$\dot{\mathbf{q}} = \{\dot{x}^A \quad \dot{y}^A \quad \dot{\theta}_2\}' \tag{3.23}$$

$$\ddot{\mathbf{q}} = \{\ddot{x}^A \quad \ddot{y}^A \quad \ddot{\theta}_2\}' \tag{3.24}$$

3.3.2 Degrees of Freedom

The minimum number of coordinates required to fully describe the configuration of a system is called the number of *degrees of freedom* (DoF). A free point or particle on a plane has two DoF—it can move along any two perpendicular axes independently. A free body on the plane has three DoF—it can translate along any two perpendicular axes, and it can also rotate about an axis perpendicular to the plane.

To determine the number of DoF of a mechanism, we must consider all the bodies in the system and the joints that connect them. For example, consider the triple pendulum shown in Figure 3.10(a). Here, no fewer than three angles, ϕ_1, ϕ_2, and ϕ_3, can uniquely determine the configuration of the pendulum. The value of any of the three angles can be varied independent of the other two angles. Therefore,

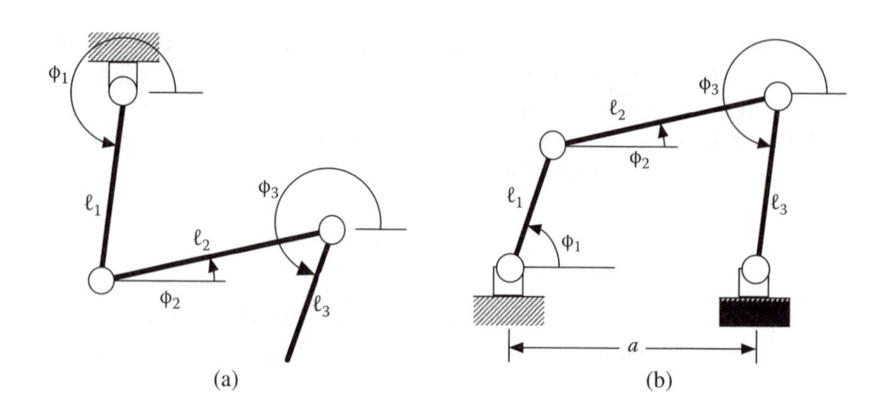

FIGURE 3.10 (a) A triple pendulum with 3 DoF and (b) a four-bar mechanism with 1 DoF.

the triple pendulum has three DoF. Similarly, for the four-bar mechanism shown in Figure 3.10(b), three angles, ϕ_1, ϕ_2, and ϕ_3, define the configuration of the system. However, the angles are not independent. As it will be seen in the upcoming chapters, there exist two algebraic equations that relate the three angles:

$$\begin{aligned} \ell_1 \cos\phi_1 + \ell_2 \cos\phi_2 + \ell_3 \cos\phi_3 - a &= 0 \\ \ell_1 \sin\phi_1 + \ell_2 \sin\phi_2 + \ell_3 \sin\phi_3 &= 0 \end{aligned} \tag{3.25}$$

The two equations can be solved for two of the angles as a function of the third; for example, ϕ_2 and ϕ_3 can be expressed as a function of ϕ_1. Therefore, ϕ_1 is the only variable needed to describe the configuration of the system, and so there is only one DoF for the four-bar mechanism.

In a mechanical system if n_{dof} is the number of DoF, then at least $n_v = n_{dof}$ coordinates are required to completely describe the configuration of the system. If the number of defined coordinates is greater than the number of DoF of the system (i.e., $n_v > n_{dof}$), then any subset of these coordinates that are independent and are equal to the number of DoF is called a set of *independent coordinates*. The remaining coordinates, which may be determined as a function of the independent coordinates, are called *dependent coordinates*. In the triple pendulum example, all the three coordinates are independent whereas in the four-bar linkage example, one of the three coordinates is independent and the other two are dependent.

3.3.3 Constraint Equations

The kinematic relationships between the independent and dependent coordinates are called *constraint equations*. The maximum number of independent constraint equations that can be written for a system is equal to the number of dependent coordinates that has been defined. For example, for the four-bar mechanism of Figure 3.10(b), the two algebraic equations of Eq. (3.25) are the constraint equations. In contrast, for the triple pendulum of Figure 3.10(a), there are no constraint equations because the three defined coordinates are independent.

A constraint equation describing a condition on the array of coordinates of a system is expressed in generic form as:

$$\Phi \equiv \Phi(\mathbf{q}) = 0 \tag{3.26}$$

The constraints on the coordinates are also referred to as *position constraints*. Assuming that there are n_c position constraints between the coordinates of a system, the constraints are expressed as:

$$\boldsymbol{\Phi} \equiv \boldsymbol{\Phi}(\mathbf{q}) = \mathbf{0} \tag{3.27}$$

Position constraints are normally nonlinear algebraic equations.

The first time derivative of the position constraints yields the *velocity constraints*. Velocity constraints provide relationships between the velocity variables that are defined for a system. The time derivative of Eq. (3.27) can be written as:

$$\dot{\boldsymbol{\Phi}} \equiv \mathbf{D}\dot{\mathbf{q}} = \mathbf{0} \tag{3.28}$$

where **D** is called the *Jacobian matrix**. The velocity constraints are linear algebraic equations.

The time derivative of the velocity constraints yields the *acceleration constraints*. The time derivative of Eq. (3.28) is written as:

$$\ddot{\boldsymbol{\Phi}} \equiv \mathbf{D}\ddot{\mathbf{q}} + \dot{\mathbf{D}}\dot{\mathbf{q}} = \mathbf{0} \tag{3.29}$$

Algebraic equality constraints in terms of the coordinates, and perhaps time**, are said to be *holonomic* constraints. In general, if constraint equations contain *inequalities* or relations between velocity components that are *not integrable* in closed form (meaning, one cannot find a position constraint whose time derivative is the given relation), they are said to be *nonholonomic* constraints. Consider, as an example, the pendulum shown in Figure 3.11(a). The rod that connects the mass B to the hinge on the ground is rigid. For the coordinates of B we can write the following holonomic constraint:

$$(x^B)^2 + (y^B)^2 - \ell^2 = 0$$

* Slightly different definitions of the Jacobian matrix may be found in literature. This term is most commonly referred to the matrix of partial derivative of the position constraints with respect to the array of coordinates which may or may not be the same as the coefficient matrix in the velocity constraints. This discrepancy arises when the velocities are not defined to be the time derivative of the coordinates.

** If time appears explicitly in a constraint, we will refer to it as a *driver* constraint. Driver constraints will be discussed separately in other chapters.

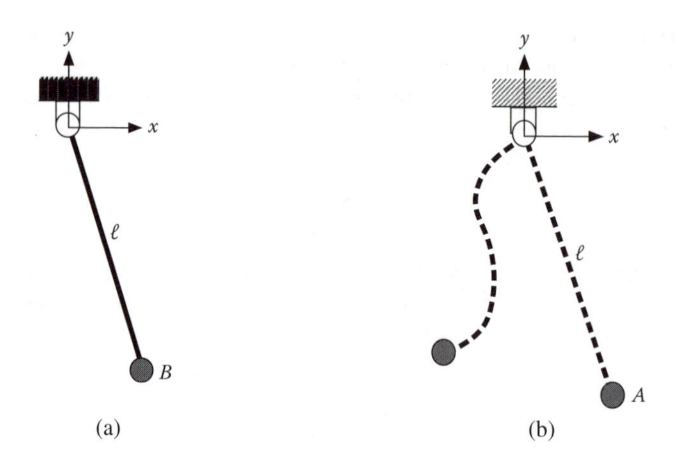

FIGURE 3.11 Example of systems that can be modeled by (a) a holonomic constraint and (b) a nonholonomic constraint.

In the system of Figure 3.11(b), the mass A is connected to the hinge on the ground by a rope. For the coordinates of A we can write the following nonholonomic constraint:

$$(x^A)^2 + (y^A)^2 - \ell^2 \leq 0$$

In this textbook, for brevity, the term *constraint* will refer to a holonomic constraint, unless specified otherwise.

Knowledge of the number of DoF of a mechanism can be useful when constraint equations are being formulated. Often, the pictorial description of a mechanism can be misleading. Several joints may restrict the same DoF and may therefore be *equivalent* or *redundant*. As an example consider the double-crank mechanism shown in Figure 3.12(a). This system has one DoF. If this system is modeled for analysis by four moving bodies and six revolute joints, the set of constraint equations will contain *redundant* equations. The reason for redundancy becomes clear when one of the coupler links is removed to obtain the mechanism shown in Figure 3.12(b). The two mechanisms are kinematically equivalent.

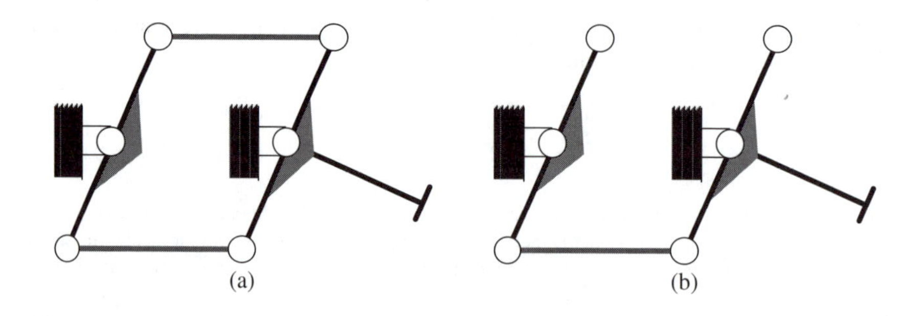

FIGURE 3.12 (a) A double-crank mechanism, and (b) its kinematically equivalent system.

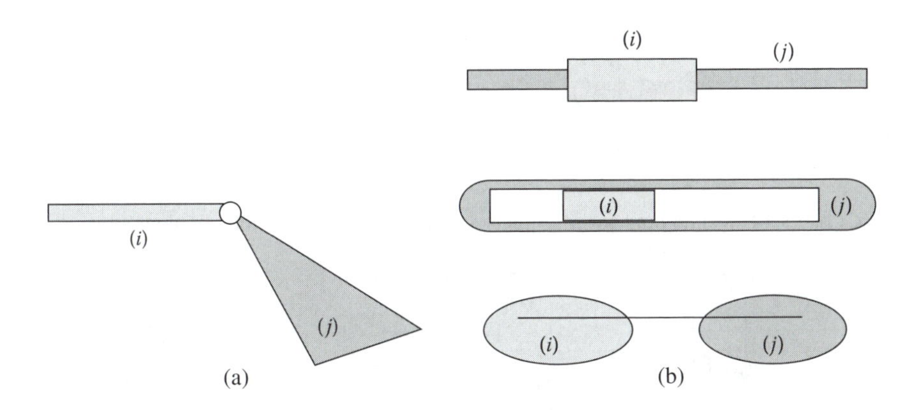

FIGURE 3.13 Schematic representation of (a) revolute joints and (b) translational joints.

In general, for a system having n_c independent constraint equations and n_v coordinates, the number of DoF is determined as:

$$n_{dof} = n_v - n_c \tag{3.30}$$

Redundant equations in a set of constraints can be identified using numerical techniques.

3.3.4 Kinematic Joints

A variety of kinematic joints appear in multibody systems. The most commonly used joints are *revolute* (or pin) and *translational* (or sliding) joints. These are among the joints categorized as the *lower-pairs*. Formulating kinematic relationships for mechanisms containing revolute and translational joints is quite simple. However, kinematic formulation for some *higher-pairs* joints, such as *gears*, *point-followers,* and *cam-followers* could be more difficult.

A revolute joint connects two bodies such that they have a common point. The schematic presentation of a revolute joint between two bodies is shown in Figure 3.13(a). A revolute joint eliminates two DoF between the two bodies.

A translational joint connects two bodies in such a way that they translate relative to each other along a common axis. The axis of relative translational motion, or any axis parallel to that, is called the *line of translation*. A translational joint removes two DoF between the two bodies. This type of joint may appear in different shapes. Schematic representations of three translational joints are shown in Figure 3.13(b).

A composite joint is defined as the combination of two joints and the body between them into a single element. For example the two revolute joints and the connecting link in the system of Figure 3.14 can be combined into a *revolute-revolute* joint connecting bodies i and j. Composite representation of joints and bodies can simplify the kinematic formulations in some problems. Another form of a composite joint is the *revolute-translational* joint shown for several examples in Figure 3.15. In the system shown in (a), body i is connected to body k by a revolute joint and body k is connected to body j by a translational joint. The revolute joint, the

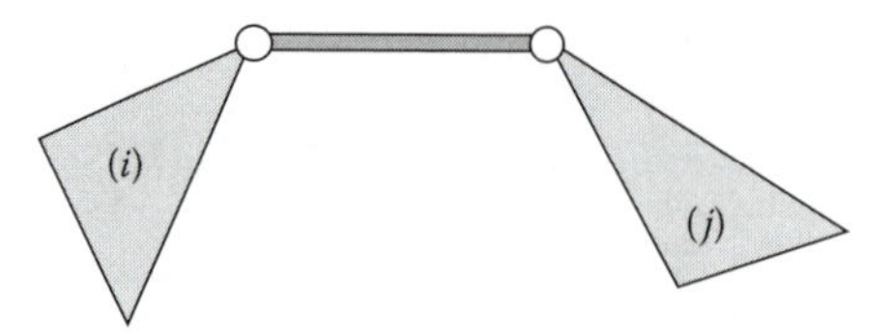

FIGURE 3.14 Combining two pin joints and a body into in revolute-revolute joint.

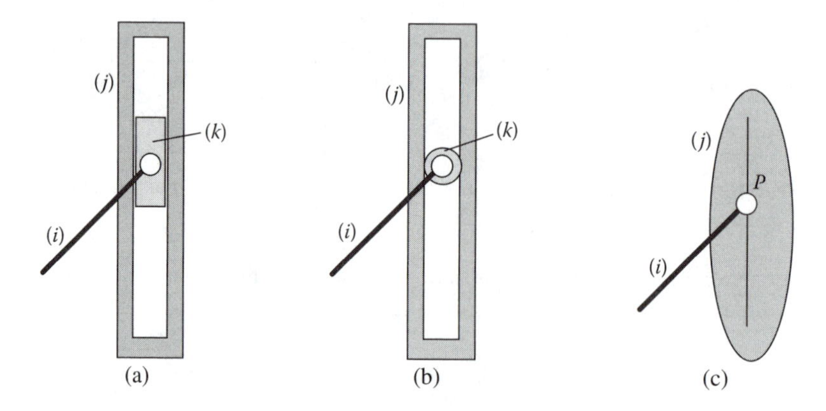

FIGURE 3.15 (a), (b) Two examples of a revolute-translational joint and (c) their schematic representation.

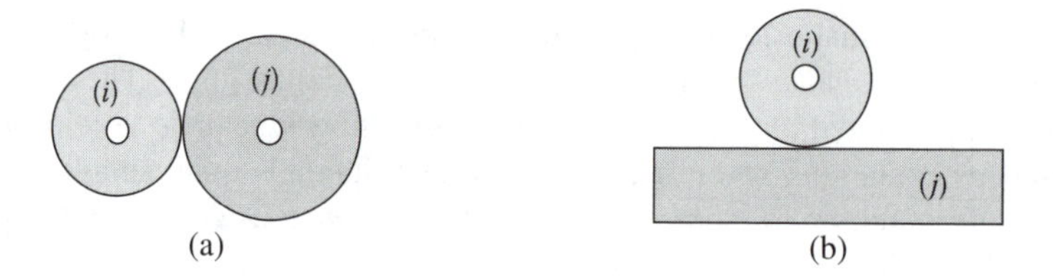

FIGURE 3.16 (a) A pair of spur gears, (b) a rack-and-pinion, and their equivalent systems.

translational joint, and body k can be combined into a composite joint in which point P on body i moves along a well defined axis on body j. A similar composite joint is described in the system shown in (b) by eliminating roller k as a body. A composite joint, such as revolute-revolute or revolute-translational, eliminates one DoF between the two bodies.

Spur gears are used to transmit motion between parallel shafts. A pair of spur gears is shown schematically in Figure 3.16(a) as two rotating discs. It is assumed that there is no slipping between the two circular discs. It should be clear that in this description of spur gears, we are not modeling the intricate contact mechanism between the two gears. A rack-and-pinion can be considered as a special form of a pair of spur gears in which the rack is a portion of a gear having an infinite pitch radius; thus its pitch circle is a straight line. A rack-and-pinion and its equivalent system, consisting of a rolling cylinder on a flat surface (a straight line), are shown in Figure 3.16(b). In

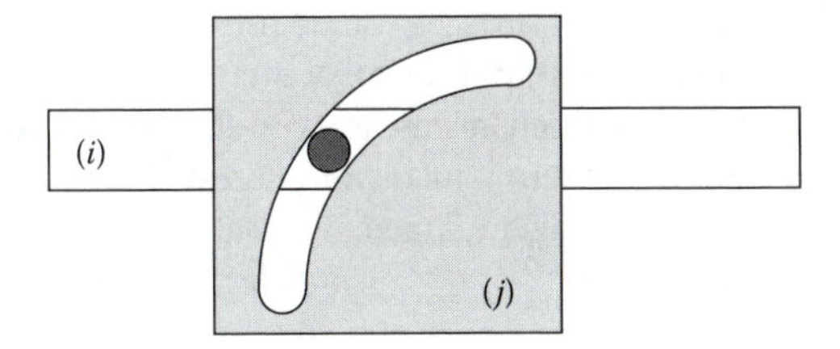

FIGURE 3.17 A point-follower joint.

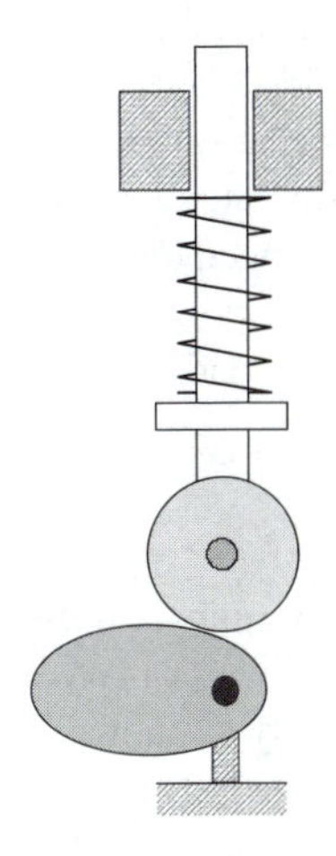

FIGURE 3.18 A cam-follower joint.

these joints, unlike a revolute or a translational joint, a third body or a spring keeps a pair of spur gears or a rack-and-pinion together. Without a third body, or some other mechanisms, the two bodies will not remain their contact to each other.

A point-follower is a joint where a pin on one body can slide in a slot that is cut into a second body. The slot is normally in the form of a curve as shown in Figure 3.17. If the slot is a straight line, then the point-follower joint becomes equivalent to a revolute-translational joint. For kinematic analysis, the shape of the curved slot must be described either analytically in a closed-form expression or numerically as a discretized lookup table.* Hence, it should be obvious that kinematic analysis with this type of joints is more complex than the pin and sliding joints. A point-follower joint eliminates one DoF between the two connecting bodies.

Cam-follower joints appear in a variety of forms where only one is shown in Figure 3.18. These joints, similar to the point-follower joints, require the description

* If the shape of the slot of a point-follower or the shape of a cam is described in a discretized form, we could transform the discretized values into slpine functions. Interested reader should refer to any textbook on numerical methods or go to MATLAB `help spline`.

of the curvature of the contacting surfaces. Sometimes, the contacting surfaces are circular shape, or one can be a line (flat surface), or even a point. These joints normally require a spring, or some other mechanisms, to keep the two bodies together. When the analysis of the chattering in these joints is of interest, then the problem becomes a dynamic analysis and it is no longer a kinematic analysis problem.

3.4 REMARKS

The kinematic definitions and the fundamental formulas for kinematics of particles and rigid bodies that have been discussed in this chapter, and the definitions and formulas for the dynamics of particles and bodies from the next chapter, will be used extensively throughout this textbook. Therefore, it is very important to understand these fundamentals. In addition, because one emphasis of this textbook is on the computational methods of kinematics and dynamics, a reader who is not familiar with programming with MATLAB is encouraged to follow the programming exercises in this and the following chapters closely. The programming exercises in the first few chapters use simple commands and they are intended for inexperienced programmers.

3.5 PROBLEMS

Solve the following problems on paper—do not use MATLAB:

3.1 Vector **s** is attached to a body as shown. Take direct measurements from the Figure and determine:
   (a) Components of the vector in the x-y frame (i.e., **s**).
   (b) Components of the vector in the ξ-η frame (i.e., $\mathbf{s}'$).
   (c) Magnitude of the vector using first the x-y components and then the ξ-η components. Are the two computed magnitudes the same or different? Explain.
   (d) Measure the angle between the x and the ξ axes. Use this angle and test the validity of Eq. (3.4).

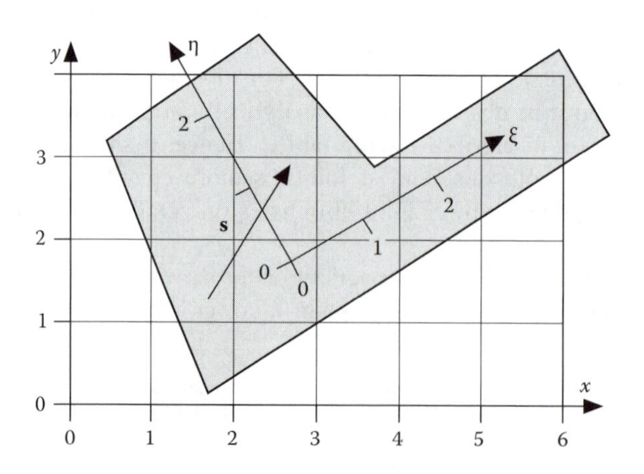

3.2 A link of a mechanism is shown. A body-fixed frame is defined on the link. Determine the local coordinates of the points $A, \cdots, E$ and two unit vectors along the axes T_1 and T_2.

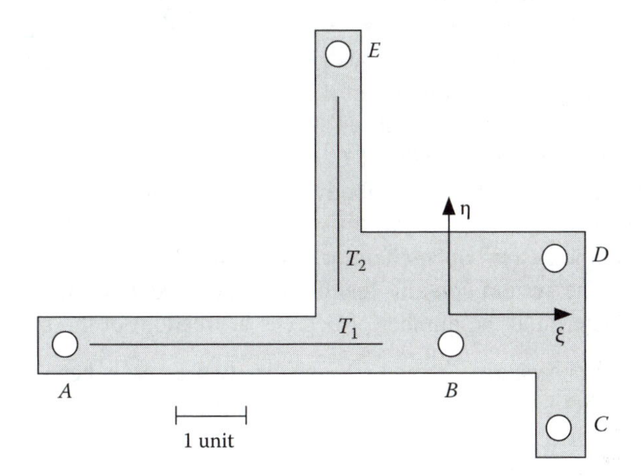

3.3 The ξ-axis of a moving reference frame with respect to a nonmoving x-y frame is shown.
(a) Draw the η-axis on the Figure.
(b) What are the x-y components of unit vectors $\mathbf{u}_{(\xi)}$ and $\mathbf{u}_{(\eta)}$ along the ξ and η axes respectively? Take direct measurements from the Figure.
(c) Construct the **A** matrix using the unit vectors from (b). Check for the orthogonality of the matrix.
(d) Measure the angle of the ξ-axis from the Figure and construct the **A** matrix using Eq. (3.5). Compare the result against that obtained in (c).

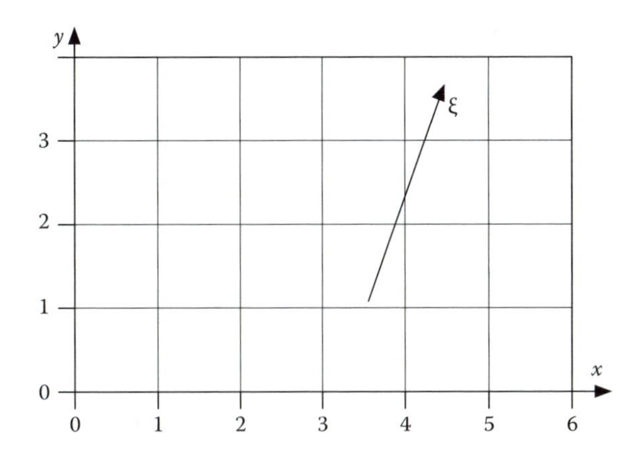

3.4 Point P_i has local and global coordinates $\mathbf{s}_i'^P = \{1.3 \quad -2.2\}'$ and $\mathbf{r}_i^P = \{-1.7 \quad 0.5\}'$. Find the translational coordinates of the body if $\phi_i = 32°$. On graph paper show the orientation of the body-fixed frame with respect to the global frame and locate point P.

3.5 The coordinates of body i are $\mathbf{r}_i = \{3.2 \quad 2.8\}'$ and $\phi_i = 80°$. Points A and B have local coordinates $\mathbf{s}_i'^A = \{-1.1 \quad -0.4\}'$ and $\mathbf{s}_i'^B = \{1.9 \quad 2.3\}'$. Point C has global coordinates $\mathbf{r}_i^C = \{5.3 \quad 4.0\}'$. Find the following:
   (a) The global coordinates of A
   (b) The global components $\mathbf{s}_i^B$
   (c) The local coordinates of C

3.6 For each of the planar multibody systems shown, answer the following questions:
   (a) Is the system open-chain or closed-chain?
   (b) If the system contains any closed chains, identify all of the loops.
   (c) Determine the number of degrees of freedom of the system.

   Note: For systems (h) and (i) consider two cases: where the wheel(s) do(es) not slip; or where the wheel(s) slip(s).

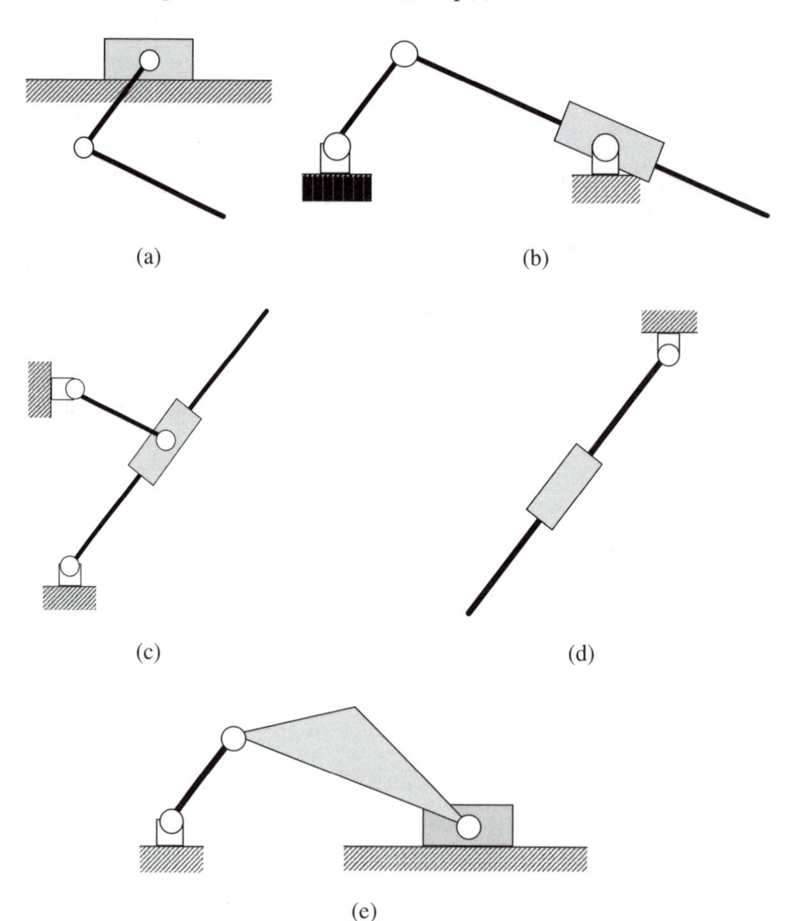

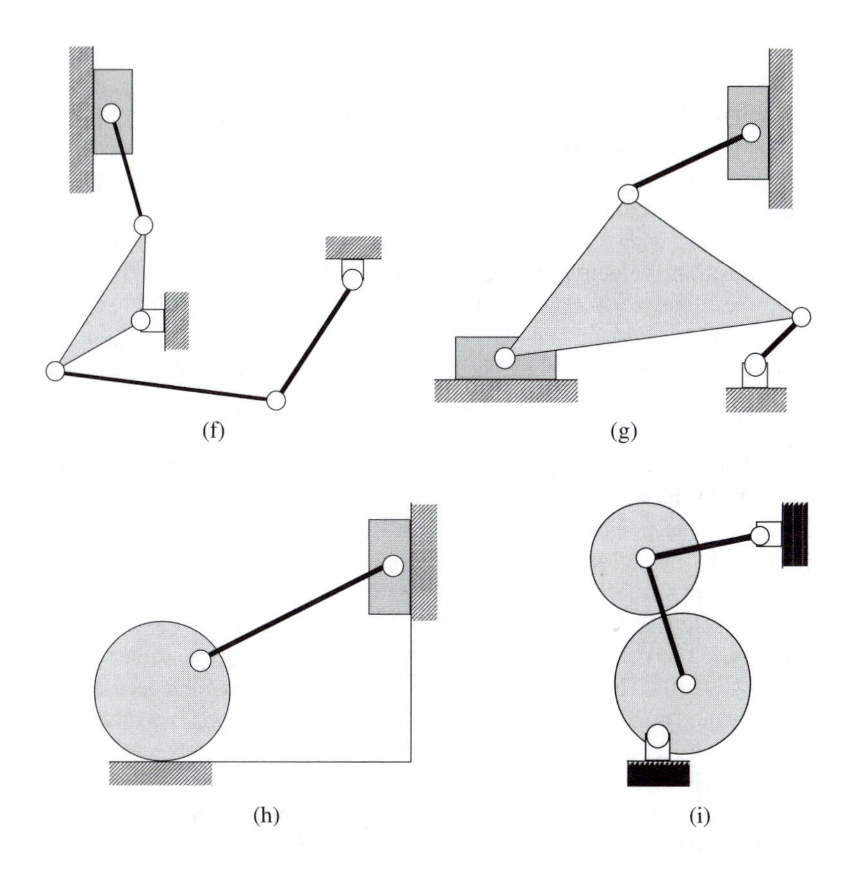

3.7 Consider the following position constraints:

$$\Phi_1 \equiv x_4 + 1.6\cos\phi_4 - 0.3\sin\phi_4 - x_1 + 0.75\sin\phi_1 = 0$$

$$\Phi_2 \equiv y_4 + 1.6\sin\phi_4 + 0.3\cos\phi_4 - y_1 - 0.75\cos\phi_1 = 0$$

(a) Construct the corresponding velocity constraints and express them in the form of Eq. (3.28).
(b) Identify the Jacobian matrix and state the order for which you have arranged the columns of the matrix.
(c) Construct the acceleration constraints and identify the quadratic velocity terms.

Solve the following problems with MATLAB:

3.8 The angle between two reference frames *x*-*y* and ξ-η, according to our convention for defining angles, is $\phi_i = 328°$. The *x*-*y* components of vector **a** are $\mathbf{a} = \{-1.5 \quad 0.6\}'$ and the ξ-η components of vector **b** are $\mathbf{b}' = \{-0.7 \quad -0.4\}'$. Compute $\mathbf{a}'$ and **b**. Make use of the function `A_matrix`.

3.9 Repeat Problem 3.8 for $\phi_i = -32°$ and compare the results. What do you conclude?

3.10 The body-fixed reference frame in Problem 3.2 is positioned in an x-y frame by the following coordinates: $\mathbf{r} = \{2 \quad -3\}'$, $\phi_i = \pi / 6$ rad. Determine the global components of all the $\mathbf{s}^P$ and $\mathbf{u}$ vectors. Then determine the global coordinates of all five points.

3.11 Point Q on a body has local coordinates $\mathbf{s}'^Q = \{1.2 \quad -0.5\}'$. At a given time the position, velocity, and acceleration of the body are: $\mathbf{r} = \{3 \quad 2\}'$, $\phi_i = \pi / 3$ rad, $\dot{\mathbf{r}} = \{-0.4 \quad 1.1\}'$, $\dot{\phi}_i = 0.2\,\pi$ rad/sec, $\ddot{\mathbf{r}} = \{0.2 \quad -0.3\}'$, $\ddot{\phi}_i = -0.1\,\pi$ rad/sec^2.

(a) Determine the absolute position, velocity, and acceleration of point Q. Make use of functions `r_Point`, `r_Point_d`, and `r_Point_dd`.

(b) On engineering paper, draw the x-y frame at a convenient position. Then draw the body-fixed frame and locate point Q. Measure the absolute position of the point and verify your computed results. Show the velocity and acceleration of this point as vectors on the Figure (choose your own scale).

3.12 The absolute position of point B attached to a body is $\mathbf{r}^B = \{5 \quad -2\}'$. The origin of the body-fixed ξ-η frame is located by the position vector $\mathbf{r} = \{4 \quad -1\}'$. The rotational coordinate between the two reference frames is measured as $\phi_i = 25°$. Determine the body-fixed components of vector $\mathbf{s}^B$ (i.e., $\mathbf{s}'^B$).

3.13 Point A on body *(1)* has local coordinates $\mathbf{s}'^A_1 = \{2 \quad -3\}'$. Point B on body *(2)* has local coordinates $\mathbf{s}'^B_2 = \{-1 \quad 2\}'$. At a given time the two bodies have the following coordinates: $\mathbf{r}_1 = \{6 \quad 4\}'$, $\phi_1 = 25°$, $\mathbf{r}_2 = \{-3 \quad 7\}'$, $\phi_2 = -85°$. Determine the distance between A and B.

3.14 The first time derivative of a fixed-length vector $\mathbf{s}$ is computed as $\dot{\mathbf{s}} = \breve{\mathbf{s}}\dot{\phi}$. The second time derivative of $\mathbf{s}$ can be determined either as $\ddot{\mathbf{s}} = \breve{\mathbf{s}}\ddot{\phi} - \mathbf{s}\dot{\phi}^2$ or as $\ddot{\mathbf{s}} = \breve{\mathbf{s}}\ddot{\phi} + \dot{\breve{\mathbf{s}}}\dot{\phi}$. The components of $\mathbf{s}$ are found to be $\mathbf{s} = \{-1 \quad 2\}'$. Numerically show that either formula for $\ddot{\mathbf{s}}$ yields the same answer. Assume $\dot{\phi} = 2.5$ rad/sec and $\ddot{\phi} = -0.8$ rad/sec^2.

4 Fundamentals of Dynamics

This chapter summarizes some of the fundamental concepts of planar dynamics. It is assumed that the reader is already familiar with most of the concepts—we present them more as a reminder. We first review the dynamics of a single particle before considering a system of particles. Then the dynamic equations of motion for a rigid body are derived from the equations of motion of a system of particles. These rigid body equations of motion form the necessary building blocks of the multibody dynamics in the following chapters. Formulations to compute the force or the moment of some commonly used force elements, such as springs and dampers, are discussed. Finally, the concept of Lagrange multipliers for presenting reaction forces within a multiparticle or a multibody system is reviewed.

4.1 NEWTON'S LAWS OF MOTION

Newton's laws of motion form the basis of dynamics. These laws can be stated as:

I. A particle remains at rest or continues to move in a straight line with constant speed if there is no force acting on it.
II. The acceleration of a particle is proportional to and in the same direction of the force acting on it.
III. The forces of action and reaction between interacting particles are equal in magnitude, collinear, and opposite in direction.

Although these laws are for particles, they can be applied to rigid bodies if interpreted properly. Newton's first law is the basis of static or static equilibrium analysis, where the second law provides the basis of dynamic analysis for one or more interacting particles or bodies. Newton's third law can be applied to analyze the *action* and *reaction* forces between contacting bodies or particles.

Newton's second law provides precise meanings to the terms *mass* and *force*. Once a unit mass is chosen, a unit force is defined to be the force necessary to give one unit of mass an acceleration of one unit magnitude.

In the discussion of Newton's laws of motion, it is presumed that position, and hence acceleration, is measured in an *inertial* reference frame. Such a reference frame should technically be defined as one fixed in the stars. For most engineering purposes, an adequate reference frame is an *earth-fixed* reference system. It is important, however, to note that for applications concerning space dynamics, or even in long-range trajectories, the rotation of the earth has a significant effect on the precision with which points can be located by means of Newton's equations of motion.

In such applications, an earth-fixed reference system may be inadequate. In this text we use the nonmoving x-y reference frame as the inertial frame.

4.1.1 Dynamics of a Particle

Newton's second law of motion for a particle relates the total force acting on the particle, **f**, the mass of the particle, m, and the acceleration, **a**, as $m\,\mathbf{a} = \mathbf{f}$ or

$$m\,\ddot{\mathbf{r}} = \mathbf{f} \tag{4.1}$$

where it is assumed that a consistent system of units is used. Equation (4.1) is expanded and matrix forms are written as

$$\begin{aligned} m\,\ddot{x} &= f_{(x)} \\ m\,\ddot{y} &= f_{(y)} \end{aligned} \qquad \begin{bmatrix} m & 0 \\ 0 & m \end{bmatrix} \begin{Bmatrix} \ddot{x} \\ \ddot{y} \end{Bmatrix} = \begin{Bmatrix} f_{(x)} \\ f_{(y)} \end{Bmatrix}$$

The condition for particle *equilibrium* (Newton's first law) may be deduced from Eq. (4.1); that is, a particle remains at rest (in equilibrium), or in a state of *constant velocity*, if the total force that acts on the particle is zero. In a state of equilibrium $\ddot{\mathbf{r}} = \mathbf{0}$, because $\mathbf{f} = \mathbf{0}$.

4.1.2 Dynamics of a System of Particles

The governing laws of dynamics of individual particles are now extended to systems of interacting particles. The equations of motion for such systems can be written simply as the collection of equations of motion for all the particles taken individually.

Consider the system of n_p particles shown in Figure 4.1 where a typical particle i has a mass m^i and is located by a position vector $\mathbf{r}^i$. External forces may act on some of the particles; for example, an external force, $\mathbf{f}^i$, acts on particle i.

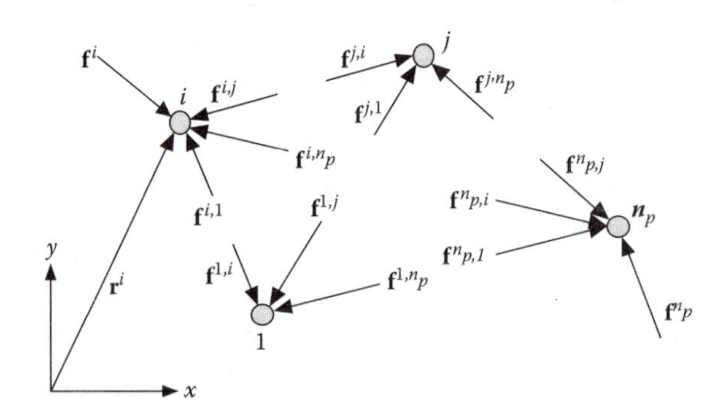

FIGURE 4.1 A system of n_p particles.

We define the total mass of the system as the sum of the individual masses,

$$m \equiv \sum_{i=1}^{n_p} m^i \tag{4.2}$$

The sum of the external forces that act on the system is denoted as

$$\mathbf{f} = \sum_{i=1}^{n_p} \mathbf{f}^i \tag{4.3}$$

We locate the *center of mass* (or *centroid*) of the system of particles, as shown in Figure 4.2, by vector $\mathbf{r}^C$. This vector is defined as

$$\mathbf{r}^C \equiv \frac{1}{m} \sum_{i=1}^{n_p} m^i \, \mathbf{r}^i \tag{4.4}$$

Then Newton's second law can describe the motion of the mass center of the system as

$$m\ddot{\mathbf{r}}^C = \mathbf{f} \tag{4.5}$$

This equation states that the resultant of all external forces on any system of mass equals the total mass of the system times the acceleration of the mass center; that is, the center of the mass moves as if it were a particle of mass m under the action of the force $\mathbf{f}$. This result will be the basis for deriving the equations of motion for a rigid body.

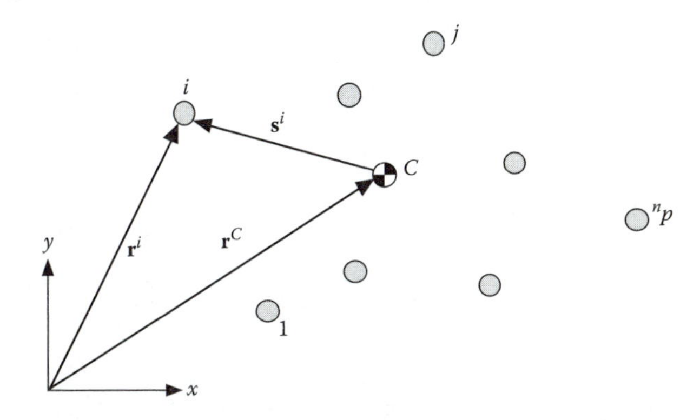

FIGURE 4.2 Center of mass for a system of particles.

Proof 4.1

The following is a derivation of the equation of motion for the mass center of a system of particles:

In addition to external forces, internal forces also act between particles of a system. Assume that the internal forces that act on a typical particle i are $\mathbf{f}^{i,j}$; $j = 1, \cdots, n_p$, where $\mathbf{f}^{i,i} = \mathbf{0}$. The total force acting on particle i is the summation of external and internal forces. Thus, for particle i, Eq. (4.1) becomes

$$m^i \ddot{\mathbf{r}}^i = \mathbf{f}^i + \sum_{j=1}^{n_p} \mathbf{f}^{i,j};\ i = 1, \cdots, n_p \tag{a}$$

This system of equations describes the motion of the system of n_p particles.

Because the forces of interaction between particles in a system must satisfy Newton's law of action and reaction, the force on particle i due to particle j must be equal to the negative of the force on particle j due to particle i; that is,

$$\mathbf{f}^{i,j} = -\mathbf{f}^{j,i};\ i, j = 1, \cdots, n_p \tag{b}$$

Note that because $\mathbf{f}^{i,i} = \mathbf{0}$, *(b)* also holds for $i = j$. If the external forces acting on each particle are known and the nature of the forces acting between particles i and j are also known, the system of equations in *(a)* may be written explicitly. However, the force of interaction between particles will generally depend on the position of the particles, and on how the particles are connected to one another.

We write *(a)* for each particle in the system and then the equations are summed to obtain

$$\sum_{i=1}^{n_p} m^i \ddot{\mathbf{r}}^i = \sum_{i=1}^{n_p} \mathbf{f}^i + \sum_{i=1}^{n_p} \sum_{j=1}^{n_p} \mathbf{f}^{i,j} \tag{c}$$

The double sum in *(c)* contains both $\mathbf{f}^{j,i}$ and $\mathbf{f}^{i,j}$, and hence from *(b)* it is found that

$$\sum_{i=1}^{n_p} \sum_{j=1}^{n_p} \mathbf{f}^{i,j} = \mathbf{0} \tag{d}$$

Thus, *(c)* reduces to

$$\sum_{i=1}^{n_p} m^i \ddot{\mathbf{r}}^i = \sum_{i=1}^{n_p} \mathbf{f}^i \tag{e}$$

Substituting Eq. (4.2), Eq. (4.3), and the second time derivative of Eq. (4.4) in *(e)* yields Eq. (4.5).

Assume that particle i is positioned from the center of mass by vector $\mathbf{s}^i$; $i = 1, \cdots, n_p$, as shown in Fig. 3.5. Therefore,

$$\mathbf{r}^i = \mathbf{r}^C + \mathbf{s}^i; \; i = 1, \cdots, n_p \tag{4.6}$$

One useful characteristic of the center of mass of a system of particles is obtained by substituting Eq. (4.6) into Eq. (4.4):

$$\sum_{i=1}^{n_p} m^i \, \mathbf{s}^i = \mathbf{0} \tag{4.7}$$

This is known as the equation of *first moments.*

EXAMPLE 4.1

The data for the vector of coordinates, the mass, and the vector of applied (external) force for a system of three particles is given as:

$$\mathbf{r}^1 = \begin{Bmatrix} 2 \\ 1 \end{Bmatrix}, \; m^1 = 2, \; \mathbf{f}^1 = \begin{Bmatrix} -6 \\ -2 \end{Bmatrix}, \; \mathbf{r}^2 = \begin{Bmatrix} 1 \\ 4 \end{Bmatrix}, \; m^2 = 5, \; \mathbf{f}^2 = \begin{Bmatrix} 0 \\ 0 \end{Bmatrix},$$

$$\mathbf{r}^3 = \begin{Bmatrix} -3 \\ 2 \end{Bmatrix}, \; m^3 = 3, \; \mathbf{f}^3 = \begin{Bmatrix} 5 \\ -1 \end{Bmatrix}$$

Determine: (a) the position of the mass center; (b) the acceleration of the mass center; (c) numerically verify Eq. (4.7).

Solution:

The following program can perform the computation.

```
% Define coordinates, masses and applied forces for the particles
    r_1 = [2; 1];     r_2 = [1; 4];     r_3 = [-3; 2];
    m_1 = 2;          m_2 = 5;          m_3 = 3;
    f_1 = [-6; -2];   f_2 = [0; 0];     f_3 = [5; -1];
% Compute the total mass
    m = m_1 + m_2 + m_3;
% Compute the coordinates of mass center
    r_C = (m_1*r_1 + m_2*r_2 + m_3*r_3)/m
```

```
% Compute the total force vector
    f = f_1 + f_2 + f_3;
% Compute the acceleration of mass center
    r_C_dd = f/m
% Compute s_i vectors
    s_1 = r_1 - r_C; s_2 = r_2 - r_C; s_3 = r_3 - r_C;
% Compute the first moment equation
    sigma_m_s = m_1*s_1 + m_2*s_2 + m_3*s_3
```

The program provides the following results for (a) and (b):

```
r_C =          r_C_dd =
          0          -0.1000
     2.8000          -0.3000
```

For (c), evaluating the first moment equation yields a zero vector within the limits of the round-off error:

```
sigma_m_s =
    1.0e-14 *
          0
     0.1776
```

4.2 DYNAMICS OF A BODY

A rigid body is regarded as a collection of a very large number of particles. In addition, from the definition of rigidity, the location of the particles in a body relative to one another remains unchanged.

The *center of mass* or the *centroid* of a body is defined in the same way as that of a system of particles, where the *summation* over all particles is replaced by an *integral* over the *area* of the planar body. If a particle is replaced by an infinitesimal mass *dm*, positioned at an arbitrary point *i* on the area of body, as shown in Figure 4.3, Eqs. (4.4) and (4.7) can be written as

$$\mathbf{r} \equiv \frac{1}{m} \int_{(area)} \mathbf{r}^i dm \tag{4.8}$$

and

$$\int_{(area)} \mathbf{s}^i \, dm = \mathbf{0} \tag{4.9}$$

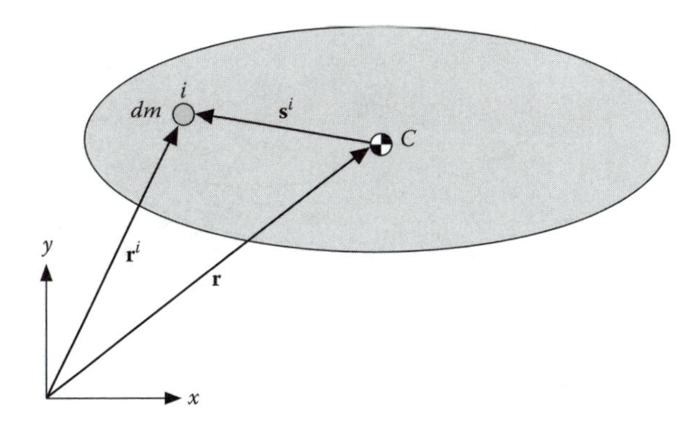

FIGURE 4.3 A body is viewed as a collection of infinitesimal masses.

The first and second time derivatives of Eq. (4.9) yield two other useful identities as

$$\int_{(area)} \dot{\mathbf{s}}^i \, dm = \mathbf{0}, \quad \int_{(area)} \ddot{\mathbf{s}}^i \, dm = \mathbf{0} \tag{4.10}$$

Because a rigid body can be considered as a collection of infinite number of particles, Newton's second law also describes how the mass center of a body accelerates under the application of a force. In addition to its tendency to move a body in the direction of its application, a force also tends to rotate a body. The measure of this tendency is known as the *moment* of the force about a point or a given axis. The moment of a force is also referred to as *torque*.

4.2.1 Moment of a Force

Consider a force, $\mathbf{f}$, acting on a body at point P as shown in Figure 4.4. The moment of this force about an axis perpendicular to the plane that passes through point O can be

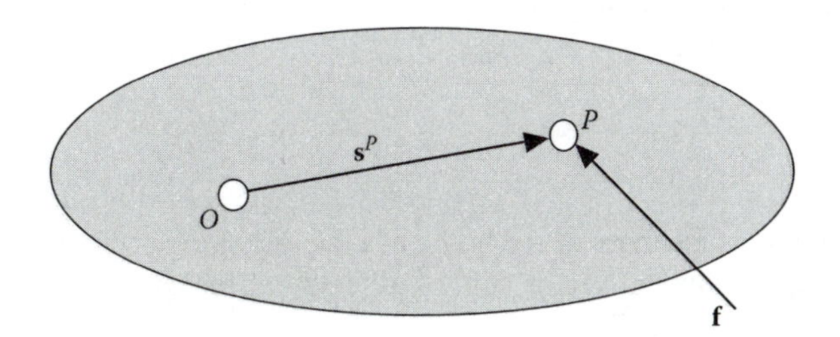

FIGURE 4.4 A force acting on point P of a body.

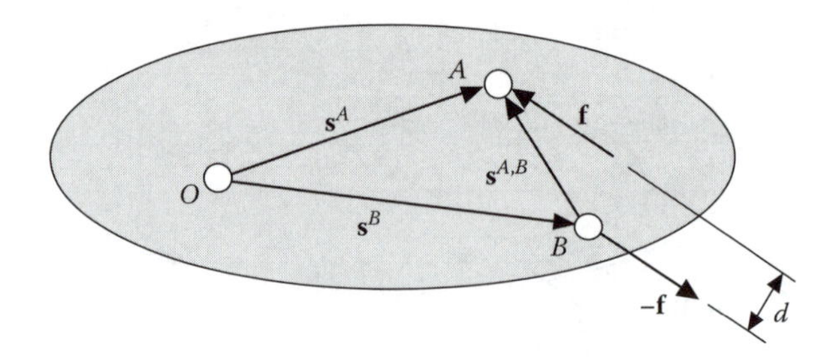

FIGURE 4.5 A couple results into a pure moment.

computed from the cross product of vectors $\mathbf{s}^P$ and $\mathbf{f}$. According to Eq. (2.13) this cross product can be obtained as

$$n = \breve{\mathbf{s}}^{P\prime}\,\mathbf{f} \tag{4.11}$$

where vector $\mathbf{s}^P$ locates point P from point O.*

The moment produced by two equal, opposite, and collinear forces is known as a *couple*. Couples have certain unique properties. The resultant force of a couple that acts on a body is zero. Figure 4.5 shows two equal and opposite forces $\mathbf{f}$ and $-\mathbf{f}$ acting on a body at points A and B. Vector $\mathbf{s}^{A,B}$ joins point A to B. These two points are located by position vectors $\mathbf{s}^A$ and $\mathbf{s}^B$ from an arbitrary point O. The combined moment of the two forces about O is

$$n = \breve{\mathbf{s}}^{A\prime}\,\mathbf{f} + \breve{\mathbf{s}}^{B\prime}(-\mathbf{f}) = (\breve{\mathbf{s}}^A - \breve{\mathbf{s}}^B)'\,\mathbf{f} = \breve{\mathbf{s}}'^{A,B}\,\mathbf{f} \tag{4.12}$$

Note that the magnitude of the moment is $n = d\,f$, where d is the distance between the lines of action of the two forces; that is, the magnitude of this moment is independent of the position of point O. The moment of a couple is a free vector; hence, sometimes it is referred to as a *pure moment*.

4.2.2 Centroidal Equations of Motion

The equation of motion for the mass center of a rigid body is directly obtained from Newton's second law as

$$m\ddot{\mathbf{r}} = \mathbf{f} \tag{4.13}$$

In this equation m is the mass of the body, $\ddot{\mathbf{r}}$ is the vector of acceleration of the mass center, and $\mathbf{f}$ represents the vector sum of all external forces acting on the body.

* Equation (4.11) simply represents the cross product of two planar vectors. In vector notation, this equation is expressed as $\mathbf{n} = \mathbf{s}^P \times \mathbf{f}$ where $\mathbf{n}$ is a vector perpendicular to the plane.

In addition to the two translational DoF that a free body exhibits on a plane, a free body also exhibits one rotational DoF. The rotational equation of motion of a body is described as

$$J\,\ddot{\phi} = n \tag{4.14}$$

where J is the *polar moment of inertia* of the body*, $\ddot{\phi}$ is the angular acceleration of the body, and n is the sum of all the moments that act on the body. These moments can be the result of forces that act on the body and also pure moments.

Equations (4.13) and (4.14) can be appended and put in matrix form:

$$\begin{bmatrix} m\mathbf{I} & \mathbf{0} \\ \mathbf{0} & J \end{bmatrix} \begin{Bmatrix} \ddot{\mathbf{r}} \\ \ddot{\phi} \end{Bmatrix} = \begin{Bmatrix} \mathbf{f} \\ n \end{Bmatrix} \quad \begin{bmatrix} m & 0 & 0 \\ 0 & m & 0 \\ 0 & 0 & J \end{bmatrix} \begin{Bmatrix} \ddot{x} \\ \ddot{y} \\ \ddot{\phi} \end{Bmatrix} = \begin{Bmatrix} f_{(x)} \\ f_{(y)} \\ n \end{Bmatrix} \tag{4.15}$$

These equations may also be presented in compact form as

$$\mathbf{M}\,\ddot{\mathbf{c}} = \mathbf{h} \tag{4.16}$$

where $\ddot{\mathbf{c}} \equiv \begin{Bmatrix} \ddot{\mathbf{r}} \\ \ddot{\phi} \end{Bmatrix}$ is the *array of accelerations*, $\mathbf{h} \equiv \begin{Bmatrix} \mathbf{f} \\ n \end{Bmatrix}$ is the *array of forces*, and $\mathbf{M} \equiv \begin{bmatrix} m\mathbf{I} & \mathbf{0} \\ \mathbf{0} & J \end{bmatrix}$ is the *mass* or *inertia matrix*. Equation (4.15) is referred to as the *centroidal equations of motion*, because the reference point on the body is the body mass center. In this text, the term *equations of motion* will be used to refer to the centroidal equations of motion unless specified otherwise.

PROOF 4.2

The following is the derivation of the rotational equation of motion for a planar rigid body as described in Eq. (4.14). For this purpose we return to the idea that a rigid body can be viewed as a system of particles. We substitute Eq. (3.17) into Eq. *(a)* from Proof 4.1 to get

$$m^i\,(\ddot{\mathbf{r}} + \breve{\mathbf{s}}^i\,\ddot{\phi} - \mathbf{s}^i\,\dot{\phi}^2) = \mathbf{f}^i + \sum_{j=1}^{n_p} \mathbf{f}^{i,j};\; i = 1, \cdots, n_p \tag{a}$$

We premultiply both sides of this equation by the transpose of $\breve{\mathbf{s}}^i$:

$$m^i\,\breve{\mathbf{s}}^{i\prime}(\ddot{\mathbf{r}} + \breve{\mathbf{s}}^i\,\ddot{\phi} - \mathbf{s}^i\,\dot{\phi}^2) = \breve{\mathbf{s}}^{i\prime}(\mathbf{f}^i + \sum_{j=1}^{n_p} \mathbf{f}^{i,j});\; i = 1, \cdots, n_p \tag{b}$$

* Refer to Appendix A for a more detailed discussion on the mass center and moment of inertia of common and uncommon geometrically shaped planar bodies.

We then sum up all the n_p equations in *(b)* to obtain

$$\sum_{i=1}^{n_p}\left(m^i\,\breve{\mathbf{s}}^{i\prime}(\ddot{\mathbf{r}}+\breve{\mathbf{s}}^i\,\ddot{\phi}-\mathbf{s}^i\,\dot{\phi}^2)\right)=\sum_{i=1}^{n_p}(\breve{\mathbf{s}}^{i\prime}\mathbf{f}^i)+\sum_{i=1}^{n_p}(\breve{\mathbf{s}}^{i\prime}\sum_{j=1}^{n_p}\mathbf{f}^{i,j}) \quad (c)$$

We now examine the terms in *(c)* individually. The first term in *(c)* can be written as

$$\sum_{i=1}^{n_p}(m^i\,\breve{\mathbf{s}}^{i\prime}\ddot{\mathbf{r}})=\left(\sum_{i=1}^{n_p}(m^i\,\breve{\mathbf{s}}^{i\prime})\right)\ddot{\mathbf{r}}=0 \quad (d)$$

where we have used Eq. (4.7). Note that Eq. (4.7) is still valid if all the vectors are rotated in the same direction by $90°$ and/or written in transposed form.

The second term in *(c)* is expressed as

$$\sum_{i=1}^{n_p}(m^i\,\breve{\mathbf{s}}^{i\prime}\,\breve{\mathbf{s}}^i\,\ddot{\phi})=\left(\sum_{i=1}^{n_p}m^i\,s^{i2}\right)\ddot{\phi}=J\ddot{\phi} \quad (e)$$

where we have applied Eq. (3.7) and defined the *moment of inertia* for the system as

$$J\equiv\sum_{i=1}^{n_p}m^i\,s^{i2} \quad (f)$$

The third term in *(c)* becomes

$$\sum_{i=1}^{n_p}(m^i\,\breve{\mathbf{s}}^{i\prime}\mathbf{s}^i\,)\,\dot{\phi}^2=0 \quad (g)$$

because the scalar product of two perpendicular vectors is zero.

The first term on the right-hand side of *(c)* is the sum of the moments caused by all external forces; that is,

$$n\equiv\sum_{i=1}^{n_p}(\breve{\mathbf{s}}^{i\prime}\mathbf{f}^i) \quad (h)$$

The second term on the right-hand side of *(c)* can be expressed as

$$\begin{aligned}\sum_{i=1}^{n_p}(\breve{\mathbf{s}}^{i\prime}\sum_{j=1}^{n_p}\mathbf{f}^{i,j}) &= \cdots+\breve{\mathbf{s}}^{i\prime}\mathbf{f}^{i,j}+\breve{\mathbf{s}}^{j\prime}\mathbf{f}^{j,i}+\cdots=\cdots+\breve{\mathbf{s}}^{i\prime}\mathbf{f}^{i,j}-\breve{\mathbf{s}}^{j\prime}\mathbf{f}^{i,j}+\cdots \\ &= \cdots+(\breve{\mathbf{s}}^{i}-\breve{\mathbf{s}}^{j})'\mathbf{f}^{i,j}+\cdots=\cdots+\breve{\mathbf{s}}^{i,j\prime}\mathbf{f}^{i,j}+\cdots=0\end{aligned} \quad (i)$$

where we have noted that vectors $\mathbf{s}^{i,j}$ and $\mathbf{f}^{i,j}$ are collinear, as shown in Figure 4.6, therefore the scalar product of $\breve{\mathbf{s}}^{i,j}$ and $\mathbf{f}^{i,j}$ is zero.

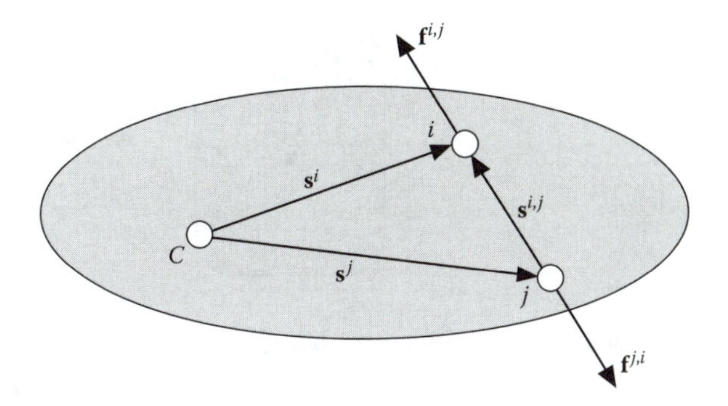

FIGURE 4.6 Reaction forces between two infinitesimal masses.

The preceding process can be repeated with an integral over area replacing the summation to obtain the rotational equation of motion for a planar body as stated in Eq. (4.14). In the process, the *polar moment of inertia* is defined as

$$J \equiv \int_{(area)} (s^i)^2 dm \qquad (j)$$

We should note that *(j)* provides the moment of inertia about an axis going through the origin regardless of whether the origin is at the body mass center or not. For objects with known geometrical shapes, lookup tables provide simple expressions to compute the moment of inertia.

Example 4.2

A body with a mass $m = 2$ is acted upon by gravity, a constant force, and a pure moment as shown in Figure 4.7. The magnitude of the pure moment is 0.6. The local coordinates of the application point of the force and the constant components of the force are

$$\boldsymbol{s}'^{P} = \begin{Bmatrix} -0.2 \\ 0.3 \end{Bmatrix}, \quad \mathbf{f}^{P} = \begin{Bmatrix} 1.2 \\ 0.5 \end{Bmatrix}$$

Determine the array of forces for the body if $\phi = 30°$ and $\mathbf{r} = \{2.1 \quad 1.6\}'$.

Solution

The weight of the body is $w = 2 \times 9.81 = 19.62$. The angular orientation of the body, ϕ, is used to determine the transformation matrix $\mathbf{A}$. Then the components

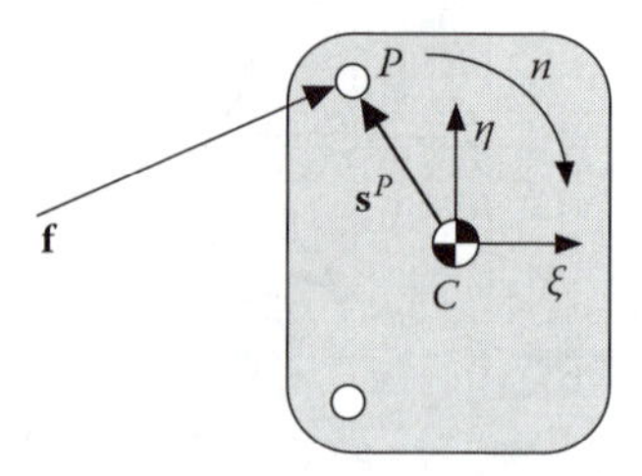

FIGURE 4.7 A force and a moment acting on a body.

of vector $\mathbf{s}^P$are computed from Eq. (3.8). The moment of the force is computed, based on Eq. (4.11), to be -0.35. Therefore, the array of forces for this body is:

$$\mathbf{g} = \begin{Bmatrix} 1.2 \\ 0.5 - 19.62 \\ -0.35 - 0.6 \end{Bmatrix} = \begin{Bmatrix} 1.2 \\ -19.12 \\ -0.95 \end{Bmatrix}$$

Example 4.3

This is an extension of Example 3.1. In addition to the coordinate data given in Example 3.1, assume that two external forces act on the link at points A and B:

$$\mathbf{f}^A = \begin{Bmatrix} 2 \\ -1 \end{Bmatrix}, \ \mathbf{f}^B = \begin{Bmatrix} -3 \\ 2 \end{Bmatrix}$$

The mass and the polar moment of inertia for the link are provided as $m = 2.5$ and $J = 1.2$. Compute the linear and angular acceleration of the body.

Solution

In the following program, Eqs. (4.11), (4.13) and (4.14) are implemented to compute the accelerations.

```
% Define coordinates
    r = [2.5; 1.2]; phi = 5.67;
% Define constant coordinates
    s_A_p = [2.18; 0]; s_B_p = [-1.8; 1.3];
% Define mass and polar moment of inertia
    m = 2.5; J = 1.2;
```

```
% Define forces at A and B
    f_A = [2; -1]; f_B = [-3; 2];
% Compute matrix A
    A = A_matrix(phi);
% Compute global coordinates of A and B
    s_A = A*s_A_p; s_B = A*s_B_p;
% Compute the sum of forces
    f = f_A + f_B;
% Compute the sum of moments
    n = (s_rot (s_A))'*f_A + (s_rot(s_B))'*f_B;
% Compute accelerations
    r_dd = f/m
    phi_dd = n/J
```

The results are:

```
r_dd =                    phi_dd =
   -0.4000                      4.6461
    0.4000
```

In this program, we have taken advantage of `function s_rot` in computing the moment of a force. Note that our programs must have access to all the necessary function M-files.

4.2.3 Noncentroidal Equations of Motion

If the reference point on a body is not defined at its mass center, the corresponding equations of motion are called *noncentroidal equations of motion*. For example, for the body shown in Figure 4.8, the mass center is at C but the reference point is defined at O. The noncentroidal equations of motion for this body are expressed in matrix form as

$$\begin{bmatrix} m\mathbf{I} & m\breve{\rho} \\ m\breve{\rho}' & J^O \end{bmatrix} \begin{Bmatrix} \ddot{\mathbf{r}}^O \\ \ddot{\phi} \end{Bmatrix} = \begin{Bmatrix} \mathbf{f}^O \\ n^O \end{Bmatrix} \qquad {}^{(n-c)}\mathbf{M}\,{}^{(n-c)}\mathbf{c} = {}^{(n-c)}\mathbf{h} \tag{4.17}$$

where ρ locates the mass center from point O and J^O is the *moment of inertia* with respect to an axis passing through O which is computed as

$$J^O = J + m\rho^2 \tag{4.18}$$

In addition a new array of forces is defined as

$$\mathbf{f}^O = \mathbf{f}^P + m\rho\,\dot{\phi}^2 \tag{4.19}$$

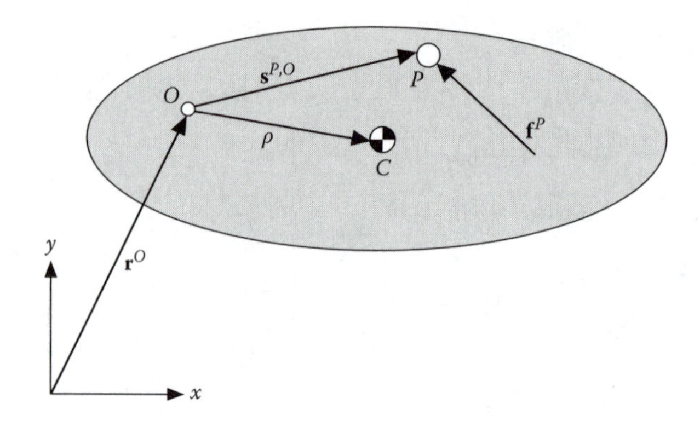

FIGURE 4.8 The body reference point is at O, not at the mass center C.

and n^O is defined as the moment acting on the body with respect to point O. This moment is computed as

$$n^O = \breve{\mathbf{s}}^{P,O\prime}\mathbf{f}^P \tag{4.20}$$

In the compact form of Eq. (4.17), the following arrays and matrix have been defined:

$$^{(n-c)}\mathbf{c} \equiv \begin{Bmatrix} \mathbf{r}^O \\ \phi \end{Bmatrix}, \quad ^{(n-c)}\mathbf{h} \equiv \begin{Bmatrix} \mathbf{f}^O \\ n^O \end{Bmatrix}, \quad ^{(n-c)}\mathbf{M} \equiv \begin{bmatrix} m\mathbf{I} & m\rho \\ m\rho' & J^O \end{bmatrix}$$

Proof 4.3

The following is the derivation of the noncentroidal equations of motion. The centroidal equations of motion can be transformed to the noncentroidal form through a simple process. We first rewrite the centroidal equations of motion, from Eq. (4.15), as

$$\begin{bmatrix} m\mathbf{I} & \mathbf{0} \\ \mathbf{0} & J^C \end{bmatrix} \begin{Bmatrix} \ddot{\mathbf{r}}^C \\ \ddot{\phi} \end{Bmatrix} = \begin{Bmatrix} \mathbf{f} \\ n^C \end{Bmatrix} \tag{a}$$

where we have emphasized that the body mass center, point C, is the reference point as shown in Figure 4.9. To simplify the process of transforming *(a)* into the noncentroidal equations of motion, we assume that a single force $\mathbf{f}^P$ acts at point P on the body. Therefore in *(a)* $\mathbf{f} = \mathbf{f}^P$ and $n^C = \breve{\mathbf{s}}^{P,C\prime}\mathbf{f}^P$. This assumption does not make the resultant equation any less general—other forces and moments can be added to the system.

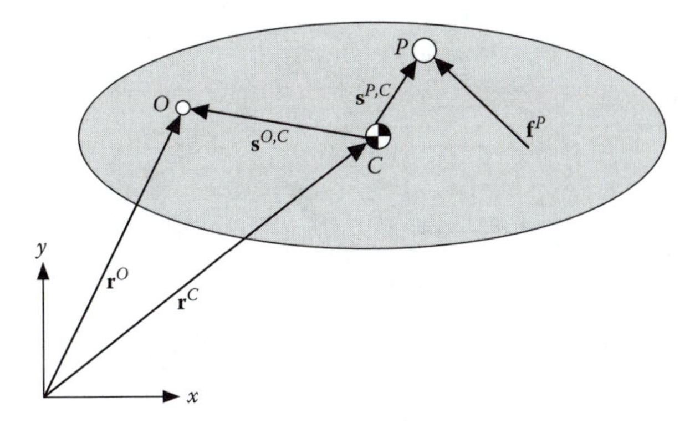

FIGURE 4.9 Vector descriptions for the centroidal and noncentroidal systems.

The acceleration of the new reference point O can be obtained from Eq. (3.17) as

$$\ddot{\mathbf{r}}^O = \ddot{\mathbf{r}}^C + \breve{\mathbf{s}}^{O,C}\,\ddot{\phi} - \mathbf{s}^{O,C}\,\dot{\phi}^2 \qquad (b)$$

By referring to Figure 4.8 and Figure 4.9, we note that $\mathbf{s}^{O,C} = -\rho$. Therefore *(b)* is rewritten as

$$\ddot{\mathbf{r}}^C = \ddot{\mathbf{r}}^O + \breve{\rho}\,\ddot{\phi} - \rho\,\dot{\phi}^2 \qquad (c)$$

This is a transformation equation for the translational acceleration of the body. The rotational acceleration of the body does not depend on the location of the body origin, or on how the body-fixed frame is attached to the body (note that we have not defined a body-fixed frame yet). Hence, a more complete transformation equation for both the translational and rotational accelerations can be written as

$$\begin{Bmatrix} \ddot{\mathbf{r}}^C \\ \ddot{\phi} \end{Bmatrix} = \begin{bmatrix} \mathbf{I} & \breve{\rho} \\ 0 & 1 \end{bmatrix} \begin{Bmatrix} \ddot{\mathbf{r}}^O \\ \ddot{\phi} \end{Bmatrix} - \begin{Bmatrix} \rho\,\dot{\phi}^2 \\ 0 \end{Bmatrix} \qquad (d)$$

We substitute *(d)* into *(a)*, then premultiply both sides of the equation by the transpose of the coefficient matrix in *(d)*:

$$\begin{bmatrix} \mathbf{I} & 0 \\ \breve{\rho}' & 1 \end{bmatrix} \begin{bmatrix} m\mathbf{I} & \mathbf{0} \\ \mathbf{0} & J^C \end{bmatrix} \left(\begin{bmatrix} \mathbf{I} & \breve{\rho} \\ 0 & 1 \end{bmatrix} \begin{Bmatrix} \ddot{\mathbf{r}}^O \\ \ddot{\phi} \end{Bmatrix} - \begin{Bmatrix} \rho\,\dot{\phi}^2 \\ 0 \end{Bmatrix} \right) = \begin{bmatrix} \mathbf{I} & 0 \\ \breve{\rho}' & 1 \end{bmatrix} \begin{Bmatrix} \mathbf{f}^P \\ \breve{\mathbf{s}}^{P,C\prime}\mathbf{f}^P \end{Bmatrix}$$

or

$$\begin{bmatrix} m\mathbf{I} & m\breve{\rho} \\ m\breve{\rho}' & J^C + m\rho^2 \end{bmatrix} \begin{Bmatrix} \ddot{\mathbf{r}}^O \\ \ddot{\phi} \end{Bmatrix} = \begin{Bmatrix} \mathbf{f}^P + m\rho\,\dot{\phi}^2 \\ \breve{\mathbf{s}}^{P,O\prime}\,\mathbf{f}^P \end{Bmatrix} \qquad (e)$$

where the identities $\breve{\rho}'\rho = 0$, $\rho^2 = \breve{\rho}'\breve{\rho}$, $\mathbf{s}^{P,O} = \mathbf{s}^{P,C} + \rho$ and have been used. Equation *(e)* is the noncentroidal equation of motion as described in Eq. (4.17).

FIGURE 4.10 An actuator in the unloading process of a dump truck.

4.3 FORCE ELEMENTS

Some mechanical systems contain such elements as springs, dampers, and actuators. These are called force elements. In our formulations, we assume that the force elements are ideal components that do not impose any kinematic constraints on a system. Therefore, we can ignore them in kinematic modeling. The actual effects of force elements will be considered in the dynamic modeling.

One example of a force element is the actuator of a dump truck shown in Figure 4.10. Another example is the combined spring-damper element of a double A-arm suspension system shown in Figure 4.11(a). If the combined spring-damper element is removed, we obtain the four-bar linkage system shown in Figure 4.11(b). The removal of the force elements from this system does not influence the kinematics of the suspension mechanism.

We may, however, encounter a system where a force element, in addition to a pair of forces, also introduces some form of constraints into the system. For kinematic analysis we can remove the force of the element but we must keep its contribution to the system as a joint. As an example consider the frontal planar view of a MacPherson suspension system shown in Figure 4.12(a). The spring can be

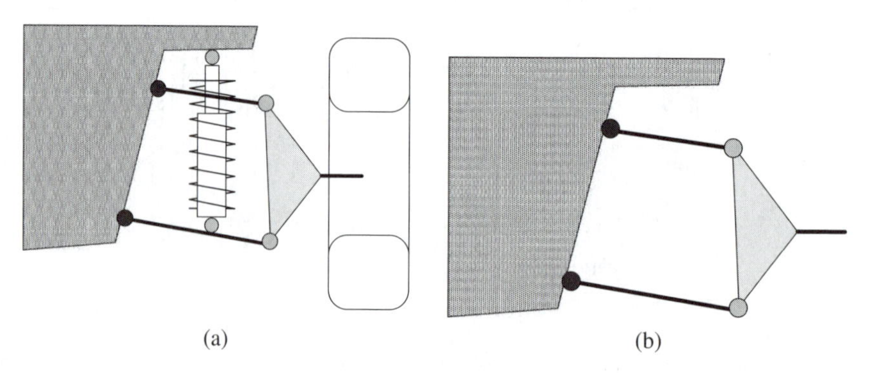

FIGURE 4.11 A double A-arm suspension system.

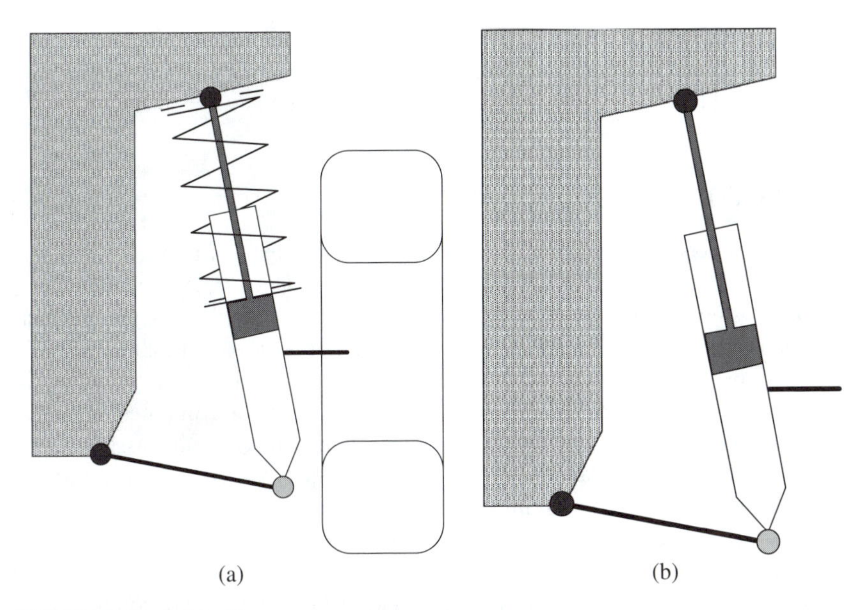

FIGURE 4.12 A MacPherson suspension system.

removed from this system without any effect on its kinematics. However, the damper (shock absorber) cannot be removed completely because it acts as a force element and as a sliding joint. We can remove the force effect of the damper but we must leave its constraints as a joint as shown in Figure 4.12(b).

In some other systems, the removal of a force element may change the kinematic behavior of the system completely. In such cases, most often if a force element is removed, the resultant system will have a higher number of degrees of freedom than the original system. As an example consider the cam-follower system shown in Figure 4.13(a). The spring keeps the cam and the follower in contact at all times (if we do not consider

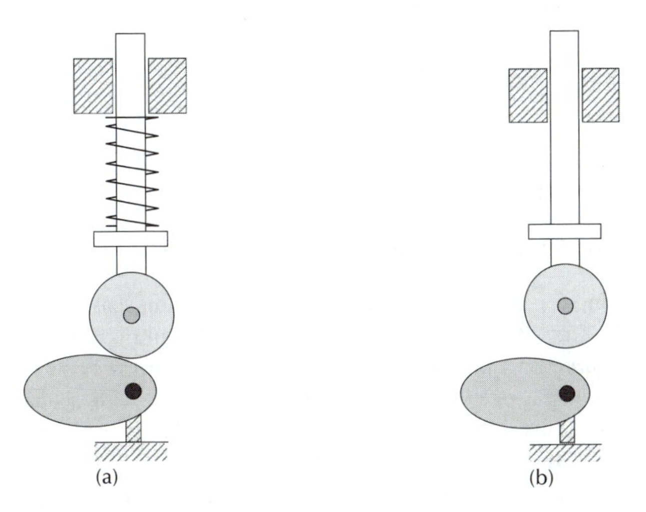

FIGURE 4.13 A cam-follower system.

possible chattering). If the spring is removed, the cam and the follower will not remain in contact as shown in Figure 4.13(b). For kinematic analysis of such systems, we must provide additional *fictitious* constraints to keep the bodies in contact.

4.4 APPLIED FORCES

The force that acts on a link may be categorized into either an applied force or a reaction force. Applied forces are produced by *force elements*. An *applied* force, also known as an *external* force, is either a constant or a function of time. Furthermore, a nonconstant force could be a function of the position coordinates and or a function of the velocities. Therefore, in general, if the positions and velocities of the links are known, the applied force between the links can be computed. Reaction forces, on the other hand, are functions of the applied forces and possibly some other factors. Computing reaction forces between bodies of a system, in general, is not as straight forward as determining the applied forces.

In this section we consider some of the commonly used force elements. Except for the gravitational force that acts on individual links, force elements act between two links. In such a case, we assume that the position of the two points and, if necessary, the velocity of the two points are known. In the following formulations we assume that the force element is connected between two points, A and B, with known positions and velocities $\mathbf{r}^A$, $\mathbf{r}^B$, $\dot{\mathbf{r}}^A$, and $\dot{\mathbf{r}}^B$.

4.4.1 Gravitational Force

One typical force that may act on a body is the gravitational force. As long as the mass of the body is known, the force of gravity can be computed. In this textbook, the gravitational field will be considered to be acting in the negative y-direction unless indicated otherwise. This choice of direction for the gravity is totally arbitrary. If w is the *weight* (mass times the gravitational constant), then the contribution of this force to the array of force of the body is

$$^{(g)}\mathbf{f} = \begin{Bmatrix} 0 \\ -w \end{Bmatrix} \tag{4.21}$$

The gravitational force acts directly at the mass center of a body and, therefore, there is no moment associated with it.

4.4.2 Point-to-Point Actuator

A point-to-point (or translational) force element, in this case an actuator, acts between points A and B on two different links, as shown in Figure 4.14. An actuator provides a *constant* or a *time-dependent* pair of forces on the two end points. The forces making up the pair are in opposite directions but with a common line of action. The sign convention for the pair of forces can be defined as *positive* when the forces *pull* on the links and *negative* when the forces *push* on the links. If the actuator force is denoted by $^{(a)}f$, then $^{(a)}f > 0$ constitutes a pull (tension) and $^{(a)}f < 0$ constitutes a push (compression).

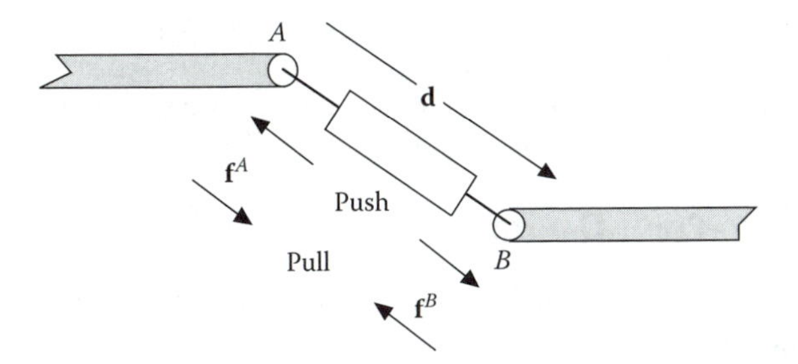

FIGURE 4.14 A force element acting between two points of two different links.

To determine the components of the pair of forces (i.e., ${}^{(a)}\mathbf{f}^A$ and ${}^{(a)}\mathbf{f}^B$), a unit vector on the line of action of the actuator must be defined. First, a vector $\mathbf{d}$ connecting the two points, from A to B, is defined:

$$\mathbf{d} = \mathbf{r}^B - \mathbf{r}^A \tag{4.22}$$

The magnitude of this vector is found as

$$\ell^2 = \mathbf{d}'\mathbf{d} = (x^B - x^A)^2 + (y^B - y^A)^2 \tag{4.23}$$

A unit vector $\mathbf{u}$ is computed as

$$\mathbf{u} = \frac{1}{\ell}\mathbf{d} \tag{4.24}$$

The unit vector $\mathbf{u}$ has the same direction as ${}^{(a)}\mathbf{f}^A$ in the case of a pull and ${}^{(a)}\mathbf{f}^B$ in the case of a push. Therefore,

$${}^{(a)}\mathbf{f}^A = {}^{(a)}f\,\mathbf{u},\ {}^{(a)}\mathbf{f}^B = -\,{}^{(a)}f\,\mathbf{u} \tag{4.25}$$

Note that because ${}^{(a)}f$ could be either positive or negative, the sign convention in Eq. (4.25) is automatically taken care of. The x- and y-components of ${}^{(a)}\mathbf{f}^A$ and ${}^{(a)}\mathbf{f}^B$ contribute to the vector of force for their corresponding bodies.

4.4.3 Point-to-Point Spring

Point-to-point (or translational) springs are the most commonly used force elements in mechanical systems. Assume that the force element in Figure 4.14 is a spring that is attached between points A and B. The force of this spring is computed as

$${}^{(s)}f = k(\ell - {}^{0}\ell) \tag{4.26}$$

where k is the *spring stiffness*, ℓ is the deformed length and ${}^{0}\ell$ is the undeformed length of the spring. The deformed length of a spring is found from Eq. (4.23).

The sign convention for a spring force is similar to that of the actuator force—positive in *tension* (pull) and negative in *compression* (push). The forces of the spring acting on two bodies are

$${}^{(s)}\mathbf{f}^A = {}^{(s)}f\,\mathbf{u},\ {}^{(s)}\mathbf{f}^B = -\,{}^{(s)}f\,\mathbf{u} \tag{4.27}$$

where the unit vector **u** is computed from Eq. (4.24). Equation (4.27) is valid in both tension and compression: if $\ell > {}^0\ell$ (for tension), ${}^{(s)}f$ is positive so the bodies are pulled toward each other; if $\ell < {}^0\ell$ (for compression), ${}^{(s)}f$ is negative so the bodies are pushed away from one another.

4.4.4 Point-to-Point Damper

Assume that the force element in Figure 4.14 is a point-to-point damper. The damping force is computed as

$$ {}^{(d)}f = d_c \dot{\ell} \tag{4.28} $$

where d_c is the damping coefficient and $\dot{\ell}$ is the time rate of change of the damper length. We can compute $\dot{\ell}$ by taking the time derivative of Eq. (4.23):

$$ 2\ell\dot{\ell} = 2\mathbf{d}'\dot{\mathbf{d}} = 2(x_j - x_i)(\dot{x}_j - \dot{x}_i) + 2(y_j - y_i)(\dot{y}_j - \dot{y}_i) $$

or

$$ \dot{\ell} = \frac{1}{\ell}\mathbf{d}'\dot{\mathbf{d}} = \frac{1}{\ell}[(x_j^P - x_i^P)(\dot{x}_j^P - \dot{x}_i^P) + (y_j^P - y_i^P)(\dot{y}_j^P - \dot{y}_i^P)] \tag{4.29} $$

The sign convention for the damper force is defined as positive for $\dot{\ell} > 0$ and negative for $\dot{\ell} < 0$. Because a damper opposes the motion between two bodies, when the bodies move away from each other (when $\dot{\ell} > 0$), the forces of the damper exhibit a pull, and when the bodies move toward each other ($\dot{\ell} < 0$), the forces of the damper exhibit a push.

By defining a unit vector **u** along **d** as in Eq. (4.24), we compute the forces acting on the bodies as

$$ {}^{(d)}\mathbf{f}^A = {}^{(d)}f\,\mathbf{u}, \quad {}^{(d)}\mathbf{f}^B = -\,{}^{(d)}f\,\mathbf{u} \tag{4.30} $$

These equations are valid for both pull and push cases.

4.4.5 Combined Elements

A spring, a damper, and an actuator can act as a combined force element between points A and B of two links. Because the point of application of the forces from each element is the same, we may combine the forces into a single force as

$$ {}^{(s,d,a)}f = {}^{(s)}f + {}^{(d)}f + {}^{(a)}f \tag{4.31} $$

Then the vector of forces are computed as

$$ {}^{(s,d,a)}\mathbf{f}^A = {}^{(s,d,a)}f\,\mathbf{u}, \quad {}^{(s,d,a)}\mathbf{f}^B = -\,{}^{(s,d,a)}f\,\mathbf{u} \tag{4.32} $$

The following MATLAB® program presents a function that computes the forces of a combined element that is connected between two points A and B. It is assumed that the characteristics of the combined element, and the coordinates and velocities of the two points are sent to the function as known quantities.

EXAMPLE 4.4

Two bodies are connected by a point-to-point combined spring-damper-actuator between point A on link 1 and point B on link 2 as shown in Figure 4.15. The two points are positioned in their respective ξ-η frame as

$$\mathbf{s}_1'^A = \begin{Bmatrix} 0.15 \\ 0 \end{Bmatrix}, \ \mathbf{s}_2'^B = \begin{Bmatrix} 0 \\ 0.1 \end{Bmatrix}$$

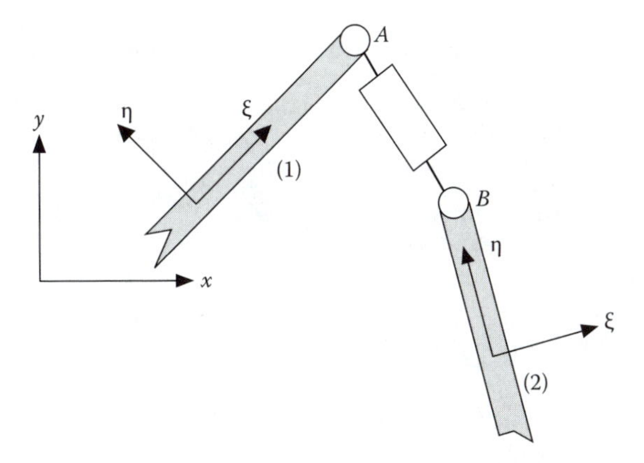

FIGURE 4.15 A combined force element acting between two bodies.

The actuator pushes on the bodies with a force of 2 units. The spring has an undeformed length ${}^0\ell = 0.15$ and a stiffness $k = 10$. The damping coefficient is $d_c = 5$. In the shown configuration, the two bodies have the following coordinates and velocities:

$$\mathbf{c}_1 = \begin{Bmatrix} -0.1 \\ 0.05 \\ 0.785 \end{Bmatrix}, \quad \mathbf{c}_2 = \begin{Bmatrix} 0.1 \\ -0.05 \\ 0.262 \end{Bmatrix}, \quad \dot{\mathbf{c}}_1 = \begin{Bmatrix} 0.1 \\ 0.2 \\ -0.25 \end{Bmatrix}, \quad \dot{\mathbf{c}}_2 = \begin{Bmatrix} -0.2 \\ 0.1 \\ 0.12 \end{Bmatrix}$$

Determine the force and the moment that act on each link.

Solution

We use Eqs. (3.8), (3.9), and (3.13) in the following MATLAB program to perform the computations. In addition, a new M-file is developed to compute the force of a combined point-to-point spring-damper-actuator element.

```
% Define body-fixed vectors
   s_A1_p = [0.15; 0]; s_B2_p = [0; 0.1];
% Define coordinates (inertial) and velocities
   r1 = [-0.1; 0.05]; phi1 = 0.785;
   r2 = [ 0.1; -0.05]; phi2 = 0.262;
```

```
    r1_d = [ 0.1; 0.2]; phi1_d = -0.25;
    r2_d = [-0.2; 0.1]; phi2_d = 0.12;
% Define spring-damper-actiator constants
    L_0 = 0.15; k = 10; d_c = 5; f_a = -2;
% Compute A matrices
    A1 = A_matrix(phi1); A2 = A_matrix(phi2);
% Compute global components of s_A1 and s_B2
    s_A1 = A1*s_A1_p; s_B2 = A2*s_B2_p;
% Compute global coordinates of A1 and B2
    r_A1 = r1 + s_A1
    r_B2 = r2 + s_B2
% Compute velocity of A1 and B2
    r_A1_d = r1_d + s_rot(s_A1)*phi1_d
    r_B2_d = r2_d + s_rot(s_B2)*phi2_d
% Compute vector d between A and B and its time derivative
    d = r_B2 - r_A1
    d_d = r_B2_d - r_A1_d
% Compute the force of the spring-damper
    f_sda = pp_sda (d, d_d, k, L_0, d_c, f_a)
    f_A1 = f_sda
    f_B2 = -f_sda
% Compute each moment
    n1 = s_rot(s_A1)'*f_A1
    n2 = s_rot(s_B2)'*f_B2
```

```
function f_sda = pp_sda(d, d_d, k, L_0, d_c, f_a)
% This function computes the force of a point-to-point combined
% spring-damper-actuator element along vector d
% Compute length of d and unit vector u
    L = sqrt(d'*d);
    u = d/L;
% Compute L_d
    L_d = d'*d_d/L;
% Compute spring and damper forces
    f_s = k*(L - L_0);
    f_d = d_c*L_d;
% Compute the sum of forces
    f = f_s + f_d + f_a;
% Compute the vector force along u
    f_sda = f * u;
```

The main program first computes the following position and velocity vectors:

```
r_A1 =              r_B2 =
    0.0061              0.0741
    0.1560              0.0466

r_A1_d =            r_B2_d =
    0.1265             -0.2116
    0.1735              0.0969
```

Then $\mathbf{d} = \mathbf{r}_2^B - \mathbf{r}_1^A$ and $\dot{\mathbf{d}} = \dot{\mathbf{r}}_2^B - \dot{\mathbf{r}}_1^A$ are computed:

```
d =                d_d =
  0.0680            -0.3381
 -0.1094            -0.0766
```

Function `pp_sda` is then used to compute the element force. This yields

```
f_sda =
  -1.4663
   2.3601
```

If we report the intermediate computational results within function `pp_sda,` we will obtain:

$$\ell = 0.129,\ \dot{\ell} = -0.113,\ {}^{(s)}f = -0.212,\ {}^{(d)}f = -0.567$$

$${}^{(s,d,a)}f = -2 - 0.217 - 0.567 = -2.779,\ \mathbf{u} = \begin{Bmatrix} 0.528 \\ -0.849 \end{Bmatrix}$$

Since vector **d** (or **u**) points in the direction of B, we get the following force vectors:

```
f_A1 =                     f_B2 =
  -1.4663                     1.4663
   2.3601                    -2.3601
```

The moments are computed to be

```
n1 =                       n2 =
   0.4059                    -0.0805
```

4.4.6 Rotational Elements

Similar to the point-to-point force elements, we can also have *rotational* (or *torsional*) force (moment) elements between two links. We assume that a rotational force element acts about the axis of a pin joint that is already defined between the two links, and it applies a pair of moments, equal in magnitude but opposite in direction, on the links. A typical force element, such as a spring, is shown in Figure 4.16.

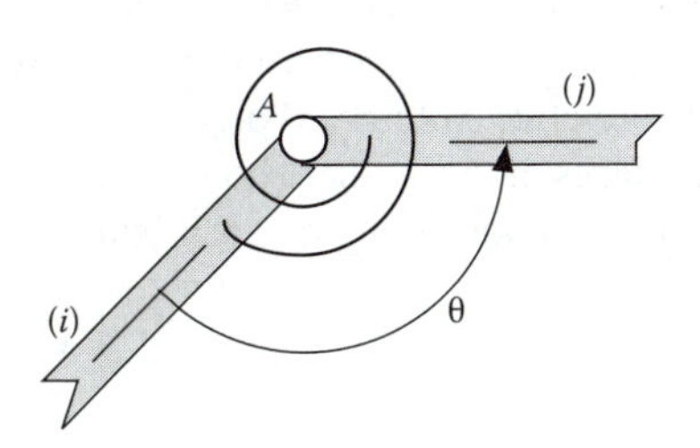

FIGURE 4.16 A rotational force element acting between two links.

A *rotational actuator* is a motor that applies known moments, ${}^{(r-a)}n$ and $-{}^{(r-a)}n$, either constant or time varying, on the connecting bodies. We can adopt a sign convention that if the applied moment on a body is counterclockwise, we consider the moment positive, otherwise it is a negative moment.

The moment associated with a rotational spring between two bodies is calculated as

$${}^{(r-s)}n = {}^{(r)}k\ (\theta - {}^{0}\theta) \tag{4.33}$$

where ${}^{(r)}k$ is the *rotational stiffness* of the spring, and θ is a measure of the relative rotational displacement between the two links. This angle may be defined as the angle between two defined vectors on the links. We observe that when $\theta > {}^{0}\theta$, the computed moment, ${}^{(r-s)}n$, is positive, which indicates the spring being in *tension*, and when $\theta < {}^{0}\theta$ the spring is in *compression*.

Similarly, for a rotational damper between two bodies the moment of the damper is computed as

$${}^{(r-d)}n = {}^{(r)}d_c\ \dot{\theta} \tag{4.34}$$

where ${}^{(r)}d_c$ is the rotational damping coefficient and $\dot{\theta}$ is the relative rotational velocity between the two links.

For a combined rotational spring-damper-actuator the individual moments can be added to obtain a combined moment as

$${}^{(r-s,d,a)}n = {}^{(r-s)}n + {}^{(r-d)}n + {}^{(r-a)}n \tag{4.35}$$

This moment can then be applied to bodies i and j as

$$n_i = {}^{(r-s,d,a)}n\ ,\quad n_j = -{}^{(r-s,d,a)}n \tag{4.36}$$

In the upcoming chapters we will discuss in more detail how to compute θ and $\dot{\theta}$ between two links. The formulation depends on whether we use points or bodies to model a multibody system.

4.4.7 Viscous Friction

Viscous friction (or damping) between two surfaces can be considered as another type of applied force because it is proportional to the relative velocity of two contacting surfaces. A point-to-point or a rotational damper provides a form of viscous friction. Knowing the coefficient of viscous friction and the relative velocity, one can compute the force due to viscous friction. This force always acts in the opposite direction of the relative velocity. For example, if the relative velocity between the contacting surfaces of two bodies shown in Figure 4.17(a) is $\mathbf{v}_{j,i} = -\mathbf{v}_{i,j}$, then the force due to viscous friction acting on each body is computed as

$${}^{(vf)}\mathbf{f}_i = -{}^{(vf)}d_c\ \mathbf{v}_{i,j}\ ,\quad {}^{(vf)}\mathbf{f}_j = -{}^{(vf)}d_c\ \mathbf{v}_{j,i} \tag{4.37}$$

where ${}^{(vf)}d_c$ is the coefficient of viscous friction and the negative sign implies that the force acts in the opposite direction of the velocity.

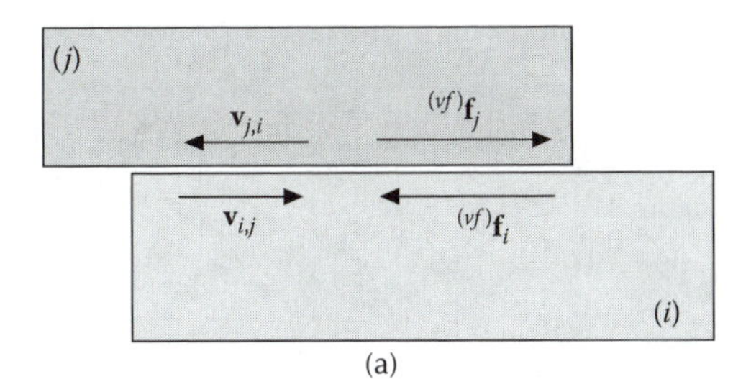

(a)

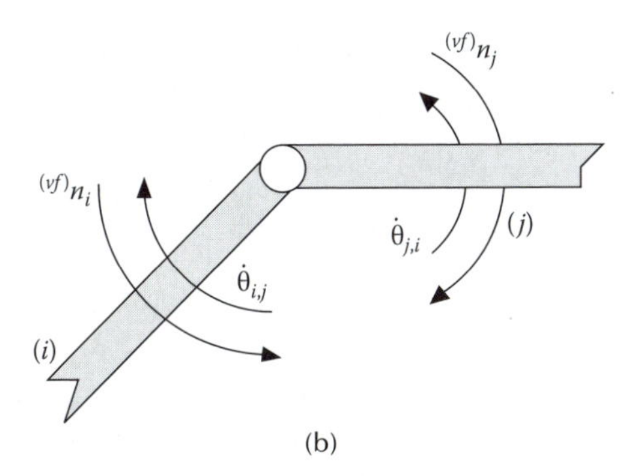

(b)

FIGURE 4.17 Viscous friction between (a) two flat and (b) two circular surfaces.

Similar formula can be written to compute the moment due to viscous friction between two circular surfaces, such as the contacting surfaces about a pin joint, as shown in Figure 4.17(b):

$$^{(vf)}n_i = -{}^{(r,vf)}d_c\,\dot{\theta}_{i,j},\quad {}^{(vf)}n_j = -{}^{(r,vf)}d_c\,\dot{\theta}_{j,i} \tag{4.38}$$

In this equation $^{(r,vf)}d_c$ is the coefficient of rotational viscous friction and $\dot{\theta}_{i,j} = -\dot{\theta}_{j,i}$ is the relative angular velocity between the two bodies.

It is common to describe the magnitude of viscous force as a linear function of the magnitude of velocity. However, some other models may describe the force as a quadratic function of the magnitude of velocity. The appropriate friction model must be described based on the characteristics of the contacting surfaces.

4.5 REACTION FORCE

When kinematic joints are present and the corresponding kinematic constraints appear explicitly in the equations of motion, the system is referred to as a constrained system. This means that reaction forces must be incorporated into the equations of motion.

An applied force is either a constant, or a function of coordinates and/or velocities. A constant force that can act on a body (or a particle) is the gravitational force. An actuator, for example, provides a known pair of forces between two bodies regardless of their positions or velocities. A spring force depends on the relative

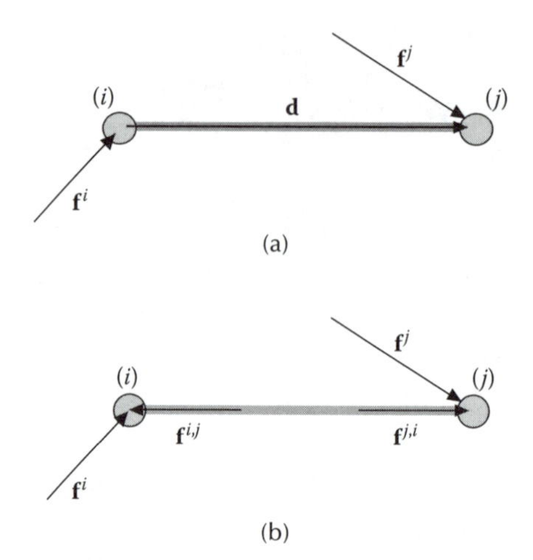

FIGURE 4.18 Two particles connected by a rod.

positions of its attachment points. The force of a damper is a function of the velocity of its attachment points. Such forces must be computed in a system of bodies or particles before determining the accelerations.

Reaction or *internal* forces, on the other hand, not only are a function of coordinates and velocities, but also a function of accelerations and applied forces. For systems in which reaction forces exist, normally the applied forces are known but the reaction forces and the accelerations are the unknowns.

4.5.1 Newton's Third Law

Newton's third law of reaction force between two particles can directly be correlated with the constraint equation representing a massless rod between two particles. Consider two particles i and j connected by a rod of length ℓ as shown in Figure 4.18(a). The particles are acted upon by external forces $\mathbf{f}^i$ and $\mathbf{f}^j$. According to Newton's third law of motion, the two particles apply reaction forces on one another equal in magnitude but opposite in direction, along the axis of the rod (collinear). These reaction forces are shown as $\mathbf{f}^{j,i}$ and $\mathbf{f}^{i,j}$ in Figure 4.18(b) with arbitrarily assigned directions. The components of $\mathbf{f}^{j,i}$ and $\mathbf{f}^{i,j}$ represent four unknowns. However, Newton's third law tells us that $\mathbf{f}^{i,j} = -\mathbf{f}^{j,i}$ and, furthermore, $\mathbf{f}^{j,i}$ and $\mathbf{f}^{i,j}$ are along the axis of the rod (i.e., vector $\mathbf{d}$). Because the reaction forces are along $\mathbf{d}$, they can be expressed as

$$\mathbf{f}^{j,i} = \lambda\,\mathbf{d} \qquad \mathbf{f}^{i,j} = -\lambda\,\mathbf{d} \tag{4.39}$$

where λ is a scale factor. Equation (4.39) shows that the four unknown components of the reaction forces can be described in term of a single unknown λ.

The discussion for two particles constrained by a rod can be extended to a system of particles or a system of bodies. The number of unknown components of reaction forces can always be reduced if we introduce the concept of Lagrange multipliers.

4.5.2 Method of Lagrange Multipliers

The reaction forces between two particles connected by a rod are described in Eq. (4.39). These forces can be stacked in an array as:

$$^{(c)}\mathbf{f} \equiv \begin{Bmatrix} \mathbf{f}^{i,j} \\ \mathbf{f}^{j,i} \end{Bmatrix} = \begin{bmatrix} -\mathbf{d} \\ \mathbf{d} \end{bmatrix} \lambda \tag{4.40}$$

As we will see in Chapters 5 and 6, the coefficient matrix of λ is the transpose of the Jacobian of the constraint between the two particles. This equation shows that the reaction forces can be expressed as the transpose of the constraint Jacobian times a scale factor. We can generalize this finding to a system of interconnected particles. If the Jacobian of all the constraints in the system is denoted as matrix $\mathbf{D}$, then the array of reaction forces acting on the particles can be expressed as:

$$^{(c)}\mathbf{f} = \mathbf{D}'\lambda \tag{4.41}$$

where λ is an array containing as many scale factors as the number of constraints. These scale factors are called *Lagrange multipliers*.

We can further generalize Eq. (4.41) to a system of bodies and express the constraint forces and moments in such a system as:

$$^{(c)}\mathbf{h} = \mathbf{D}'\lambda \tag{4.42}$$

where $\mathbf{D}$ is the Jacobian of the constraints between the bodies.

In the upcoming chapters, Eqs. (4.41) and (4.42) will be applied to a variety of kinematic joints and will be further discussed. For a few kinematic joints the corresponding equation of reaction force will be derived and analyzed in detail. However, it is outside the scope of this textbook to derive Eqs. (4.41) and (4.42) in a general form because that requires more advanced knowledge of some other fundamental principles.

4.5.3 Coulomb Friction

The *Coulomb* friction, also known as the *dry* friction, is the resistive force between two contacting surfaces that oppose the motion. This force may exist even if the two surfaces do not move relative to each other. The force associated with the Coulomb friction is normally computed as a function of coefficient of friction times the magnitude of the reaction force normal to the two surfaces. The friction force is applied in the opposite direction in which the surfaces move, or would tend to move, relative to one another as shown in Figure 4.19. If there is no relative motion between the two surfaces, we use the so-called *coefficient of static friction* otherwise we use the *coefficient of kinetic friction*. We will denote the coefficient of Coulomb friction as $^{(Cf)}d_c$.

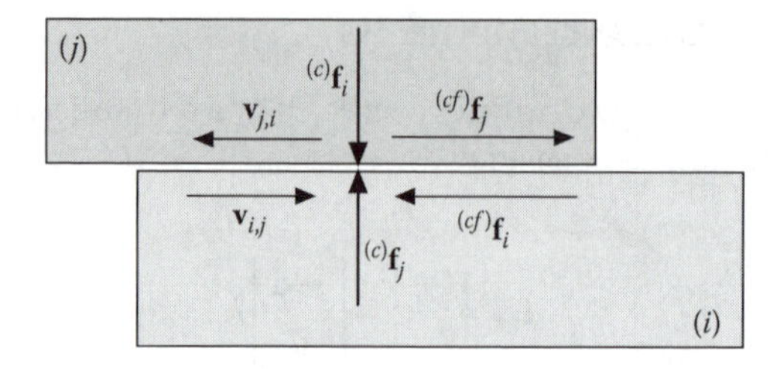

FIGURE 4.19 Coulomb friction acting between two flat surfaces.

Implementation of Coulomb friction in the equations of motion of a multibody system is not difficult. However, inclusion of Coulomb friction makes the process of solving the equations not to be a straight forward process. The difficulty arises from the fact that the friction force depends on the reaction force between two surfaces. In turn the reaction forces depend on the accelerations whereas the accelerations depend on the applied and reaction forces. This fact suggests that the Coulomb friction forces should be determined simultaneously with the reaction forces. In Chapter 14 we will discuss how to model and analyze systems containing Coulomb friction.

4.6 REMARKS

The kinematic definitions and the fundamental formulas for kinematics of particles and rigid bodies that were discussed in the previous chapter, and the definitions and formulas for the dynamics of particles and bodies from this chapter, will be used extensively throughout this textbook. Therefore, it is important to understand these fundamentals before moving to the next chapters. The following six chapters apply these fundamentals to derive the kinematic and dynamic equations of motion for multibody systems based on three different formulations.

4.7 PROBLEMS

4.1 Derive the first moment equation given in Eq. (4.7).

4.2 The data for a system of four particles are given as:

$$\mathbf{r}^1 = \begin{Bmatrix} -1 \\ 3 \end{Bmatrix},\ \mathbf{r}^2 = \begin{Bmatrix} 2 \\ -2 \end{Bmatrix},\ \mathbf{r}^3 = \begin{Bmatrix} 1 \\ 2 \end{Bmatrix},\ \mathbf{r}^4 = \begin{Bmatrix} -4 \\ -1 \end{Bmatrix},$$

$$\dot{\mathbf{r}}^1 = \begin{Bmatrix} 2 \\ -0.5 \end{Bmatrix},\ \dot{\mathbf{r}}^2 = \begin{Bmatrix} -1.5 \\ 1.2 \end{Bmatrix},\ \dot{\mathbf{r}}^3 = \begin{Bmatrix} 1 \\ -1.2 \end{Bmatrix},\ \dot{\mathbf{r}}^4 = \begin{Bmatrix} -2.1 \\ 0 \end{Bmatrix}$$

$$m^1 = 2,\ m^2 = 5,\ m^3 = 1.6,\ m^4 = 2.7$$

$$\mathbf{f}^1 = \begin{Bmatrix} 1 \\ -2 \end{Bmatrix},\ \mathbf{f}^2 = \begin{Bmatrix} 3 \\ 0 \end{Bmatrix},\ \mathbf{f}^3 = \begin{Bmatrix} 0 \\ 0 \end{Bmatrix},\ \mathbf{f}^4 = \begin{Bmatrix} -3 \\ 4 \end{Bmatrix}$$

Revise the MATLAB program in Example 4.1 to determine the followings:

(a) Position of the mass center
(b) Velocity of the mass center
(c) Acceleration of the mass center
(d) Position vector for each particle from the mass center

4.3 Two forces act on a body at points A and B. The position vector and the components of each force with respect to the body reference frame are given as

$$\mathbf{s}'^{A} = \begin{Bmatrix} 1 \\ -1 \end{Bmatrix},\ \mathbf{f}'^{A} = \begin{Bmatrix} 3 \\ 1 \end{Bmatrix},\ \mathbf{s}'^{B} = \begin{Bmatrix} -3 \\ 2 \end{Bmatrix},\ \mathbf{f}'^{B} = \begin{Bmatrix} -2 \\ 1 \end{Bmatrix}$$

The rotational coordinate of the body is $\phi = -35°$.

(a) Without transforming the position and force vectors to the global frame, compute the sum of the moments as the result of the two forces.
(b) Transform the position and force vectors to the global frame, and then compute the sum of the moments as the result of the two forces.
(c) Compare the results from (a) and (b).

4.4 The rod shown has a length of 0.3 units and is attached to the ground by a spring. Write the equations of motion for the rod. Assume that the mass center is at the rod's geometric center, $m = 4$, $J = 3$, $k = 40$, and ${}^{0}\ell = 0.15$. Take direct measurements from the figure if needed.

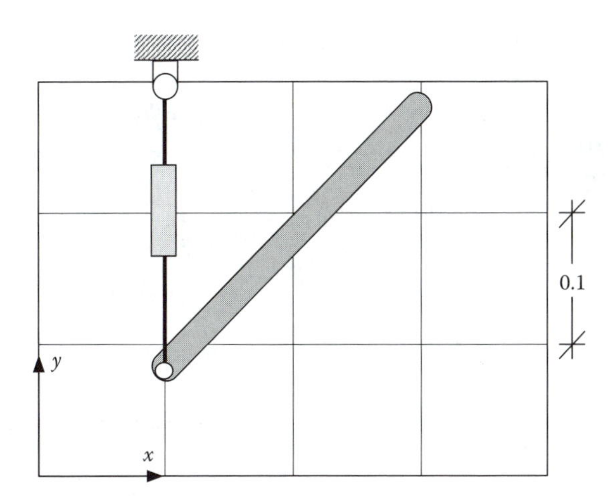

4.5 The single body shown has a mass of 5 and a moment of inertia of 3.5 (SI units). Gravity, a constant force at point P, and a pure moment act upon the body as shown. Point P has the local coordinates $\mathbf{s}'^{P} = \{-0.8 \quad 0.4\}'$. The components of the force are $\mathbf{f} = \{1.2 \quad 0.5\}'$ and the magnitude of the pure moment is 0.8. At a given time the body has the following coordinates: $\mathbf{r} = \{2.0 \quad 1.6\}'$, $\phi = 30°$.

(a) Construct the mass matrix and the force array according to Eq. (4.16).
(b) Compute the accelerations.

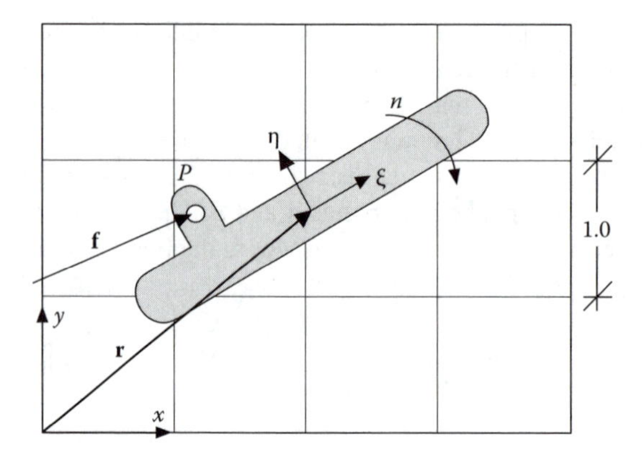

4.6 The single body in Problem 4.5 is revised by removing the gravity, the external constant force and the pure moment. A spring is attached between point P and point A as shown. The coordinates of point A are $\mathbf{r}^A = \{0 \quad 2\}'$. The spring has an undeformed length of 1.4 and a stiffness of 6.

(a) Compute the array of forces.
(b) Compute the accelerations.

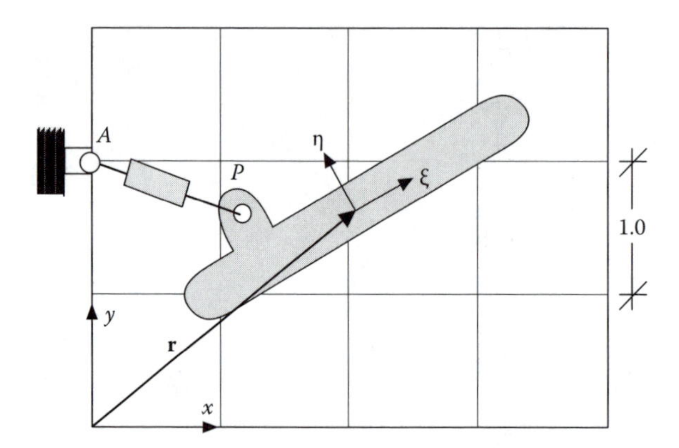

4.7 A rod, body (1), is 1.5 units long and is pinned to the ground at one end. A spring-damper-actuator connects the other end of the rod to the ground as shown. The characteristics of this force element are: $k = 100$, ${}^0\ell = 2$, $d_c = 200$, ${}^{(a)}f = 30$. In the shown configuration, the rod has an angular velocity of 1 rad/sec CW.

(a) Determine the magnitude and the direction of the force of this element.
(b) Determine the array of forces for body (1); assume the gravity is not present.

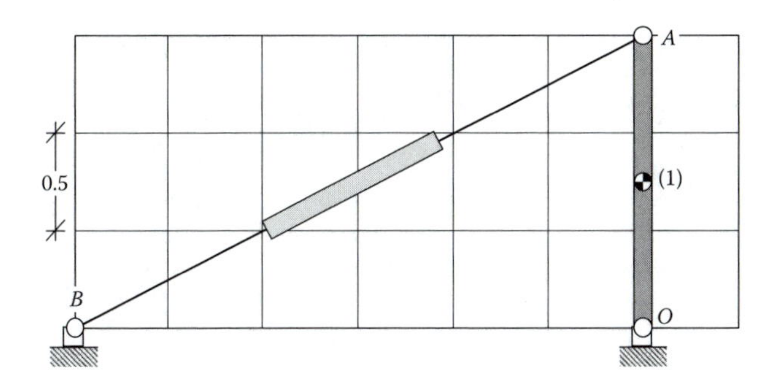

4.8 For the objects shown determine the location of the mass center and the moment of inertia. Assume that the mass is distributed evenly.

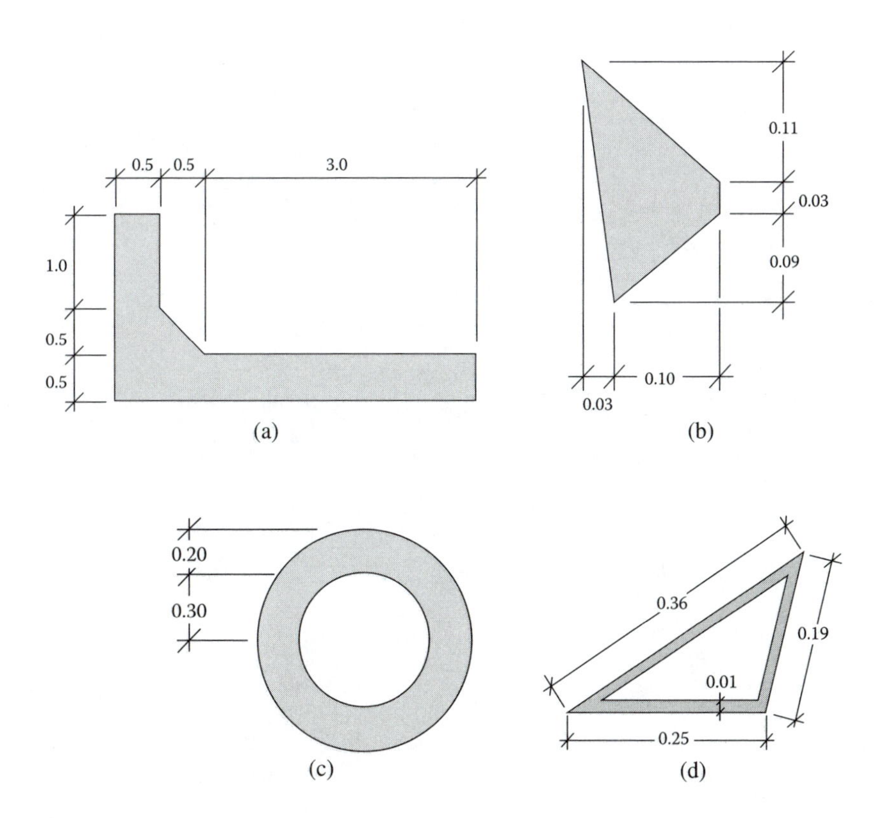

5 Point-Coordinates: Kinematics

Although a mechanical system is a collection of interconnected links, as it will be seen in this chapter, a mechanical system can also be described as a collection of interconnected points. In other words a multibody system can be represented as a multiparticle system. This process does not create an approximated representation of the original multibody system—the kinematics of the system can fully be preserved if the process is performed correctly. Furthermore, it will turn out that the kinematic equations of the representing multiparticle system have a simpler form than the equations of the equivalent multibody system.

In this chapter we learn how to represent a planar multibody system by a minimum number of points defined at proper locations on each body. We construct the necessary algebraic relationships between the coordinates of these points. We also see how to define additional points to describe additional kinematic information about the system. We first learn the concept of multiparticle presentation of a system through several simple examples. Then the process is described in a more general form. In our discussions, we will use the terms *points* and *particles* interchangeably.

Programs: In this chapter we develop several new MATLAB® programs. First we need to organize some of the functions that we developed in Chapters 3 and 4 in a folder/directory named `Basics`, as shown in Table 5.1. We create three more folders naming them `PC_Basics`, `PC_AA`, and `PC_MP`. We put these four folders inside another newly created folder named `Point_Coord`. These folders and their contents, as we develop them in this chapter and in Chapter 6, are shown in Table 5.2. In programming exercises of this and the next chapter, we make the folder `Point_Coord` the Current Directory of MATLAB.

5.1 MULTIPOINT REPRESENTATION

The process of describing a multibody system as a collection of interconnected points is illustrated in this section. Most of the points are defined at/on the joints that connect the links of a system and, therefore, their locations can be determined easily. We can apply the following as rule-of-thumb in defining most or possibly all of the necessary points:

Pin joints: A point must be defined at the center of a pin joint connecting two links.

Sliding joints: Two or more points must be defined on the axis of a sliding joint that connects two links—at least one point on each link.

TABLE 5.1

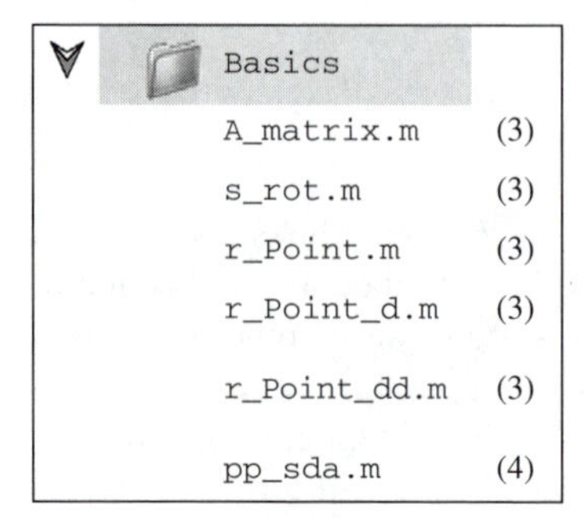

Basics	
A_matrix.m	(3)
s_rot.m	(3)
r_Point.m	(3)
r_Point_d.m	(3)
r_Point_dd.m	(3)
pp_sda.m	(4)

TABLE 5.2

Point_Coord		
PC_AA.m		(5) (6)
PC_MP.m		(5) (6)
PC_FS_A.m		(5) (6)
Basics		(3) (4)
PC_Basics		
	PC_Phi_L.m	(5)
	PC_Phi_n.m	(5)
	PC_Phi_p.m	(5)
	PC_Phi_theta.m	(5)
	PC_jacob_L.m	(5)
	PC_gamma_L.m	(5)
	PC_jacob_p_3.m	(5)
	PC_gamma_p_3.m	(5)
	PC_secondary.m	(5)
	PC_mass_3_approx.m	(6)
	PC_mass_4_exact.m	(6)
	PC_mass_3_exact.m	(6)
PC_AA		
	PC_AA_Phi.m	(5)
	PC_AA_jacob.m	(5)
	PC_AA_gamma.m	(5)
	PC_AA_h.m	(6)
PC_MP		
	PC_MP_Phi.m	(5)
	PC_MP_jacob.m	(5)
	PC_MP_gamma.m	(5)
	PC_MP_h.m	(6)

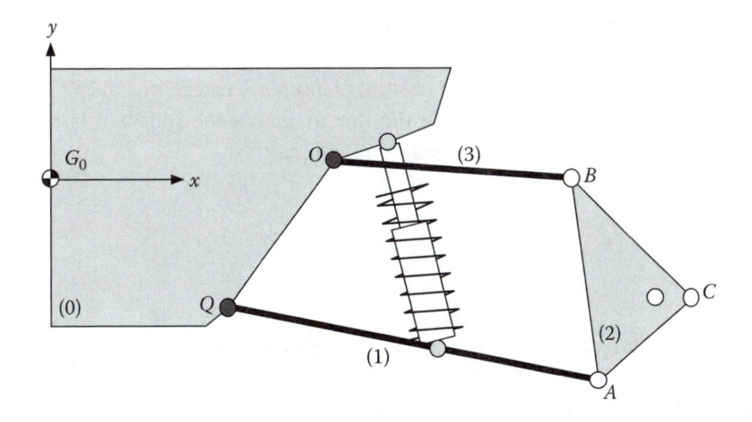

FIGURE 5.1 The double A-arm suspension system adopted for the point coordinate formulation.

Besides these points, we may need to define other points depending on the kinematic characteristics of the system. We show the process of defining points through several examples.

5.1.1 Double A-Arm Suspension

The description and the data for this system are provided in Section 1.5.1. The links and the joints of this double A-arm suspension system form a four-bar mechanism. The spring and damper elements do not contribute to the kinematics of the system—they are assumed to be ideal force elements—and therefore they will be considered in the dynamic equations of motion.

For kinematic modeling with point-coordinate formulation, because there are four pin joints in this mechanism, we define four points O, Q, A, and B—one for each pin joint—as shown in Figure 5.1. It is assumed that the frame is not moving and we are only interested in the kinematics of the suspension links with respect to the frame (i.e., the ground). We define a nonmoving x-y frame as shown. Points O and Q are nonmoving—their coordinates in the x-y frame remain constant as

$$\mathbf{r}^O = \begin{Bmatrix} 0.32 \\ 0.02 \end{Bmatrix}, \quad \mathbf{r}^Q = \begin{Bmatrix} 0.20 \\ -0.14 \end{Bmatrix} \tag{a}$$

Points A and B are moving points; that is, their coordinates in x-y remain variables. Therefore, for these two points we define their coordinates as

$$\mathbf{r}^A = \begin{Bmatrix} x^A \\ y^A \end{Bmatrix}, \quad \mathbf{r}^B = \begin{Bmatrix} x^B \\ y^B \end{Bmatrix} \tag{b}$$

Although at any given configuration these coordinates have certain values, as the links move, these values change. We refer to moving points such as A and B as *primary* points. The primary and the nonmoving points describe the four pin joints of this four-bar mechanism.

In addition to the stationary points O and Q, points A and B are the two necessary and sufficient primary points to define the kinematics of the four-bar. However, we define a third primary point at C which is positioned as:

$$\mathbf{r}^C = \begin{Bmatrix} x^C \\ y^C \end{Bmatrix} \tag{c}$$

The coordinates of A, B and C are not independent of one another—the three points must move in such a way that the distances between them and O and Q remain as defined. Therefore, five algebraic equations can be written indicating what the lengths should be:

$$\begin{aligned} (x^A - 0.20)^2 + (y^A + 0.14)^2 - 0.42^2 = 0 \\ (x^B - x^A)^2 + (y^B - y^A)^2 - 0.22^2 = 0 \\ (x^B - 0.32)^2 + (y^B - 0.02)^2 - 0.26^2 = 0 \\ (x^C - x^A)^2 + (y^C - y^A)^2 - 0.14^2 = 0 \\ (x^C - x^B)^2 + (y^C - y^B)^2 - 0.18^2 = 0 \end{aligned} \tag{d}$$

These algebraic relations are called *position constraints* or, more specifically, *length constraints*. The three primary points are represented by six variable coordinates and Eq. (d) provides five constraints. This agrees with the fact that the system has 6 – 5 = 1 DoF.

5.1.2 MacPherson Suspension

This example is described in more detail in Section 1.5.2. Unlike the double A-arm suspension system, the spring-damper element cannot be removed completely for kinematic modeling because the damper, besides providing a damping force, also acts as a sliding joint. Therefore, in addition to the primary points that we position at the pin joints, we also define points to describe the axis of the sliding joint between the links.

Because we are interested in investigating the kinematics of the suspension with respect to the frame, we assume that the frame is the ground. We define a nonmoving x-y frame as shown in Figure 5.2. We also define two fixed and three primary points, O, Q, A, B and C having the following coordinates:

$$\mathbf{r}^Q = \begin{Bmatrix} 0.12 \\ -0.41 \end{Bmatrix},\ \mathbf{r}^O = \begin{Bmatrix} 0.41 \\ 0.13 \end{Bmatrix},\ \mathbf{r}^A = \begin{Bmatrix} x^A \\ y^A \end{Bmatrix},\ \mathbf{r}^B = \begin{Bmatrix} x^B \\ y^B \end{Bmatrix},\ \mathbf{r}^C = \begin{Bmatrix} x^C \\ y^C \end{Bmatrix} \tag{a}$$

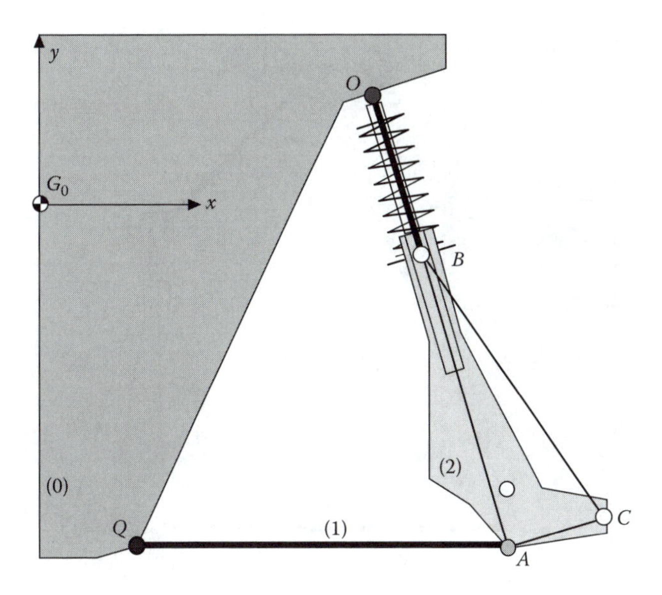

FIGURE 5.2 Point coordinate presentation of the MacPherson suspension system.

Points O, Q, and A are defined at the pin joints. The primary point B is defined on link (2) along the strut axis. This point could be defined anywhere on link (2) as long as it is positioned on the strut axis. In our example, we define point B at the base of the spring. Point C is defined as a primary point, similar to the previous example, to simplify the dynamics formulation in the next chapter.

The following constant length constraints can be written:

$$\begin{aligned}
&(x^A - 0.12)^2 + (y^A + 0.41)^2 - 0.45^2 = 0 \\
&(x^A - x^B)^2 + (y^A - y^B)^2 - 0.36^2 = 0 \\
&(x^C - x^A)^2 + (y^C - y^A)^2 - 0.12^2 = 0 \\
&(x^C - x^B)^2 + (y^C - y^B)^2 - 0.39^2 = 0
\end{aligned} \tag{b}$$

When the system moves, the distance between B and O varies and, therefore, we cannot write a length constraint for it. The four constraints in *(b)* are necessary but not sufficient. With these constraints alone, if we allow the three primary points to move, the system may end up in a configuration shown in Figure 5.3. To avoid such situations, we must guarantee that points O, B, and A remain on the same axis. For this purpose, we construct two vectors as

$$\mathbf{d}^{A,B} = \mathbf{r}^A - \mathbf{r}^B = \begin{Bmatrix} x^A - x^B \\ y^A - y^B \end{Bmatrix}, \ \mathbf{d}^{B,O} = \mathbf{r}^B - \mathbf{r}^O = \begin{Bmatrix} x^B - 0.41 \\ y^B - 0.13 \end{Bmatrix} \tag{c}$$

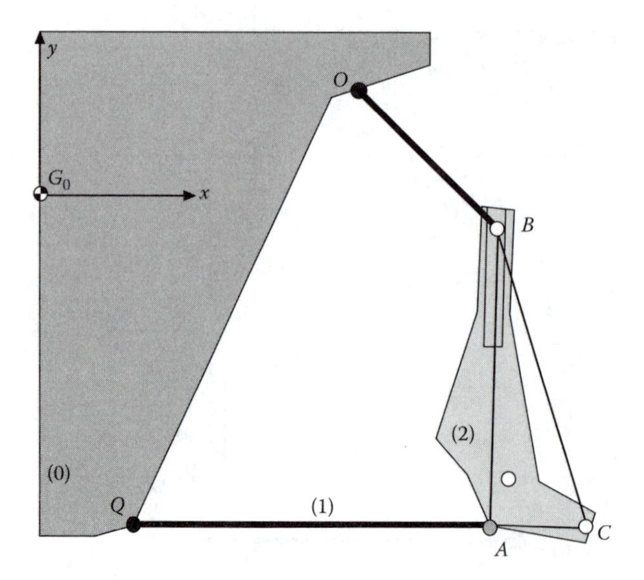

FIGURE 5.3 An undesirable configuration for the MacPherson suspension model.

These two vectors need to remain collinear (parallel) at all times. This can be accomplished, using Eq. (2.15), by the following constraint:

$$\breve{\mathbf{d}}^{B,O\prime}\mathbf{d}^{A,B} = (y^A - y^B)(x^B - 0.41) - (x^A - x^B)(y^B - 0.13) = 0 \qquad (d)$$

Now we have five constraints in Eqs. (*b*) and (*d*). Because there are three primary points, the model provides correct number of DoF; that is, 6 – 5 = 1.

5.1.3 Filmstrip Advancer

A highly simplified mechanism to advance a filmstrip in a movie projector is described in Section 1.5.3. Similar to the double A-arm system, because this mechanism is a four-bar, we define two stationary points, O and Q, and two primary points, A and B, as shown in Figure 5.4(a). In a defined nonmoving x-y frame these points are positioned as

$$\mathbf{r}^O = \begin{Bmatrix} 0 \\ 0 \end{Bmatrix}, \mathbf{r}^Q = \begin{Bmatrix} 3.0 \\ -4.5 \end{Bmatrix}, \mathbf{r}^A = \begin{Bmatrix} x^A \\ y^A \end{Bmatrix}, \mathbf{r}^B = \begin{Bmatrix} x^B \\ y^B \end{Bmatrix} \qquad (a)$$

Three length constraints can be written between the coordinates of A, B, O and Q:

$$\begin{aligned} &(x^A - 3.0)^2 + (y^A + 4.5)^2 - 1.5^2 = 0 \\ &(x^B - x^A)^2 + (y^B - y^A)^2 - 5.0^2 = 0 \\ &(x^B)^2 + (y^B)^2 - 3.0^2 = 0 \end{aligned} \qquad (c)$$

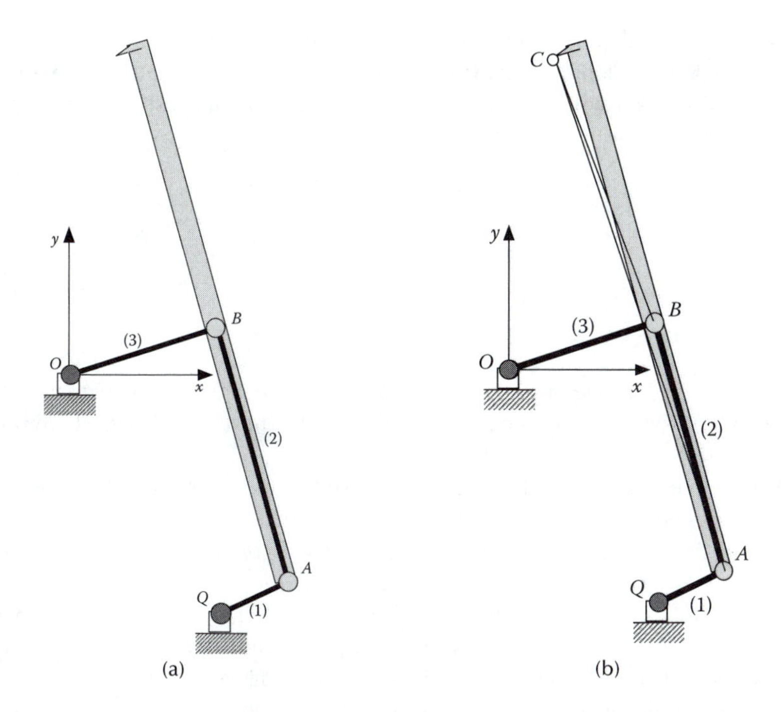

FIGURE 5.4 (a) Representation of the filmstrip mechanism with primary and nonmoving points; (b) inclusion of one more primary point at C.

Because point C is a coupler point, its kinematics can be described as a function of the kinematics of the other points that have already been defined on the same link. We discuss two methods for describing the kinematics of point C. The first method is to introduce another primary point at C even though C is not a pin joint. Considering C as a primary point requires its coordinates to be defined as additional variables; that is,

$$\mathbf{r}^C = \begin{Bmatrix} x^C \\ y^C \end{Bmatrix} \tag{d}$$

As shown in Figure 5.4(b), the distances between C and A, and C and B must remain unchanged. This requires two more constraints:

$$\begin{aligned} (x^C - x^A)^2 + (y^C - y^A)^2 - (10.5^2 + 0.5^2) = 0 \\ (x^C - x^B)^2 + (y^C - y^B)^2 - (5.5^2 + 0.5^2) = 0 \end{aligned} \tag{e}$$

Now the system is defined by three primary points—that is, six variable coordinates, and five constraints (6 – 5 = 1 DoF).

The second method is to include point C into the system as a *secondary* point—that is, to express its coordinates explicitly in terms of the coordinates of the primary points A and B. As it will be seen in Section 5.5, the coordinates of point C can be computed from the following expression:

$$\begin{Bmatrix} x^C \\ y^C \end{Bmatrix} = -1.1 \begin{Bmatrix} x^A \\ y^A \end{Bmatrix} + 2.1 \begin{Bmatrix} x^B \\ y^B \end{Bmatrix} + 0.1 \begin{Bmatrix} -y^B + y^A \\ x^B - x^A \end{Bmatrix} \tag{f}$$

5.2 STATIONARY AND PRIMARY POINTS

In the examples of the previous section, we observed that in describing the kinematics of a multibody system, a minimum number of points must be defined. These points can be either nonmoving or moving.

A point that is defined with two constant coordinates is referred to as a *stationary* (*fixed* or *nonmoving*) point, such as point O in Figure 5.5 having coordinates $\mathbf{r}^O = \{2.0 \quad 1.0\}'$. We can define as many fixed points in a problem as needed without increasing the problem size computationally. A *primary* point, on the other hand, has two variable coordinates, such as point A in Figure 5.5 with coordinates $\mathbf{r}^A = \{x^A \quad y^A\}'$. When the mechanical system undergoes motion, the value of both coordinates may change. Increasing the number of primary points in a problem increases the computational effort during an analysis. The vector of coordinates for a typical primary point i is denoted as

$$\mathbf{r}^i = \begin{Bmatrix} x^i \\ y^i \end{Bmatrix} \tag{5.1}$$

If n_p (number of points) primary points are defined for a system, then an *array of coordinates* can be constructed as

$$\mathbf{r} \equiv \begin{Bmatrix} \mathbf{r}^1 \\ \vdots \\ \mathbf{r}^{n_p} \end{Bmatrix} \tag{5.2}$$

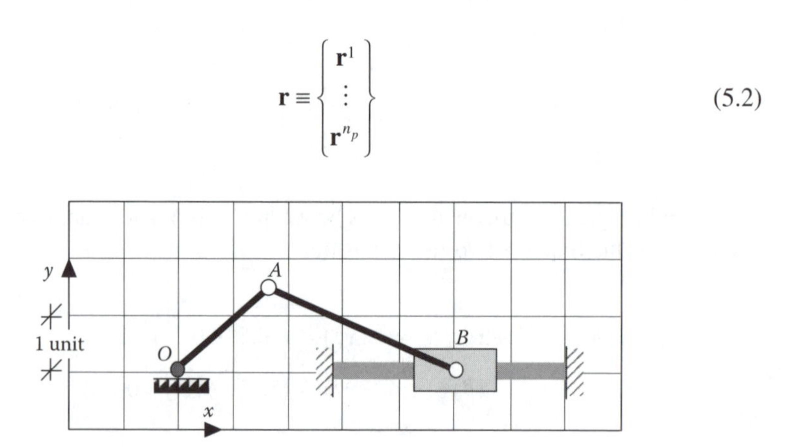

FIGURE 5.5 Three types of points: O, A, and B.

The number of elements in the array of coordinates, $\mathbf{r}$, is denoted as n_v (number of variables). If each of the primary points that define a system has been assigned two variable coordinates, then $n_v = 2 \times n_p$. However, this may not be true in general because, as we will see next, some of the primary points may have only one variable coordinate.

Another type of moving point can be defined with one fixed and one variable coordinate. For example, point B in Figure 5.5 can only move on an axis parallel to the x-axis—its y-coordinate remains a constant at all times. This type of point is considered as a primary point that only contributes one variable to the array of coordinates. We can deal with this type of points in two ways. For example, for point B we can either define the vector of coordinates as $\mathbf{r}^B = \{x^B \quad 1.0\}'$ which contains only one variable. Or we define the vector of coordinates as $\mathbf{r}^B = \{x^B \quad y^B\}'$ which contains two variables. Then, we define a constraint as $y^B = 1.0$.

5.3 CONSTRAINTS

The defined primary and nonmoving points that describe a multibody system, in general, are not completely free—they are dependent on one another through algebraic relationships called *constraints*. In the examples of Section 5.1 we saw that the most common constraint between two points is a constant length constraint. In one particular case we also used another type of constraint to keep two vectors parallel. In this section we state these and several other commonly used constraints in a general form.

A constraint equation is denoted by the *lightface* character Φ. If there is more than one algebraic equation in a constraint, we denote it by the *boldface* character $\boldsymbol{\Phi}$. The constraint symbol may carry a left superscript in parentheses with two entries, such as ${}^{(a,b)}\boldsymbol{\Phi}$. The first entry, a, states the constraint *type*—for example, ℓ for *length*, p for *parallel*, and n for *normal*. The second entry, b, denotes the *number* of algebraic equations in a constraint. In the point-coordinate formulation b is most likely equal to one.

An algebraic constraint equation removes one degree of freedom from a system. The number of constraint equations, n_c, must be equal to the difference between the number of coordinates and the number of DoF; that is, $n_c = n_v - n_{dof}$.

If the array of coordinates is defined as $\mathbf{r}$, as stated in Eq. (5.2), then the position constraints are expressed in a general form as

$$\boldsymbol{\Phi} \equiv \boldsymbol{\Phi}(\mathbf{r}) = \mathbf{0} \tag{5.3}$$

In point-coordinate formulation, most constraints describe a condition on a vector connecting two points. Therefore, the first task before defining a constraint is to construct a vector between the two points. Between two typical points i and j, with coordinate vectors $\mathbf{r}^i$ and $\mathbf{r}^j$, a position vector is constructed as*

$$\mathbf{d}^{j,i} = \mathbf{r}^j - \mathbf{r}^i \tag{5.4}$$

* In general, $\mathbf{d}^{i,j} = -\mathbf{d}^{j,i}$.

The two points i and j, can be both primary points or one primary and one fixed point. If both points are fixed, there is no need to construct a vector between them because the points do not move and, therefore, there is no need to define a constraint between them.

5.3.1 Length Constraint

To keep the distance between two points a known constant, a *length* constraint must be constructed. Assuming that both points, such as i and j, are primary points, a constraint is constructed as

$$(x^j - x^i)^2 + (y^j - y^i)^2 - (\ell^{j,i})^2 = 0$$

where $\ell^{j,i}$ is the distance between the two points. This equation can also be written as

$$^{(\ell,1)}\Phi \equiv \tfrac{1}{2}[(x^j - x^i)^2 + (y^j - y^i)^2 - (\ell^{j,i})^2] = 0 \qquad (5.5)$$

In this equation we have introduced the coefficient "$\frac{1}{2}$" to eliminate numerous "2" coefficients from the derivatives of this equation as it will be seen in the velocity constraint equations. Equation (5.5) is expressed in expanded form. The compact form of this equation is

$$^{(\ell,1)}\Phi \equiv \tfrac{1}{2}[\mathbf{d}^{j,i\prime}\mathbf{d}^{j,i} - (\ell^{j,i})^2] = 0 \qquad (5.6)$$

5.3.2 Angle Constraints

Two vectors may be enforced to remain parallel or perpendicular, or to keep a known constant angle between them. Assume that the two vectors are

$$\mathbf{d}^{j,i} = \mathbf{r}^j - \mathbf{r}^i,\ \mathbf{d}^{l,k} = \mathbf{r}^l - \mathbf{r}^k$$

To keep these two vectors perpendicular (*normal*) to each other, as shown in Figure 5.6(a), their scalar product must be set to zero. This constraint is written in both expanded and compact forms as

$$^{(n,1)}\Phi \equiv (x^j - x^i)(x^l - x^k) + (y^j - y^i)(y^l - y^k)$$

$$= \mathbf{d}^{j,i\prime}\mathbf{d}^{l,k} = 0 \qquad (5.7)$$

Similarly, to enforce two vectors to remain *parallel*, as shown in Figure 5.6(b), we set their vector product to zero; that is,

$$^{(p,1)}\Phi \equiv -(y^j - y^i)(x^l - x^k) + (x^j - x^i)(y^l - y^k)$$

$$= \breve{\mathbf{d}}^{j,i\prime}\mathbf{d}^{l,k} = 0 \qquad (5.8)$$

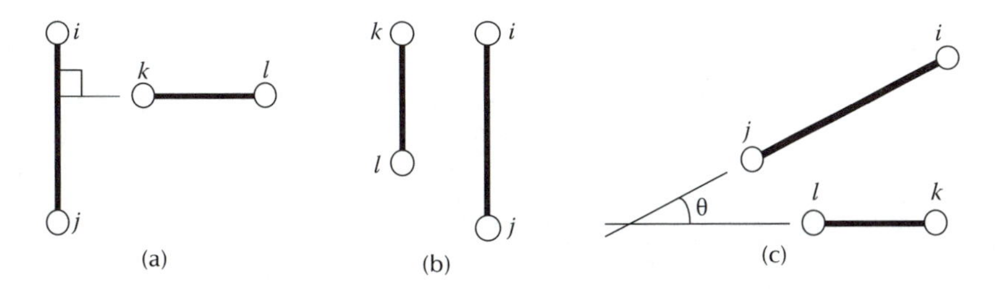

FIGURE 5.6 Angle conditions between two links: (a) normal; (b) parallel; (c) any other angle.

To keep the *angle* between two vectors a known *constant*, as shown in Figure 5.6(c), we use their scalar product as

$$^{(\theta,1)}\Phi \equiv (x^j - x^i)(x^l - x^k) + (y^j - y^i)(y^l - y^k) - \ell^{j,i}\ell^{l,k}\cos\theta$$

$$= \mathbf{d}^{j,i\prime}\mathbf{d}^{l,k} - \ell^{j,i}\ell^{l,k}\cos\theta = 0 \qquad (5.9)$$

The vectors used in any of the angle constraints could be between three points instead of four. This is the more common case in most of the examples that we will encounter. The general form of these constraints, especially in their compact form, will not be different whether we have three or four points.

5.3.3 Simple Constraints

A simple constraint is a condition that we impose on a single coordinate. For example, the y-coordinate of primary point B in the mechanism of Figure 5.5 must be set to $y^B = 1.0$. These constraints are usually defined when we want to keep a particular coordinate to appear as a variable in the formulation where in reality that coordinate is a known constant. The advantage of doing this is that we can apply a constraint equation in its general form on a vector regardless of the coordinates of its end points being all variables or not. This normally simplifies the programming effort. However, a minor drawback is that we unnecessarily increase the number of variables and the number of algebraic constraints by one for every constant coordinate that is treated in this way.

Example 5.1

For the inverted slider-crank mechanism shown in Figure 5.7(a) define the necessary primary points and write the constraint equations.

Solution:

This problem can be solved with either three or two primary points.

Solution 1: We define point A at the pin joint on the slider, point B at the tip of link 2, and point C on the slider block as shown in Figure 5.7(b). We write three length

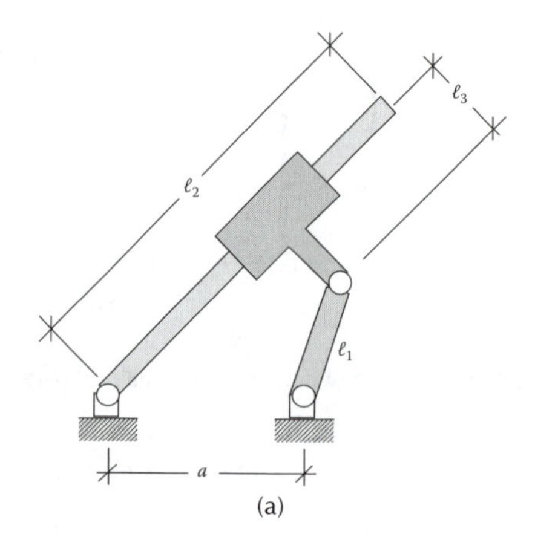

(a)

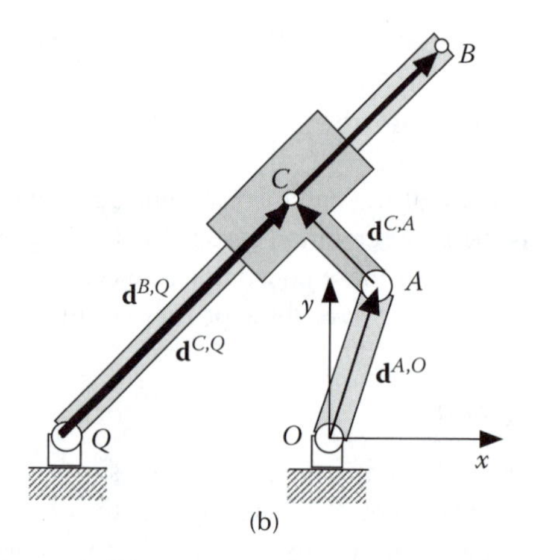

(b)

FIGURE 5.7 (a) An inverted slider-crank mechanism and (b) its point coordinate representation.

constraints, a fourth constraint to keep vectors $\mathbf{d}^{C,Q}$ and $\mathbf{d}^{C,A}$ perpendicular, and a fifth constraint to keep vectors $\mathbf{d}^{C,Q}$ and $\mathbf{d}^{B,Q}$ parallel:

$$\begin{aligned}
&\tfrac{1}{2}\left((x^A)^2+(y^A)^2-\ell_1^2\right)=0\\
&\tfrac{1}{2}\left((x^B+a)^2+(y^B)^2-\ell_2^2\right)=0\\
&\tfrac{1}{2}\left((x^C-x^A)^2+(y^C-y^A)^2-\ell_3^2\right)=0\\
&(x^C-x^A)(x^C+a)+(y^C-y^A)y^C=0\\
&(x^C+a)y^B-(x^B+a)y^C=0
\end{aligned} \qquad (a)$$

Solution 2: In this solution we only use points A and C. Hence we only need two length constraints and a third constraint to keep vectors $\mathbf{d}^{C,Q}$ and $\mathbf{d}^{C,A}$ perpendicular:

$$\begin{aligned} &\tfrac{1}{2}\left((x^A)^2 + (y^A)^2 - \ell_1^2\right) = 0 \\ &\tfrac{1}{2}\left((x^C - x^A)^2 + (y^C - y^A)^2 - \ell_3^2\right) = 0 \\ &(x^C - x^A)(x^C + a) + (y^C - y^A)y^C = 0 \end{aligned} \qquad (b)$$

Example 5.2

In the circular-cam-roller-follower system shown in Figure 5.8(a), the cam rotates about a pin joint attached to the chassis. The roller is kept in contact with the cam with a loaded spring. All the geometric data are shown on the Figure. Define the necessary primary points and then write the position constraints.*

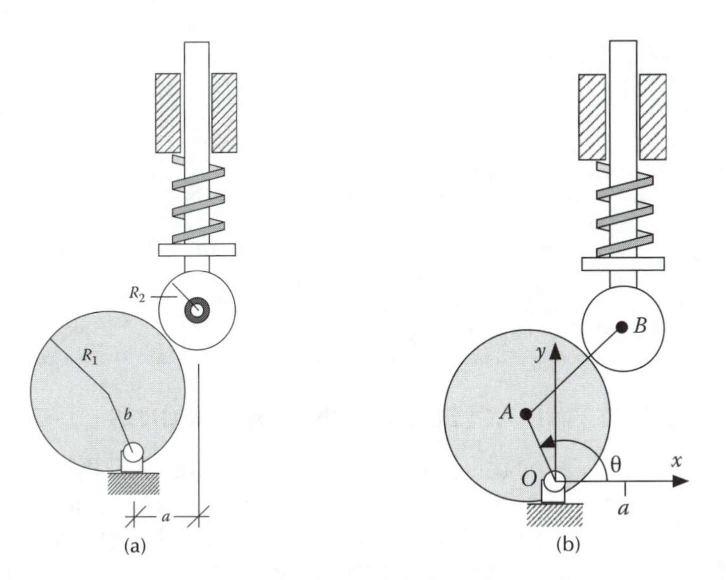

FIGURE 5.8 (a) A cam-follower mechanism and (b) its point coordinate representation.

* Because cam-followers are higher-pair joints, their kinematic formulation is much more complex than the mechanisms with pin and sliding joints. If the shape of the cam or the contacting part of the follower is circular, the kinematic relationships can be obtained easily. However, as soon as the shape of the contact part becomes more complex than a circle or a straight line, the formulation becomes complicated.

Solution:

We define two primary points at the centers of the two circles, points A and B. The distance between A and B is $R_1 + R_2$ as shown in Figure 5.8(b). We write two length constraints, noting that $x^B = a$, as:

$$(x^A)^2 + (y^A)^2 - b^2 = 0$$

$$(x^A - a)^2 + (y^A - y^B)^2 - (R_1 + R_2)^2 = 0$$

The second constraint assures that the two circles remain in contact at only one point.

Example 5.3

In this exercise four function M-files are presented to evaluate the constraints stated in Eqs. (5.6) to (5.9). In these functions it is assumed that the components of all the vectors have already been updated and all the necessary constants are available. These function M-files will be used in some of the upcoming exercises.

```
function Phi = PC_Phi_L(d, L)
% Length constraint between two points
     Phi = (d'*d - L^2)/2;
```

```
function Phi = PC_Phi_n(d1, d2)
% Normal constraint between two vectors
     Phi = d1'*d2;
```

```
function Phi = PC_Phi_p(d1, d2)
% Parallel constraint between two vectors
     Phi = (s_rot(d1))'*d2;
```

```
function Phi = PC_Phi_theta(d1, d2, L1, L2, theta)
% Angle constraint between two vectors
     Phi = d1'*d2 - L1*L2*cos(theta);
```

5.4 VELOCITY AND ACCELERATION CONSTRAINTS

Velocity constraints are obtained by taking the time derivative of the position constraints. These constraints provide algebraic relationships between the coordinates and velocities of the points that appear in a constraint. For example, the time derivative of Eq. (5.5) provides the velocity *length-constraint* between two points as:

$$^{(\ell,1)}\dot{\Phi} \equiv (x^j - x^i)(\dot{x}^j - \dot{x}^i) + (y^j - y^i)(\dot{y}^j - \dot{y}^i)$$

$$= \begin{bmatrix} -(x^j - x^i) & -(y^j - y^i) & (x^j - x^i) & (y^j - y^i) \end{bmatrix} \begin{Bmatrix} \dot{x}^i \\ \dot{y}^i \\ \dot{x}^j \\ \dot{y}^j \end{Bmatrix} = 0 \tag{5.10}$$

This velocity constraint is shown in two expanded forms where, in the second form, the coefficients of the velocities are organized in a matrix that will become useful later on. Equation (5.10) in compact form can be obtained directly from the time derivative of Eq. (5.6) as

$$
\begin{aligned}
{}^{(\ell,1)}\dot{\Phi} &\equiv \mathbf{d}^{j,i\prime}\dot{\mathbf{d}}^{j,i} \\
&= \begin{bmatrix} -\mathbf{d}^{j,i\prime} & \mathbf{d}^{j,i\prime} \end{bmatrix} \begin{Bmatrix} \dot{\mathbf{r}}^{i} \\ \dot{\mathbf{r}}^{j} \end{Bmatrix} = 0 \\
&= \begin{bmatrix} -\mathbf{d}^{j,i\prime} \end{bmatrix} \dot{\mathbf{r}}^{i} + \begin{bmatrix} \mathbf{d}^{j,i\prime} \end{bmatrix} \dot{\mathbf{r}}^{j} = 0
\end{aligned} \tag{5.11}
$$

At this point it should be clear why we introduced the coefficient "$\frac{1}{2}$" in Eqs. (5.5) and (5.6).

The velocity constraints for the angle constraints, or any other constraints, can be obtained similarly to the length constraint. The velocity constraints for a system can be expressed in a general form as

$$
\dot{\mathbf{\Phi}} \equiv \mathbf{D}\dot{\mathbf{r}} = \mathbf{0} \tag{5.12}
$$

where $\mathbf{D}$ is referred to as the *Jacobian* matrix. The velocity constraints are *linear* in the velocities, and the *right-hand side array of the velocity equations* for the kinematic constraints is always a zero array.

The time derivative of the velocity constraints yields the acceleration constraints. For example the time derivative of Eq. (5.10) yields the acceleration length-constraint as

$$
{}^{(\ell,1)}\ddot{\Phi} \equiv (x^j - x^i)(\ddot{x}^j - \ddot{x}^i) + (y^j - y^i)(\ddot{y}^j - \ddot{y}^i) + (\dot{x}^j - \dot{x}^i)^2 + (\dot{y}^j - \dot{y}^i)^2
$$

$$
= [-(x^j - x^i) \quad -(y^j - y^i) \quad (x^j - x^i) \quad (y^j - y^i)] \begin{Bmatrix} \ddot{x}^i \\ \ddot{y}^i \\ \ddot{x}^j \\ y^j \end{Bmatrix} + (\dot{x}^j - \dot{x}^i)^2 + (\dot{y}^j - \dot{y}^i)^2 = 0 \tag{5.13}
$$

This equation in compact form is expressed as

$$
\begin{aligned}
{}^{(\ell,1)}\ddot{\Phi} &\equiv \mathbf{d}^{j,i\prime}\ddot{\mathbf{d}}^{j,i} + \dot{\mathbf{d}}^{j,i\prime}\dot{\mathbf{d}}^{j,i} = 0 \\
&= \begin{bmatrix} -\mathbf{d}^{j,i\prime} \end{bmatrix} \ddot{\mathbf{r}}^{i} + \begin{bmatrix} \mathbf{d}^{j,i\prime} \end{bmatrix} \ddot{\mathbf{r}}^{j} + \dot{\mathbf{d}}^{j,i\prime}\dot{\mathbf{d}}^{j,i} = 0
\end{aligned} \tag{5.14}
$$

or,

$$\left[-\mathbf{d}^{j,i\prime}\right]\ddot{\mathbf{r}}^{i}+\left[\mathbf{d}^{j,i\prime}\right]\ddot{\mathbf{r}}^{j}={}^{(\ell,1)}\gamma \tag{5.15}$$

where

$${}^{(\ell,1)}\gamma=-\dot{\mathbf{d}}^{j,i\prime}\dot{\mathbf{d}}^{j,i} \tag{5.16}$$

The acceleration equations for a system are written in a general form as

$$\ddot{\mathbf{\Phi}}\equiv\mathbf{D}\ddot{\mathbf{r}}+\dot{\mathbf{D}}\dot{\mathbf{r}}=\mathbf{0} \tag{5.17}$$

or,

$$\mathbf{D}\ddot{\mathbf{r}}=\boldsymbol{\gamma} \tag{5.18}$$

where

$$\boldsymbol{\gamma}=-\dot{\mathbf{D}}\dot{\mathbf{r}} \tag{5.19}$$

The array $\boldsymbol{\gamma}$ is called the *right-hand side array of the acceleration equations*. This array is quadratic in the velocities. The acceleration constraints are *linear* in the accelerations and the coefficient matrix in the acceleration constraints is the same coefficient matrix as in the velocity constraints.

The coefficient matrix $\mathbf{D}$ in the velocity and acceleration constraint is called the *Jacobian* matrix. This matrix can also be obtained by taking the partial derivative of the position constraints, $\mathbf{\Phi}$, with respect to the array of coordinates, $\mathbf{r}$. This matrix appears in the kinematic constraints at different levels, as well as in the dynamics equations of motion as it will be seen in the next chapter.

Example 5.4

In this example two function M-files to evaluate the Jacobian matrix and the right-hand side of the acceleration constraint for the length constraint of Eq. (5.6) are developed. These M-files will be used in some of the upcoming exercises.

```
function [jac_i, jac_j] = PC_jacob_L(d)
% Jacobian for the length constraint
    jac_i = -d'; jac_j = d';
```

Note that the output from this function contains two submatrices. In the upcoming examples we will see how we distribute these submatrices within a larger matrix.

```
function gamma = PC_gamma_L(d_d)
% r-h-s of acceleration array for the length constraint
    gamma = -d_d'*d_d;
```

EXAMPLE 5.5

Simplify the parallel constraint of Eq. (5.8) for two vectors that share one of the primary points as shown in Figure 5.9. Derive the corresponding sub-Jacobian matrices and the right-hand side of the acceleration constraint.

Solution:

The Constraint equation is simplified to

$$\breve{\mathbf{d}}^{j,i\prime}\mathbf{d}^{k,j} = 0 \qquad (a)$$

Then the velocity constraint is obtained as

$$\breve{\mathbf{d}}^{j,i\prime}\dot{\mathbf{d}}^{k,j} - \breve{\mathbf{d}}^{k,j\prime}\dot{\mathbf{d}}^{j,i} =$$

$$\begin{bmatrix} \breve{\mathbf{d}}^{k,j\prime} & -(\breve{\mathbf{d}}^{k,j\prime} + \breve{\mathbf{d}}^{j,i\prime}) & \breve{\mathbf{d}}^{j,i\prime} \end{bmatrix} \begin{Bmatrix} \dot{\mathbf{r}}^{i} \\ \dot{\mathbf{r}}^{j} \\ \dot{\mathbf{r}}^{k} \end{Bmatrix} = 0$$

The sub-Jacobian matrices are extracted from this equation:

$$\mathbf{D}_i = \left[\breve{\mathbf{d}}^{k,j\prime}\right], \mathbf{D}_j = \left[-(\breve{\mathbf{d}}^{k,j\prime} + \breve{\mathbf{d}}^{j,i\prime})\right], \mathbf{D}_k = \left[\breve{\mathbf{d}}^{j,i\prime}\right] \qquad (b)$$

The time derivative of the velocity constraint yields the right-hand side of the acceleration constraint as

$$\gamma = 2\dot{\breve{\mathbf{d}}}^{k,j\prime}\dot{\mathbf{d}}^{j,i} \qquad (c)$$

FIGURE 5.9 Two collinear links sharing one primary point.

EXAMPLE 5.6

The following two function M-files construct the Jacobian matrix and the right-hand side of the acceleration constraint for the simplified parallel constraint of Example 5.5. Note that there is no need to develop a new M-file function because we can use the `PC_const_p` function for this constraint as well.

```
function [jac_i, jac_j, jac_k] = PC_jacob_ p_3(d_ji, d_kj)
% Jacobian for the parallel constraint between
%    two vectors sharing one primary point
     sr_ji = s_rot(d_ji); sr_kj = s_rot(d_kj);
     jac_i = sr_kj'; jac_j = -(sr_ji + sr_kj)'; jac_k = sr_ji';
```

```
function gamma = PC_gamma_p_3(d_ji_d, d_d)
% r-h-s of acceleration constraints for the parallel constraint
%    between two vectors sharing one primary point
     gamma = 2*(s_rot(d_kj_d))'*d_ji_d;
```

5.5 SECONDARY POINTS

In contrast to a primary point, the coordinates of a secondary point do not appear in any constraint equations and, therefore, these coordinates are not included in the array of coordinates. In some problems however, as we have seen in the examples of Section 5.1, we may upgrade a secondary point to a primary point for various reasons. We may do this to obtain the kinematics of all the points of interest directly through the solution of the constraints, or to make the formulation of the equations of motion more accurate in describing the dynamics of the system (we will see this in the next chapter). Upgrading a secondary point to a primary point increases the dimension of the array of coordinates, the number of constraints, and hence the computational effort during an analysis. Therefore, unless necessary, it is better to describe the coordinates of a secondary point in terms of the coordinates of the primary points as expressions and not as constraints.

In planar kinematics, the coordinates of a secondary point can be determined as a function of the coordinates of at least two primary points that are positioned on the same body. Consider the link shown in Figure 5.10(a). Assume that points A and B are primary points with coordinates $\mathbf{r}^A = \{x^A \quad y^A\}'$ and $\mathbf{r}^B = \{x^B \quad y^B\}'$. Point P is defined on the axis of the link with a known distance α from A as shown. If the coordinates of A and B are known, then the coordinates of P can be computed. To derive the necessary *expression* for this purpose, we first define a positive direction

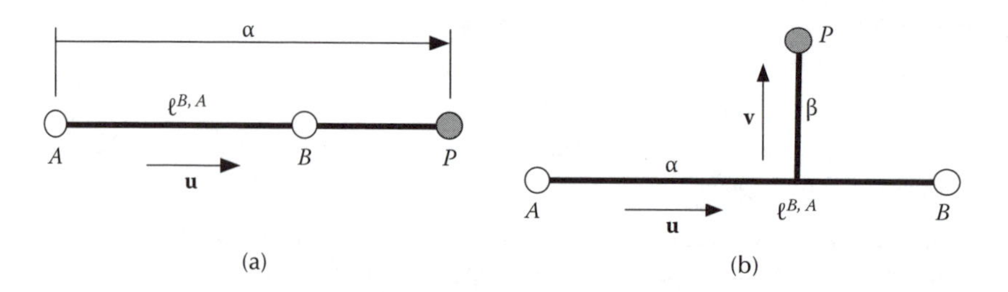

FIGURE 5.10 A secondary point P can be defined on a link either (a) collinear or (b) noncollinear.

along the axis of the link; for example, from A to B. A unit vector along this axis is computed as

$$\mathbf{u} = \frac{1}{\ell^{B,A}}(\mathbf{r}^B - \mathbf{r}^A) \tag{5.20}$$

The coordinates of P can now be computed from the following expression:

$$\begin{aligned}\mathbf{r}^P &= \mathbf{r}^A + \alpha\mathbf{u} \\ &= (1 - \frac{\alpha}{\ell^{B,A}})\mathbf{r}^A + \frac{\alpha}{\ell^{B,A}}\mathbf{r}^B\end{aligned} \tag{5.21}$$

It should be noted that α could be either a positive or a negative quantity depending on the position of P on the link axis relative to A and the direction of the unit vector.

The velocity and acceleration of P are computed from the first and second time derivatives of Eq. (5.21); that is,

$$\dot{\mathbf{r}}^P = (1 - \frac{\alpha}{\ell^{B,A}})\dot{\mathbf{r}}^A + \frac{\alpha}{\ell^{B,A}}\dot{\mathbf{r}}^B \tag{5.22}$$

$$\ddot{\mathbf{r}}^P = (1 - \frac{\alpha}{\ell^{B,A}})\ddot{\mathbf{r}}^A + \frac{\alpha}{\ell^{B,A}}\ddot{\mathbf{r}}^B \tag{5.23}$$

In a second case, the secondary point P is not on the axis of the primary points A and B, as shown in Figure 5.10(b). The position of the secondary point on the body can be defined by two known distances α and β, along the unit vectors $\mathbf{u}$ and $\mathbf{v}$ respectively, where $\mathbf{v}$ is constructed by rotating $\mathbf{u}$ by 90° in the positive sense (i.e., $\mathbf{v} = \breve{\mathbf{u}}$). We compute the components of the unit vector $\mathbf{u}$ from Eq. (5.20); then the coordinates of point P are determined from the following expression:

$$\begin{aligned}\mathbf{r}^P &= \mathbf{r}^A + \alpha\mathbf{u} + \beta\breve{\mathbf{u}} \\ &= (1 - \frac{\alpha}{\ell^{B,A}})\mathbf{r}^A + \frac{\alpha}{\ell^{B,A}}\mathbf{r}^B + \frac{\beta}{\ell^{B,A}}(\breve{\mathbf{r}}^B - \breve{\mathbf{r}}^A)\end{aligned} \tag{5.24}$$

Then, the velocity and acceleration expressions are written as

$$\dot{\mathbf{r}}^P = (1 - \frac{\alpha}{\ell^{B,A}})\dot{\mathbf{r}}^A + \frac{\alpha}{\ell^{B,A}}\dot{\mathbf{r}}^B + \frac{\beta}{\ell^{B,A}}(\breve{\dot{\mathbf{r}}}^B - \breve{\dot{\mathbf{r}}}^A) \tag{5.25}$$

$$\ddot{\mathbf{r}}^P = (1 - \frac{\alpha}{\ell^{B,A}})\ddot{\mathbf{r}}^A + \frac{\alpha}{\ell^{B,A}}\ddot{\mathbf{r}}^B + \frac{\beta}{\ell^{B,A}}(\breve{\ddot{\mathbf{r}}}^B - \breve{\ddot{\mathbf{r}}}^A) \tag{5.26}$$

In Eqs. (5.21) and (5.24), it is assumed that the values of α and β are available. This is normally the case in most problems. However, in some problems the values of α and β may not be available and, therefore, they need to be determined. One possible scenario is to assume that at a given configuration the coordinates of the primary and the secondary points are known. These coordinates can be used to determine α and β.

Assume that for the linkage of Figure 5.10(b), the coordinates of points A, B, and P are known initially. To determine the values of α and β, we write Eq. (5.24) as

$$\mathbf{r}^P - \mathbf{r}^A = \alpha \mathbf{u} + \beta \breve{\mathbf{u}} \tag{5.27}$$

Then we premultiply this equation by $\mathbf{u}'$ to get

$$\mathbf{u}'(\mathbf{r}^P - \mathbf{r}^A) = \alpha \mathbf{u}'\mathbf{u} + \beta \mathbf{u}'\breve{\mathbf{u}}$$

Because $\mathbf{u}'\mathbf{u} = 1$ and $\mathbf{u}'\breve{\mathbf{u}} = 0$, we obtain an expression to compute α as

$$\begin{aligned}\alpha &= \mathbf{u}'(\mathbf{r}^P - \mathbf{r}^A) \\ &= \frac{1}{\ell^{B,A}}(\mathbf{r}^B - \mathbf{r}^A)'(\mathbf{r}^P - \mathbf{r}^A)\end{aligned} \tag{5.28}$$

Similarly multiplication of Eq. (5.27) by $\breve{\mathbf{u}}'$ yields an expression for computing β as

$$\beta = \frac{1}{\ell^{B,A}}(\breve{\mathbf{r}}^B - \breve{\mathbf{r}}^A)'(\mathbf{r}^P - \mathbf{r}^A) \tag{5.29}$$

These values of α and β can be used in Eqs. (5.24) to (5.26) as if they were obtained from direct measurements.

Example 5.7

For the four-bar mechanism shown in Figure 5.11, an x-y frame is defined. In this reference frame the coordinates of the joints and the coupler point are measured accurately to be

$$\mathbf{r}^O = \begin{Bmatrix} 0 \\ 0 \end{Bmatrix},\ \mathbf{r}^Q = \begin{Bmatrix} 2.00 \\ 0.50 \end{Bmatrix},\ \mathbf{r}^A = \begin{Bmatrix} -0.42 \\ 0.91 \end{Bmatrix},\ \mathbf{r}^B = \begin{Bmatrix} 1.99 \\ 2.70 \end{Bmatrix},\ \mathbf{r}^P = \begin{Bmatrix} 0.07 \\ 2.27 \end{Bmatrix}$$

Construct the constraint equations and the expression for the kinematics of the coupler-point P only at the position level.

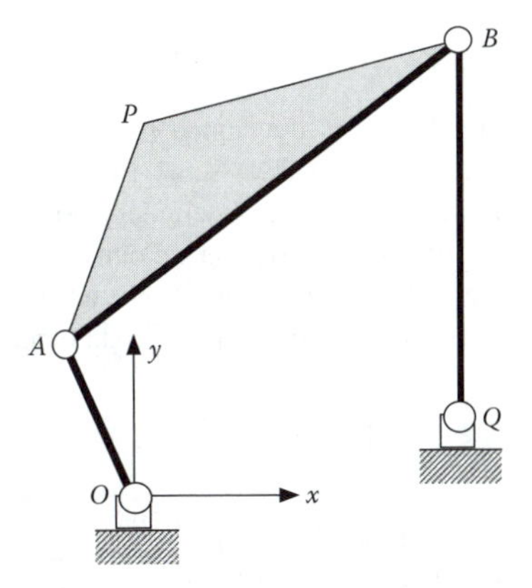

FIGURE 5.11 A coupler point P of a four-bar mechanism.

Solution:

The fixed points are at O and Q, and the primary points are at A and B. Because the link lengths are not given, we need to compute them first. Substituting the known coordinates in Eq. (5.5) for the three links yields

$$\ell^{A,O} = \left((-0.42)^2 + (0.91)^2\right)^{\frac{1}{2}} = 1.0$$
$$\ell^{B,A} = \left((1.99+0.42)^2 + (2.70-0.91)^2\right)^{\frac{1}{2}} = 3.0$$
$$\ell^{B,Q} = \left((1.99-2.00)^2 + (2.70-0.50)^2\right)^{\frac{1}{2}} = 2.2$$

Now we can write the constraint equations in expanded form as

$$^{(\ell,1)}\Phi^{A,O} \equiv \tfrac{1}{2}[(x^A)^2 + (y^A)^2 - 1.0^2] = 0$$

$$^{(\ell,1)}\Phi^{B,A} \equiv \tfrac{1}{2}[(x^B - x^A)^2 + (y^B - y^A)^2 - 3.0^2] = 0$$

$$^{(\ell,1)}\Phi^{B,Q} \equiv \tfrac{1}{2}[(x^B - 2.00)^2 + (y^B - 0.50)^2 - 2.2^2] = 0$$

Similarly, because the values for α and β for positioning the coupler point P are not given, we use Eqs. (5.28) and (5.29) to get

$$\alpha = \frac{1}{3.0}\left(\begin{Bmatrix}1.99\\2.70\end{Bmatrix} - \begin{Bmatrix}-0.42\\0.91\end{Bmatrix}\right)'\left(\begin{Bmatrix}0.07\\2.27\end{Bmatrix} - \begin{Bmatrix}-0.42\\0.91\end{Bmatrix}\right) = 1.2$$

$$\beta = \frac{1}{3.0}\left(\begin{Bmatrix}-2.70\\1.99\end{Bmatrix} - \begin{Bmatrix}-0.91\\-0.42\end{Bmatrix}\right)'\left(\begin{Bmatrix}0.07\\2.27\end{Bmatrix} - \begin{Bmatrix}-0.42\\0.91\end{Bmatrix}\right) = 0.8$$

The expression for point P is written in compact form based on Eq. (5.24) as

$$\mathbf{r}^P = \left(1 - \frac{1.2}{3.0}\right)\mathbf{r}^A + \frac{1.2}{3.0}\mathbf{r}^B + \frac{0.8}{3.0}(\breve{\mathbf{r}}^B - \breve{\mathbf{r}}^A)$$

EXAMPLE 5.8

In this exercise a function M-file is developed to compute the coordinates of a secondary point based on Eq. (5.24). The same function M-file can be used to compute the velocity and acceleration of a secondary point because Eqs. (5.24) to (5.26) are similar; that is, if we provide the velocity or the acceleration of the primary points, we receive, in turn, the velocity or the acceleration of the secondary point. The main program starts with an `addpath`* statement.

```
    addpath Basics PC_Basics
% Coordinates and constant
    r_A = [-0.42; 0.91]; r_B = [1.99; 2.70];
    alpha = 1.2; beta = 0.8; L = 3;
% Compute kinematics of the secondary point
    r_P = PC_secondary(r_A, r_B, alpha, beta, L)
```

```
function r_P = PC_secondary(r_A, r_B, alpha, beta, L)
% Coordinates (or velocity or acceleration) of a secondary
%  point based on the coordinates of two primary points
    r_P = (1 -alpha/L)*r_A + alpha/L*r_B + beta/L*s_rot (r_B - r_A);
```

Executing the program yields the following results:

```
    r_P =
          0.0667
          2.2687
```

5.6 EXAMPLE PROGRAMS

In Sections 5.2 to 5.5 we learned how to define primary points and how to construct the position, velocity, and acceleration constraints. We also discussed, if necessary, how to construct expressions for the kinematics of any secondary point. In this section we revisit these equations and extend them for complete kinematics of our three application examples.

5.6.1 DOUBLE A-ARM SUSPENSION

The expanded form of the position constraints for the kinematics of a double A-arm suspension were derived in Section 5.1.1. In this section the constraints are presented in compact form. We derive the necessary arrays and matrices for constructing the

* The `addpath` command dynamically establishes the path to the listed subdirectories. The subdirectories contain all the necessary M-files used by the program. Consult the MATLAB `help` to learn about other methods for establishing the path for a program. Also refer to Table 5.2 for the recommended subdirectory setup for various functions.

corresponding velocity and acceleration constraints, and then we develop the following MATLAB program to compute some of these entities.

```
% file name: PC_AA
% Double A-arm suspension
    clear all
    addpath Basics  PC_Basics  PC_AA
% Constant geometric data
    r_Q = [0.20; -0.14]; r_O = [0.32; 0.02];
  L_AQ = 0.42; L_BA = 0.22; L_BO = 0.26; L_CA = 0.14; L_CB = 0.18;
        els = [L_AQ L_BA L_BO L_CA L_CB];
% Coordinates
    r_A = [0.610; -0.233]; r_B = [0.578; -0.013];
    r_C = [0.709; -0.132];
% Velocities
    r_A_d = [0.263; 1.163]; r_B_d = [0.146; 1.146];
    r_C_d = [0.209; 1.216];
% Compute position vectors
    d_AQ = r_A - r_Q; d_BA = r_B - r_A; d_BO = r_B - r_O;
    d_CA = r_C - r_A; d_CB = r_C - r_B;
        dees = [d_AQ  d_BA  d_BO  d_CA  d_CB];
% Evaluate constraints
    Phi = PC_AA_Phi(dees, els);
% Evaluate Jacobian
    D = PC_A_jacob(dees);
% Compute velocity vectors
    d_AQ_d = r_A_d; d_BA_d = r_B_d - r_A_d; d_BO_d = r_B_d;
    d_CA_d = r_C_d - r_A_d; d_CB_d = r_C_d - r_B_d;
        dees_d = [d_AQ_d d_BA_d d_BO_d d_CA_d d_CB_d];
% Construct r-h-s array of acceleration constraints
    gamma = PC_AA_gamma(dees_d);
```

The first statement in this program is `clear all`* followed by an `addpath` statement. Then, the constant data are provided to the program: the coordinates of the two fixed points, $\mathbf{r}^O$ and $\mathbf{r}^Q$, and the five constant lengths. The constant lengths are arranged in an array, `els`. Following that, we have provided our best estimates for the coordinates and the velocities of the three primary points.

The five length constraints in compact form can be expressed as:

$$\begin{aligned}
\Phi^{A,Q} &\equiv \tfrac{1}{2}[\mathbf{d}^{A,Q\prime}\mathbf{d}^{A,Q} - 0.42^2] = 0 \\
\Phi^{B,A} &\equiv \tfrac{1}{2}[\mathbf{d}^{B,A\prime}\mathbf{d}^{B,A} - 0.22^2] = 0 \\
\Phi^{B,O} &\equiv \tfrac{1}{2}[\mathbf{d}^{B,O\prime}\mathbf{d}^{B,O} - 0.26^2] = 0 \\
\Phi^{C,A} &\equiv \tfrac{1}{2}[\mathbf{d}^{C,A\prime}\mathbf{d}^{C,A} - 0.14^2] = 0 \\
\Phi^{C,B} &\equiv \tfrac{1}{2}[\mathbf{d}^{C,B\prime}\mathbf{d}^{C,B} - 0.18^2] = 0
\end{aligned} \qquad (a)$$

* It is a good practice to start a MATLAB program with a `clear all` statement. This command clears the memory from any previously computed data.

To evaluate these constraints, we need to compute the **d** vectors. These vectors are evaluated in the main program and then arranged in a matrix called `dees`. By referencing different columns of this matrix, different vectors can be recalled. Matrix `dees` and array `els` are sent to the following function to evaluate the length constraints:

```
function Phi = PC_AA_Phi(dees, els)
% Evaluate constraints
    Phi_AQ = PC_Phi_L(dees(:,1), els(1));
    Phi_BA = PC_Phi_L(dees(:,2), els(2));
    Phi_BO = PC_Phi_L(dees(:,3), els(3));
    Phi_CA = PC_Phi_L(dees(:,4), els(4));
    Phi_CB = PC_Phi_L(dees(:,5), els(5));
        Phi = [Phi_AQ; Phi_BA; Phi_BO; Phi_CA; Phi_CB];
```

The velocity constraints are written based on Eq. (5.11) as:

$$\dot{\Phi}^{A,Q} \equiv \mathbf{d}^{A,Q\prime}\dot{\mathbf{d}}^{A,Q} = \mathbf{d}^{A,Q\prime}\dot{\mathbf{r}}^{A} = 0$$

$$\dot{\Phi}^{B,A} \equiv \mathbf{d}^{B,A\prime}\dot{\mathbf{d}}^{B,A} = \begin{bmatrix} \mathbf{d}^{B,A\prime} & -\mathbf{d}^{B,A\prime} \end{bmatrix} \begin{Bmatrix} \dot{\mathbf{r}}^{B} \\ \dot{\mathbf{r}}^{A} \end{Bmatrix} = 0$$

$$\dot{\Phi}^{B,O} \equiv \mathbf{d}^{B,O\prime}\dot{\mathbf{d}}^{B,O} = \mathbf{d}^{B,O\prime}\dot{\mathbf{r}}^{B} = 0$$

$$\dot{\Phi}^{C,A} \equiv \mathbf{d}^{C,A\prime}\dot{\mathbf{d}}^{C,A} = \begin{bmatrix} \mathbf{d}^{C,A\prime} & -\mathbf{d}^{C,A\prime} \end{bmatrix} \begin{Bmatrix} \dot{\mathbf{r}}^{C} \\ \dot{\mathbf{r}}^{A} \end{Bmatrix} = 0$$

$$\dot{\Phi}^{C,B} \equiv \mathbf{d}^{C,B\prime}\dot{\mathbf{d}}^{C,B} = \begin{bmatrix} \mathbf{d}^{C,B\prime} & -\mathbf{d}^{C,B\prime} \end{bmatrix} \begin{Bmatrix} \dot{\mathbf{r}}^{C} \\ \dot{\mathbf{r}}^{B} \end{Bmatrix} = 0$$

The velocity constraints can be organized into matrix form as

$$\begin{bmatrix} \mathbf{d}^{A,Q\prime} & \mathbf{0} & \mathbf{0} \\ -\mathbf{d}^{B,A\prime} & \mathbf{d}^{B,A\prime} & \mathbf{0} \\ \mathbf{0} & \mathbf{d}^{B,O\prime} & \mathbf{0} \\ -\mathbf{d}^{C,A\prime} & \mathbf{0} & \mathbf{d}^{C,A\prime} \\ \mathbf{0} & -\mathbf{d}^{C,B\prime} & \mathbf{d}^{C,B\prime} \end{bmatrix} \begin{Bmatrix} \dot{\mathbf{r}}^{A} \\ \dot{\mathbf{r}}^{B} \\ \dot{\mathbf{r}}^{C} \end{Bmatrix} = \begin{Bmatrix} 0 \\ 0 \\ 0 \\ 0 \\ 0 \end{Bmatrix} \qquad (b)$$

The choice in the order of the array of coordinates, and consequently the arrays of velocities and accelerations, is totally ours as long as we are consistent. A different choice would be equivalent of reordering the columns of the matrix. The 5 × 6 Jacobian matrix is obtained directly from the velocity constraints. The following function evaluates the Jacobian. Note how the sub-Jacobians for each constraint are assembled to form the larger matrix. Because some of the points are nonmoving, such as O and Q, they do not contribute to the columns of the Jacobian.

```
function D = PC_AA_jacob(dees)
% Construct 5x6 Jacobian
    [jac_i_AQ, jac_j_AQ]  = PC_jacob_L(dees(:,1));
    [jac_i_BA, jac_j_BA]  = PC_jacob_L(dees(:,2));
    [jac_i_BO, jac_j_BO]  = PC_jacob_L(dees(:,3));
    [jac_i_CA, jac_j_CA]  = PC_jacob_L(dees(:,4));
    [jac_i_CB, jac_j_CB]  = PC_jacob_L(dees(:,5));
    z2      = [0 0];
         D = [ jac_j_AQ     z2          z2
               jac_i_BA     jac_j_BA    z2
               z2           jac_j_BO    z2
               jac_i_CA     z2          jac_j_CA
               z2           jac_i_CB    jac_j_CB ];
```

The acceleration constraints are simply obtained from the Jacobian matrix and Eq. (5.16):

$$\begin{bmatrix} \mathbf{d}^{A,Q\prime} & \mathbf{0} & \mathbf{0} \\ -\mathbf{d}^{B,A\prime} & \mathbf{d}^{B,A\prime} & \mathbf{0} \\ \mathbf{0} & \mathbf{d}^{B,O\prime} & \mathbf{0} \\ -\mathbf{d}^{C,A\prime} & \mathbf{0} & \mathbf{d}^{C,A\prime} \\ \mathbf{0} & -\mathbf{d}^{C,B\prime} & \mathbf{d}^{C,B\prime} \end{bmatrix} \begin{Bmatrix} \ddot{\mathbf{r}}^{A} \\ \ddot{\mathbf{r}}^{B} \\ \ddot{\mathbf{r}}^{C} \end{Bmatrix} = \begin{Bmatrix} -\dot{\mathbf{d}}^{A,Q\prime}\dot{\mathbf{d}}^{A,Q} \\ -\dot{\mathbf{d}}^{B,A\prime}\dot{\mathbf{d}}^{B,A} \\ -\dot{\mathbf{d}}^{B,O\prime}\dot{\mathbf{d}}^{B,O} \\ -\dot{\mathbf{d}}^{C,A\prime}\dot{\mathbf{d}}^{C,A} \\ -\dot{\mathbf{d}}^{C,B\prime}\dot{\mathbf{d}}^{C,B} \end{Bmatrix} \quad (c)$$

The right-hand-side array is a function of $\dot{\mathbf{d}}$ arrays. These arrays are computed in the main program and then arranged in a matrix called `dees_d`. This matrix is then sent to the following function to evaluate the right-hand-side array of the acceleration constraints.

```
function gamma = PC_AA_gamma(dees_d)
% r-h-s array of acceleration constraints
    gam_AQ = PC_gamma_L(dees_d(:,1));
    gam_BA = PC_gamma_L(dees_d(:,2));
    gam_BO = PC_gamma_L(dees_d(:,3));
    gam_CA = PC_gamma_L(dees_d(:,4));
    gam_CB = PC_gamma_L(dees_d(:,5));
         gamma = [gam_AQ; gam_BA; gam_BO; gam_CA; gam_CB];
```

After executing the program, in the Command Window we type `Phi`, or `gamma`, or `D` to view the results:

```
Phi =                      gamma =
   1.0e-03 *                  -1.4217
    0.1745                    -0.0140
    0.5120                    -1.3346
    0.0265                    -0.0057
    0.2010                    -0.0089
   -0.5390
```

```
D =
    0.4100   -0.0930         0         0         0         0
    0.0320   -0.2200   -0.0320    0.2200         0         0
         0         0    0.2580   -0.0330         0         0
   -0.0990   -0.1010         0         0    0.0990    0.1010
         0         0   -0.1310    0.1190    0.1310   -0.1190
```

The computed values for `Phi` must be as close to zeros as possible. These values indicate that the coordinate values were not provided arbitrarily. In Chapter 14 we will discuss several methods for determining accurate values for coordinates and velocities that would satisfy the constraints.

> As an exercise, determine whether the provided velocity values satisfy the velocity constraints. For this purpose, we need to evaluate Eq. (*b*). In the Command Window construct the velocity array as:
>
> ```
> >> r_d = [r_A_d; r_B_d; r_C_d]
> ```
>
> Then type in:
>
> ```
> >> D*r_d
> ```
>
> Are the values small enough to be considered zeros?

5.6.2 MacPherson Suspension

The position constraints in expanded form for the kinematics of MacPherson suspension were derived in Section 5.1.2. In this section we state these constraints, the corresponding Jacobian and the right-hand-side array of the acceleration constraints in compact form. These entities are evaluated by the following MATLAB program. The main program is organized in the same form as the main program in Section 5.6.1.

```
% file name: PC_MP
% MacPherson suspension
    clear all
    addpath  Basics  PC_Basics  MP_PC
% Constant geometric data
    r_Q = [0.12; -0.41]; r_O = [0.41; 0.13];
    L_AQ = 0.45; L_AB = 0.36; L_CA = 0.12; L_CB = 0.39;
        els = [L_AQ  L_AB  L_CA  L_CB];
% Estimated coordinates
  r_A = [0.57; -0.41]; r_B = [0.468; -0.065]; r_C = [0.688; -0.387];
% Assume the following velocities
    r_A_d = [0; 1.130]; r_B_d = [-0.197; 1.072];
    r_C_d = [-0.013; 1.197];
% Compute position vectors
    d_AQ = r_A - r_Q; d_AB = r_A - r_B; d_CA = r_C - r_A;
    d_CB = r_C - r_B; d_BO = r_B - r_O;
        dees = [d_AQ  d_AB  d_CA  d_CB  d_BO];
% Evaluate constraints
    Phi = PC_M_Phi(dees, els);
% Construct Jacobian
    D = PC_MP_jacob(dees);
% Compute velocity vectors
   d_AQ_d = r_A_d; d_AB_d = r_A_d - r_B_d; d_CA_d = r_C_d - r_A_d;
    d_CB_d = r_C_d - r_B_d; d_BO_d = r_B_d;
        dees_d = [d_AQ_d d_AB_d d_CA_d d_CB_d d_BO_d];
% Construct r-h-s array of acceleration constraints
    gamma = PC_MP_gamma(dees_d);
```

The four length constraints and one parallel constraint from Section 5.1.2 are expressed in compact form as:

$$
\begin{aligned}
\Phi^{A,Q} &\equiv \tfrac{1}{2}[\mathbf{d}^{A,Q\prime}\mathbf{d}^{A,Q} - 0.45^2] = 0 \\
\Phi^{A,B} &\equiv \tfrac{1}{2}[\mathbf{d}^{A,B\prime}\mathbf{d}^{A,B} - 0.36^2] = 0 \\
\Phi^{C,A} &\equiv \tfrac{1}{2}[\mathbf{d}^{C,A\prime}\mathbf{d}^{C,A} - 0.12^2] = 0 \\
\Phi^{C,B} &\equiv \tfrac{1}{2}[\mathbf{d}^{C,B\prime}\mathbf{d}^{C,B} - 0.39^2] = 0 \\
\Phi_5 &\equiv \breve{\mathbf{d}}^{B,O\prime}\mathbf{d}^{A,B} = 0
\end{aligned}
\quad (a)
$$

The following function evaluates the constraints. This function uses the `PC_Phi_p` function from the parallel constraint in Example 5.3.

```
function Phi = PC_MP_Phi(dees, els)
% Evaluate constraints
    Phi_AQ = PC_Phi_L(dees(:,1), els(1));
    Phi_AB = PC_Phi_L(dees(:,2), els(2));
    Phi_CA = PC_Phi_L(dees(:,3), els(3));
    Phi_CB = PC_Phi_L(dees(:,4), els(4));
    Phi_p  = PC_Phi_p(dees(:,5), dees(:,2));
        Phi = [Phi_AQ; Phi_AB; Phi_CA; Phi_CB; Phi_p];
```

The velocity constraints, obtained from the time derivative of the position constraints, are written in matrix form as:

$$\begin{bmatrix} \mathbf{d}^{A,Q'} & \mathbf{0} & \mathbf{0} \\ \mathbf{d}^{A,B'} & -\mathbf{d}^{A,B'} & \mathbf{0} \\ -\mathbf{d}^{C,A'} & \mathbf{0} & \mathbf{d}^{C,A'} \\ \mathbf{0} & -\mathbf{d}^{C,B'} & \mathbf{d}^{C,B'} \\ \breve{\mathbf{d}}^{B,O'} & -\breve{\mathbf{d}}^{A,B'} - \breve{\mathbf{d}}^{B,O'} & \mathbf{0} \end{bmatrix} \begin{Bmatrix} \dot{\mathbf{r}}^A \\ \dot{\mathbf{r}}^B \\ \dot{\mathbf{r}}^C \end{Bmatrix} = \begin{Bmatrix} 0 \\ 0 \\ 0 \\ 0 \\ 0 \end{Bmatrix} \tag{b}$$

The 5×6 Jacobian matrix can be obtained directly from the velocity constraints and it is constructed by the following function. For the parallel constraint we use `PC_jacob_p_3` function.

```
function D = PC_MP_jacob(dees)
% Construct 5x6 Jacobian
     [jac_i_AQ jac_j_AQ] = PC_jacob_L(dees(:,1));
     [jac_i_AB jac_j_AB] = PC_jacob_L(dees(:,2));
     [jac_i_CA jac_j_CA] = PC_jacob_L(dees(:,3));
     [jac_i_CB jac_j_CB] = PC_jacob_L(dees(:,4));
     [jac_i_p, jac_j_p, jac_k_p] = PC_jacob_p_3(dees(:,5), dees(:,2));
     z2     = [0 0];
        D = [ jac__AQ            z2                  z2
              jac_j_AB           jac_i_AB            z2
              jac_i_CA           z2                  jac_j_CA
              z2                 jac_i_CB            jac_j_CB
              jac_k_p            jac_j_p             z2 ];
```

The acceleration constraints are obtained from the velocity constraints as:

$$\begin{bmatrix} \mathbf{d}^{A,Q'} & \mathbf{0} & \mathbf{0} \\ \mathbf{d}^{A,B'} & -\mathbf{d}^{A,B'} & \mathbf{0} \\ -\mathbf{d}^{C,A'} & \mathbf{0} & \mathbf{d}^{C,A'} \\ \mathbf{0} & -\mathbf{d}^{C,B'} & \mathbf{d}^{C,B'} \\ \breve{\mathbf{d}}^{B,O'} & -\breve{\mathbf{d}}^{A,B'} - \breve{\mathbf{d}}^{B,O'} & \mathbf{0} \end{bmatrix} \begin{Bmatrix} \ddot{\mathbf{r}}^A \\ \ddot{\mathbf{r}}^B \\ \ddot{\mathbf{r}}^C \end{Bmatrix} = \begin{Bmatrix} -\dot{\mathbf{d}}^{A,Q'}\dot{\mathbf{d}}^{A,Q} \\ -\dot{\mathbf{d}}^{A,B'}\dot{\mathbf{d}}^{A,B} \\ -\dot{\mathbf{d}}^{C,A'}\dot{\mathbf{d}}^{C,A} \\ -\dot{\mathbf{d}}^{C,B'}\dot{\mathbf{d}}^{C,B} \\ -2\dot{\breve{\mathbf{d}}}^{B,O'}\dot{\mathbf{d}}^{A,B} \end{Bmatrix} \tag{c}$$

The right-hand-side array is constructed by the following function. For the parallel constraint it uses the `PC_gamma_p_3` function.

```
function gamma = MP_PC_gamma(dees_d)
% r-h-s array of acceleration constraints
     gam_AQ = PC_gamma_L(dees_d(:,1));
     gam_AB = PC_gamma_L(dees_d(:,2));
     gam_CA = PC_gamma_L(dees_d(:,3));
     gam_CB = PC_gamma_L(dees_d(:,4));
     gam_p   = PC_gamma_p_3(dees_d(:,5), dees_d(:,2));
         gamma = [gam_AQ; gam_AB; gam_CA; gam_CB; gam_p];
```

Executing the program provides the following results:

```
Phi =                    gamma =
    1.0e-03 *
    -0.0000                 -1.2769
    -0.0855                 -0.0422
     0.0265                 -1.1880
    -0.0080                 -0.0047
    -0.1200                  0.4452
```

```
D =
    0.4500         0         0         0         0         0
    0.1020   -0.3450   -0.1020    0.3450         0         0
   -0.1180   -0.0230         0         0    0.1180    0.0230
         0         0   -0.2200    0.3220    0.2200   -0.3220
    0.1950    0.0580   -0.5400   -0.1600         0         0
```

5.6.3 Filmstrip Advancer

The position constraints for this mechanism were derived in expanded form in Section 5.1.3. We can transform these constraints into compact form and then find the corresponding Jacobian and the right-hand-side array of the acceleration constraints. However, by comparing the two Figures for this mechanism and the double A-arm, we conclude that there is no need to derive any equations. Both systems are four-bars and, furthermore, the stationary and primary points have been labeled identically. The only geometric difference is in the lengths. Therefore, we can simply change the data in the program of Section 5.6.1 and evaluate the required entities for the film-advancer mechanism.

The data portion of the main program for this mechanism is listed as:

```
% fine name:  P_FS_A
% Film Strip Advancer
     clear all
     addpath  Basics  PC_Basics  AA_PC
% Constant geometric data
     r_O = [0; 0]; r_Q = [3.0; -4.5];
   L_AQ = 1.5; L_BA = 5.0; L_BO = 3.0; L_CA = 10.5119; L_CB = 5.5227;
          els = [L_AQ  L_BA  L_BO  L_CA  L_CB];
% Coordinates
    r_A = [3.963; -3.35]; r_B = [2.61; 1.45]; r_C = [0.81; 6.67];
% Velocities
     r_A_d = [-11; 24]; r_B_d = [-8; 25]; r_C_d = [-4; 26];
% Compute position vectors
     ...
```

The main program uses the same functions that were developed for the double A-arm. This mechanism can also be modeled with only two primary points at A and B—point C can be modeled as a secondary point.

5.7 REMARKS

In this chapter we discussed how to describe a multibody system as a multiparticle system and construct the necessary kinematic constraints between the particles. This methodology will be extended to the dynamic equations of motion for multiparticle systems in the next chapter. In these two chapters we only discuss how to formulate the necessary equations—we need to wait until Chapters 11 to 14 to learn how to use these equations in different types of analyses. However, a reader may skip over the next five chapters and go directly to Chapter 11 and learn how to perform kinematic analysis with the point coordinate formulation.

5.8 PROBLEMS

5.1 Consider the simple mechanical systems shown in the Figure for kinematic modeling with the point-coordinate method. For each system:
  (a) Define a nonmoving reference frame at a convenient position and at least two primary points for each system.
  (b) Write the position, velocity, and acceleration constraints in expanded form.
  (c) Determine the Jacobian matrix and the right-hand-side array of the acceleration constraints.
  (d) Construct the array of coordinates.

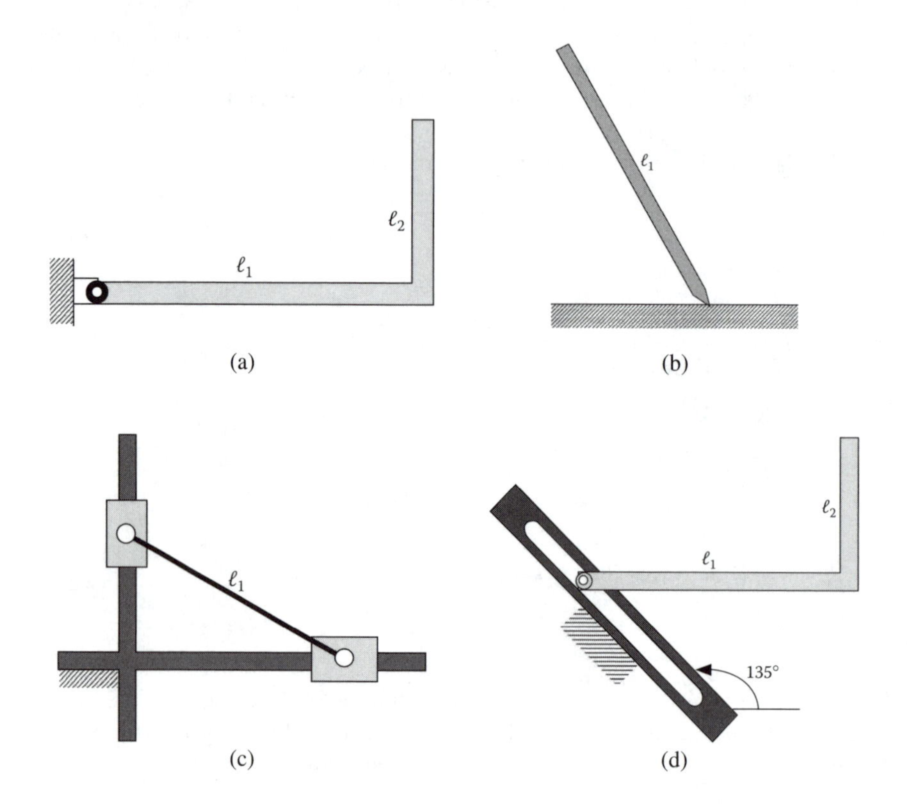

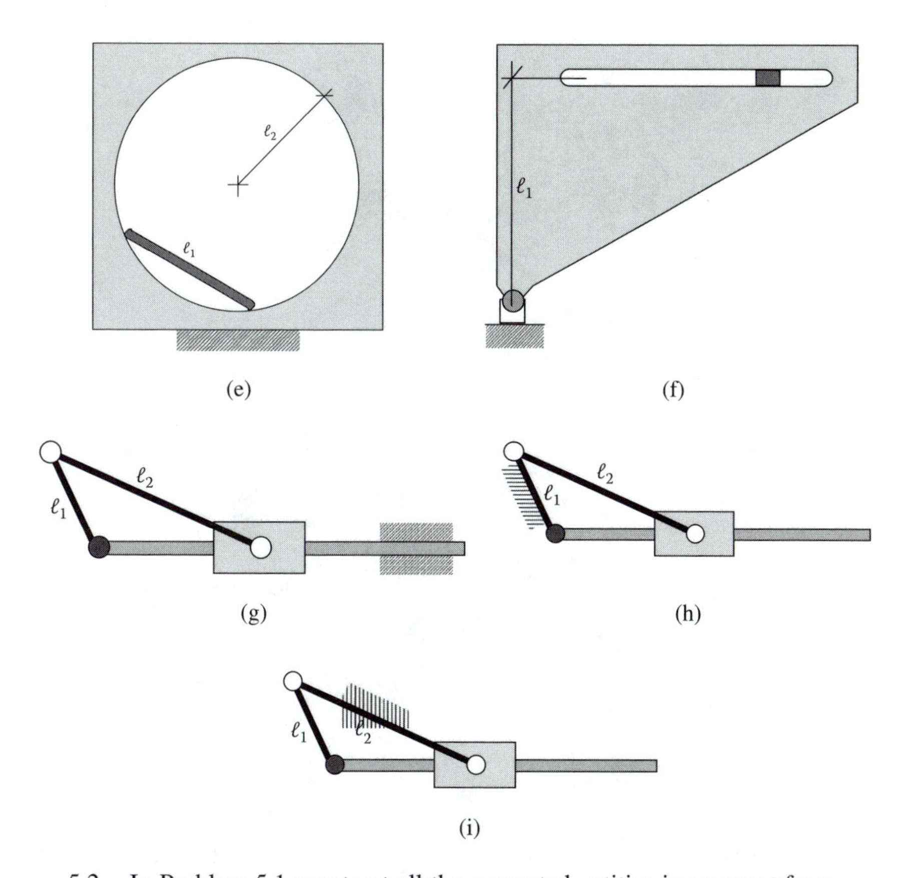

(e) (f)

(g) (h)

(i)

5.2 In Problem 5.1 construct all the requested entities in compact form.

5.3 Use the point-coordinate method to derive the necessary constraints for the systems shown.

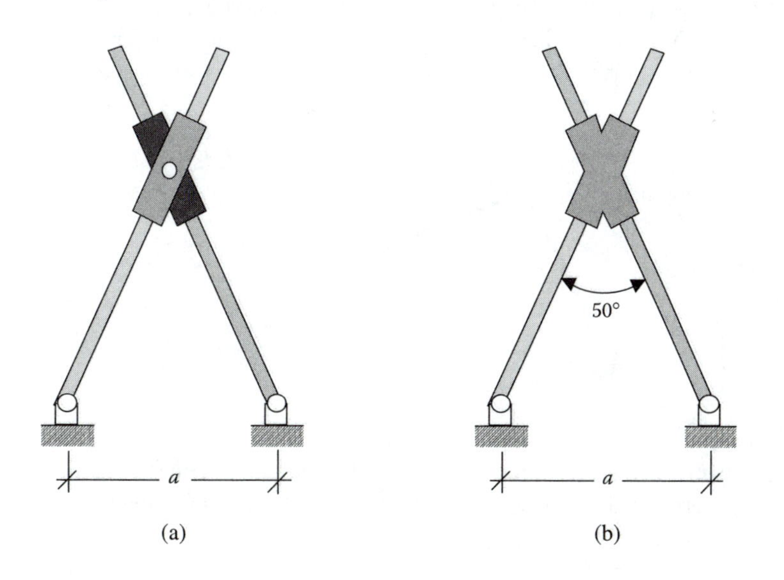

(a) (b)

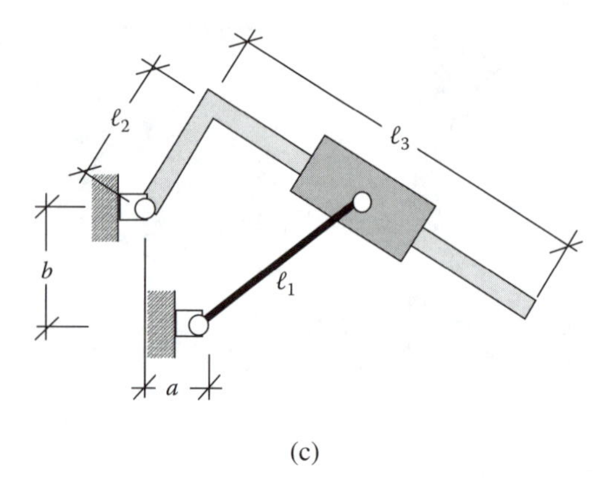

(c)

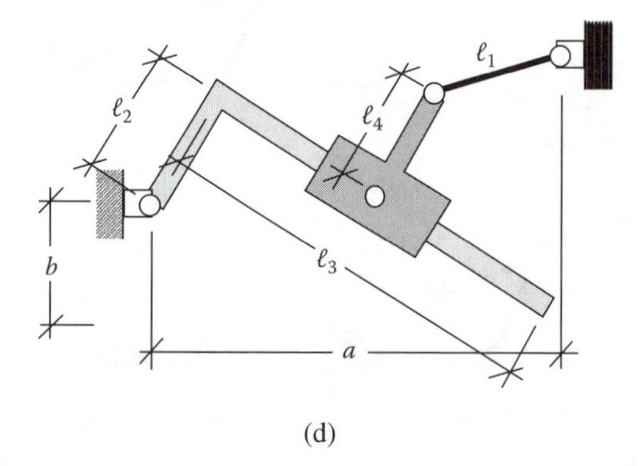

(d)

5.4 Use the point-coordinate method to derive the necessary constraints for the rolling disc shown.
(a) Assume the disc can slip relative to the ground.
(b) Assume the disc can roll without slip relative to the ground.

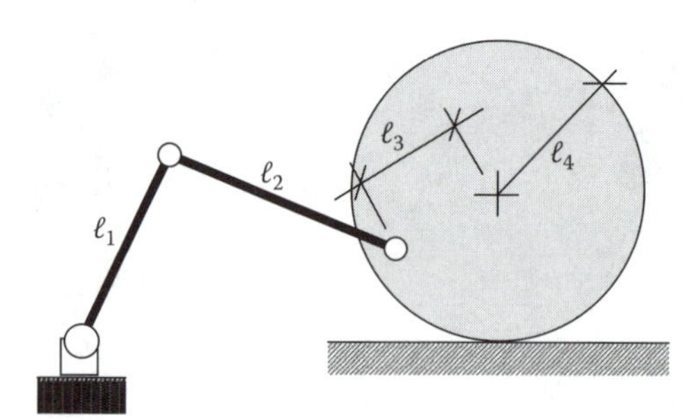

5.5 If the mechanism shown is modeled by the point-coordinate formulation, determine the minimum number of coordinates and constraints that are needed.

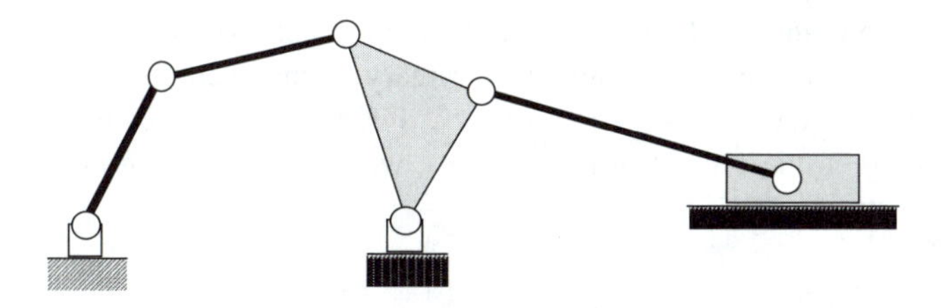

5.6 Use the point-coordinate method to derive the necessary constraints for the six-bar mechanism shown.

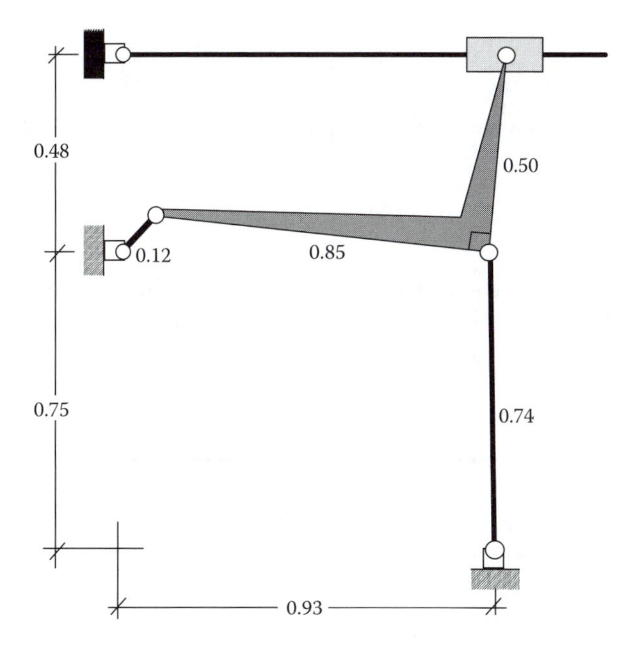

5.7 For two parallel vectors we can set $\theta = 0$ in Eq. (5.9) to obtain

$$\mathbf{d}^{i,j\prime}\mathbf{d}^{k,l} - \ell^{i,j}\ell^{k,l} = 0$$

Can we use this equation as a constraint for two parallel vectors instead of Eq. (5.8)? Explain your reasoning.

5.8 A link is described by two primary points A and B. A secondary point P is positioned on this link by an angle θ and a length ℓ as shown. Derive expressions similar to those in Eqs. (5.24) to (5.26) to compute the position, velocity, and acceleration of P.

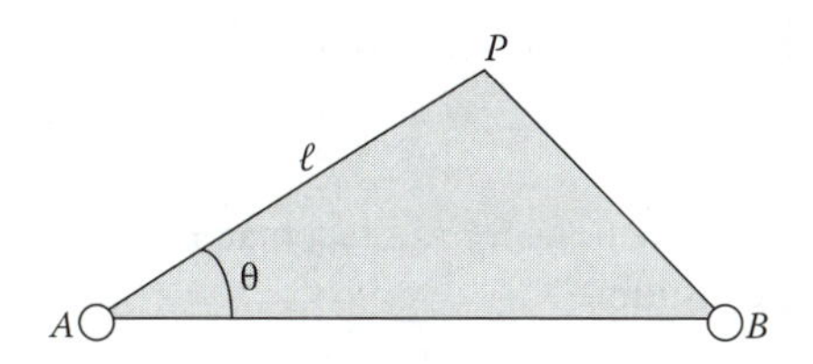

5.9 Three points A, B, and C are defined along the axis of a link as shown. If all three points are defined as primary points, then three vectors must be constructed as $\mathbf{d}^{A,B} = \mathbf{r}^A - \mathbf{r}^B$, $\mathbf{d}^{A,C} = \mathbf{r}^A - \mathbf{r}^C$, and $\mathbf{d}^{B,C} = \mathbf{r}^B - \mathbf{r}^C$. Between these three primary points, we can define the following three constraints:

$$\mathbf{d}^{A,B\prime}\mathbf{d}^{A,B} - (\ell^{A,B})^2 = 0 \qquad (a)$$

$$\mathbf{d}^{A,C\prime}\mathbf{d}^{A,C} - (\ell^{A,C})^2 = 0 \qquad (b)$$

$$\breve{\mathbf{d}}^{A,C\prime}\mathbf{d}^{A,B} = 0 \qquad (c)$$

We can also write the following constraint and use it instead of (c):

$$\mathbf{d}^{B,C\prime}\mathbf{d}^{B,C} - (\ell^{B,C})^2 = 0 \qquad (d)$$

However, we recommend not using (d) instead of (c). Explain the reason! (Hint: Consider the Jacobian matrix for the three constraints and show that the rows are dependent.)

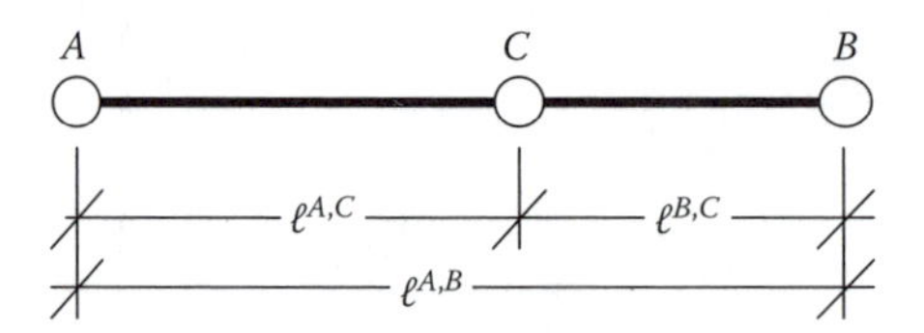

5.10 What are the advantages and disadvantages of upgrading a secondary point to a primary point?

5.11 Points A and B are primary points. Find a formulation to express point P in terms of the coordinates of A and B. What are the expressions for the velocity and acceleration of P?

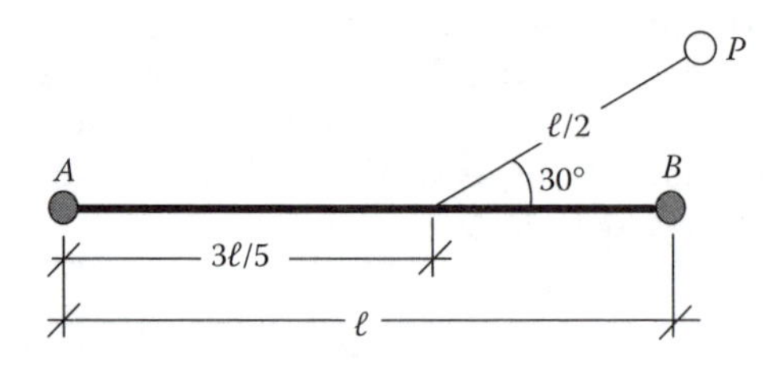

5.12 Construct M-file functions for the Jacobian matrix and the gamma array for the angle constraint.

5.13 Develop M-file functions for the Jacobian and the gamma array corresponding to the parallel constraint between four points as described in Eq. (5.8). Name these functions `PC_jacob_p_4` and `PC_gamma_p_4` respectively.

5.14 Develop an M-file function to compute parameters α and β for a secondary point based on Eqs. (5.28) and (5.29).

5.15 Transform the programs from Sections 5.6.1, 5.6.2, and 5.6.3 to a form similar to that of the programs in Section 7.6.

5.16 Revise the filmstrip program `PC_FS_A.m` from Section 5.6.3. In the revised program consider only two primary points—consider point *C* as a secondary point. Use the function `PC_secondary.m` to compute the coordinate and velocity of point *C* at the end of the program. Name the file `PC_FS_B.m`. Assume that the crank rotates at 1800 rpm, CW. At three different orientations of the crank (your choice of orientations between 0° and 360°), come up with your best estimates for the coordinates and velocities of points *A* and *B*. Execute the program at each orientation and determine the constraint violations at the position and velocity levels, and the coordinates and velocity of point *C*.

6 Point-Coordinates: Dynamics

In this chapter the kinematic modeling and formula derivation of the previous chapter is extended to deriving the dynamic equations of motion for multiparticle systems. We will learn how to construct the mass matrix, the array of applied forces, and the array of reaction forces for mechanical systems that are represented by a collection of primary points. First, the dynamics of systems of particles when only force elements are present is reviewed. Then, we consider the dynamics of systems of particles interconnected by force elements and kinematic constraints. The topics are described through examples before general formulations are presented.

Programs: In this chapter we add new statements to the programs of Chapter 5 and we develop several new function M-files. We will work with the same folders/directories as shown in Table 5.2. In programming exercises of this chapter, we make the folder `Point_Coord` the Current Directory of MATLAB®.

6.1 SYSTEM OF UNCONSTRAINED PARTICLES

A system of particles interconnected by force elements, without the presence of kinematic joints or constraints, is referred to as a system of *unconstrained particles*. The particles can be free or interconnected by force elements such as springs and dampers, and acted upon by gravitational force. In the process that leads to the construction of the equations of motion, it will be assumed that the coordinates and velocities of the particles that form the system are known. Because the number of coordinates is equal to the number of system's DoF for any unconstrained system, the coordinates and the velocities can be assigned any arbitrary or desired values. We first demonstrate the process of constructing the equations of motion for an unconstrained system of particles through a simple example before representing the general form of the equations.

6.1.1 A Two-Particle System

A system of two particles is considered in this example. As shown in Figure 6.1(a), two particles A and B are connected to each other by a spring, and particle A is connected to the ground by a spring-damper at O. The following data is provided:

$$m^A = 3,\ m^B = 4,\ k_1 = 50,\ {}^0\ell_1 = 1.1,\ d_{c,1} = 20,\ k_2 = 100,\ {}^0\ell_2 = 1.2$$

We also know that at a given instant, the particles have the following coordinates and velocities:

$$\mathbf{r}^A = \begin{Bmatrix} 0.5 \\ -1.0 \end{Bmatrix}, \dot{\mathbf{r}}^A = \begin{Bmatrix} -0.2 \\ 0.4 \end{Bmatrix}, \mathbf{r}^B = \begin{Bmatrix} 2.0 \\ -1.0 \end{Bmatrix}, \dot{\mathbf{r}}^B = \begin{Bmatrix} -0.1 \\ 0.3 \end{Bmatrix}$$

where the data is provided in an *x-y* coordinate system positioned at *O*.

The following program computes all the necessary arrays and matrices for the equations of motion.

```
  clear all
     addpath Basics   PC_Basics
% Define particle masses and spring-damper characteristics
    m_A = 3;   m_B = 4;
    k_1 = 50; L_1_0 = 1.1; d_c_1 = 20; k_2 = 100; L_2_0 = 1.2;
% Define coordinates and velocities
    r_A = [0.5; -1]; r_A_d = [-0.2; 0.4]; r_O = [0; 0];
    r_B = [2; -1];    r_B_d = [-0.1; 0.3];
% compute d1, d1_d, d2 and d2_d vectors
    d1  = r_A - r_O; d1_d = r_A_d;
    d2  = r_B - r_A; d2_d = r_B_d - r_A_d;
% Compute forces for element 1
    f_sda_1 = pp_sda(d1, d1_d, k_1, L_1_0, d_c_1, 0);
    f_O_1 =   f_sda_1   % the force on point O
    f_A_1 = -f_sda_1   % the force on point A
% Compute forces for element 2
    f_sda_2 = pp_sda(d2, d2_d, k_2, L_2_0, 0, 0);
    f_A_2 =   f_sda_2   % the force on point A
    f_B_2 = -f_sda_2   % the force on point B
% Compute total force acting on A and B
    f_A = f_A_1 + f_A_2   % sum up the two forces on A
    f_B = f_B_2
% Construct the 4x1 array of forces
    h = [f_A; f_B]
% Define a 2x2 identity matrix and a 2x2 zero matrix
    I2 = eye(2); Z2 = zeros (2,2);
% Construct the 4x4 mass matrix
  M = [m_A*I2 Z2; Z2 m_B*I2]
```

The program first determines the forces that act on each particle. For the force element between *A* and *O* the program applies the formulas from Sections 4.4.3 and 4.4.4 to obtain

```
f_O_1 =                     f_A_1 =
   -3.5967                      3.5967
    7.1935                     -7.1935
```

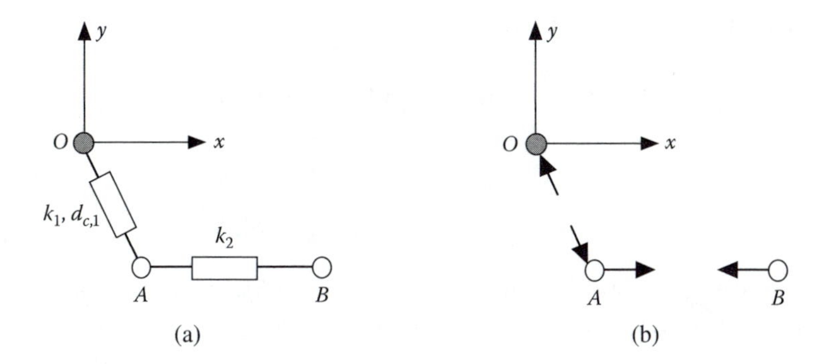

FIGURE 6.1 (a) Two unconstrained particles connected by spring-damper elements; (b) the corresponding element forces.

Similarly, for the spring element between A and B we obtain:

```
f_A_2 =                              f_B_2 =
    30.0000                              -30.0000
     0.0000                                0.0000
```

For clarification purposes, these forces are shown schematically in Figure 6.1(b). For the two particles, we find the following force vectors as the result of the force elements:

```
f_A =                                f_B =
   33.5967                              -30.0000
   -7.1935                                0.0000
```

The system force array is now obtained to be

```
h =
  33.5967
  -7.1935
 -30.0000
   0.0000
```

The mass matrix is computed to be

```
M =
     3     0     0     0
     0     3     0     0
     0     0     4     0
     0     0     0     4
```

We construct the equations of motion for the system as

$$\begin{bmatrix} 3 & 0 & 0 & 0 \\ 0 & 3 & 0 & 0 \\ 0 & 0 & 4 & 0 \\ 0 & 0 & 0 & 4 \end{bmatrix} \begin{Bmatrix} \ddot{x}^A \\ \ddot{y}^A \\ \ddot{x}^B \\ \ddot{y}^B \end{Bmatrix} = \begin{Bmatrix} 33.60 \\ -7.19 \\ -30.00 \\ 0.00 \end{Bmatrix}$$

If desired, the acceleration of the particles can be computed at this point because we have as many equations as unknown accelerations.

6.1.2 Unconstrained Particles—General

For a system of n_p unconstrained particles, assume that the array of coordinates is denoted as $\mathbf{r}$, and the arrays of velocities and accelerations are denoted as $\dot{\mathbf{r}}$ and $\ddot{\mathbf{r}}$. If the mass matrix and the array of forces are defined as $\mathbf{M}$ and $\mathbf{h}$, then the equations of motion are simply written as

$$\mathbf{M}\ddot{\mathbf{r}} = \mathbf{h} \tag{6.1}$$

where

$$\ddot{\mathbf{r}} = \begin{Bmatrix} \ddot{\mathbf{r}}^1 \\ \vdots \\ \ddot{\mathbf{r}}^{n_p} \end{Bmatrix}, \quad \mathbf{M} = \begin{bmatrix} \mathbf{M}^1 & \cdots & \mathbf{0} \\ \vdots & \ddots & \vdots \\ \mathbf{0} & \cdots & \mathbf{M}^{n_p} \end{bmatrix}, \quad \mathbf{h} = \begin{Bmatrix} \mathbf{f}^1 \\ \vdots \\ \mathbf{f}^{n_p} \end{Bmatrix} \tag{6.2}$$

The mass matrix $\mathbf{M}$ is made of 2×2 submatrices for each particle; that is,

$$\mathbf{M}^i = \begin{bmatrix} m^i & 0 \\ 0 & m^i \end{bmatrix} = m^i\,\mathbf{I};\ \ i = 1, \cdots, n_p \tag{6.3}$$

where m^i is the mass of the i-th particle. The array of forces, $\mathbf{h}$, can have contributions from gravitational force, springs, dampers and other elements. We note that Eq. (6.1) contains as many equations as the number of accelerations. This means that these equations can be solved as a set of linear algebraic equations to determine the accelerations.

6.2 SYSTEM OF CONSTRAINED PARTICLES

When kinematic constraints are present between particles, the system is referred to as a system of *constrained particles*. Because kinematic constraints result in reaction forces between particles, these forces must be incorporated into the equations of motion.

To construct the equations of motion for a constrained system of particles, we follow a two-step process. First, we ignore the constraints completely—we assume that we are dealing with an unconstrained system of particles. For this assumed unconstrained system, we construct the mass matrix and the array of forces due to the force elements as in Eq. (6.1). Then, in a second step, we return to the original system and include the reaction forces due to the constraints. The reaction forces will be described in terms of the Jacobian matrix of the constraints and an array of Lagrange multipliers. Construction of the equations of motion is first demonstrated through several simple examples before generalization.

6.2.1 A Two-Particle System

This example is a variation of the example in Section 6.1.1. As shown in Figure 6.2(a), the spring element between particles A and B is replaced by a massless rod having a length of $\ell = 1.2$; otherwise the rest of the system remains the same. We also know that at a given instant, the particles have the following coordinates and velocities:

$$\mathbf{r}^A = \begin{Bmatrix} 0.5 \\ -1.0 \end{Bmatrix},\ \dot{\mathbf{r}}^A = \begin{Bmatrix} -0.2 \\ 0.4 \end{Bmatrix},\ \mathbf{r}^B = \begin{Bmatrix} 1.7 \\ -1.0 \end{Bmatrix},\ \dot{\mathbf{r}}^B = \begin{Bmatrix} -0.2 \\ 0.3 \end{Bmatrix}$$

To construct the equations of motion, we first ignore the rod and view the system as unconstrained as depicted in Figure 6.2(b). For this system, we determine the forces that act on each particle to be:

$$\mathbf{f}^A = \begin{Bmatrix} 3.60 \\ -7.19 \end{Bmatrix},\ \mathbf{f}^B = \begin{Bmatrix} 0 \\ 0 \end{Bmatrix}$$

The equations of motion for this unconstrained system is constructed as

$$\begin{bmatrix} 3 & 0 & 0 & 0 \\ 0 & 3 & 0 & 0 \\ 0 & 0 & 4 & 0 \\ 0 & 0 & 0 & 4 \end{bmatrix} \begin{Bmatrix} \ddot{x}^A \\ \ddot{y}^A \\ \ddot{x}^B \\ \ddot{y}^B \end{Bmatrix} = \begin{Bmatrix} 3.60 \\ -7.19 \\ 0 \\ 0 \end{Bmatrix}$$

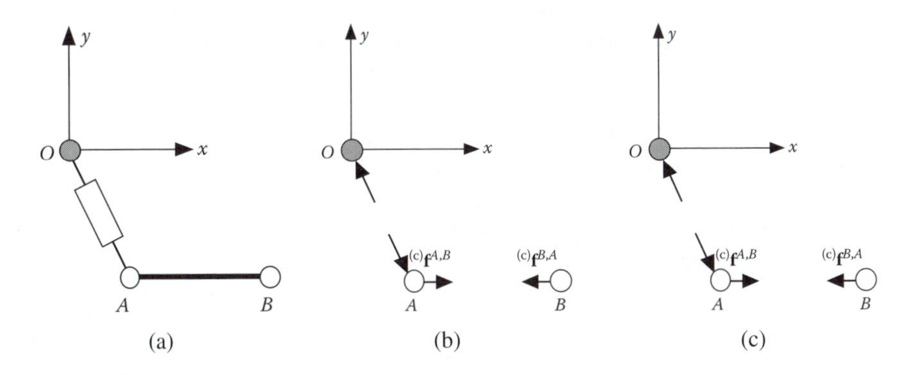

FIGURE 6.2 (a) Two constrained particles, (b) its unconstrained representation, and (c) reaction forces associated with the constraint.

When we put the rod back, we must consider the reaction forces between the particles and modify the equations of motion accordingly. The reaction forces with arbitrary directions are shown in Figure 6.2(c). We know, according to Newton's third law, that the two particles apply reaction forces on one another that are equal in magnitude and opposite in direction. At this point, we employ the Lagrange multiplier technique to describe these reaction forces.

The constant length constraint for the rod is written as

$$\frac{1}{2}[\mathbf{d}^{B,A\prime}\,\mathbf{d}^{B,A} - \ell^2] = 0 \qquad (a)$$

This yields the following velocity constraint:

$$\mathbf{d}^{B,A\prime}\,\dot{\mathbf{d}}^{B,A} = \begin{bmatrix} -\mathbf{d}^{B,A\prime} & \mathbf{d}^{B,A\prime} \end{bmatrix} \begin{Bmatrix} \dot{\mathbf{r}}^A \\ \dot{\mathbf{r}}^B \end{Bmatrix} = 0 \qquad (b)$$

The Jacobian matrix is extracted from *(b)* to be:

$$\mathbf{D} = [-\mathbf{d}^{B,A\prime} \quad \mathbf{d}^{B,A\prime}] = [-1.2 \quad 0 \quad 1.2 \quad 0] \qquad (c)$$

The transpose of this matrix is used to describe the reaction forces as

$$\begin{Bmatrix} {}^{(c)}\mathbf{f}^{A,B} \\ {}^{(c)}\mathbf{f}^{B,A} \end{Bmatrix} = \begin{bmatrix} -\mathbf{d}^{B,A} \\ \mathbf{d}^{B,A} \end{bmatrix} \lambda = \begin{bmatrix} -1.2 \\ 0 \\ 1.2 \\ 0 \end{bmatrix} \lambda \qquad (d)$$

The values of the coordinates and velocities that are given for this system are not arbitrary—these values must satisfy the constraints in (*a*) and (*b*). The coordinate values yield $\mathbf{d}^{B,A} = \mathbf{r}^B - \mathbf{r}^A = \{1.2 \quad 0\}'$. Substituting in *(a)* yields

$$\frac{1}{2}[\{1.2 \quad 0\}\begin{Bmatrix} 1.2 \\ 0 \end{Bmatrix} - 1.2^2] = 0$$

which is valid and involves no errors. Similarly, velocity values yield $\dot{\mathbf{d}}^{B,A} = \dot{\mathbf{r}}^B - \dot{\mathbf{r}}^A = \{0 \quad -0.1\}'$. Then *(b)* yields

$$\{1.2 \quad 0\}\begin{Bmatrix} 0 \\ -0.1 \end{Bmatrix} = 0$$

which also does not involve any errors.

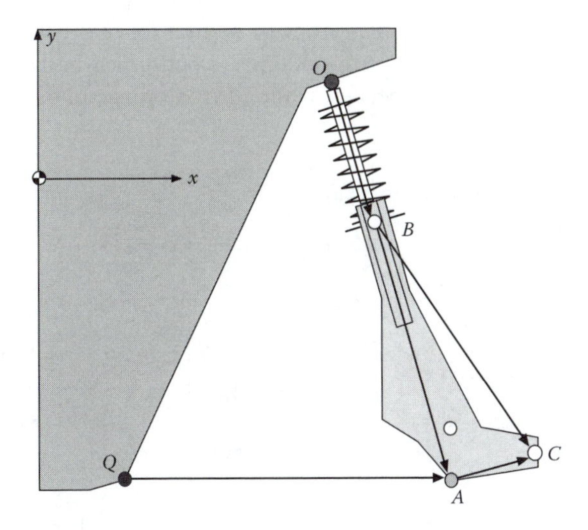

FIGURE 6.3 Point coordinate presentation of the MacPherson suspension system.

This expression contains only one Lagrange multiplier because we have only one constraint. Now the equations of motion can be written as

$$\begin{bmatrix} 3 & 0 & 0 & 0 \\ 0 & 3 & 0 & 0 \\ 0 & 0 & 4 & 0 \\ 0 & 0 & 0 & 4 \end{bmatrix} \begin{Bmatrix} \ddot{x}^A \\ \ddot{y}^A \\ \ddot{x}^B \\ \ddot{y}^B \end{Bmatrix} = \begin{Bmatrix} 3.60 \\ -7.19 \\ 0 \\ 0 \end{Bmatrix} + \begin{bmatrix} -1.2 \\ 0 \\ 1.2 \\ 0 \end{bmatrix} \lambda \qquad (e)$$

This equation contains five unknowns—four accelerations and one Lagrange multiplier.

6.2.2 MacPherson Suspension

This example has been modeled kinematically in Sections 5.1.2 and 5.6.2. For the three primary points A, B and C we are given the following masses (we will see in the upcoming sections how these masses are determined based on the mass of the bodies):

$$m^A = 13.98,\ m^B = 3.66,\ m^C = 3.36$$

In the configuration shown in Figure 6.3, the coordinates and velocity of the primary points are:

$$\mathbf{r}^A = \begin{Bmatrix} 0.570 \\ -0.410 \end{Bmatrix},\ \dot{\mathbf{r}}^A = \begin{Bmatrix} 0.000 \\ 1.130 \end{Bmatrix},\ \mathbf{r}^B = \begin{Bmatrix} 0.468 \\ -0.065 \end{Bmatrix},$$

$$\dot{\mathbf{r}}^B = \begin{Bmatrix} -0.197 \\ 1.072 \end{Bmatrix},\ \mathbf{r}^C = \begin{Bmatrix} 0.688 \\ -0.387 \end{Bmatrix},\ \dot{\mathbf{r}}^C = \begin{Bmatrix} -0.013 \\ 1.197 \end{Bmatrix}$$

Our objective is to construct the equations of motion for this system at this given configuration. All the necessary computations are performed in the following program. This program must be appended to the MATLAB program of Section 5.6.2.

```
% Append to the file P_MP
% ...
% Masses, gravitational constant, and spring-damper data
   m_A = 13.98; m_B = 3.66; m_C = 3.36; g = 9.81;
   k = 22600; L_0 = 0.34; d_c = 1270;
% Construct h array
   h = PC_MP_h(d_BO, d_BO_d, m_A, m_B, m_C, g, k, L_0, d_c);
% Construct mass matrix
      M_array = [m_A m_A m_B m_B m_C m_C];
  M_diag = diag(M_array);
```

In this part of the program, we have provided the mass and spring-damper data. The data is sent to the following function to evaluate the force array. In addition, because the spring-damper is connected between points O and B, the components of vectors $\mathbf{d}^{B,O}$ and $\dot{\mathbf{d}}^{B,O}$ are also supplied to the function. In this function, we consider the system as an unconstrained system of three particles and a spring-damper; that is, we can ignore the constraints completely. This function first computes the force of the spring-damper through function `pp_sda`; then it includes the gravity acting in the negative y-direction on all three particles.

```
function f = PC_MP_h(d_BO, d_BO_d, m_A, m_B, m_C, g, k, L_0, d_c)
% Compute spring-damper force
    f_sda = pp_sda(d_BO, d_BO_d, k, L_0, d_c, 0);
    f_sda_B = -f_sda;
% Construct gravitational vector of forces
    f_g_A = [0; -m_A*g];
    f_g_B = [0; -m_B*g];
    f_g_C = [0; -m_C*g];
% Construct array of forces
    h = [f_g_A; (f_sda_B + f_g_B); f_g_C];
```

The last entity that is constructed by the main program is the mass matrix. The program constructs the diagonal elements of the mass matrix in an array called `M_array`. Then the program uses a MATLAB function, `diag()`, to convert the array into a diagonal matrix.

Executing the program provides the following force vector to be applied at B as the result of the spring-damper element:

```
f_sda_B =
    1.0e+03 *
      1.2722
     -4.2773
```

Inclusion of the gravitational force provides the following three vectors:

```
f_g_A =                    f_g_B =                   f_g_C =
          0                          0                         0
  -137.1438                   -35.9046                  -32.9616
```

The array of forces is obtained by summing the forces that act on each point:

```
h =
    1.0e+03  *
          0
    -0.1371
     1.2722
    -4.3132
          0
    -0.0330
```

The mass matrix finds the following values:

```
M_diag =
   13.9800         0         0         0         0         0
         0   13.9800         0         0         0         0
         0         0    3.6600         0         0         0
         0         0         0    3.6600         0         0
         0         0         0         0    3.3600         0
         0         0         0         0         0    3.3600
```

6.2.3 Constrained Particles—General

For a system of n_p constrained particles, the mass matrix and the array of applied forces are defined as $\mathbf{M}$ and $\mathbf{h}$, exactly as those in Eq. (6.1). However, we must revise Eq. (6.1) by including the reaction forces as

$$\mathbf{M}\ddot{\mathbf{r}} = \mathbf{h} + {}^{(c)}\mathbf{h} \tag{6.4}$$

where

$${}^{(c)}\mathbf{h} = \begin{Bmatrix} {}^{(c)}\mathbf{f}^1 \\ \vdots \\ {}^{(c)}\mathbf{f}^{n_p} \end{Bmatrix} \tag{6.5}$$

This is the array of reaction (constraint) forces acting on the particles.

Assume that the position constraints between the n_p particles form n_c constraint equations that are expressed in a general form as

$$\mathbf{\Phi}(\mathbf{r}) = \mathbf{0} \tag{6.6}$$

The velocity and acceleration constraints are expressed as

$$\dot{\mathbf{\Phi}} \equiv \mathbf{D}\dot{\mathbf{r}} = \mathbf{0} \tag{6.7}$$

$$\ddot{\mathbf{\Phi}} \equiv \mathbf{D}\ddot{\mathbf{r}} + \dot{\mathbf{D}}\dot{\mathbf{r}} = \mathbf{0} \tag{6.8}$$

In these equations **D** is the Jacobian matrix. According to the method of Lagrange multipliers, the array of reaction forces of Eq. (6.5) can be represented as

$$^{(c)}\mathbf{h} = \mathbf{D}'\boldsymbol{\lambda} \tag{6.9}$$

where

$$\boldsymbol{\lambda} = \begin{Bmatrix} \lambda_1 \\ \vdots \\ \lambda_{n_c} \end{Bmatrix} \tag{6.10}$$

is an array of n_c Lagrange multipliers. Hence Eq. (6.4) is rewritten as*:

$$\mathbf{M}\ddot{\mathbf{r}} = \mathbf{h} + \mathbf{D}'\boldsymbol{\lambda} \tag{6.11}$$

We note that Eq. (6.11) contains n_p equations and $n_p + n_c$ unknowns (accelerations and Lagrange multipliers). This means that these equations cannot be solved in their present form as a set of linear algebraic equations to determine the accelerations. We will learn how to solve these equations to determine the unknowns in the analysis chapters.

Example 6.1

Construct the equations of motion for the MacPherson suspension system in numerical form.

Solution

The mass matrix and the array of forces for this system were obtained numerically in Section 6.2.2. Numerical values for the entries of the Jacobian matrix were obtained

---

* The equations of motion can also be written as $\mathbf{M}\ddot{\mathbf{r}} = \mathbf{h} - \mathbf{D}'\boldsymbol{\lambda}$. This can be interpreted that the reaction forces are described as $^{(c)}\mathbf{h} = -\mathbf{D}'\boldsymbol{\lambda}$ and not as $^{(c)}\mathbf{h} = \mathbf{D}'\boldsymbol{\lambda}$. The preference in using this revised form of equations of motion could be that we will have a symmetric coefficient matrix when we perform forward dynamic analysis.

in Section 5.6.2. These entities yield the equation of motion for the MacPherson suspension system, with the specified coordinates and velocities, as

$$
\left[\begin{array}{cc:cc:cc}
13.98 & 0 & 0 & 0 & 0 & 0 \\
0 & 13.98 & 0 & 0 & 0 & 0 \\
\hdashline
0 & 0 & 3.66 & 0 & 0 & 0 \\
0 & 0 & 0 & 3.66 & 0 & 0 \\
\hdashline
0 & 0 & 0 & 0 & 3.36 & 0 \\
0 & 0 & 0 & 0 & 0 & 3.36
\end{array}\right]
\left\{\begin{array}{c}
\ddot{x}^A \\ \ddot{y}^A \\ \hdashline \ddot{x}^B \\ \ddot{y}^B \\ \hdashline \ddot{x}^C \\ \ddot{y}^C
\end{array}\right\} =
$$

$$
\left\{\begin{array}{c}
0 \\ -137.1 \\ \hdashline 1272.2 \\ -4313.2 \\ \hdashline 0 \\ -33.0
\end{array}\right\} +
\left[\begin{array}{ccccc}
0.450 & 0.102 & -0.118 & 0 & -0.195 \\
0 & -0.345 & -0.023 & 0 & -0.058 \\
\hdashline
0 & -0.102 & 0 & -0.220 & 0.540 \\
0 & 0.345 & 0 & 0.322 & 0.160 \\
\hdashline
0 & 0 & 0.118 & 0.220 & 0 \\
0 & 0 & 0.023 & -0.322 & 0
\end{array}\right]
\left\{\begin{array}{c}
\lambda_1 \\ \lambda_2 \\ \lambda_3 \\ \lambda_4 \\ \lambda_5
\end{array}\right\}
$$

6.3 FORCE AND MASS DISTRIBUTION

Applied forces on a body must be distributed to the particles that represent the body. If the point of application of a force on a body coincides with a primary point (i.e., one of the particles that represents the body), then the force can be applied directly to that particle. However, if the point of application of a force does not coincide with one of the particles, then that force must properly be distributed to the particles. Similarly, the mass of a body must be distributed to the particles that represent that body. In this section we show how to distribute an applied force that acts on a body and its mass to primary points. We state the necessary conditions in a general form for both distributions.

We assume that n_p primary points represent a body. Each primary point is positioned from a reference point O by a vector $\mathbf{s}^i$. To simplify our discussion, we consider the body mass center as the reference point. A body and its equivalent primary point representation are shown in Figure 6.4(a) and (b).

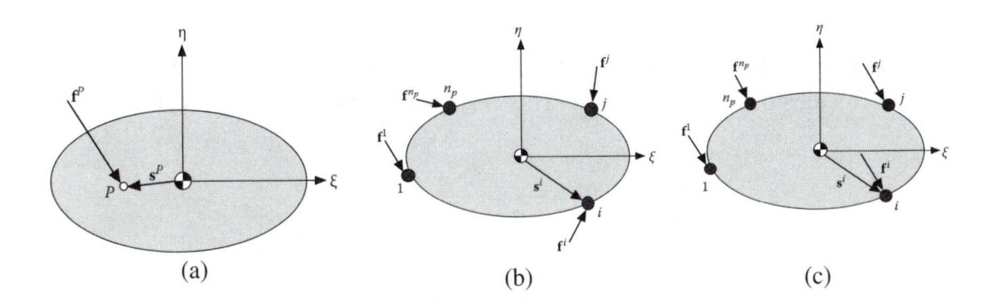

FIGURE 6.4 Equivalency of forces for a body and its point representation.

For the force distribution, assume a force $\mathbf{f}^P$ acts on a body at point P as shown in Figure 6.4(a). Point P is positioned on the body from a reference point O by vector $\mathbf{s}^P$. In order for the two force systems shown in Figure 6.4(a) and (b) to be equivalent, two conditions must be met:

1. The sum of forces must be equal:

$$\sum_{i=1}^{n_p} \mathbf{f}^i = \mathbf{f}^P \tag{6.12}$$

2. The sum of moments about any arbitrary point must be equal:

$$\sum_{i=1}^{n_p} \breve{\mathbf{s}}^{i\prime}\, \mathbf{f}^i = \breve{\mathbf{s}}^{P\prime}\, \mathbf{f}^P \tag{6.13}$$

Equations (6.12) and (6.13) represent three algebraic equations: one for the x-component of the force; one for the y-component of the force; and one for the moment about an axis perpendicular to the plane. Therefore, to obtain a unique solution these equations could contain only three unknowns. Because the equations contain more than three unknowns, one possible solution would be to assume that the distributed forces on the particles are all along the axis of the original force; that is,

$$\mathbf{f}^i = \alpha^i\, \mathbf{f}^P;\ i = 1, \cdots, n_p \tag{6.14}$$

where α^i 's are positive or negative scale factors. Such a distribution is shown in Figure 6.4(c). Substituting Eq. (6.14) into Eqs. (6.12) and (6.13) yields

$$\sum_{i=1}^{n_p} \alpha^i\, \mathbf{f}^P = \mathbf{f}^P$$

$$\sum_{i=1}^{n_p} \alpha^i\, \breve{\mathbf{s}}^{i\prime}\, \mathbf{f}^P = \breve{\mathbf{s}}^{P\prime}\, \mathbf{f}^P$$

Because $\mathbf{f}^P$ is a vector having arbitrary magnitude and direction, it can be eliminated from both sides of these equations. This results into the following equations:

$$\begin{gathered} \sum_{i=1}^{n_p} \alpha^i = 1 \\ \sum_{i=1}^{n_p} \alpha^i\, \mathbf{s}^i = \mathbf{s}^P \quad \text{or} \quad \sum_{i=1}^{n_p} \alpha^i\, \not{\mathbf{s}}^i = \not{\mathbf{s}}^P \end{gathered} \tag{6.15}$$

Equation (6.15) represents three algebraic equations; therefore we can only have three unknowns to obtain a unique solution. We keep three of the unknown α^i's; (i.e., three forces) and express Eq. (6.15) in expanded form as

$$\begin{bmatrix} 1 & 1 & 1 \\ s^i_{(\xi)} & s^j_{(\xi)} & s^k_{(\xi)} \\ s^i_{(\eta)} & s^j_{(\eta)} & s^k_{(\eta)} \end{bmatrix} \begin{Bmatrix} \alpha^i \\ \alpha^j \\ \alpha^k \end{Bmatrix} = \begin{Bmatrix} 1 \\ s^P_{(\xi)} \\ s^P_{(\eta)} \end{Bmatrix} \tag{6.16}$$

For the mass distribution, assume the body has a mass m. If the mass of each primary point is denoted as m^i; $i = 1, \cdots, n_p$, then the following two conditions must be satisfied in order for this system of mass points to *approximate* the inertial characteristics of its corresponding body:

1. The sum of the mass particles must be equal to the mass of the body:

$$\sum_{i=1}^{n_p} m^i = m \tag{6.17}$$

2. The mass center of the system of particles must be at the same position as the mass center of the body:

$$\sum_{i=1}^{n_p} m^i \, \mathbf{s}^i = \mathbf{0} \tag{6.18}$$

These two conditions represent three algebraic equations. This means that with only three unknown masses we can have a unique solution. With three unknown masses, these two conditions can be put together in expanded form as

$$\begin{bmatrix} 1 & 1 & 1 \\ s^i_{(\xi)} & s^j_{(\xi)} & s^k_{(\xi)} \\ s^i_{(\eta)} & s^j_{(\eta)} & s^k_{(\eta)} \end{bmatrix} \begin{Bmatrix} m^i \\ m^j \\ m^k \end{Bmatrix} = \begin{Bmatrix} m \\ 0 \\ 0 \end{Bmatrix} \tag{6.19}$$

Satisfying the conditions of Eqs. (6.17) and (6.18) provides an approximate mass distribution, because we did not state a condition between the moment of inertia of the system of particles and that of the body. We will consider an exact mass distribution in Section 6.4.

6.3.1 Two Primary Points

A force acting on a rod can be distributed to only two primary points. In other words, there is no need to define three primary points on the rod for the purpose of distributing a force. This is due to the fact that vectors $\mathbf{s}^i$, $\mathbf{s}^j$, $\mathbf{s}^k$, and $\mathbf{s}^P$ become collinear for a rod and, as a result, Eq. (6.15) turns into two algebraic equations; therefore we can have only two unknowns.

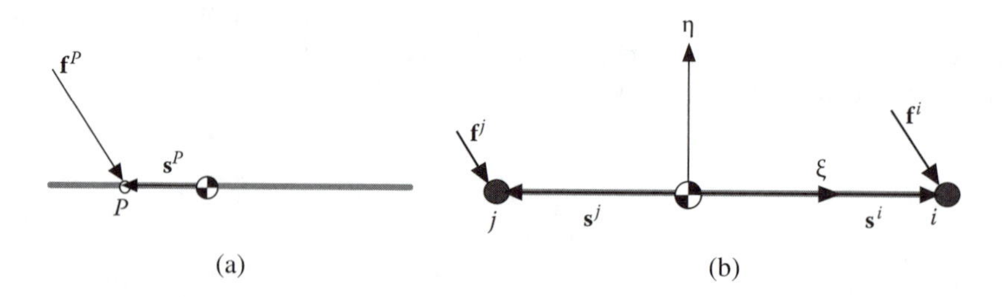

FIGURE 6.5 Force equivalency for a rod represented by two primary points.

The force distribution for a rod represented by two primary points is shown in Figure 6.5. For this system, Eq. (6.19) can be written for two unknowns as

$$\begin{bmatrix} 1 & 1 \\ s^i_{(\xi)} & s^j_{(\xi)} \end{bmatrix} \begin{Bmatrix} \alpha^i \\ \alpha^j \end{Bmatrix} = \begin{Bmatrix} 1 \\ s^P_{(\xi)} \end{Bmatrix} \tag{6.20}$$

This leads to

$$\alpha^i = \frac{s^P_{(\xi)} - s^j_{(\xi)}}{s^i_{(\xi)} - s^j_{(\xi)}}, \quad \alpha^j = \frac{s^P_{(\xi)} - s^i_{(\xi)}}{s^j_{(\xi)} - s^i_{(\xi)}}$$

Then the two forces are determined as

$$\mathbf{f}^i = \alpha^i \mathbf{f}^P, \mathbf{f}^j = \alpha^j \mathbf{f}^P \tag{6.21}$$

Example 6.2

Consider the lower A-arm of the double A-arm suspension from Section 1.5.1 as shown in Figure 6.6. Distribute the spring-damper force to the two points at the ends of the link.

Solution

For the lower A-arm, link *(1)*, we position the origin of the reference frame at E which is the point at which the spring-damper force is applied. Then, Eq. (6.20) becomes

$$\begin{bmatrix} 1 & 1 \\ 0.18 & -0.24 \end{bmatrix} \begin{Bmatrix} \alpha^A \\ \alpha^Q \end{Bmatrix} = \begin{Bmatrix} 1 \\ 0 \end{Bmatrix}$$

This equation yields $\alpha^A = 0.571$ and $\alpha^Q = 0.429$.

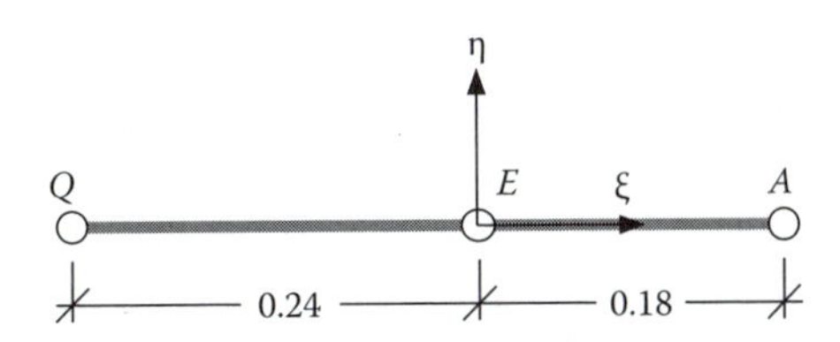

FIGURE 6.6 The lower A-arm of the double A-arm suspension.

For this system of two primary points, in a similar process as that of the force distribution, the mass distribution of Eq. (6.19) is simplified and written as

$$\begin{bmatrix} 1 & 1 \\ s^i_{(\xi)} & s^j_{(\xi)} \end{bmatrix} \begin{Bmatrix} m^i \\ m^j \end{Bmatrix} = \begin{Bmatrix} m \\ 0 \end{Bmatrix} \tag{6.22}$$

This equation yields

$$m^i = \frac{s^j_{(\xi)}}{s^j_{(\xi)} - s^i_{(\xi)}} m, \; m^j = \frac{s^i_{(\xi)}}{s^i_{(\xi)} - s^j_{(\xi)}} m$$

Example 6.3

A slender rod with a uniform mass distribution has its mass center at its geometric center. If the mass and the length of the rod are m and ℓ respectively, we know that the moment of inertia of the rod with respect to its mass center is $J = m\ell^2/12$. Distribute the mass to two primary points at the ends of the rod and compute the moment of inertia of this system with respect to the mass center. What do you conclude?

Solution

The two primary points A and B are positioned from the mass center as $s^A_{(\xi)} = \ell/2$ and $s^B_{(\xi)} = -\ell/2$. Then, Eq. (6.22) is expressed as

$$\begin{bmatrix} 1 & 1 \\ \ell/2 & -\ell/2 \end{bmatrix} \begin{Bmatrix} m^A \\ m^B \end{Bmatrix} = \begin{Bmatrix} m \\ 0 \end{Bmatrix}$$

This equation results in $m^A = m^B = m/2$. The moment of inertia for this system of two particles about its mass center is computed as

$$J^{A+B} = m^A(\ell/2)^2 + m^B(-\ell/2)^2 = m\ell^2/4$$

As we see $J^{A+B} = m\ell^2/4$ is different from the correct value $m\ell^2/12$. This is due to the fact that we did not consider the moment of inertia of the body as another condition, in addition to Eqs. (6.17) and (6.18). Therefore, this is only an approximated mass distribution.

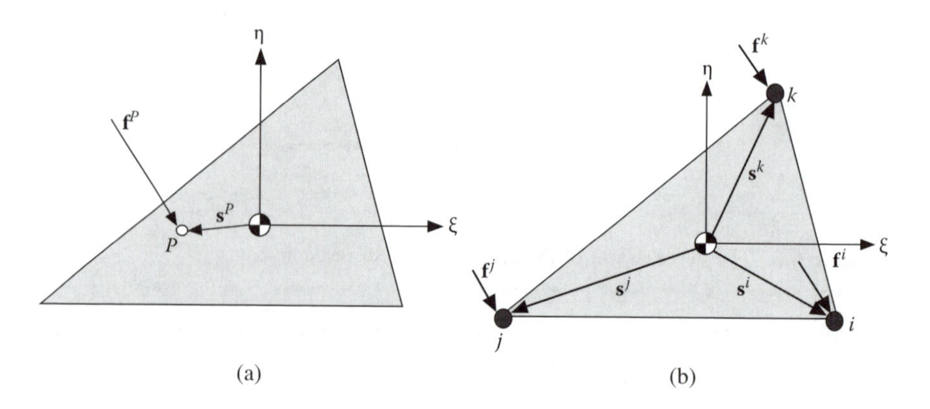

FIGURE 6.7 (a) A force acting on a plate and (b) its equivalent distribution to three primary points.

6.3.2 Three Primary Points

A force $\mathbf{f}^P$ acts at point P on a body that is represented by three primary points as shown in Figure 6.7. For this system Eq. (6.16) can be applied directly to obtain the distribution coefficients. Then we can determine the three forces as

$$\mathbf{f}^i = \alpha^i\,\mathbf{f}^P,\ \mathbf{f}^j = \alpha^j\,\mathbf{f}^P,\ \mathbf{f}^k = \alpha^k\,\mathbf{f}^P \tag{6.23}$$

To distribute the mass to the three primary points, we can directly apply Eq. (6.19).

Example 6.4

For the double A-arm suspension system example, the primary points were defined in Section 5.1.1. Consider body (*2*) of this system as shown in Figure 6.8. Distribute the mass of this body to the three primary points A, B, and C.

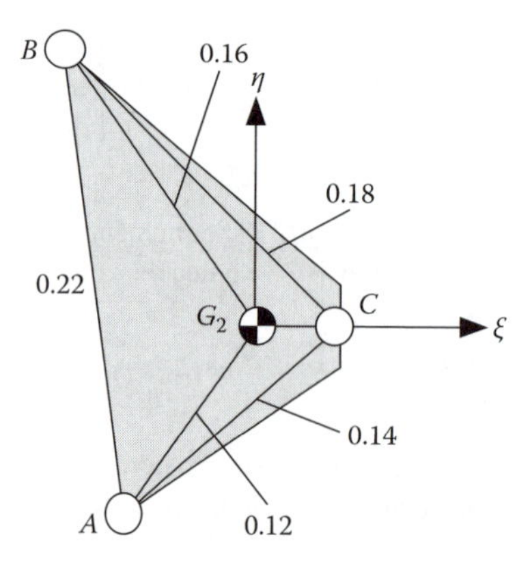

FIGURE 6.8 The knuckle of the double A-arm suspension.

Solution

In the attached ξ-η reference frame, the coordinates of the three primary points are measured:

$$\mathbf{s}^A = \begin{Bmatrix} -0.061 \\ -0.079 \end{Bmatrix}, \ \mathbf{s}^B = \begin{Bmatrix} -0.089 \\ 0.126 \end{Bmatrix}, \ \mathbf{s}^C = \begin{Bmatrix} 0.037 \\ 0 \end{Bmatrix}$$

This data and the mass of the body, $m = 30$, are used in the following program to determine the three masses:

```
% Define body mass and three position vectors
    m = 30;
  s_A = [-0.061; -0.079]; s_B = [-0.089; 0.126]; s_C = [0.037; 0];
% Distribute the mass
    m_ps = PC_mass_3_approx(m, s_A, s_B, s_C)
```

```
function m_ps = PC_mass_3_approx(m, s_A, s_B, s_C)
% Distribute mass of the body to three particles (approximate)
    P = [1      1      1
           s_A   s_B s_C];
    rhs = [m; 0; 0];
    m_ps = P\rhs;
```

Executing the program provides the following masses:

```
m_ps =
     6.2712
     3.9319
    19.7969
```

That is: $m^A = 6.2712$, $m^B = 3.9319$ and $m^C = 19.7969$.

EXAMPLE 6.5

We repeat the process of Example 6.4 for link *(2)* of the MacPherson suspension system (refer to Section 1.5.2) where the primary points are defined in Section 5.1.3. The mass of the link is $m = 20$. In an attached ξ-η frame, the coordinates of the primary points are measured as:

$$\mathbf{s}^A = \begin{Bmatrix} 0.0 \\ -0.07 \end{Bmatrix}, \quad \mathbf{s}^B = \begin{Bmatrix} -0.11 \\ 0.28 \end{Bmatrix}, \quad \mathbf{s}^C = \begin{Bmatrix} 0.12 \\ -0.035 \end{Bmatrix}$$

Revising the data in the preceding program yields the following masses:

```
m_A =            m_B =        m_C =
  12.9771            3.6641        3.3588
```

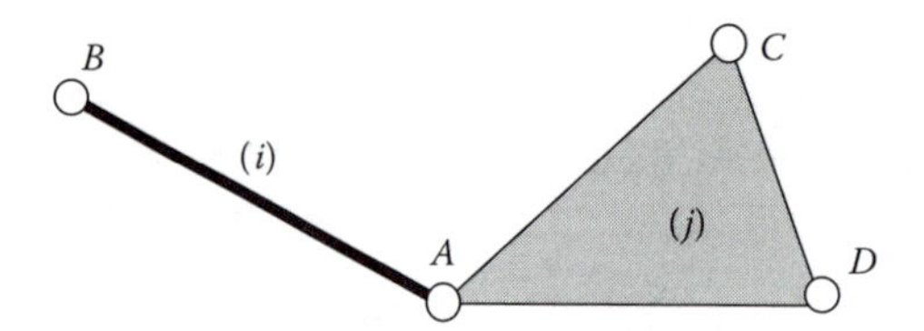

FIGURE 6.9 Primary point A receives masses from two links.

6.3.3 Mass Collection

A primary point could receive mass contribution from more than one link. This is the case when a primary point is shared between more than one body. For example, the primary point A is defined on a pin joint connecting links *(i)* and *(j)* as shown in Figure 6.9. The mass of link *(i)*, m_i, is distributed to its two primary points as m_i^A and m_i^B. The mass of link *(j)*, m_j, is distributed to its primary points as m_j^A, m_j^C, and m_j^D. Then, the mass of point A received contributions from both links as

$$m^A = m_i^A + m_j^A \tag{6.24}$$

6.3.4 Double A-Arm Suspension

This example has been modeled kinematically in Sections 5.1.1 and 5.6.1. In this section we compute the necessary arrays and matrices for constructing the equations of motion. For easy reference, the system is shown again in Figure 6.10.

In Example 6.4 we distributed the mass of link (2) to the three primary points. The three masses were found to be

$$m_2^A = 6.2712,\ m_2^B = 3.9319,\ m_2^C = 19.7969$$

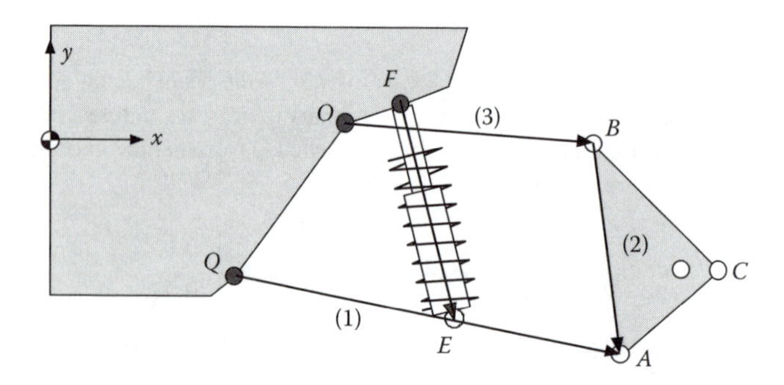

FIGURE 6.10 The double A-arm suspension system adopted for the point coordinate formulation.

Primary points A and B also receive mass distributions from links (1) and (3) respectively. Because the positions of the mass center for these links are not provided, it would be logical to assume uniform mass distributions for these links. Therefore the mass distribution from these two links yields:

$$m_1^A = 1.0, \ m_3^B = 0.5$$

Now the mass of each primary point is determined to be:

$$m^A = m_1^A + m_2^A = 7.2712, \ m^B = m_2^B + m_3^B = 4.4319, \ m^C = m_2^C = 19.7969$$

The spring-damper is connected between points F and E. Because E is not a primary point, the force that acts on E must be distributed to Q and A. In Example 6.2, we determined that the parameters for this distribution are $\alpha^A = 0.571$ and $\alpha^Q = 0.429$.

All the necessary computations are performed in the following program which must be appended to the main program of Section 5.1.1.

```
% Append to the file PC_AA
% ...
% Constant geometric data
    L_EQ = 0.24; r_F = [0.38; 0.04];
% Masses, gravitational constant, and spring-damper characteristics
    m_A = 7.27; m_B = 4.43; m_C = 19.80; g = 9.81;
    k = 91600; L_0 = 0.3; d_c = 1433;
    alpha_A = 0.57; alpha_Q = 0.43;
% Construct f array
    % Compute coordinate and velocity of secondary point E
    r_E   = PC_secondary(r_Q, r_A, L_EQ, 0, L_AQ);
    r_E_d = PC_secondary([0; 0], r_A_d, L_EQ, 0, L_AQ);
    d_EF = r_E - r_F; d_EF_d = r_E_d;
  h = PC_AA_h(d_EF, d_EF_d, m_A, m_B, m_C, g, k, L_0, d_c, alpha_A);
% Construct mass matrix
        M_diag = [m_A m_A m_B m_B m_C m_C];
     M = diag(M_diag);
```

Following the input data, the program computes the coordinates and velocity of E as a secondary point by using the function `secondary`. This yields the following values:

```
r_E =                              r_E_d =
       0.4343                             0.1503
      -0.1931                             0.6646
```

This information is then used to compute the force of the spring damper that is connected between E and the nonmoving point F. The spring-damper force and the gravitational forces are computed by the following function.

```
function h = PC_AA_h(d_EF, d_EF_d, m_A, m_B, m_C, g, ...
                    k, L_0, d_c, alpha_A)
% Compute spring-damper force
   f_sda = pp_sda(d_EF, d_EF_d, k, L_0, d_c, 0);
% Distribute the spring-damper force from E to B and Q
   f_sda_E = -f_sda;
   f_sda_A = alpha_A * f_sda_E;
% Construct gravitational vector of forces
    f_g_A = [0; -m_A*g]; f_g_B = [0; -m_B*g]; f_g_C = [0; -m_C*g];
% Construct array of forces
   h = [(f_sda_A + f_g_A); f_g_B; f_g_C];
```

The spring-damper force that acts on E is distributed to A using parameter α^A. The array of forces for the three particles is reported to be:

```
h =
    1.0e+03 *
     0.8314
    -3.6418
          0
    -0.0435
          0
    -0.1942
```

The mass matrix is constructed next and reported as:

```
M =
          7.2700         0         0         0         0         0
               0    7.2700         0         0         0         0
               0         0    4.4300         0         0         0
               0         0         0    4.4300         0         0
               0         0         0         0   19.8000         0
               0         0         0         0         0   19.8000
```

6.4 EXACT MASS DISTRIBUTION

In Section 6.3 we learned how to distribute the mass of a planar body to the primary points that represent that body. The distribution was based on the criteria to satisfying the total mass and the center of mass conditions. This distribution of mass did not consider any condition for the body moment of inertia and, therefore, it provided only approximate mass distribution formulas.

In this section we expand on the mass distribution formulas of Section 6.3 and present *exact* (nonapproximate) formulas for distributing the mass of a body to the representing primary points. As in Section 6.3, we assume that the body has a mass m and the mass of each primary point is denoted as m^i ; $i = 1, \cdots, n_p$. Then the following three conditions must be enforced in order for this system of mass points to *exactly* represent the inertial characteristics of their corresponding body:

1. The sum of the masses of the particles must be equal to the mass of the body—Eq. (6.17).

2. The mass center of the system of particles must be at the mass center of the body—Eq. (6.18).
3. The moment of inertia of the system of particles about any point must be equal to the moment of inertia of the body about the same point. To simplify the formulation, we select that point to be the mass center:

$$\sum_{i=1}^{n_p} m^i \, (s^i)^2 = J \tag{6.25}$$

These three conditions together represent four algebraic equations, which means that we need only four unknown masses. We put Eqs. (6.17), (6.18), and (6.25) together in expanded form for four unknown masses:

$$\begin{bmatrix} 1 & 1 & 1 & 1 \\ s^i_{(\xi)} & s^j_{(\xi)} & s^k_{(\xi)} & s^o_{(\xi)} \\ s^i_{(\eta)} & s^j_{(\eta)} & s^k_{(\eta)} & s^o_{(\eta)} \\ (s^i)^2 & (s^j)^2 & (s^k)^2 & (s^o)^2 \end{bmatrix} \begin{Bmatrix} m^i \\ m^j \\ m^k \\ m^o \end{Bmatrix} = \begin{Bmatrix} m \\ 0 \\ 0 \\ J \end{Bmatrix} \tag{6.26}$$

This equation can slightly be simplified if we place point o at the mass center; that is, $\mathbf{s}^o = \mathbf{0}$:

$$\begin{bmatrix} 1 & 1 & 1 & 1 \\ s^i_{(\xi)} & s^j_{(\xi)} & s^k_{(\xi)} & 0 \\ s^i_{(\eta)} & s^j_{(\eta)} & s^k_{(\eta)} & 0 \\ (s^i)^2 & (s^j)^2 & (s^k)^2 & 0 \end{bmatrix} \begin{Bmatrix} m^i \\ m^j \\ m^k \\ m^o \end{Bmatrix} = \begin{Bmatrix} m \\ 0 \\ 0 \\ J \end{Bmatrix} \tag{6.27}$$

In the following subsections, we apply this equation to a body that is represented by two or three primary points.

6.4.1 Two Primary Points

Application of Eq. (6.27) to a slender rod requires four points to be defined along the axis of the rod. Because for these four points $\mathbf{s}^i$, $\mathbf{s}^j$, $\mathbf{s}^k$, and $\mathbf{s}^o$ are collinear, one of the equations (second or third row) in Eq. (6.27) becomes redundant. This means that for a slender rod we need only three primary points to obtain an exact mass distribution. If we eliminate point k from the system, as shown in Figure 6.11, Eq. (6.27) becomes

$$\begin{bmatrix} 1 & 1 & 1 \\ s^i_{(\xi)} & s^j_{(\xi)} & 0 \\ (s^i_{(\xi)})^2 & (s^j_{(\xi)})^2 & 0 \end{bmatrix} \begin{Bmatrix} m^i \\ m^j \\ m^o \end{Bmatrix} = \begin{Bmatrix} m \\ 0 \\ J \end{Bmatrix} \tag{6.28}$$

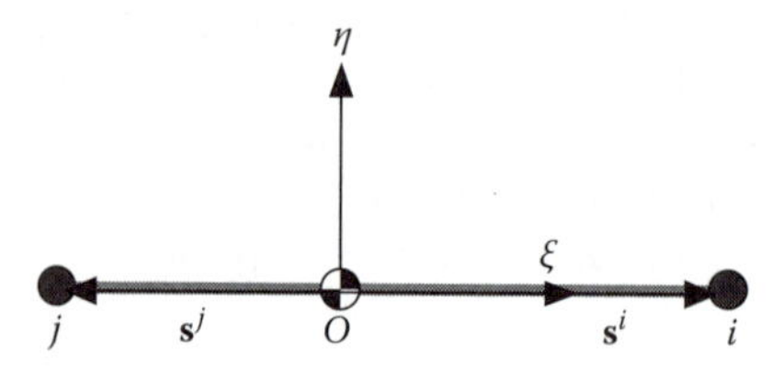

FIGURE 6.11 Mass distribution of a rod to three and then to two points.

Solution of these equations provides values for three masses, m^i , m^j, and m^o. Then, for this system of three particles, a diagonal mass matrix can be constructed as

$$\mathbf{M} = \begin{bmatrix} m^i\mathbf{I} & 0 & 0 \\ 0 & m^j\mathbf{I} & 0 \\ 0 & 0 & m^o\mathbf{I} \end{bmatrix} \tag{a}$$

Next, we try to eliminate particle o from the formulation. For this purpose, the coordinates of particle o are described in terms of the coordinates of the other two particles, based on Eq. (4.4), as

$$(m^i + m^j + m^o)\mathbf{r}^0 = m^i\mathbf{r}^i + m^j\mathbf{r}^j + m^o\mathbf{r}^0$$

or

$$\mathbf{r}^0 = \mu^i\mathbf{r}^i + \mu^j\mathbf{r}^j \tag{b}$$

where

$$\mu^i = \frac{m^i}{m^i + m^j},\ \mu^j = \frac{m^j}{m^i + m^j}$$

We can write the following transformation expression:

$$\begin{Bmatrix} \mathbf{r}^i \\ \mathbf{r}^j \\ \mathbf{r}^0 \end{Bmatrix} = \begin{bmatrix} \mathbf{I} & \mathbf{0} \\ \mathbf{0} & \mathbf{I} \\ \mu^i\mathbf{I} & \mu^j\mathbf{I} \end{bmatrix} \begin{Bmatrix} \mathbf{r}^i \\ \mathbf{r}^j \end{Bmatrix} \tag{c}$$

The coefficient matrix in *(c)* is called the transformation matrix:

$$\mathbf{B} = \begin{bmatrix} \mathbf{I} & \mathbf{0} \\ \mathbf{0} & \mathbf{I} \\ \mu^i\mathbf{I} & \mu^j\mathbf{I} \end{bmatrix} \tag{d}$$

With this matrix we can transform the mass matrix in *(a)* to a smaller size matrix as*:

$$\mathbf{M}^{(exact)} = \begin{bmatrix} \mathbf{M}^{i,i} & \mathbf{M}^{i,j} \\ \mathbf{M}^{j,i} & \mathbf{M}^{j,j} \end{bmatrix} = \mathbf{B}'\mathbf{M}\mathbf{B}$$

or

$$\mathbf{M}^{(exact)} = \begin{bmatrix} \left(m^i + m^o(\mu^i)^2\right)\mathbf{I} & m^o\mu^i\mu^j\mathbf{I} \\ m^o\mu^j\mu^i\mathbf{I} & \left(m^j + m^o(\mu^j)^2\right)\mathbf{I} \end{bmatrix} \tag{6.29}$$

In this process we have transformed a 6×6 matrix (for three particles) to a 4×4 matrix (for two particles). We have eliminated particle o from the system but have kept its contribution to the mass matrix. The resultant matrix is no longer diagonal.

For a rod, in most cases, a symmetric mass distribution results into $m^i = m^j$. This in turn simplifies the mass matrix of Eq. (6.29) to:

$$\mathbf{M}^{(exact)} = \begin{bmatrix} \mathbf{M}^{i,i} & \mathbf{M}^{i,j} \\ \mathbf{M}^{j,i} & \mathbf{M}^{j,j} \end{bmatrix} = \begin{bmatrix} (m^i + \frac{m^o}{4})\mathbf{I} & \frac{m^o}{4}\mathbf{I} \\ \frac{m^o}{4}\mathbf{I} & (m^j + \frac{m^o}{4})\mathbf{I} \end{bmatrix} \tag{6.30}$$

Example 6.6

Distribute the mass of a slender rod to two primary points that are positioned at its ends, as shown in Figure 6.12, by applying the conditions of: (a) Section 6.3; and (b) Section 6.4. Assume that the mass and the moment of inertia for the rod are m and $J = \frac{1}{12}m\ell^2$, where ℓ is the length of the rod.

Solution

The mass center for a slender rod is at its geometric center. The approximated process of Section 6.3 distributes the mass to the two points as $m^A = m^B = m/2$. This yields the following mass matrix for the two particle system:

$$\mathbf{M}^{(approx.)} = \frac{m}{2}\begin{bmatrix} \mathbf{I} & \mathbf{0} \\ \mathbf{0} & \mathbf{I} \end{bmatrix}$$

---

* This transformation process is part of a transformation process of the equations of motion from three particles to two particles. If we construct the equations of motion for three particles, they must contain three constraints: one length constraint between i and j; and another constraint (this contains two algebraic constraints) similar to *(b)*. The transformation expression of *(c)* is written for the accelerations and is substituted in the equations of motion. Then the entire equation is premultiplied by $\mathbf{B}'$. This eliminates particle o and its corresponding constraints from the equations of motion.

For an exact distribution, Eq. (6.28) is written as

$$\begin{bmatrix} 1 & 1 & 1 \\ -\ell/2 & \ell/2 & 0 \\ (-\ell/2)^2 & (\ell/2)^2 & 0 \end{bmatrix} \begin{Bmatrix} m^A \\ m^B \\ m^o \end{Bmatrix} = \begin{Bmatrix} m \\ 0 \\ m\ell^2/12 \end{Bmatrix}$$

The solution to this equation yields $m^A = m^B = m/6$ and $m^o = 2m/3$. Then Eq. (6.30) provides the following mass matrix:

$$\mathbf{M}^{(exact)} = \frac{m}{6}\begin{bmatrix} 2\mathbf{I} & \mathbf{I} \\ \mathbf{I} & 2\mathbf{I} \end{bmatrix}$$

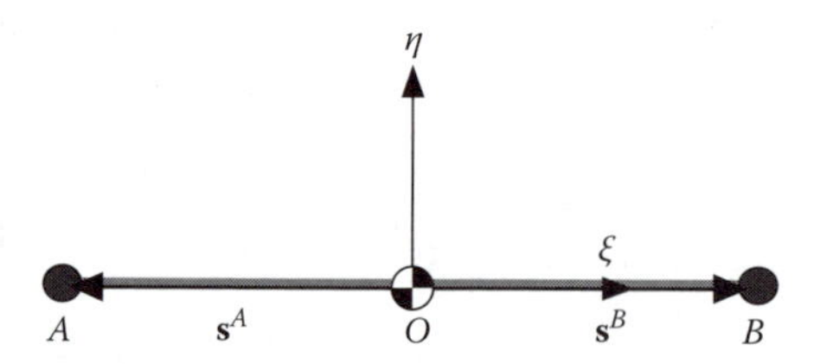

FIGURE 6.12 A rod with uniform mass distribution.

Example 6.7

Consider the lower A-arm of the double A-arm suspension from Section 1.5.1 as shown in Figure 6.13. Assume that the mass center for this link is located at its geometric center. Distribute the mass of this link to the two primary points that are positioned at the ends of the link by applying: (a) the two conditions of Section 6.3; and (b) the three conditions of Section 6.4. The mass and the moment of inertia for this link are provided in Section 1.5.1 as m = 2.0 and J = 0.5. Note that the moment of inertia for this link is not $J = \frac{1}{12}m\ell^2$ due to the shape of the link.

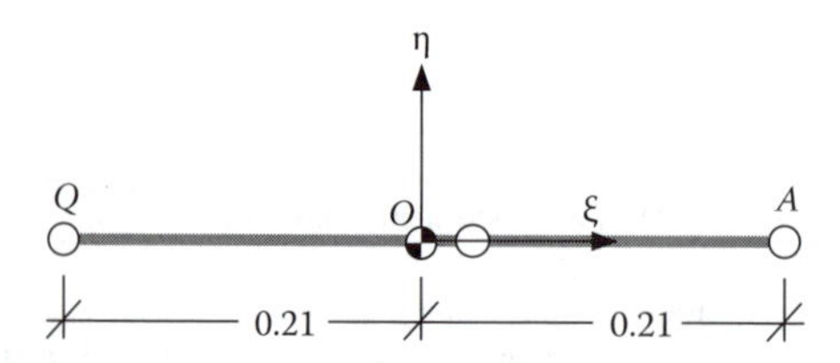

FIGURE 6.13 Lower arm of the double A-arm suspension.

Solution

The approximated process of Section 6.3 distributes the mass to the two points as $m^A = m^Q = m/2 = 1.0$. This yields a mass matrix for the two particle system as

$$\mathbf{M}^{(approx.)} = \begin{bmatrix} \mathbf{I} & \mathbf{0} \\ \mathbf{0} & \mathbf{I} \end{bmatrix}$$

For an exact distribution, Eq. (6.28) is written as

$$\begin{bmatrix} 1 & 1 & 1 \\ 0.21 & -0.21 & 0 \\ (0.21)^2 & (-0.21)^2 & 0 \end{bmatrix} \begin{Bmatrix} m^A \\ m^Q \\ m^o \end{Bmatrix} = \begin{Bmatrix} 2.0 \\ 0 \\ 0.5 \end{Bmatrix}$$

which yields $m^A = m^Q = 5.67$ and $m^o = -9.34$. Substituting these values in the mass matrix of Eq. (6.30) provides the following mass matrix:

$$\mathbf{M}^{(exact)} = \begin{bmatrix} 3.33\mathbf{I} & -2.33\mathbf{I} \\ -2.33\mathbf{I} & 3.33\mathbf{I} \end{bmatrix}$$

It may look strange that the absolute value of each element in this matrix is larger than the total mass of the link and, furthermore, that the off-diagonal terms are negative. But, this is an exact distribution of the mass! As a check, if we add the four elements of this matrix, we find that it is equal to the mass of the body.

6.4.2 Three Primary Points

A planar body represented by four particles, *i, j, k,* and *o*, where *o* is positioned at the body mass-center, is shown in Figure 6.14. The mass of the four particles can be computed using Eq. (6.27). Then, through a transformation process similar

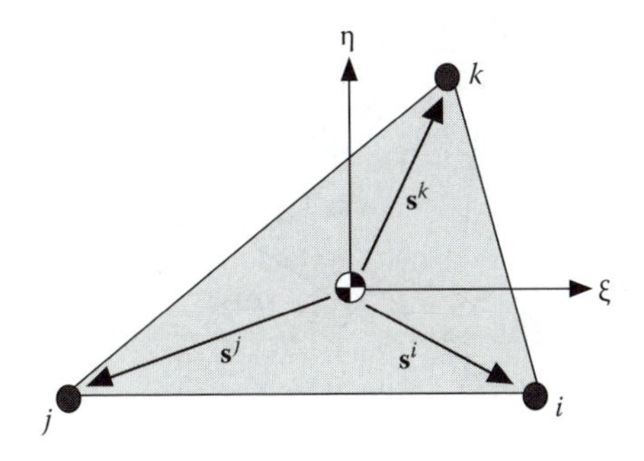

FIGURE 6.14 Mass distribution of a plate to four then to three points.

to that of Section 6.4.1, particle o is eliminated and a smaller size mass matrix is obtained as:

$$\mathbf{M}^{(exact)} = \begin{bmatrix} \mathbf{M}^{i,i} & \mathbf{M}^{i,j} & \mathbf{M}^{i,k} \\ \mathbf{M}^{j,i} & \mathbf{M}^{j,j} & \mathbf{M}^{j,k} \\ \mathbf{M}^{k,i} & \mathbf{M}^{k,j} & \mathbf{M}^{k,k} \end{bmatrix} = \begin{bmatrix} \left(m^i + m^o(\mu^i)^2\right)\mathbf{I} & m^o\mu^i\mu^j\mathbf{I} & m^o\mu^i\mu^k\mathbf{I} \\ m^o\mu^j\mu^i\mathbf{I} & \left(m^j + m^o(\mu^j)^2\right)\mathbf{I} & m^o\mu^j\mu^k\mathbf{I} \\ m^o\mu^k\mu^i\mathbf{I} & m^o\mu^k\mu^j\mathbf{I} & \left(m^k + m^o(\mu^k)^2\right)\mathbf{I} \end{bmatrix} \tag{6.31}$$

where

$$\mu^i = \frac{m^i}{m^i + m^j + m^k}, \quad \mu^j = \frac{m^j}{m^i + m^j + m^k}, \quad \mu^k = \frac{m^k}{m^i + m^j + m^k}$$

In special cases where the mass is distributed symmetrically with respect to the mass center (e.g., where $m^i = m^j = m^k$), the mass matrix of Eq. (6.31) can be simplified.

Example 6.8

Determine the mass matrix for link (2) of the double A-arm suspension system, shown in Figure 6.15, based on the mass distribution of Section 6.4. All the necessary data can be found in Section 1.5.2.

Solution

The origin of a ξ-η reference frame is positioned at the mass center. The coordinates of the primary points, the mass, and the moment of inertia are used in the following program to compute the mass matrix.

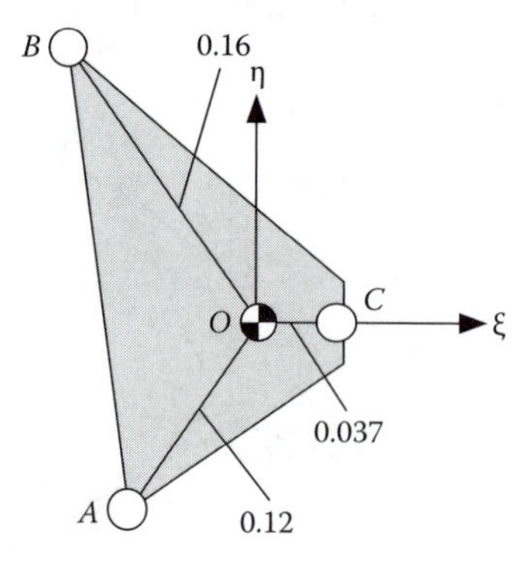

FIGURE 6.15 Knuckle of the double A-arm suspension.

```
% Body mass, moment of inertia, and three position vectors
% Double A-arm
    m = 30; J = 2.5;
    s_A = [-0.061; -0.079]; s_B = [-0.089; 0.126]; s_C = [0.037; 0];
% MacPherson
%   m = 20; J = 2.5;
%   s_A = [0.0; -0.07]; s_B = [-0.11; 0.28]; s_C = [0.12; -0.035];
% Distribute the mass to four particles
    m_ps = PC_mass_4_exact(m, s_A, s_B, s_C, J)
% Construct the exact mass matrix for three points
    M_3 = PC_mass_3_exact(m_ps)
```

```
function m_ps = PC_mass_4_exact(m, s_A, s_B, s_C, J)
% Distribute mass of a body to four particles (exact)
    s_A_2 = s_A'*s_A; s_B_2 = s_B'*s_B; s_C_2 = s_C'*s_C;
    P = [1      1      1      1
         s_A    s_B    s_C    [0; 0]
         s_A_2  s_B_2  s_C_2  0];
    rhs = [m; 0; 0; J];
    m_ps = P\rhs;
```

```
function M_exact = PC_mass_3_exact(m_ps)
% Construct the exact mass matrix for a plate as three particles
    m_ijk = m_ps(1) + m_ps(2) + m_ps(3);
    mu_i = m_ps(1)/m_ijk; mu_j = m_ps(2)/m_ijk; mu_k = m_ps(3)/m_ijk;
    mii = (m_ps(1) + m_ps(4)*mu_i^2)*eye(2);
    mjj = (m_ps(2) + m_ps(4)*mu_j^2)*eye(2);
    mkk = (m_ps(3) + m_ps(4)*mu_k^2)*eye(2);
mij = m_ps(4)*mu_i*mu_j*eye(2);
    mik = m_ps(4)*mu_i*mu_k*eye(2);
    mjk = m_ps(4)*mu_j*mu_k*eye(2);
   M_exact = [ mii   mij   mik
               mij   mjj   mjk
               mik   mjk   mkk ];
```

This program and its two function M-files yield the following masses and mass matrix:

```
m_ps =
    85.6047
    53.6728
   270.2369
  -379.5144

M_3 =
   69.0209          0  -10.3978          0  -52.3519          0
         0    69.0209         0  -10.3978         0  -52.3519
  -10.3978          0   47.1535          0  -32.8238          0
         0   -10.3978         0   47.1535         0  -32.8238
  -52.3519          0  -32.8238          0  104.9725          0
         0   -52.3519         0  -32.8238         0  104.9725
```

6.4.3 Mass Addition

A primary point that is shared between two links receives its mass from both links. If the mass of each link is distributed to its primary points without approximation, the mass matrices of two links that share a primary point overlap. As an example we consider the two links of Figure 6.9. The mass of links *(i)* and *(j)* are distributed to their primary points according to Eqs. (6.30) and (6.31) respectively as

$$\mathbf{M}_i^{(exact)} = \begin{bmatrix} \mathbf{M}_i^{B,B} & \mathbf{M}_i^{B,A} \\ \mathbf{M}_i^{A,B} & \mathbf{M}_i^{A,A} \end{bmatrix}, \mathbf{M}_j^{(exact)} = \begin{bmatrix} \mathbf{M}_j^{A,A} & \mathbf{M}_j^{A,C} & \mathbf{M}_j^{A,D} \\ \mathbf{M}_j^{C,A} & \mathbf{M}_j^{C,C} & \mathbf{M}_j^{C,D} \\ \mathbf{M}_j^{D,A} & \mathbf{M}_j^{D,C} & \mathbf{M}_j^{D,D} \end{bmatrix}$$

Then, the mass matrix for the system is constructed as

$$\mathbf{M}^{(exact)} = \begin{bmatrix} \mathbf{M}_i^{B,B} & \mathbf{M}_i^{B,A} & \mathbf{0} & \mathbf{0} \\ \mathbf{M}_i^{A,B} & \mathbf{M}_i^{A,A} + \mathbf{M}_j^{A,A} & \mathbf{M}_j^{A,C} & \mathbf{M}_j^{A,D} \\ \mathbf{0} & \mathbf{M}_j^{C,A} & \mathbf{M}_j^{C,C} & \mathbf{M}_j^{C,D} \\ \mathbf{0} & \mathbf{M}_j^{D,A} & \mathbf{M}_j^{D,C} & \mathbf{M}_j^{D,D} \end{bmatrix}. \quad (6.32)$$

Similar process is used to construct the overall mass matrix for any multibody system when primary points are shared between the links.

6.5 REMARKS

In this chapter we discussed how to construct the equations of motion for a multi-particle system. The subjects discussed in this chapter were complimentary to the kinematic formulation of Chapter 5. At this point a reader may skip over the next four chapters and go directly to Chapters 11 to 14 and learn how to perform different types of analyses with the point-coordinate formulation. However, it is recommended that a reader should learn the body-coordinate formulation in Chapters 7 and 8 as well before moving to the analyses chapters.

6.6 PROBLEMS

6.1 Two particles, *A* and *B*, can only slide along the *x*-axis. A spring-damper connects the two particles and each particle is acted upon by an external force. Derive the equations of motion for the system. Consider the following data: $m^A = 1$, $m^B = 2$, $k = 800$, ${}^0\ell = 0.2$, $d_c = 15$, $\mathbf{f}^A = 35$, $\mathbf{f}^B = 15$. In the given configuration, the coordinates and velocities are; $x^A = 0.15$, $\dot{x}^A = 30$, $x^B = 0.33$, $\dot{x}^B = 40$.

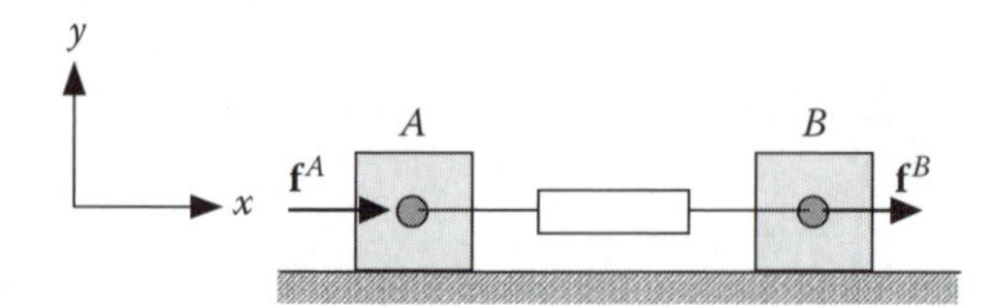

6.2 Two particles, A and B, can slide along the x- and y-axes respectively. A spring-damper connects the two particles and each particle is acted upon by an external force. Derive the equations of motion for the system. Consider the following data: $m^A = 10$, $m^B = 20$, $k = 150$, ${}^0\ell = 6$, $d_c = 300$, $\mathbf{f}^A_{(x)} = -15$, $\mathbf{f}^B_{(y)} = 5$. In the given configuration the coordinates and velocities are $x^A = 4$, $\dot{x}^A = 0.1$, $y^B = 3$, $\dot{y}^B = -0.2$.

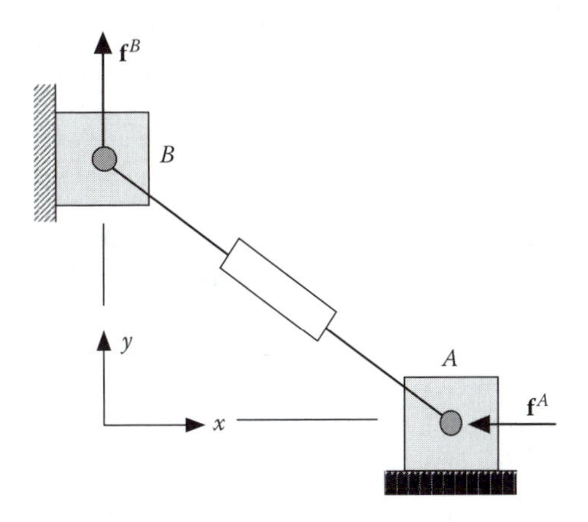

6.3 Derive the equations of motion in closed form for three particles, i, j, and k, connected by two massless rods having lengths $\ell^{i,j}$ and $\ell^{j,k}$. Assume that the particles have masses m^i, m^j, and m^k. Consider external forces $\mathbf{f}^i$, $\mathbf{f}^j$, and $\mathbf{f}^k$ acting on the particles.

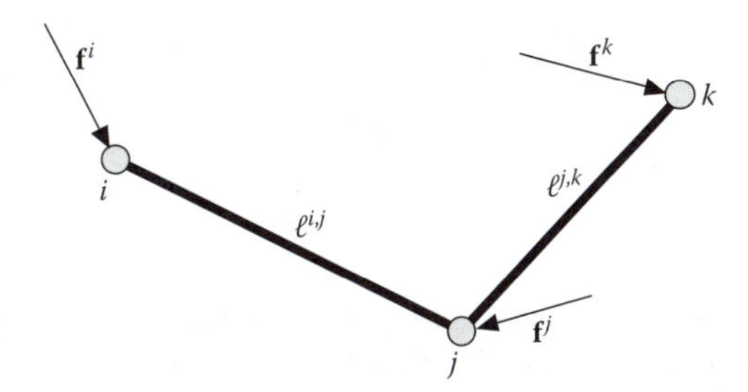

6.4 Three particles with masses $m^A = 5$, $m^B = 8$, and $m^C = 6$ are constrained by two rods with lengths $\ell^{A,B} = \sqrt{2}$ and $\ell^{B,C} = \sqrt{5}$ as shown. One spring, S_1, is connected between particles A and C, and one spring, S_2, connects particle B to the ground at Q. The following data are available for the springs: $k_1 = 60$, ${}^0\ell_1 = 2$, $k_2 = 80$, ${}^0\ell_2 = 2$. In the configuration shown, the coordinates are $\mathbf{r}^A = \{1.0 \quad 1.5\}'$, $\mathbf{r}^B = \{2.0 \quad 0.5\}'$, $\mathbf{r}^C = \{3.0 \quad 2.5\}'$, and all the velocities are zero. The spring attachment point to the ground is at $\mathbf{r}^Q = \{3.5 \quad 0.5\}'$. Determine all the necessary terms of the equations of motion in numerical form.

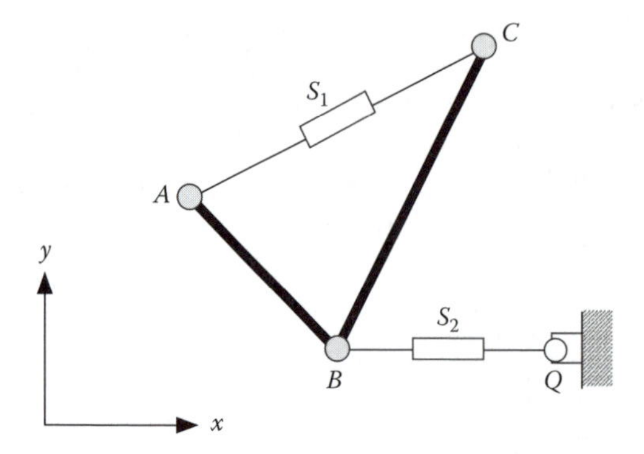

6.5 Two particles A and B having masses $m^A = 1.0$ and $m^B = 2.0$ are connected by a massless rod with a length $\ell^{A,B} = 1.0$. An external force $\mathbf{f}^A = \{0.3 \quad 0.4\}'$ acts on particle A. Derive the equations of motion for this system in each of the following two conditions:

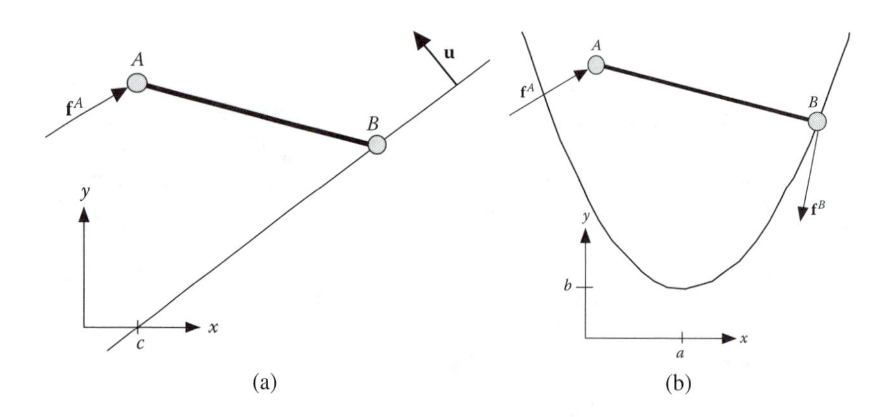

(a) (b)

(a) Particle B is constrained to slide along a straight line as shown. It is assumed that the line intersects the x-axis at $x = c = 0.2$ and a unit vector $\mathbf{u}$ normal to the line is described as $\mathbf{u} = \{u_{(x)} \quad u_{(y)}\}' = \{-0.6 \quad 0.8\}'$.

(b) Particle B slides on a curve described by $y = b + (x - a)^2 = 0.1 + (x - 0.2)^2$ and an external force $\mathbf{f}^B = \{-0.2 \quad -0.7\}'$ acts on this particle.

In each case assume zero initial velocities and determine a reasonable set of values for the coordinates.

6.6 A massless rod of length $\ell^{A,Q} = 3.0$ connects particle A to the ground at Q that is positioned at $a = 0.6$ and $b = 0.6$. Particle B can slide along the rod as shown. The particles have masses $m^A = 1.0$ and $m^B = 2.0$, and an external forces $\mathbf{f}^A = \{0.6 \quad 0.1\}'$ and $\mathbf{f}^B = \{-0.1 \quad 0.6\}'$ act on the particles. Derive the equations of motion for this system. Assume a reasonable set of values for the coordinates and consider zero initial velocities.

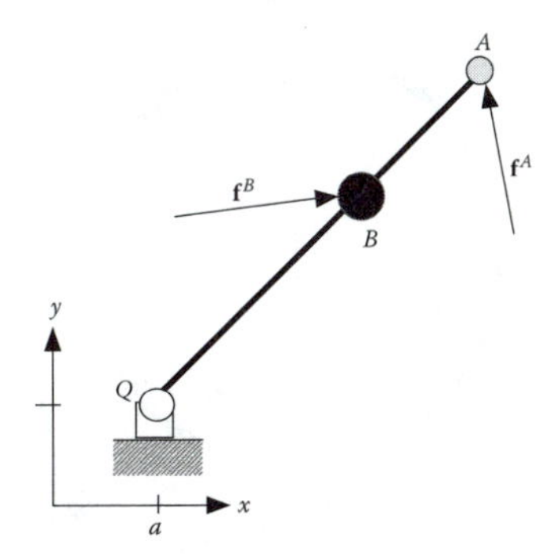

6.7 The mass center of a uniform rod is at its geometric center. Two forces with magnitudes $f_1 = 100$ and $f_2 = 200$ act on the rod as shown. The rod is modeled by two primary points A and B. Find the force distribution at A and B. Assume the mass of the rod is $m = 5$ and the gravitational constant is 9.81.

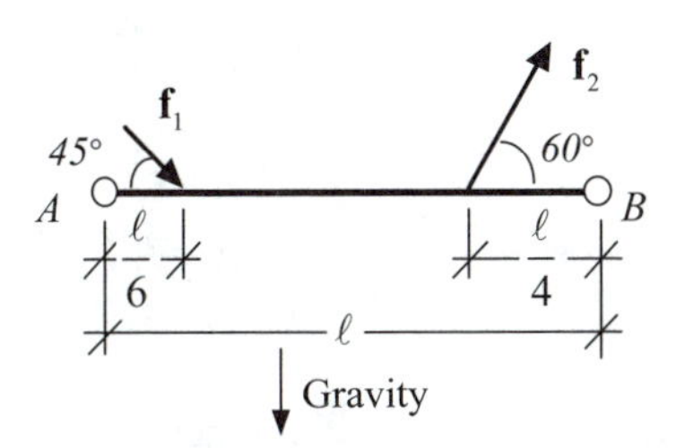

6.8 The radius of a solid disk is $R = 1$ and its mass is $m = 1$. Two primary points at A and B are used to model the disk. Determine the exact mass distribution for the disk.

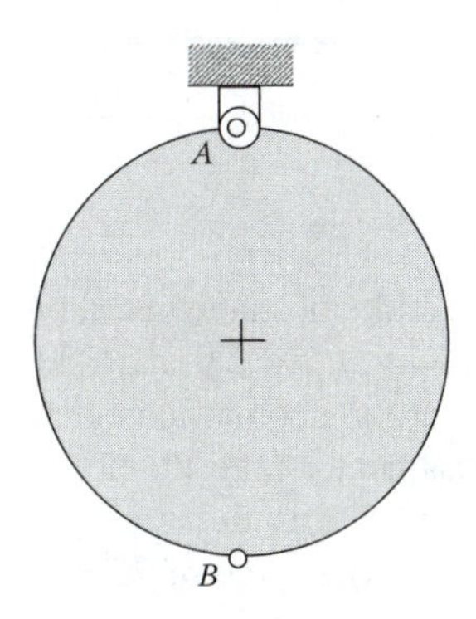

6.9 Repeat Problem 6.8 for a ring.

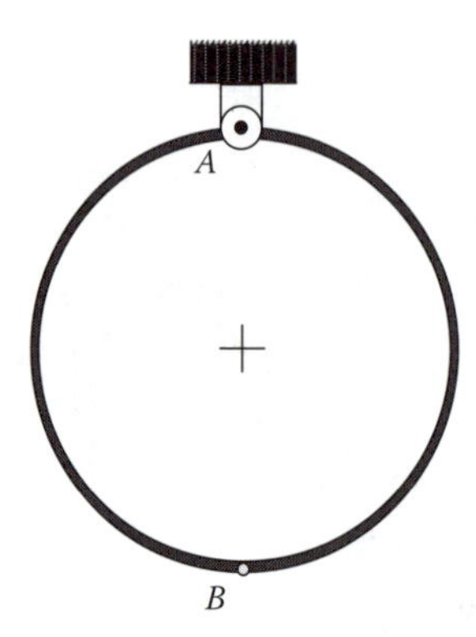

6.10 An L-shaped rod with uniform mass distribution is modeled by three primary points as shown. Assume $\ell_1 = 2$, $\ell_2 = 1$, and $m = 18$. For this system determine:
(a) An approximate mass matrix
(b) An exact mass matrix

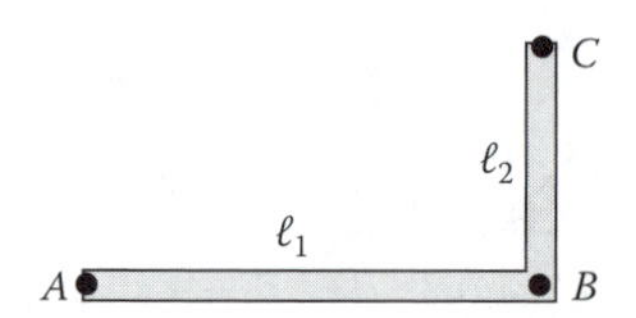

6.11 For a slender rod with mass m and moment of inertia $J = m\ell^2 / 12$, two particles may be positioned symmetrically with respect to the mass center, but not at the ends. Find the distance d between the two particles for which all the three inertial conditions are satisfied.

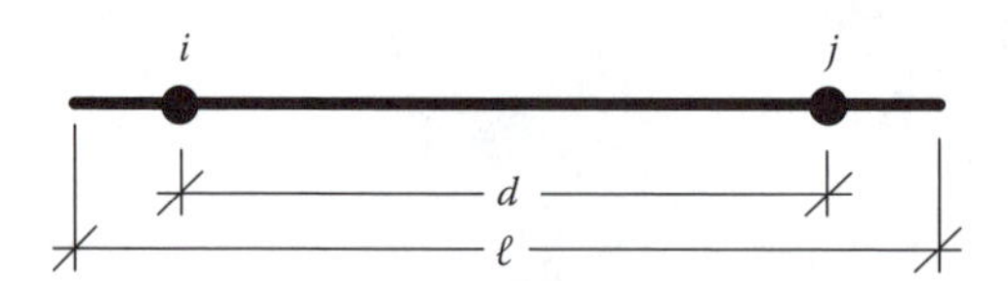

6.12 Consider the slider-crank mechanisms shown. The link lengths and masses are $\ell_1 = 2.0$, $\ell_2 = 4.0$, $a = 2.0$, $m_1 = 3.0$, $m_2 = 6.0$, and $m_3 = 24.0$. The mass centers are at the geometric centers. Link (1) is connected to the ground at its mass center. The moments of inertia with respect to the link's mass center are determined to be $J_1 = 1.0$, $J_2 = 2.0$, and $J_3 = 2.5$. In system (a) a force of 20 units acts on the slider block;

in system (b) a moment of 28 units acts on link (2); in system (c) a moment of 16 units acts on link (1). For each system:

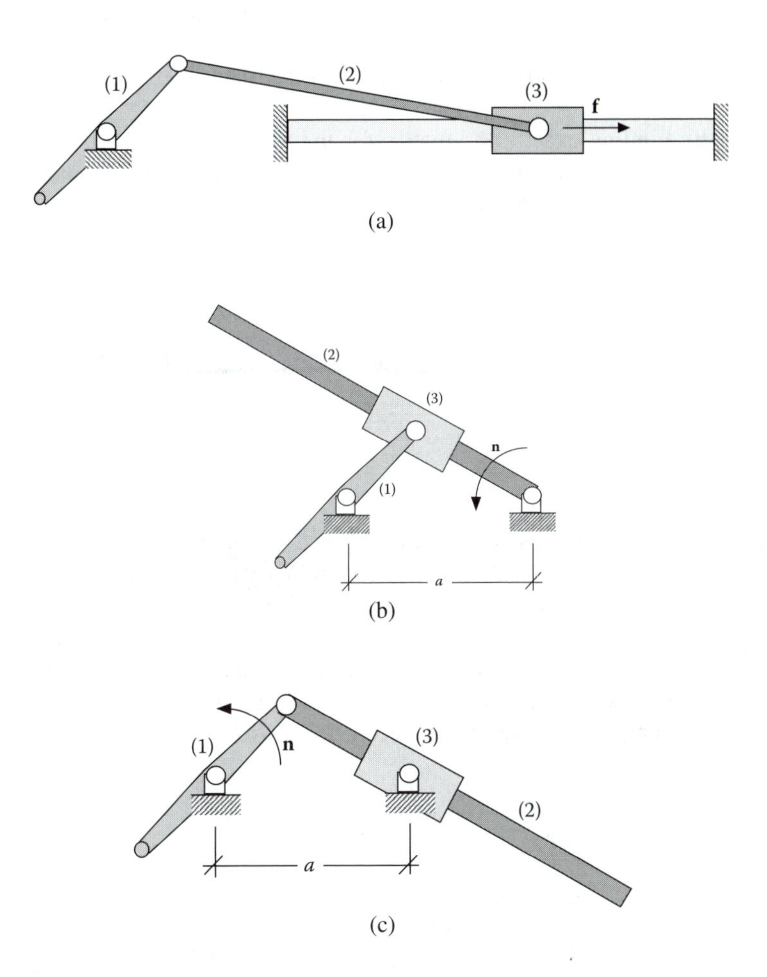

(a) Represent the system as a collection of particles.
(b) Construct the necessary kinematic entities.
(c) Construct the mass matrix;
 (c.1) using approximate mass distribution
 (c.2) using exact mass distribution
(d) Construct the array of forces (no gravity). [Hint: convert the moment to an equivalent pair of forces.]
(e) Construct the equations of motion.

6.13 In Problem 6.12, for each system assume that link (1) makes a 45° angle with the ground and its angular velocity is 1 rad/sec CCW. Use these values to determine (or approximate) the coordinates and velocities of the particles. Develop a MATLAB program to evaluate the entities of the equations of motion for each system.

6.14 The rod shown is described by two primary points i and j, where j is at the mass center of the rod. A force **f** with variable axis acts at point P. The mass of the rod is m and its moment of inertia is $J = m\ell^2 / 12$.

(a) Find the coefficients α^i and α^j to distribute the force to the two particles.

(b) Find an exact mass matrix for the system of two particles. *Hint*: Define a third particle at P, and then follow a process similar to that of in Section 6.4.1 to eliminate P.

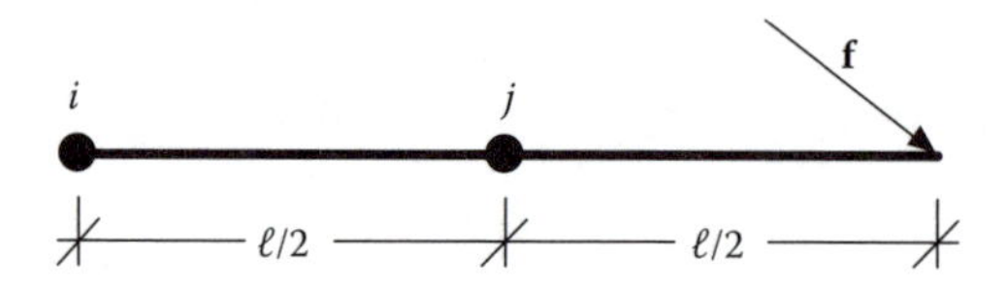

6.15 Find the exact mass distribution for link *(2)* of the McPherson suspension (refer to Section 1.5.2). Consider the primary points at A, B and C. Use MATLAB functions `mass_4_exact` and `mass_3_exact`.

6.16 For the multibody system shown the following data are given: $m_1 = 5.0$, $m_2 = 3.0$, $k = 20$, and ${}^0\ell = 2.0$. Define primary points and distribute the masses to the primary points by the approximation method. Develop a program to construct the necessary arrays and matrices for future analysis.

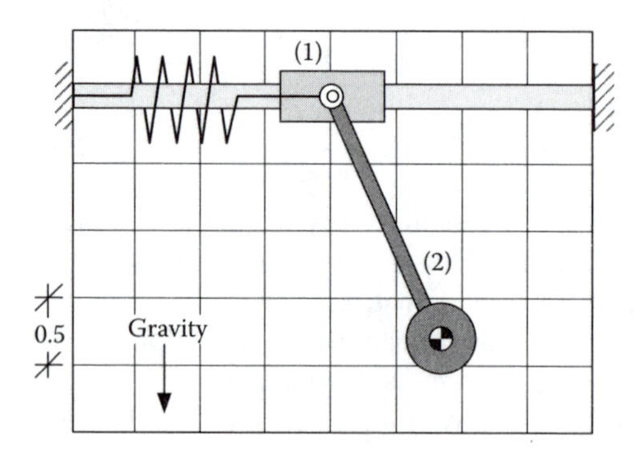

6.17 This multibody system is a slight variation of the system in Problem 6.16. Define primary points and distribute the masses to the primary points by the approximation method. Develop a program to construct the necessary arrays and matrices for future analysis.

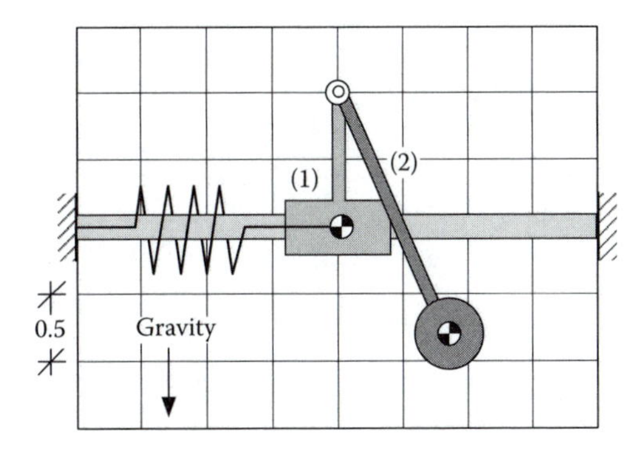

6.18 For the multibody system shown the following data are given: $m_1 = 5.0$, $m_2 = 1.0$, $m_3 = 3.0$, $k = 20$, and ${}^0\ell = 2.0$. Define primary points and distribute the masses to the primary points by the approximation method. Develop a program to construct the necessary arrays and matrices for future analysis.

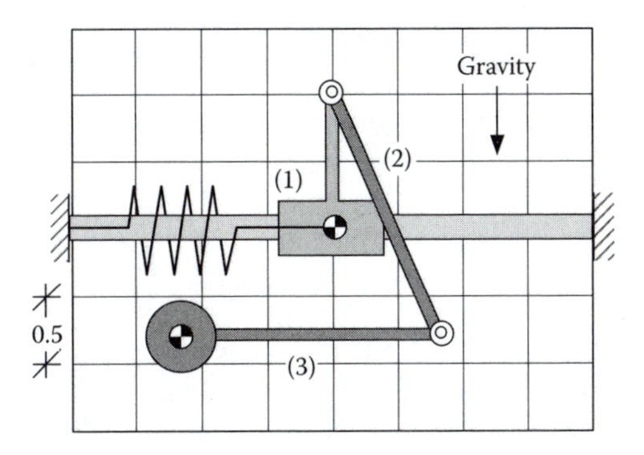

6.19 For the multibody system shown the following data are given: $m_1 = 4.0$, $m_2 = 1.5$, and $m_3 = 1.5$. Define primary points and distribute the masses to the primary points by the approximation method. Develop a program to construct the necessary arrays and matrices for future analysis.

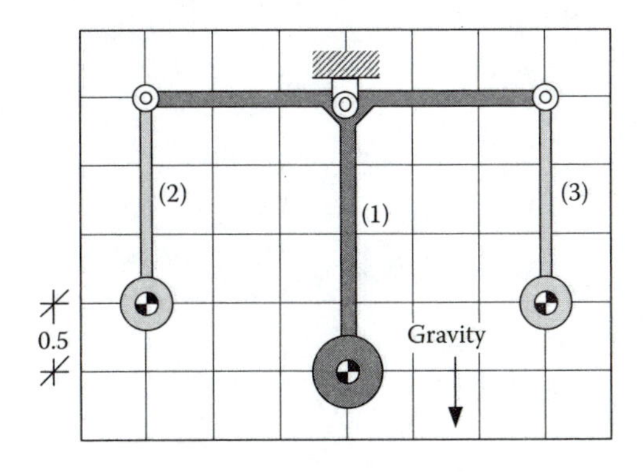

7 Body-Coordinates: Kinematics

Body-coordinate formulation is a method suitable for implementation in a computer program; the method is not recommended for the pencil-and-paper approach. The method can systematically generate the kinematic and dynamic equations for a wide variety of mechanical systems. The number of equations generated by this method is by far much larger than those generated by other formulations for the same multibody system.

The fundamental difference between the body-coordinate and the joint-coordinate formulations is the following. In the point-coordinate formulation, a multibody system is defined as a collection of points representing the joints in the system, where each point is assigned two coordinates. Then, kinematic constraints are constructed between the coordinates. In contrast, in the body-coordinate formulation, equal number of coordinates is defined for each rigid body regardless of the kinematic joints that exist in the system. Then, kinematic constraints describing the joints are defined between the body coordinates. In short, in the point-coordinate formulation, the coordinates represent the joints and the constraints represent the bodies, whereas in the body-coordinate formulation the coordinates represent the bodies and the constraints represent the joints.

Programs: In this chapter we develop several MATLAB programs. The file organization for this chapter and for Chapter 8 is shown in Table 7.1. The main folder is named `Body_Coord` which will be the Current Directory of MATLAB®. In this folder we must provide a copy of the folder `Basics`. We need to construct the following new folders within `Body_Coord`: `BC_Basics`, `BC_AA`, `BC_MP_A`, `BC_MP_B`, `BC_MP_C`, and `BC_FS`. Except for the first folder, the last five folders contain files with identical names. By setting the path properly, we can direct a program to the correct folder to access its contents.

7.1 GENERAL PROCEDURE

The process of describing a multibody system as a collection of interconnected bodies can be illustrated with a simple generic example. Assume the planar multibody system shown in Figure 7.1. Because at this stage the system is considered only for kinematic modeling, no force elements are shown in the figure. We apply the following steps as a rule of thumb in the kinematic modeling process.

- Assign indices (numbers) to each body. If the ground is part of the system, assign it index (0). Number the other bodies consecutively as (1), (2), (3), and so forth.

TABLE 7.1

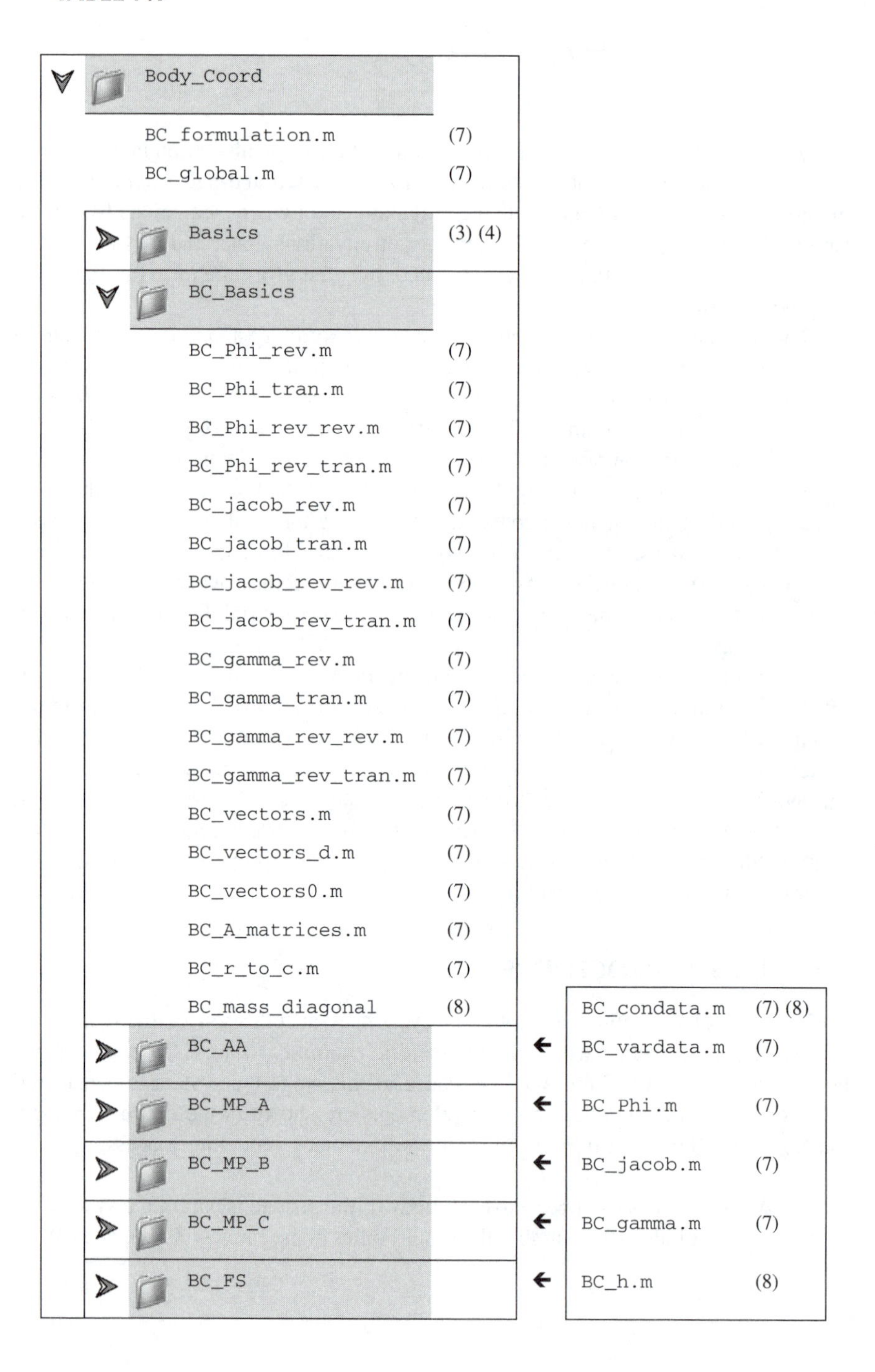

Folder / File	Chapter		Model folder file	Chapter
Body_Coord				
BC_formulation.m	(7)			
BC_global.m	(7)			
Basics	(3) (4)			
BC_Basics				
BC_Phi_rev.m	(7)			
BC_Phi_tran.m	(7)			
BC_Phi_rev_rev.m	(7)			
BC_Phi_rev_tran.m	(7)			
BC_jacob_rev.m	(7)			
BC_jacob_tran.m	(7)			
BC_jacob_rev_rev.m	(7)			
BC_jacob_rev_tran.m	(7)			
BC_gamma_rev.m	(7)			
BC_gamma_tran.m	(7)			
BC_gamma_rev_rev.m	(7)			
BC_gamma_rev_tran.m	(7)			
BC_vectors.m	(7)			
BC_vectors_d.m	(7)			
BC_vectors0.m	(7)			
BC_A_matrices.m	(7)			
BC_r_to_c.m	(7)			
BC_mass_diagonal	(8)		BC_condata.m	(7) (8)
BC_AA		←	BC_vardata.m	(7)
BC_MP_A		←	BC_Phi.m	(7)
BC_MP_B		←	BC_jacob.m	(7)
BC_MP_C		←	BC_gamma.m	(7)
BC_FS		←	BC_h.m	(8)

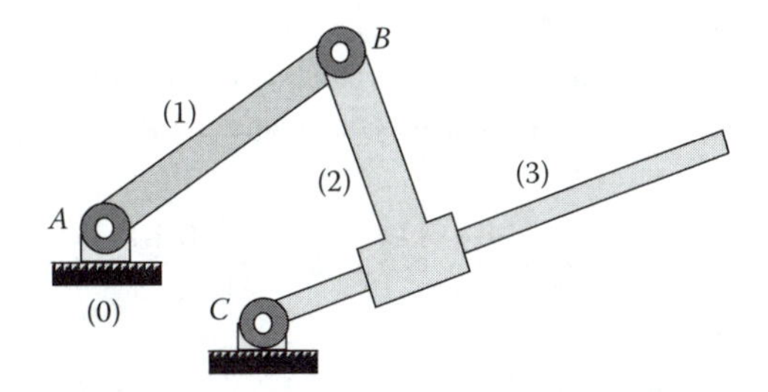

FIGURE 7.1 A generic planar system containing three moving bodies and the ground.

- Separate the bodies at the kinematic joints and position them in any convenient orientation, such as the bodies shown in Figure 7.2.
- Attach a reference frame *x*-*y* to body (0) at a convenient location; in this example the origin is attached to the pin joint at *C*.
- Attach body-fixed ξ-η frames to each moving body. The origin of these frames can be located anywhere on the corresponding bodies. However, because for dynamic analysis, as it will be discussed in Chapter 8, the origin of these frames must be positioned at the mass center of each body, it is a good practice to do the same in kinematic analysis.* The ξ-axis of each frame can be oriented in any desired direction. It is a good practice to orient the ξ-axis (or the η-axis) along a geometric or an existing joint axis on the body.

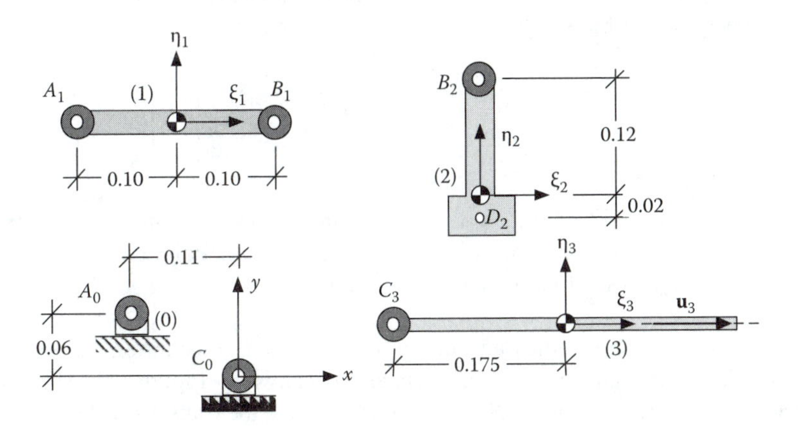

FIGURE 7.2 Individual bodies of the system from Figure 7.1.

* The position of the mass center on a body is either provided in a problem statement, or we need to find it based on the shape of the body and any information we might have on the mass distribution (using a CAD software would be ideal). If none of these are available, we can use our best judgment in locating the mass center.

- Define the location of each joint on its corresponding bodies. In our example, the pin joint A is connected between bodies (0) and (1). The coordinates of A on these two bodies (i.e., the coordinates of A_0 and A_1) are

$$\mathbf{r}_0^A = \mathbf{s}_0^A = \begin{Bmatrix} -0.11 \\ 0.06 \end{Bmatrix}, \quad \mathbf{s}_1^{\prime A} = \begin{Bmatrix} -0.10 \\ 0 \end{Bmatrix}$$

Because A_0 is a point on the ground, we state its x-y coordinates, but A_1 is attached to a moving body, therefore we state its ξ-η coordinates.

For the pin joint at B, we determine the coordinates of points B_1 and B_2 on their corresponding bodies:

$$\mathbf{s}_1^{\prime B} = \begin{Bmatrix} 0.10 \\ 0 \end{Bmatrix}, \quad \mathbf{s}_2^{\prime B} = \begin{Bmatrix} 0 \\ 0.12 \end{Bmatrix}$$

Similarly, for the pin joint at C we have

$$\mathbf{r}_0^C = \mathbf{s}_0^C = \begin{Bmatrix} 0 \\ 0 \end{Bmatrix}, \quad \mathbf{s}_3^{\prime C} = \begin{Bmatrix} -0.175 \\ 0 \end{Bmatrix}$$

For a sliding joint we define one point and one unit vector on the axis of the joint on one of the bodies, and one point on the same axis on the other body. In our example, we define point D_2 on body (2) and point C_3 and vector $\mathbf{u}_3$ on body (3):

$$\mathbf{s}_2^{\prime D} = \begin{Bmatrix} 0 \\ -0.02 \end{Bmatrix}, \quad \mathbf{s}_3^{\prime C} = \begin{Bmatrix} -0.175 \\ 0 \end{Bmatrix}, \quad \mathbf{u}_3^{\prime} = \begin{Bmatrix} 1.0 \\ 0 \end{Bmatrix}$$

Points D_2 and C_3 could have been defined anywhere on the axis. We chose D_2 at a location that was most convenient. Point C_3 was chosen because it was already defined for the pin joint.

The three moving bodies and the ground, without considering the existence of the joints, form an unconstrained system as shown in Figure 7.3. The x- and y-coordinates of the origin of each ξ-η frame describe the translational coordinates of that body. The angle between each ξ-axis and the x-axis describes the rotational coordinate of that body. These coordinates are arranged in three arrays as

$$\mathbf{c}_1 = \begin{Bmatrix} x_1 \\ y_1 \\ \phi_1 \end{Bmatrix}, \quad \mathbf{c}_2 = \begin{Bmatrix} x_2 \\ y_2 \\ \phi_2 \end{Bmatrix}, \quad \mathbf{c}_3 = \begin{Bmatrix} x_3 \\ y_3 \\ \phi_3 \end{Bmatrix}$$

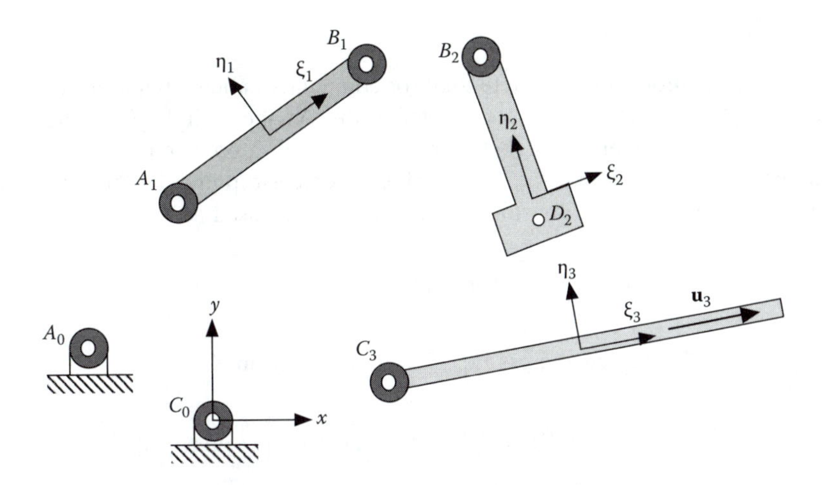

FIGURE 7.3 The mechanism of Figure 7.1 viewed as an unconstrained system.

These nine coordinates are variables; that is, when the bodies move, the coordinates find different values. Obviously, if we enforce the presence of the kinematic joints, these variables no longer remain independent from one another. In the next sections, we learn how to describe the kinematic constraints representing pin and sliding joints.

7.2 KINEMATIC JOINTS

In this section we present the body-coordinate formulation of constraint equations for several commonly used kinematic joints. Similar to the discussion on constraints with point-coordinates in Section 5.3, for the body-coordinate formulation we adopt a similar notation. A constraint equation is denoted by the *lightface* character Φ. If there is more than one algebraic equation in a constraint, we denote it by the *boldface* character $\boldsymbol{\Phi}$. The constraint symbol may carry a left superscript in parentheses with two entries, such as ${}^{(a,b)}\boldsymbol{\Phi}$. The first entry, a, states the constraint *type*—for example, r for *revolute* and t for *translational*. The second entry, b, denotes the *number* of algebraic equations in a constraint. If the array of coordinates is defined as $\mathbf{c}$, then the position constraints are expressed in a general form as

$$\boldsymbol{\Phi} \equiv \boldsymbol{\Phi}(\mathbf{c}) = \mathbf{0} \tag{7.1}$$

Most of the kinematic constraints are nonlinear in the coordinates, $\mathbf{c}$. For a system of n_b bodies, the array of coordinates $\mathbf{c}$ contains n_b 3-arrays as

$$\mathbf{c} = \begin{Bmatrix} \mathbf{c}_1 \\ \vdots \\ \mathbf{c}_{n_b} \end{Bmatrix} \tag{7.2}$$

where $\mathbf{c}_i = \{x_i \quad y_i \quad \phi_i\}'$; $i = 1, \cdots, n_b$.

7.2.1 Revolute (Pin) Joint

Constraint equations for a revolute joint (or commonly called a pin joint) in body-coordinate formulation can be constructed easily. As shown in Figure 7.4(a), it is desired to attach a pin joint between points P_i and P_j on bodies i and j. The kinematic condition is that the x-y coordinates of these points be the same; that is, $\mathbf{r}_j^P = \mathbf{r}_i^P$. Therefore the constraint equation are expressed as

$$^{(r,2)}\mathbf{\Phi} = \mathbf{r}_j^P - \mathbf{r}_i^P = \mathbf{0} \tag{7.3}$$

Using Eq. (3.10), we can express Eq. (7.3) in matrix form as

$$\begin{Bmatrix} x_j \\ y_j \end{Bmatrix} + \begin{bmatrix} \cos\phi_j & -\sin\phi_j \\ \sin\phi_j & \cos\phi_j \end{bmatrix} \begin{Bmatrix} \xi_j^P \\ \eta_j^P \end{Bmatrix} - \begin{Bmatrix} x_i \\ y_i \end{Bmatrix} - \begin{bmatrix} \cos\phi_i & -\sin\phi_i \\ \sin\phi_i & \cos\phi_i \end{bmatrix} \begin{Bmatrix} \xi_i^P \\ \eta_i^P \end{Bmatrix} = \begin{Bmatrix} 0 \\ 0 \end{Bmatrix}$$

We can also express the equation in expanded form as

$$^{(r,2)}\mathbf{\Phi} = \begin{Bmatrix} x_j + \xi_j^P \cos\phi_j - \eta_j^P \sin\phi_j - x_i - \xi_i^P \cos\phi_i + \eta_i^P \sin\phi_i \\ y_j + \xi_i^P \sin\phi_j + \eta_j^P \cos\phi_j - y_i - \xi_i^P \sin\phi_i - \eta_i^P \cos\phi_i \end{Bmatrix} = \begin{Bmatrix} 0 \\ 0 \end{Bmatrix}$$

Enforcing these conditions forms a pin joint between the two bodies as shown in Figure 7.4(b). The two constraints in Eq. (7.3) reduce the DoF between the two bodies by 2.

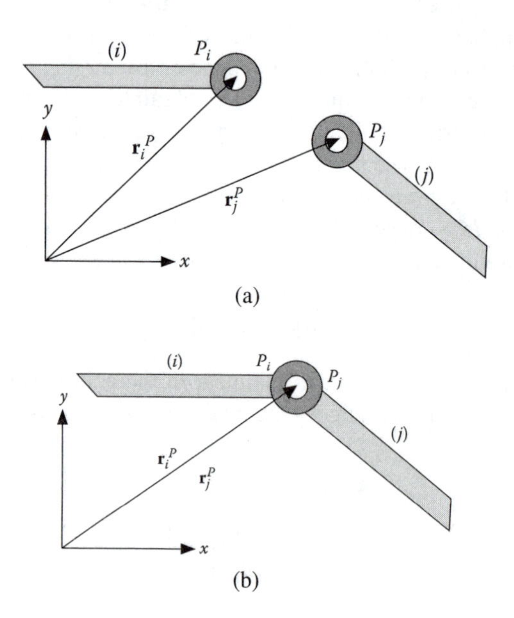

FIGURE 7.4 Two bodies (a) before and (b) after the pin joint constraints are enforced.

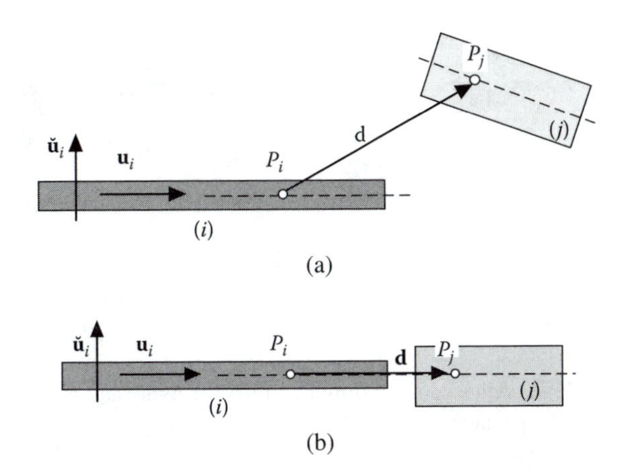

FIGURE 7.5 Two bodies (a) before and (b) after the sliding joint constraints are enforced.

7.2.2 Translational (Sliding) Joint

A translational (also known as sliding) joint allows two bodies to slide relative to one another along the axis of the joint. To construct a translational joint between bodies i and j, we describe one point and one unit vector along the axis of translation on body i and one point along the same axis on body j as shown in Figure 7.5(a). These are points P_i and P_j and unit vector $\mathbf{u}_i$. Vector $\mathbf{d}$ is constructed by connecting P_i to P_j; that is, $\mathbf{d} = \mathbf{r}_j^P - \mathbf{r}_i^P$. The necessary and sufficient conditions for the two bodies for not rotating relative to one another and also for the vectors $\mathbf{u}_i$ and $\mathbf{d}$ to remain parallel at all times are

$$^{(t,2)}\mathbf{\Phi} = \begin{Bmatrix} \phi_j - \phi_i - {}^c\phi \\ \breve{\mathbf{u}}_i' \, \mathbf{d} \end{Bmatrix} = \begin{Bmatrix} 0 \\ 0 \end{Bmatrix} \tag{7.4}$$

where ${}^c\phi$ is the constant angle between ξ_j and ξ_i axes. These equations can also be expressed in terms of the x-y components of the vectors as

$$^{(t,2)}\mathbf{\Phi} = \begin{Bmatrix} \phi_j - \phi_i - {}^c\phi \\ -u_{i(y)}\, d_{(x)} + \mathrm{u}_{i(x)}\, d_{(y)} \end{Bmatrix} = \begin{Bmatrix} 0 \\ 0 \end{Bmatrix}$$

If these constraints are enforced, a sliding joint is formed between the two bodies as illustrated in Figure 7.5(b). A sliding joint removes 2 DoF between the connecting bodies.

7.2.3 Revolute-Revolute Joint

The total number of coordinates and constraint equations describing a mechanical system can be reduced if some of the joints and bodies are combined and represented as composite joints. This technique simplifies the analytical formulation without

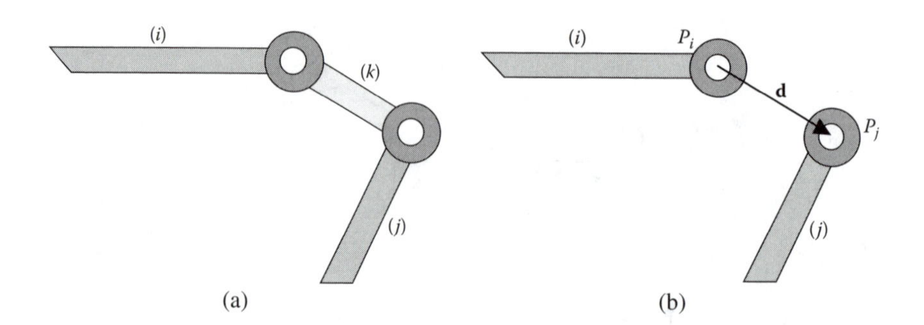

FIGURE 7.6 (a) Three bodies connected by two revolute joints and (b) the composite revolute-revolute joint equivalence.

changing the kinematic characteristics of a system. For example, consider the three bodies connected by two revolute joints shown in Figure 7.6(a). This system requires nine coordinates (three per body) and four constraint equations (two per joint). Therefore, this system has 9 – 4 = 5 DoF. This system may be represented by the kinematically equivalent system shown in Figure 7.6(b) where body k and the two revolute joints are combined to form a *revolute-revolute* joint (an imaginary link). This configuration requires six coordinates for bodies i and j, and one constraint equation. Therefore, this equivalent system has 6 – 1 = 5 DoF—the same as the original system.

To formulate the constraint equation for a revolute-revolute joint, we take advantage of the length of the imaginary link being a known constant ℓ. As shown in Figure 7.6(b), the centers of the two joints are denoted as points P_i and P_j. Vector $\mathbf{d} = \mathbf{r}_j^P - \mathbf{r}_i^P$ connecting these two points must keep a constant length. Therefore the constraint equation is written as*

$$^{(r-r,1)}\mathbf{\Phi} = \frac{1}{2}(\mathbf{d}'\,\mathbf{d} - \ell^2) = 0 \tag{7.5}$$

This equation can be expressed in terms of the x-y components of vector $\mathbf{d}$ as

$$^{(r-r,1)}\mathbf{\Phi} = \frac{1}{2}(d_{(x)}^2 + d_{(y)}^2 - \ell^2) = 0$$

This joint eliminates one DoF between the two bodies.

7.2.4 Revolute-Translational Joint

A second type of composite joint is the *revolute-translational* joint, or *pin-in-a-slot* joint. Figure 7.7(a) illustrates schematically three bodies connected by a revolute and a translational joint. Similar to the revolute-revolute joint formulation, this

* The reason for introducing the factor $^1/_2$ in front of the expression is the same as what was discussed for the length constraint in the point-coordinate formulation.

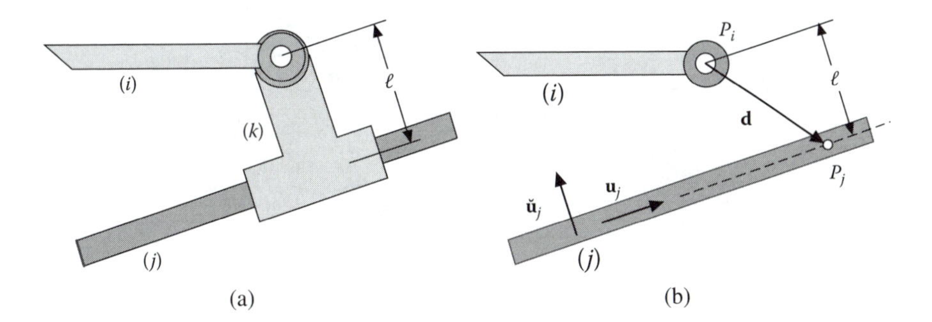

FIGURE 7.7 (a) Three bodies connected by a revolute and a translational joint and (b) the composite revolute-translational joint equivalence.

system can be modeled as a revolute-translational joint between bodies i and j as shown in Figure 7.7(b). Because there are two relative DoF between the two bodies, we only need one constraint equation. To construct the constraint equation for this composite joint, the center of the pin joint is denoted as point P_i, and point P_j and unit vector $\mathbf{u}_j$ are defined anywhere along the translational axis on body j. A constraint equation for this joint is found by requiring that the projection of vector $\mathbf{d} = \mathbf{r}_j^P - \mathbf{r}_i^P$ onto an axis perpendicular to $\mathbf{u}_j$ must keep a length equal to ℓ; that is,

$$^{(r\text{-}t,1)}\Phi = \breve{\mathbf{u}}_j'\,\mathbf{d} - \ell = 0 \tag{7.6}$$

This equation can be expressed in terms of the x-y components of its vectors as

$$^{(r\text{-}t,1)}\Phi = -u_{j(y)}\,d_{(x)} + \mathrm{u}_{j(x)}\,d_{(y)} - \ell = 0$$

This joint eliminates one DoF between the two bodies.

In Eq. (7.6), the length ℓ must be considered as a directional scalar; that is, it may be a positive or a negative quantity. For example, if we examine the vectors in Figure 7.7(b), ℓ must be considered as a negative quantity. Where as in the system of Figure 7.8(a), ℓ must be considered as a positive quantity. In general, the sign of ℓ depends on how the two vectors, $\mathbf{u}_j$ and $\mathbf{d}$, are defined. If the pin joint slides directly on the axis of the sliding joint, as shown in Figure 7.8(b), then $\ell = 0$ and therefore the direction of the vectors is no longer an issue.

7.2.5 Rigid Joint

The connection between two bodies that are fixed to each other at one point is called a *rigid* joint. We can think of a rigid joint as a pin joint between two bodies and an additional constraint eliminating the relative rotation between the

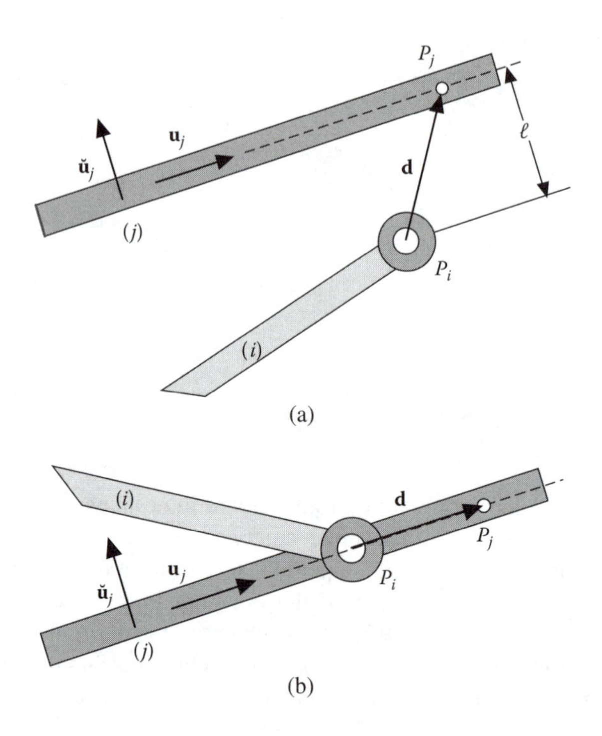

FIGURE 7.8 Two other possible configurations for a revolute-translational joint.

two bodies. Consider the two bodies shown in Figure 7.9 that are joined together at points P_i and P_j. The rigid joint constraints between the two bodies can be expressed as

$$^{(rigid,3)}\boldsymbol{\Phi} = \begin{Bmatrix} \mathbf{r}_j^P - \mathbf{r}_i^P \\ \phi_j - \phi_i - {}^c\phi \end{Bmatrix} = \begin{Bmatrix} \mathbf{0} \\ 0 \end{Bmatrix} \tag{7.7}$$

where ${}^c\phi$ is a constant angle between ξ_j and ξ_i axes. Obviously, such a joint eliminates all three relative degrees of freedom between the two bodies.

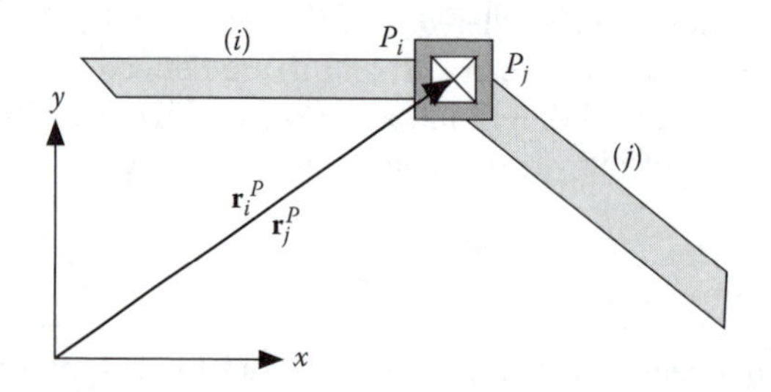

FIGURE 7.9 A rigid (also called bracket or welded) joint between two bodies.

7.2.6 Simple Constraints

A simple constraint is a condition that we impose on one or more coordinates of a body. For example, the y-coordinate of a typical body *(i)* must remain equal to 2.5. Therefore, we write a simple constraint as $y_i - 2.5 = 0$. Because a body has three coordinates, we can impose the following simple constraints:

$$
\begin{aligned}
{}^{(x,1)}\Phi &= x_i - c_1 = 0 \\
{}^{(y,1)}\Phi &= y_i - c_2 = 0 \\
{}^{(\phi,1)}\Phi &= \phi_i - c_3 = 0
\end{aligned} \tag{7.8}
$$

where c_1, c_2 and c_3 are constants.

Example 7.1

In this exercise four function M-files are presented to evaluate the constraints stated in Eqs. (7.3) to (7.6). In these functions it is assumed that the components of all the vectors have already been updated and all the necessary constants are available. These function M-files will be used in some of the upcoming exercises.

```
function Phi = BC_Phi_rev(r_P_i, r_P_j)
% Revolute joint constraints
    Phi = r_P_j - r_P_i;
```

```
function Phi = BC_Phi_tran(u_i, phi_i, phi_j, d, phi_c)
% Translational joint constraints
    Phi = [phi_j - phi_i - phi_c; s_rot(u_i)'*d];
```

```
function Phi = BC_Phi_rev_rev(d, L)
% Revolute-revolute joint constraint
    Phi = (d'*d - L^2)/2;
```

```
function Phi = BC_Phi_rev_tran(u_j, d, L)
% Revolute-translational joint constraint
    Phi = s_rot(u_j)'*d - L;
```

7.3 EXAMPLES

In this section we construct the kinematic constraints with body-coordinate formulation for some of the application examples from Section 1.5. These same examples were considered in Chapter 5 with the point-coordinate formulation. This should provide the reader the means to compare different formulations when applied to the same system.

7.3.1 Double A-Arm Suspension

In this section we consider the double A-arm suspension system for kinematic modeling with the body-coordinate formulation. The system is shown in its assembled and unassembled forms in Figure 7.10. The bodies are numbered as (0), (1), (2) and (3). A nonmoving *x-y* frame and three body-fixed frames are defined. If the mass centers are known, the origin of the body frames are positioned at the mass centers otherwise a frame is positioned at a point reasonably close to that body's mass center. The pin joints are positioned with respect to their corresponding frames as:

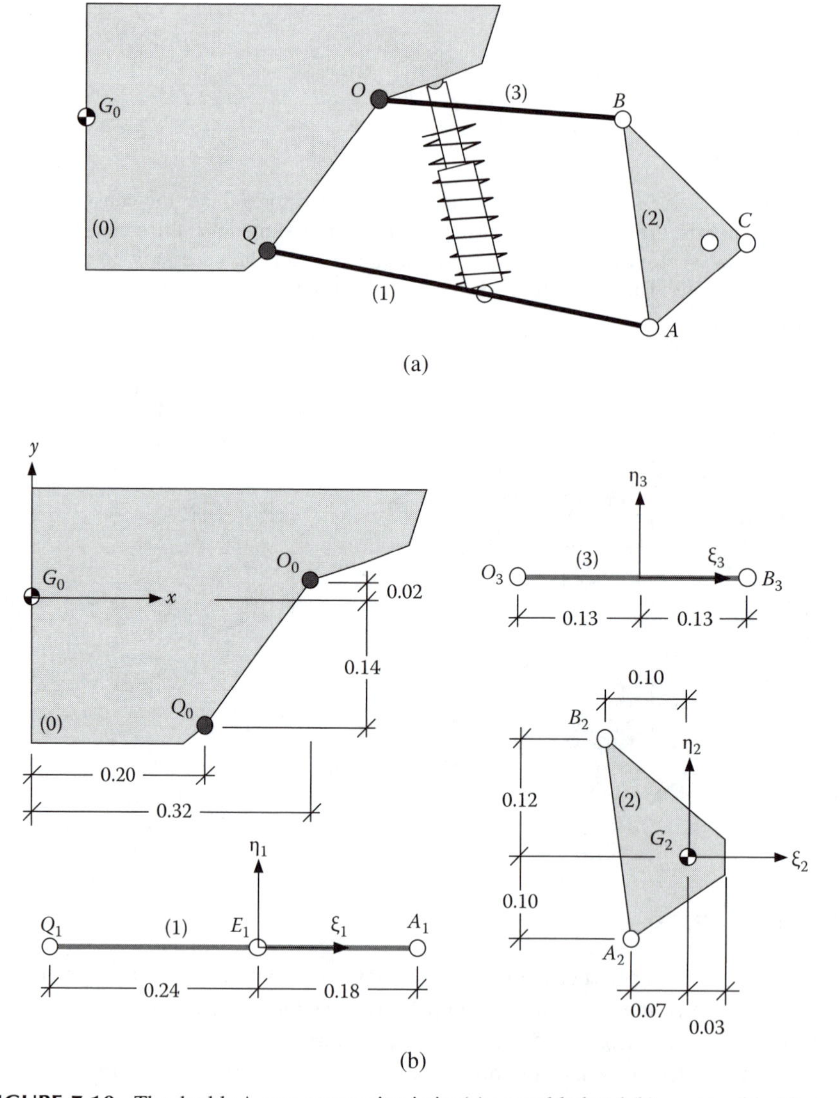

FIGURE 7.10 The double A-arm suspension in its (a) assembled and (b) unassembled forms.

$$\text{Pin joint } Q\text{: } \mathbf{r}_0^Q = \begin{Bmatrix} x_0^Q \\ y_0^Q \end{Bmatrix} = \begin{Bmatrix} 0.20 \\ -0.14 \end{Bmatrix}, \quad \mathbf{s}_1'^Q = \begin{Bmatrix} \xi_1^Q \\ \eta_1^Q \end{Bmatrix} = \begin{Bmatrix} -0.24 \\ 0 \end{Bmatrix}$$

$$\text{Pin joint } A\text{: } \mathbf{s}_1'^A = \begin{Bmatrix} \xi_1^A \\ \eta_1^A \end{Bmatrix} = \begin{Bmatrix} 0.18 \\ 0 \end{Bmatrix}, \quad \mathbf{s}_2'^A = \begin{Bmatrix} \xi_2^A \\ \eta_2^A \end{Bmatrix} = \begin{Bmatrix} -0.07 \\ -0.10 \end{Bmatrix}$$

$$\text{Pin joint } B\text{: } \mathbf{s}_2'^B = \begin{Bmatrix} \xi_2^B \\ \eta_2^B \end{Bmatrix} = \begin{Bmatrix} -0.10 \\ 0.12 \end{Bmatrix}, \quad \mathbf{s}_3'^B = \begin{Bmatrix} \xi_3^B \\ \eta_3^B \end{Bmatrix} = \begin{Bmatrix} 0.13 \\ 0 \end{Bmatrix}$$

$$\text{Pin joint } O\text{: } \mathbf{s}_3'^O = \begin{Bmatrix} \xi_3^O \\ \eta_3^O \end{Bmatrix} = \begin{Bmatrix} -0.13 \\ 0 \end{Bmatrix}, \quad \mathbf{r}_0^O = \begin{Bmatrix} x_0^O \\ y_0^O \end{Bmatrix} = \begin{Bmatrix} 0.32 \\ 0.02 \end{Bmatrix}$$

The constraint equations for the four pin joints are written in compact form, based on Eq. (7.3) as

$$\begin{aligned} \mathbf{\Phi}^Q &= \mathbf{r}_1^Q - \mathbf{r}_0^Q = \mathbf{0} \\ \mathbf{\Phi}^A &= \mathbf{r}_2^A - \mathbf{r}_1^A = \mathbf{0} \\ \mathbf{\Phi}^B &= \mathbf{r}_3^B - \mathbf{r}_2^B = \mathbf{0} \\ \mathbf{\Phi}^O &= \mathbf{r}_0^O - \mathbf{r}_3^O = \mathbf{0} \end{aligned} \tag{7.9}$$

These constraints can be expanded and slightly simplified into the following form:

$$\mathbf{\Phi}^Q = \begin{Bmatrix} -0.20 + x_1 - 0.24\cos\phi_1 \\ 0.14 + y_1 - 0.24\sin\phi_1 \end{Bmatrix} = \mathbf{0}$$

$$\mathbf{\Phi}^A = \begin{Bmatrix} -x_1 - 0.18\cos\phi_1 + x_2 - 0.07\cos\phi_2 + 0.10\sin\phi_2 \\ -y_1 - 0.18\sin\phi_1 + y_2 - 0.07\sin\phi_2 - 0.10\cos\phi_2 \end{Bmatrix} = \mathbf{0}$$

$$\mathbf{\Phi}^B = \begin{Bmatrix} -x_2 + 0.10\cos\phi_2 + 0.12\sin\phi_2 + x_3 + 0.13\cos\phi_3 \\ -y_2 + 0.10\sin\phi_2 - 0.12\cos\phi_2 + y_3 + 0.13\sin\phi_3 \end{Bmatrix} = \mathbf{0}$$

$$\mathbf{\Phi}^O = \begin{Bmatrix} -x_3 + 0.13\cos\phi_3 + 0.32 \\ -y_3 + 0.13\sin\phi_3 + 0.02 \end{Bmatrix} = \mathbf{0}$$

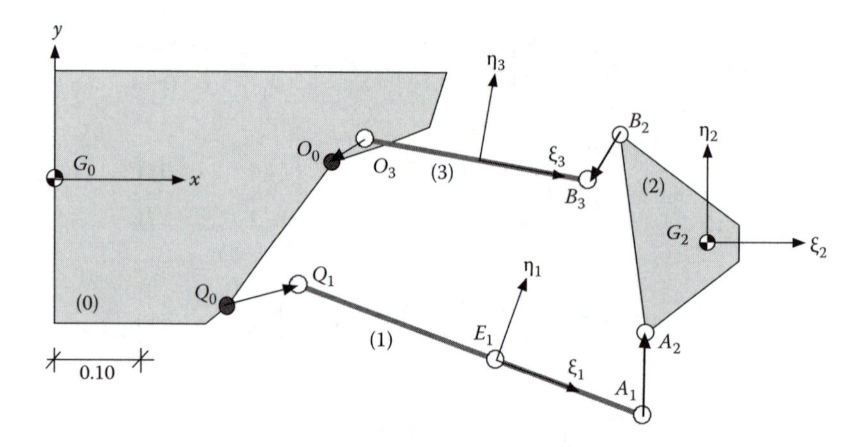

FIGURE 7.11 Constraint violations at the pin joints are shown with thick arrows.

Although this conversion from compact to expanded form is not necessary, it is done here to reveal the contents of the constraints.

Assume that the following values are provided for the body coordinates:

$$\mathbf{r}_1 = \begin{Bmatrix} 0.51 \\ -0.20 \end{Bmatrix},\ \phi_1 = 340°,\ \mathbf{r}_2 = \begin{Bmatrix} 0.75 \\ -0.07 \end{Bmatrix},\ \phi_2 = 0,\ \mathbf{r}_3 = \begin{Bmatrix} 0.49 \\ 0.02 \end{Bmatrix},\ \phi_3 = 350°$$

With these values we can evaluate the eight constraints. This evaluation can be performed either by hand (not recommended) or by a computer program to obtain*

```
Phi_Q =           Phi_A =           Phi_B =           Phi_O =
    0.0845            0.0009           -0.0320           -0.0420
    0.0221            0.0916           -0.0526           -0.0226
```

It is obvious that the constraints are not satisfied. This means that the values given for the coordinates are not correct. If we construct the system according to the given values, as shown in Figure 7.11, it becomes clear why the constraints are violated. In the analysis chapters we will learn how to correct the coordinate values to ensure that the constraints are not violated.

7.3.2 MacPherson Suspension

The MacPherson suspension system from Section 1.5.2, as shown in Figure 7.12, is considered here for kinematic modeling with the body-coordinate formulation. We construct three different kinematic models for this system to demonstrate how to take advantage of composite joints to reduce the size of a problem. The models contain three, two, and one moving body respectively. Although these models are

* For this computation we have used the program from Section 7.6.1.

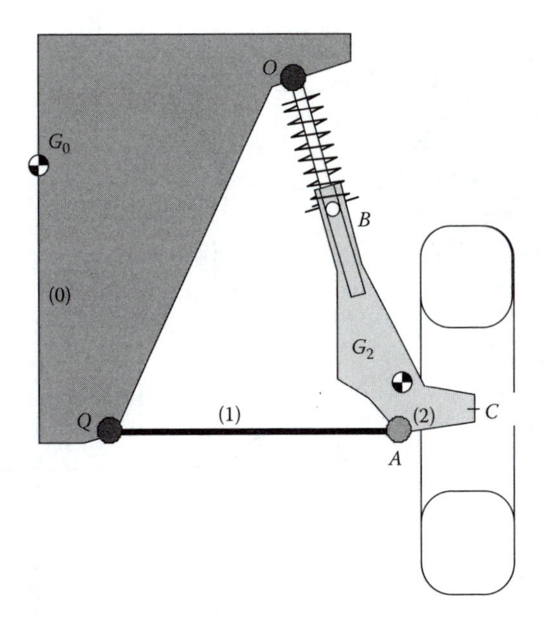

FIGURE 7.12 The MacPherson suspension.

constructed with different number of bodies and joints, they are kinematically identical. They each have one DoF.

Model A: In this model we consider one nonmoving and three moving bodies where in unassembled form it is shown in Figure 7.13. The moving bodies are numbered (1), (2), and (3). There are three pin joints at O, Q, and A, and one sliding joint between bodies (2) and (3) along the strut axis. For the three pin joints the following data are obtained from the figure:

$$\mathbf{r}_0^Q = \begin{Bmatrix} 0.12 \\ -0.41 \end{Bmatrix}, \mathbf{s}_1'^Q = \begin{Bmatrix} -0.225 \\ 0 \end{Bmatrix}; \mathbf{s}_1'^A = \begin{Bmatrix} 0.225 \\ 0 \end{Bmatrix}, \mathbf{s}_2'^A = \begin{Bmatrix} 0 \\ -0.07 \end{Bmatrix};$$

$$\mathbf{r}_0^O = \begin{Bmatrix} 0.41 \\ 0.13 \end{Bmatrix}, \mathbf{s}_3'^O = \begin{Bmatrix} -0.15 \\ 0 \end{Bmatrix}$$

On the axis of the sliding joint we need two points and one unit vector:

$$\mathbf{s}_3'^O = \begin{Bmatrix} -0.15 \\ 0 \end{Bmatrix}, \quad \mathbf{u}_3' = \begin{Bmatrix} 1 \\ 0 \end{Bmatrix}; \quad \mathbf{s}_2'^B = \begin{Bmatrix} -0.17 \\ 0.25 \end{Bmatrix}$$

Points O_3 and B_2 are used to construct vector $\mathbf{d} = \mathbf{r}_2^B - \mathbf{r}_3^O$.

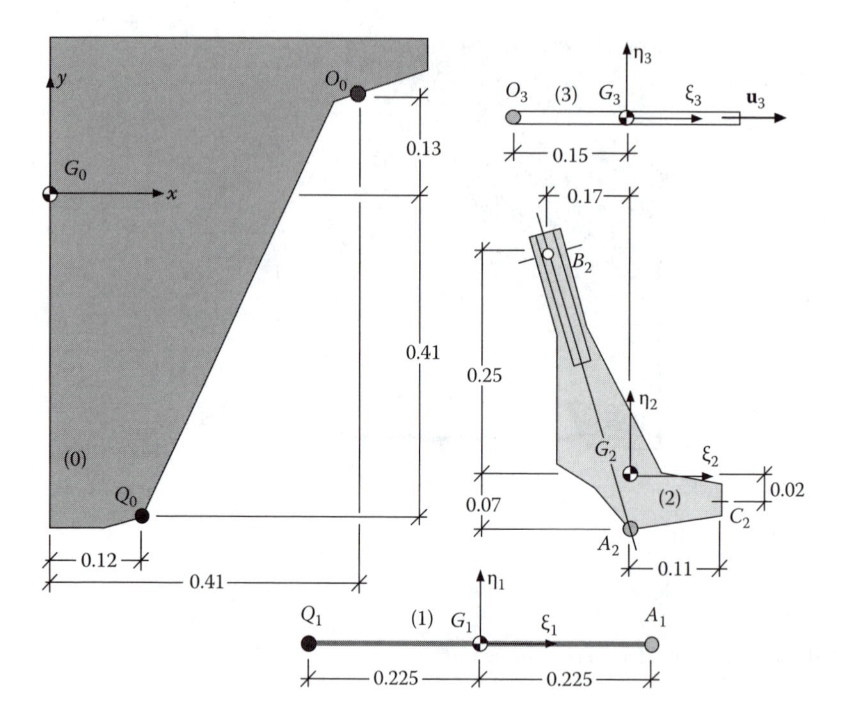

FIGURE 7.13 The MacPherson suspension modeled with three moving bodies.

The constraints for the three pin joints are constructed as

$$\begin{aligned} \mathbf{r}_1^Q - \mathbf{r}_0^Q &= \mathbf{0} \\ \mathbf{r}_2^A - \mathbf{r}_1^A &= \mathbf{0} \\ \mathbf{r}_3^O - \mathbf{r}_0^O &= \mathbf{0} \end{aligned} \tag{7.10}$$

For the sliding joint the following two constraints are constructed:

$$\begin{aligned} \phi_2 - \phi_3 - {}^c\phi &= 0 \\ \breve{\mathbf{u}}_3' \, \mathbf{d} &= 0 \end{aligned} \tag{7.11}$$

where ${}^c\phi = 61.9° = 1.08$ rad is the angle between the ξ_2 and ξ_3 axes. These are eight algebraic constraints between the nine coordinates (three per moving body); that is, 9 − 8 = 1 DoF.

Model B: In this model we combine link (3), the pin joint at O, and the sliding joint into a revolute-translational joint. Therefore, we will have two moving bodies, two pin joints at Q and A, and a revolute-translational joint between bodies (0) and

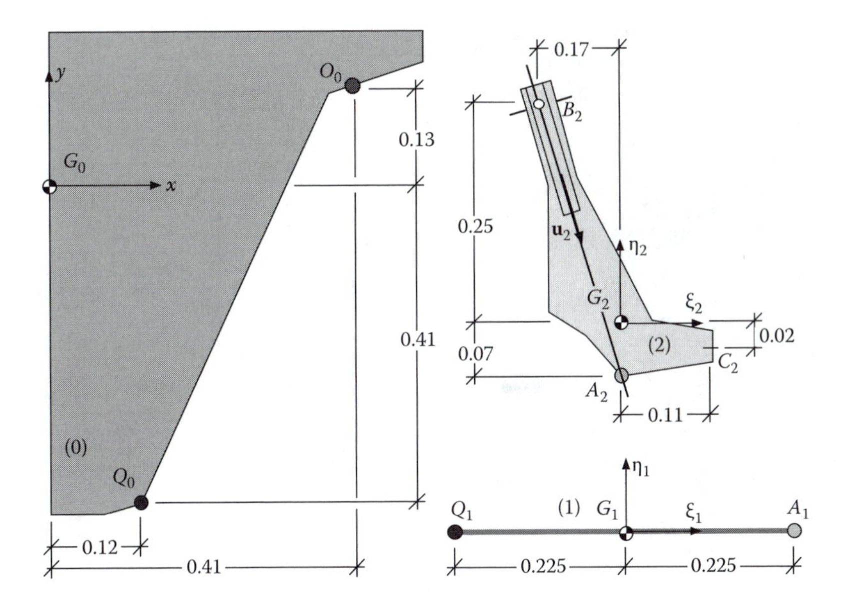

FIGURE 7.14 The MacPherson suspension modeled with two moving bodies.

(2) as shown in Figure 7.14. For the pin joints the data is the same as in Model A. The following constraints can be written for the pin joints:

$$\begin{aligned} \mathbf{r}_1^Q - \mathbf{r}_0^Q &= \mathbf{0} \\ \mathbf{r}_2^A - \mathbf{r}_1^A &= \mathbf{0} \end{aligned} \tag{7.12}$$

For the revolute-translational joint—the pin joint at O_0 and the sliding joint along the axis of the strut passing through A_2 —the following data are needed:

$$\mathbf{r}_0^O = \mathbf{s}_0^O = \begin{Bmatrix} 0.41 \\ 0.13 \end{Bmatrix}; \quad \mathbf{s}_2'^B = \begin{Bmatrix} -0.17 \\ 0.25 \end{Bmatrix}, \quad \mathbf{u}_2' = \begin{Bmatrix} 0.47 \\ -0.88 \end{Bmatrix}$$

Points O_0 and B_2 will be used to construct vector $\mathbf{d} = \mathbf{r}_2^B - \mathbf{r}_0^O$. For this joint, we write the following constraint based on Eq. (7.6):

$$\breve{\mathbf{u}}_2' \mathbf{d} = 0 \tag{7.13}$$

where $\ell = 0$ (the distance between the pin joint and the sliding axis). In this model we have constructed five algebraic constraints between the six coordinates; that is, 6 – 5 = 1 DoF.

Model C: In this model, in addition to the revolute-translational joint, we introduce a second composite joint to replace the rod between Q and A. The only

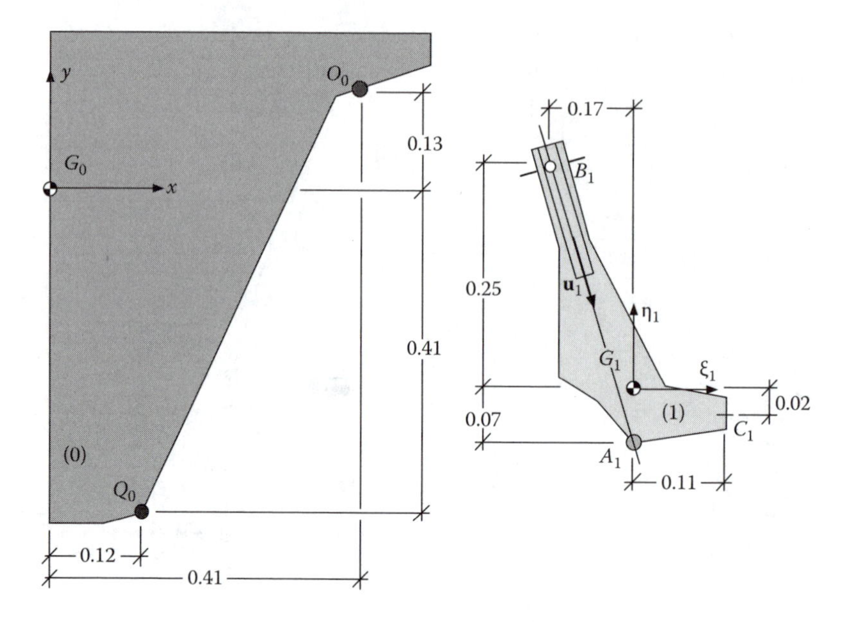

FIGURE 7.15 The MacPherson suspension modeled with one moving body.

body in the system is the strut assembly. As shown in Figure 7.15, this body is renumbered as link (1). For the revolute-translational joint, the data is the same as in Model B except for a change in the body index:

$$\mathbf{r}_0^O = \mathbf{s}_0^O = \begin{Bmatrix} 0.41 \\ 0.13 \end{Bmatrix}, \quad \mathbf{s}_1^{\prime B} = \begin{Bmatrix} -0.17 \\ 0.25 \end{Bmatrix}, \quad \mathbf{u}_1^{\prime} = \begin{Bmatrix} 0.47 \\ -0.88 \end{Bmatrix}$$

Points O_0 and B_1 will be used to construct vector $\mathbf{d}^{B,O} = \mathbf{r}_1^B - \mathbf{r}_0^O$. For the revolute-revolute joint between Q and A we need the following data:

$$\mathbf{r}_0^Q = \mathbf{s}_0^Q = \begin{Bmatrix} 0.12 \\ -0.41 \end{Bmatrix}, \quad \mathbf{s}_1^{\prime A} = \begin{Bmatrix} 0 \\ -0.07 \end{Bmatrix}, \quad \ell = 0.25$$

Points Q_0 and A_1 will be used to construct vector $\mathbf{d}^{A,Q} = \mathbf{r}_1^A - \mathbf{r}_0^Q$.

The following constraints can be written for the two composite joints:

$$\begin{gathered} \breve{\mathbf{u}}_1^{\prime} \, \mathbf{d}^{B,O} = 0 \\ \frac{1}{2}(\mathbf{d}^{A,Q\prime} \, \mathbf{d}^{A,Q} - \ell^2) = 0 \end{gathered} \tag{7.14}$$

With two algebraic constraints between the three coordinates we have 3 – 2 = 1 DoF.

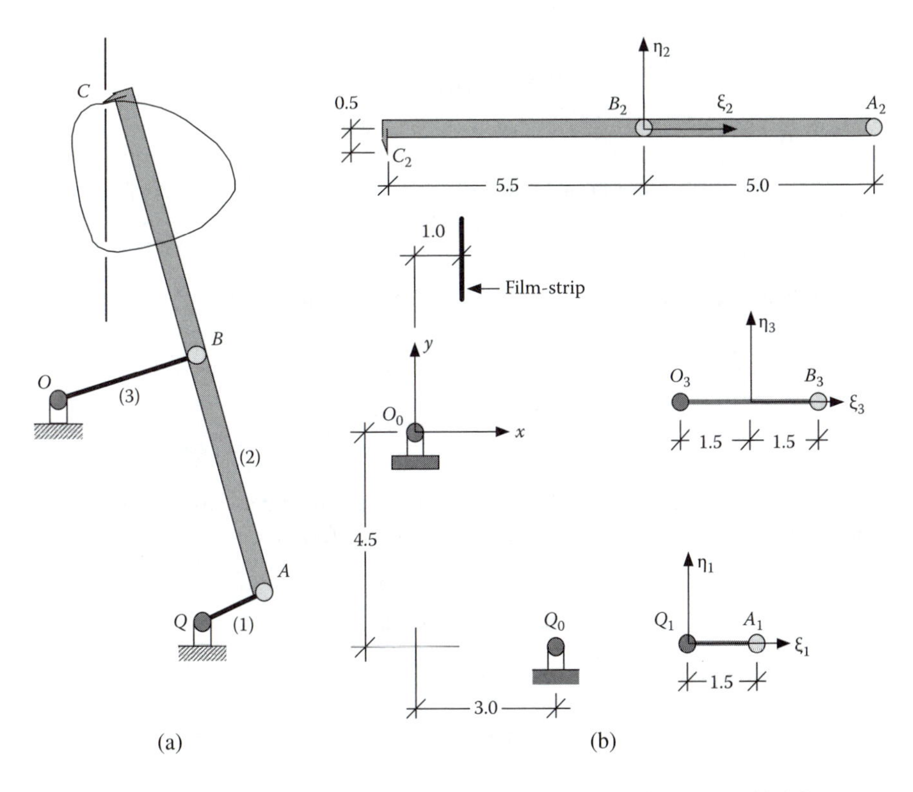

FIGURE 7.16 The film-strip mechanism in (a) assembled and (b) unassembled forms.

7.3.3 Filmstrip Advancer

The filmstrip advancer mechanism from Section 1.5.3 is shown in Figure 7.16 in the assembled and unassembled forms. The ground and the three moving bodies are numbered as (0), (1), (2), (3). A nonmoving *x-y* frame is positioned at *O* and three body-fixed frames are defined at convenient locations on the bodies. Because we will consider this system only for kinematic analysis, there is no need to position the origin of the body frames at the mass centers.

For each pin joint the following constant coordinates are extracted from the figure:

$$\text{Pin joint } Q\text{: } \mathbf{r}_0^Q = \mathbf{s}_0^Q = \begin{Bmatrix} 3.0 \\ -4.5 \end{Bmatrix}, \quad \mathbf{s}_1^{\prime Q} = \begin{Bmatrix} 0 \\ 0 \end{Bmatrix}$$

$$\text{Pin joint } A\text{: } \mathbf{s}_1^{\prime A} = \begin{Bmatrix} 1.5 \\ 0 \end{Bmatrix}, \quad \mathbf{s}_2^{\prime A} = \begin{Bmatrix} 5.0 \\ 0 \end{Bmatrix}$$

$$\text{Pin joint } B\text{: } \mathbf{s}_2'^B = \begin{Bmatrix} 0 \\ 0 \end{Bmatrix}, \quad \mathbf{s}_3'^B = \begin{Bmatrix} 1.5 \\ 0 \end{Bmatrix}$$

$$\text{Pin joint } O\text{: } \mathbf{r}_0^O = \mathbf{s}_0^O = \begin{Bmatrix} 0 \\ 0 \end{Bmatrix}, \quad \mathbf{s}_3'^O = \begin{Bmatrix} -1.5 \\ 0 \end{Bmatrix}$$

The coordinates of point C on body (2) are also determined to be

$$\text{Point } C\text{: } \mathbf{s}_2'^C = \begin{Bmatrix} -5.5 \\ -0.5 \end{Bmatrix}$$

The constraint equations for the pin joints can be constructed based on Eq. (7.3). In this exercise we express the equations in compact form as:

$$\begin{aligned} \mathbf{\Phi}^Q &= \mathbf{r}_1^Q - \mathbf{r}_0^Q = \mathbf{0} \\ \mathbf{\Phi}^A &= \mathbf{r}_2^A - \mathbf{r}_1^A = \mathbf{0} \\ \mathbf{\Phi}^B &= \mathbf{r}_3^B - \mathbf{r}_2^B = \mathbf{0} \\ \mathbf{\Phi}^O &= \mathbf{r}_0^O - \mathbf{r}_3^O = \mathbf{0} \end{aligned} \tag{7.15}$$

The x-y coordinates of point C on body (2) are determined based on Eq. (3.10) as

$$\mathbf{r}_2^C = \mathbf{r}_2 + \mathbf{s}_2^C \tag{7.16}$$

This expression can be converted into expanded form as

$$\begin{aligned} x_2^C &= x_2 - 5.5\cos\phi_2 + 0.5\sin\phi_2 \\ y_2^C &= y_2 - 5.5\sin\phi_2 - 0.5\cos\phi_2 \end{aligned}$$

These are algebraic expressions—not constraints—that can be evaluated after numerical values for the coordinates of body (2) have been determined.

7.4 VELOCITY AND ACCELERATION CONSTRAINTS

Velocity and acceleration constraints are obtained by taking the first and second time derivatives of the position constraints. The arrays of velocities and accelerations associated with the array of coordinates of Eq. (7.2) are denoted as

$$\dot{\mathbf{c}} = \begin{Bmatrix} \dot{\mathbf{c}}_1 \\ \vdots \\ \dot{\mathbf{c}}_{n_b} \end{Bmatrix}, \quad \ddot{\mathbf{c}} = \begin{Bmatrix} \ddot{\mathbf{c}}_1 \\ \vdots \\ \ddot{\mathbf{c}}_{n_b} \end{Bmatrix} \tag{7.17}$$

The time derivative of the constraint equations in Eq. (7.1) yields the velocity constraints in a general form. These constraints are *linear* in the velocities and they are expressed as

$$\dot{\mathbf{\Phi}} = \mathbf{D}\,\dot{\mathbf{c}} = \mathbf{0} \tag{7.18}$$

The coefficient matrix $\mathbf{D}$ is the *Jacobian*.

The acceleration constraints are expressed in their general form as

$$\ddot{\mathbf{\Phi}} = \mathbf{D}\,\ddot{\mathbf{c}} + \dot{\mathbf{D}}\,\dot{\mathbf{c}} = \mathbf{0} \tag{7.19}$$

These constraints are linear in the accelerations. The coefficient matrix of the acceleration array is the same Jacobian. The product $\dot{\mathbf{D}}\,\dot{\mathbf{c}}$ contains quadratic terms in velocities. The acceleration constraints can be written in another useful form as

$$\mathbf{D}\,\ddot{\mathbf{c}} = \boldsymbol{\gamma} \tag{7.20}$$

where

$$\boldsymbol{\gamma} = -\dot{\mathbf{D}}\,\dot{\mathbf{c}} \tag{7.21}$$

The array $\boldsymbol{\gamma}$ is called the *right-hand side* array of acceleration constraints.

In the following subsections we derive the velocity and acceleration constraints for the commonly used kinematic joints. In these derivations, several equations from Chapter 3 are used repeatedly. These are Eqs. (3.12), (3.13), (3.16), and (3.17) that are restated here for easy reference:

$$\dot{\mathbf{s}} = \breve{\mathbf{s}}\,\dot{\phi} \tag{7.22}$$

$$\dot{\mathbf{r}}^P = \dot{\mathbf{r}} + \dot{\mathbf{s}}^P = \dot{\mathbf{r}} + \breve{\mathbf{s}}^P\dot{\phi} \tag{7.23}$$

$$\ddot{\mathbf{s}} = \breve{\mathbf{s}}\,\ddot{\phi} + \dot{\breve{\mathbf{s}}}\,\dot{\phi} = \breve{\mathbf{s}}\,\ddot{\phi} - \mathbf{s}\,\dot{\phi}^2 \tag{7.24}$$

$$\ddot{\mathbf{r}}^P = \ddot{\mathbf{r}} + \breve{\mathbf{s}}^P\ddot{\phi} + \dot{\breve{\mathbf{s}}}^P\dot{\phi} = \ddot{\mathbf{r}} + \breve{\mathbf{s}}^P\ddot{\phi} - \mathbf{s}^P\dot{\phi}^2 \tag{7.25}$$

Other useful equations are the first and second time derivative of $\mathbf{d} = \mathbf{r}_j^P - \mathbf{r}_i^P$:

$$\begin{aligned}\dot{\mathbf{d}} &= \dot{\mathbf{r}}_j^P - \dot{\mathbf{r}}_i^P = \dot{\mathbf{r}}_j + \dot{\mathbf{s}}_j^P - \dot{\mathbf{r}}_i - \dot{\mathbf{s}}_i^P \\ &= \dot{\mathbf{r}}_j + \breve{\mathbf{s}}_j^P \dot{\phi}_j - \dot{\mathbf{r}}_i - \breve{\mathbf{s}}_i^P \dot{\phi}_i\end{aligned} \tag{7.26}$$

$$\begin{aligned}\ddot{\mathbf{d}} &= \ddot{\mathbf{r}}_j^P - \ddot{\mathbf{r}}_i^P = \ddot{\mathbf{r}}_j + \ddot{\mathbf{s}}_j^P - \ddot{\mathbf{r}}_i - \ddot{\mathbf{s}}_i^P \\ &= \ddot{\mathbf{r}}_j + \breve{\mathbf{s}}_j^P \ddot{\phi}_j + \dot{\breve{\mathbf{s}}}_j^P \dot{\phi}_j - \ddot{\mathbf{r}}_i - \breve{\mathbf{s}}_i^P \ddot{\phi}_i - \dot{\breve{\mathbf{s}}}_i^P \dot{\phi}_i \\ &= \ddot{\mathbf{r}}_j + \breve{\mathbf{s}}_j^P \ddot{\phi}_j - \mathbf{s}_j^P \dot{\phi}_j^2 - \ddot{\mathbf{r}}_i - \breve{\mathbf{s}}_i^P \ddot{\phi}_i + \mathbf{s}_i^P \dot{\phi}_j^2\end{aligned} \tag{7.27}$$

Some of the equations are shown in more than one form. In the derivations of the following subsections, we will use the form that yields the simplest representation.

7.4.1 Revolute Joint

The velocity constraints for a revolute joint are obtained from the time derivative of Eq. (7.3) as

$$^{(r,2)}\dot{\mathbf{\Phi}} = \dot{\mathbf{r}}_j^P - \dot{\mathbf{r}}_i^P = \mathbf{0}$$

Using Eq. (7.23), this equation is transformed into its components:

$$^{(r,2)}\dot{\mathbf{\Phi}} = \dot{\mathbf{r}}_j + \breve{\mathbf{s}}_j^P \dot{\phi}_j - \dot{\mathbf{r}}_i - \breve{\mathbf{s}}_i^P \dot{\phi}_i = \mathbf{0}$$

or

$$^{(r,2)}\dot{\mathbf{\Phi}} = \left[-\mathbf{I} \quad -\breve{\mathbf{s}}_i^P \;\middle|\; \mathbf{I} \quad \breve{\mathbf{s}}_j^P \right] \begin{Bmatrix} \dot{\mathbf{r}}_i \\ \dot{\phi}_i \\ \hline \dot{\mathbf{r}}_j \\ \dot{\phi}_j \end{Bmatrix} = \mathbf{0} \tag{7.28}$$

This form of expressing the velocity constraints clearly shows that the constraints are *linear* in the velocities. In expanded form this equation becomes

$$^{(r,2)}\dot{\mathbf{\Phi}} = \left[\begin{array}{ccc|ccc} -1 & 0 & (\xi_i^P \sin\phi_i + \eta_i^P \cos\phi_i) & 1 & 0 & -(\xi_j^P \sin\phi_j + \eta_j^P \cos\phi_j) \\ 0 & -1 & -(\xi_i^P \cos\phi_i - \eta_i^P \sin\phi_i) & 0 & 1 & (\xi_i^P \cos\phi_j - \eta_j^P \sin\phi_j) \end{array} \right] \begin{Bmatrix} \dot{x}_i \\ \dot{y}_i \\ \dot{\phi}_i \\ \hline \dot{x}_j \\ \dot{y}_j \\ \dot{\phi}_j \end{Bmatrix} = \begin{Bmatrix} 0 \\ 0 \end{Bmatrix}$$

However, there is no need to expand a velocity constraint in this form. For MATLAB programming, the compact matrix form of Eq. (7.28) is preferred.

The time derivative of Eq. (7.28) yields the acceleration constraints:

$$^{(r,2)}\ddot{\mathbf{\Phi}} = \begin{bmatrix} -\mathbf{I} & -\breve{\mathbf{s}}_i^P & \mathbf{I} & \breve{\mathbf{s}}_j^P \end{bmatrix} \begin{Bmatrix} \ddot{\mathbf{r}}_i \\ \ddot{\phi}_i \\ \ddot{\mathbf{r}}_j \\ \ddot{\phi}_j \end{Bmatrix} + \begin{bmatrix} \mathbf{0} & -\dot{\breve{\mathbf{s}}}_i^P & \mathbf{0} & \dot{\breve{\mathbf{s}}}_j^P \end{bmatrix} \begin{Bmatrix} \dot{\mathbf{r}}_i \\ \dot{\phi}_i \\ \dot{\mathbf{r}}_j \\ \dot{\phi}_j \end{Bmatrix} = \mathbf{0}$$

The quadratic velocity terms are moved to the right-hand side to obtain:

$$\begin{bmatrix} -\mathbf{I} & -\breve{\mathbf{s}}_i^P & \mathbf{I} & \breve{\mathbf{s}}_j^P \end{bmatrix} \begin{Bmatrix} \ddot{\mathbf{r}}_i \\ \ddot{\phi}_i \\ \ddot{\mathbf{r}}_j \\ \ddot{\phi}_j \end{Bmatrix} = \dot{\breve{\mathbf{s}}}_i^P \dot{\phi}_i - \dot{\breve{\mathbf{s}}}_j^P \dot{\phi}_j \tag{7.29}$$

The Jacobian submatrices and the right-hand side of the acceleration constraints for a revolute joint are restated here for easy reference:

$$^{(r,2)}\mathbf{D}_i = \begin{bmatrix} -\mathbf{I} & -\breve{\mathbf{s}}_i^P \end{bmatrix}, \; ^{(r,2)}\mathbf{D}_j = \begin{bmatrix} \mathbf{I} & \breve{\mathbf{s}}_j^P \end{bmatrix} \tag{7.30}$$

$$^{(r,2)}\boldsymbol{\gamma} = \dot{\breve{\mathbf{s}}}_i^P \dot{\phi}_i - \dot{\breve{\mathbf{s}}}_j^P \dot{\phi}_j \tag{7.31}$$

EXAMPLE 7.2

In this example two function M-files to evaluate the Jacobian matrix and the right-hand side of the acceleration constraints for a revolute joint are developed.

```
function [jac_i, jac_j] = BC_jacob_rev(s_P_i, s_P_j)
% Jacobian submatrices for a revolute joint
    jac_i = [-eye(2) -s_rot(s_P_i)]; jac_j = [eye(2) s_rot(s_P_j)];
```

```
function gamma = BC_gamma_rev(s_P_i_d, phi_i_d, s_P_j_d, phi_j_d)
% Gamma array for a revolute joint
    gamma = s_rot(s_P_i_d*phi_i_d - s_P_j_d*phi_j_d);
```

7.4.2 Translational Joint

For a translational joint, the time derivative of Eq. (7.4) provides the velocity constraints as

$$
{}^{(t,2)}\dot{\boldsymbol{\Phi}} = \begin{cases} \dot{\phi}_j - \dot{\phi}_i = 0 \\ \dot{\breve{\mathbf{u}}}_i'\,\mathbf{d} + \breve{\mathbf{u}}_i'\,\dot{\mathbf{d}} = 0 \end{cases}
$$

The terms in this equation are rearranged as (refer to Eq. (2.14))

$$
{}^{(t,2)}\dot{\boldsymbol{\Phi}} = \begin{cases} \dot{\phi}_j - \dot{\phi}_i = 0 \\ -\breve{\mathbf{d}}'\dot{\mathbf{u}}_i + \breve{\mathbf{u}}_i'\,\dot{\mathbf{d}} = 0 \end{cases}
$$

Using Eqs. (7.22), (7.23), (7.26), and (2.16), we expand the constraint into another form:

$$
{}^{(t,2)}\dot{\boldsymbol{\Phi}} = \left[\begin{array}{cc:cc} \mathbf{0} & -1 & \mathbf{0} & 1 \\ -\breve{\mathbf{u}}_i' & -\mathbf{u}_i'(\mathbf{d}+\mathbf{s}_i^P) & \breve{\mathbf{u}}_i' & \mathbf{u}_i'\,\mathbf{s}_j^P \end{array}\right] \begin{Bmatrix} \dot{\mathbf{r}}_i \\ \dot{\phi}_i \\ \hline \dot{\mathbf{r}}_j \\ \dot{\phi}_j \end{Bmatrix} = \begin{Bmatrix} 0 \\ 0 \end{Bmatrix} \tag{7.32}
$$

The time derivative of Eq. (7.32) yields the acceleration constraints as

$$
{}^{(t,2)}\ddot{\boldsymbol{\Phi}} = \left[\begin{array}{cc:cc} \mathbf{0} & -1 & \mathbf{0} & 1 \\ -\breve{\mathbf{u}}_i' & -\mathbf{u}_i'(\mathbf{d}+\mathbf{s}_i^P) & \breve{\mathbf{u}}_i' & \mathbf{u}_i'\,\mathbf{s}_j^P \end{array}\right] \begin{Bmatrix} \ddot{\mathbf{r}}_i \\ \ddot{\phi}_i \\ \hline \ddot{\mathbf{r}}_j \\ \ddot{\phi}_j \end{Bmatrix} +
$$

$$
\left[\begin{array}{cc:cc} \mathbf{0} & 0 & \mathbf{0} & 0 \\ -\dot{\breve{\mathbf{u}}}_i' & -\dot{\mathbf{u}}_i'(\mathbf{d}+\mathbf{s}_i^P) - \mathbf{u}_i'(\dot{\mathbf{d}}+\dot{\mathbf{s}}_i^P) & \dot{\breve{\mathbf{u}}}_i' & \dot{\mathbf{u}}_i'\,\mathbf{s}_j^P + \mathbf{u}_i'\,\dot{\mathbf{s}}_j^P \end{array}\right] \begin{Bmatrix} \dot{\mathbf{r}}_i \\ \dot{\phi}_i \\ \hline \dot{\mathbf{r}}_j \\ \dot{\phi}_j \end{Bmatrix} = \begin{Bmatrix} 0 \\ 0 \end{Bmatrix}
$$

Taking advantage of the fact that $\dot{\phi}_i = \dot{\phi}_j$, we can simplify the acceleration constraints to obtain

$$
\left[\begin{array}{cc:cc} \mathbf{0} & -1 & \mathbf{0} & 1 \\ -\breve{\mathbf{u}}_i' & -\mathbf{u}_i'(\mathbf{d}+\mathbf{s}_i^P) & \breve{\mathbf{u}}_i' & \mathbf{u}_i'\,\mathbf{s}_j^P \end{array}\right] \begin{Bmatrix} \ddot{\mathbf{r}}_i \\ \ddot{\phi}_i \\ \hline \ddot{\mathbf{r}}_j \\ \ddot{\phi}_j \end{Bmatrix} = \begin{Bmatrix} 0 \\ \dot{\mathbf{u}}_i'(\dot{\breve{\mathbf{d}}} - \dot{\breve{\mathbf{r}}}_i + \dot{\breve{\mathbf{r}}}_j + \mathbf{d}\dot{\phi}_i) \end{Bmatrix} \tag{7.33}
$$

The Jacobian submatrices and the right-hand side of the acceleration constraints for a translational joint are restated here as

$$^{(t,2)}\mathbf{D}_i = \begin{bmatrix} \mathbf{0} & -1 \\ -\breve{\mathbf{u}}_i' & -\mathbf{u}_i'(\mathbf{d}+\mathbf{s}_i^P) \end{bmatrix},\ ^{(t,2)}\mathbf{D}_j = \begin{bmatrix} \mathbf{0} & 1 \\ \breve{\mathbf{u}}_i' & \mathbf{u}_i'\mathbf{s}_j^P \end{bmatrix} \tag{7.34}$$

$$^{(t,2)}\boldsymbol{\gamma} = \begin{Bmatrix} 0 \\ \dot{\mathbf{u}}_i'(\dot{\breve{\mathbf{d}}} - \dot{\breve{\mathbf{r}}}_i + \dot{\breve{\mathbf{r}}}_j + \mathbf{d}\dot{\phi}_i) \end{Bmatrix} \tag{7.35}$$

Example 7.3

In this example two function M-files to evaluate the Jacobian matrix and the right-hand side of the acceleration constraints for a translational joint are developed.

```
function [jac_i, jac_j] = BC_jacob_tran(u_i, d, s_P_i, s_P_j)
% Jacobian submatrices for a translational joint
     z2 = [0 0]; u_i_r = s_rot(u_i);
     jac_i = [z2  -1; -u_i_r'  -u_i'*(d + s_P_i)];
     jac_j = [z2   1;  u_i_r'   u_i'*s_P_j       ];
```

```
function gamma = BC_gamma_tran(u_i_d, r_i_d, r_j_d, d, d_d, phi_i_d)
% Gamma array for a translational joint
    gamma = [0; u_i_d'*(s_rot(d_d - r_i_d + r_j_d) + d*phi_i_d)];
```

7.4.3 Revolute-Revolute Joint

The velocity constraint for a revolute-revolute joint is obtained by taking the time derivative of Eq. (7.5):

$$^{(r-r,1)}\dot{\Phi} = \mathbf{d}'\dot{\mathbf{d}} = 0$$

In matrix form we have

$$^{(r-r,1)}\dot{\Phi} = \begin{bmatrix} -\mathbf{d}' & -\mathbf{d}'\breve{\mathbf{s}}_i^P & \mathbf{d}' & \mathbf{d}'\breve{\mathbf{s}}_j^P \end{bmatrix} \begin{Bmatrix} \dot{\mathbf{r}}_i \\ \dot{\phi}_i \\ \dot{\mathbf{r}}_j \\ \dot{\phi}_j \end{Bmatrix} = 0 \tag{7.36}$$

The time derivative of Eq. (7.36) yields the acceleration constraint as

$$
{}^{(r-r,1)}\ddot{\boldsymbol{\Phi}} = \left[-\mathbf{d}' \quad -\mathbf{d}'\breve{\mathbf{s}}_i^P \;\middle|\; \mathbf{d}' \quad \mathbf{d}'\breve{\mathbf{s}}_j^P \right] \begin{Bmatrix} \ddot{\mathbf{r}}_i \\ \ddot{\phi}_i \\ \hline \ddot{\mathbf{r}}_j \\ \ddot{\phi}_j \end{Bmatrix} +
$$

$$
\left[-\dot{\mathbf{d}}' \quad -\dot{\mathbf{d}}'\breve{\mathbf{s}}_i^P - \mathbf{d}'\dot{\breve{\mathbf{s}}}_i^P \;\middle|\; \dot{\mathbf{d}}' \quad \dot{\mathbf{d}}'\breve{\mathbf{s}}_j^P + \mathbf{d}'\dot{\breve{\mathbf{s}}}_j^P \right] \begin{Bmatrix} \dot{\mathbf{r}}_i \\ \dot{\phi}_i \\ \hline \dot{\mathbf{r}}_j \\ \dot{\phi}_j \end{Bmatrix} = 0
$$

After simplification, this equation is written as

$$
\left[-\mathbf{d}' \quad -\mathbf{d}'\breve{\mathbf{s}}_i^P \;\middle|\; \mathbf{d}' \quad \mathbf{d}'\breve{\mathbf{s}}_j^P \right] \begin{Bmatrix} \ddot{\mathbf{r}}_i \\ \ddot{\phi}_i \\ \hline \ddot{\mathbf{r}}_j \\ \ddot{\phi}_j \end{Bmatrix} = -\dot{\mathbf{d}}'\dot{\mathbf{d}} - \mathbf{d}'(\dot{\breve{\mathbf{s}}}_j^P \dot{\phi}_j - \dot{\breve{\mathbf{s}}}_i^P \dot{\phi}_i) \tag{7.37}
$$

The Jacobian submatrices and the right-hand side of the acceleration constraints for a revolute-revolute joint are restated here as

$$
{}^{(r-r,1)}\mathbf{D}_i = \left[-\mathbf{d}' \quad -\mathbf{d}'\breve{\mathbf{s}}_i^P \right],\; {}^{(r-r,1)}\mathbf{D}_j = \left[\mathbf{d}' \quad \mathbf{d}'\breve{\mathbf{s}}_j^P \right] \tag{7.38}
$$

$$
{}^{(r-r,1)}\gamma = -\dot{\mathbf{d}}'\dot{\mathbf{d}} - \mathbf{d}'(\dot{\breve{\mathbf{s}}}_j^P \dot{\phi}_j - \dot{\breve{\mathbf{s}}}_i^P \dot{\phi}_i) \tag{7.39}
$$

Example 7.4

In this example two function M-files to evaluate the Jacobian matrix and the right-hand side of the acceleration constraint for a revolute-revolute joint are developed.

```
function [jac_i, jac_j] = BC_jacob_rev_rev(d, s_P_i, s_P_j)
% Jacobian submatrices for a revolute-revolute joint
   jac_i = [-d' -d'*s_rot(s_P_i)]; jac_j = [ d'  d'*s_rot(s_P_j)];
```

```
function gamma = BC_gamma_rev_rev(d,d_d, s_P_i_d, s_P_j_d, ...
                                         phi_i_d, phi_j_d)
% Gamma array for a revolute-revolute joint
   gamma = -d_d'*d_d - d'*s_rot(s_P_j_d*phi_j_d - s_P_i_d*phi_i_d);
```

7.4.4 Revolute-Translational Joint

For the revolute-translational joint, we obtain the velocity constraint from the time derivative of Eq. (7.6):

$$^{(r\text{-}t,1)}\dot{\Phi} = \breve{\mathbf{u}}'_j\,\dot{\mathbf{d}} + \mathbf{d}'\,\dot{\breve{\mathbf{u}}}_j = 0$$

In matrix form we have

$$^{(r-t,1)}\dot{\Phi} = \left[-\breve{\mathbf{u}}'_j \quad -\mathbf{u}'_j\mathbf{s}_i^P \;\vdots\; \breve{\mathbf{u}}'_j \quad \mathbf{u}'_j(\mathbf{s}_j^P - \mathbf{d})\right]\begin{Bmatrix}\dot{\mathbf{r}}_i \\ \dot{\phi}_i \\ \dot{\mathbf{r}}_j \\ \dot{\phi}_j\end{Bmatrix} = 0 \tag{7.40}$$

The time derivative of Eq. (7.40) after simplifications and rearrangement of terms yields the acceleration constraint as

$$\left[-\breve{\mathbf{u}}'_j \quad -\mathbf{u}'_j\mathbf{s}_i^P \;\vdots\; \breve{\mathbf{u}}'_j \quad \mathbf{u}'_j(\mathbf{s}_j^P - \mathbf{d})\right]\begin{Bmatrix}\ddot{\mathbf{r}}_i \\ \ddot{\phi}_i \\ \ddot{\mathbf{r}}_j \\ \ddot{\phi}_j\end{Bmatrix} = \dot{\mathbf{u}}'_j(\mathbf{d}\dot{\phi}_j + 2\dot{\breve{\mathbf{d}}}) - \breve{\mathbf{u}}'_j(\dot{\breve{\mathbf{s}}}_j^P\dot{\phi}_j - \dot{\breve{\mathbf{s}}}_i^P\dot{\phi}_i) \tag{7.41}$$

The Jacobian submatrices and the right-hand side of the acceleration constraints for a revolute-translational joint are restated here as

$$^{(r-t,1)}\mathbf{D}_i = \left[-\breve{\mathbf{u}}'_j \quad -\mathbf{u}'_j\mathbf{s}_i^P\right],\; ^{(r-t,1)}\mathbf{D}_j = \left[\breve{\mathbf{u}}'_j \quad \mathbf{u}'_j(\mathbf{s}_j^P - \mathbf{d})\right] \tag{7.42}$$

$$^{(r-t,1)}\gamma = \dot{\mathbf{u}}'_j(\mathbf{d}\dot{\phi}_j + 2\dot{\breve{\mathbf{d}}}) - \breve{\mathbf{u}}'_j(\dot{\breve{\mathbf{s}}}_j^P\dot{\phi}_j - \dot{\breve{\mathbf{s}}}_i^P\dot{\phi}_i) \tag{7.43}$$

Example 7.5

In this example two function M-files to evaluate the Jacobian matrix and the right-hand side of the acceleration constraint for a revolute-translational joint are developed.

```
function [jac_i, jac_j] = BC_jacob_rev_tran(u_j, d, s_P_i, s_P_j)
% Jacobian submatrices for a revolute-translational Joint
     u_j_r = s_rot(u_j);
   jac_i = [-u_j_r' -u_j'*s_P_i];jac_j = [ u_j_r'  u_j'*(s_P_j - d)];
```

```
function gamma = BC_gamma_rev_tran(u_j, u_j_d, d, d_d, s_P_i_d, ...
                                   s_P_j_d, phi_i_d, phi_j_d)
% Gamma array for a revolute-translational joint
     gamma = u_j_d'*(d*phi_j_d + 2*s_rot(d_d)) - ...
            s_rot(u_j)'*s_rot(s_P_j_d*phi_j_d - s_P_i_d*phi_i_d);
```

7.4.5 Simple Constraints

The velocity and acceleration constraints for the simple constraints of Eq. (7.8) have trivial forms:

$$
\begin{aligned}
&{}^{(x,1)}\dot{\Phi} = \dot{x}_i = 0, \quad {}^{(x,1)}\ddot{\Phi} = \ddot{x}_i = 0 \\
&{}^{(y,1)}\dot{\Phi} = \dot{y}_i = 0, \quad {}^{(y,1)}\ddot{\Phi} = \ddot{y}_i = 0 \\
&{}^{(\phi,1)}\dot{\Phi} = \dot{\phi}_i = 0, \quad {}^{(\phi,1)}\ddot{\Phi} = \ddot{\phi}_i = 0
\end{aligned}
\tag{7.44}
$$

It should be obvious that these equations can also be expressed in expanded matrix form. For example, the first velocity constraint can be expressed as

$$
{}^{(x,1)}\dot{\Phi} = [1 \quad 0 \quad 0]\begin{Bmatrix} \dot{x}_i \\ \dot{y}_i \\ \dot{\phi}_i \end{Bmatrix} = 0
$$

These constraints yield Jacobians and right-hand-side acceleration arrays as

$$
\begin{aligned}
&{}^{(x,1)}\mathbf{D}_i = \begin{bmatrix} 1 & 0 & 0 \end{bmatrix} = 0, \quad {}^{(x,1)}\gamma = 0 \\
&{}^{(y,1)}\mathbf{D}_i = \begin{bmatrix} 0 & 1 & 0 \end{bmatrix} = 0, \quad {}^{(y,1)}\gamma = 0 \\
&{}^{(\phi,1)}\mathbf{D}_i = \begin{bmatrix} 0 & 0 & 1 \end{bmatrix} = 0, \quad {}^{(\phi,1)}\gamma = 0
\end{aligned}
\tag{7.45}
$$

7.4.6 System Jacobian

The Jacobian matrix for a multibody system is constructed from the Jacobian submatrices of its joints. In this subsection we show the process in a generic form with an example. Consider the velocity constraint for a revolute joint from Eq. (7.28). We rewrite this equation in the following form by separating the joint Jacobian into two submatrices:

$$
\begin{bmatrix} -\mathbf{I} & -\breve{\mathbf{s}}_i^P \end{bmatrix}\begin{Bmatrix} \dot{\mathbf{r}}_i \\ \dot{\phi}_i \end{Bmatrix} + \begin{bmatrix} \mathbf{I} & \breve{\mathbf{s}}_j^P \end{bmatrix}\begin{Bmatrix} \dot{\mathbf{r}}_j \\ \dot{\phi}_j \end{Bmatrix} = \mathbf{0} \tag{a}
$$

Each submatrix is 2×3 —two rows corresponding to the two algebraic constraints and three columns corresponding to the three velocities (or coordinates) of a body. We further simply *(a)* into a more generic form as

$$
\begin{bmatrix} \times\times \end{bmatrix}\begin{Bmatrix} \dot{\mathbf{r}}_i \\ \dot{\phi}_i \end{Bmatrix} + \begin{bmatrix} \times\times \end{bmatrix}\begin{Bmatrix} \dot{\mathbf{r}}_j \\ \dot{\phi}_j \end{Bmatrix} = \mathbf{0} \tag{b}
$$

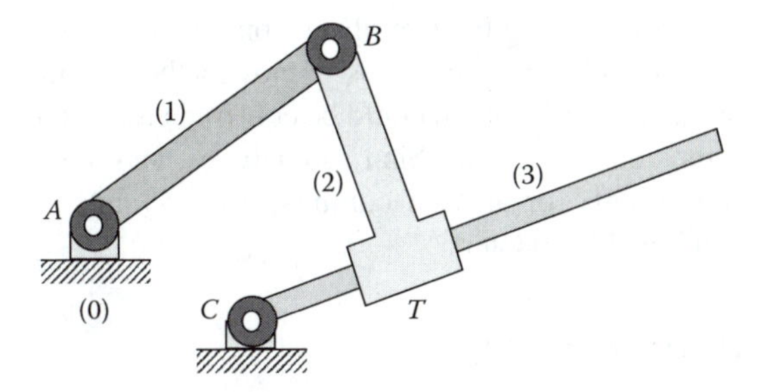

FIGURE 7.17 A generic multibody system.

where ×'s represent possible nonzero entries in a matrix. With this representation we can concentrate on the structure of a Jacobian matrix and not on its detailed contents.

We consider the multibody system from Figure 7.1 which is repeated here in Figure 7.17. This system contains three pin joints and one sliding joint which yield eight algebraic constraints (2 per joints). For the moment let us assume that body (0) is a moving body; therefore, we have twelve coordinates (3 per body) and therefore the system Jacobian will be 8×12. Assume that the joint constraints are assembled in the following order: *A*, *B*, *T* and *C*. A Jacobian matrix for this system can be constructed in generic form as:

	(0)	(1)	(2)	(3)
A	××	××		
B		××	××	
T			××	××
C	××			××

(*c*)

where each generic row represents two actual rows and each generic column represents three actual columns. The blanks represent zero entries. Note how the submatrices of each joint are placed in proper rows and columns. Furthermore note that the sliding joint has the same generic form for its Jacobian as the pin joint. So far, we have considered body (0) as a moving body to make the process of constructing the overall Jacobian more easily understood. Because body (0) is a nonmoving body (i.e., its velocities and accelerations are always zero), we can remove the columns associated with body (0) from the Jacobian. This yields an 8×9 system Jacobian as

	(1)	(2)	(3)
A	××		
B	××	××	
T		××	××
C			××

(*d*)

If the process of constructing the overall Jacobian is well understood, there will be no need to first include the ground as a moving body and then remove its corresponding columns—the Jacobian could be constructed in its final form directly. For convenience, the constraints, Jacobian submatrices, and the right-hand side of the acceleration constraints for all four commonly used joints are summarized in Table 7.4, at the end of this chapter.

7.5 PROGRAMMING NOTE

In the next section we develop several programs to construct all the necessary kinematic entities for several application examples. In Chapter 8 these programs will be extended to include the formulation for the array of forces and the mass matrix. These programs will further be extended in Chapters 9 and 10 from body-coordinates to joint-coordinates. And finally, these programs will be adopted for different types of analyses in Chapters 11 to 14. Changing a program from one formulation to another, revising it from one type of analysis to another type, and then switching between different application examples raise several programming issues. The main issues are: program size; ease of adaptation from one form to another; and flow of data. These were issues that we did not pay any attention to in the programs that we developed in Chapters 5 and 6 because we did not want to introduce the reader to too many ideas all at once. The reader will be directed through problem statements later on to rewrite the programs in Chapters 5 and 6.

When the size of a program becomes too large, debugging the program becomes difficult. Therefore, it is recommended to split a large program, if possible, into several segments, saved in separate files. It is always easier to debug smaller segments individually. Splitting a program into segments, not only makes the debugging process easier, it also makes the program adaptable to other forms. For example, a program that performs kinematic analysis of the double A-arm suspension with the body coordinate formulation, and a program that performs forward dynamic analysis of the same system formulated with the joint coordinates, can share most of their segments. In MATLAB, we can split a program into several smaller programs, or M-files. There are two kinds of M-files: scripts and functions.*

When a program is split into many M-files, organizing the flow of data between M-files deserves special attention. If there are a few pieces of data sent and received from a function, the flow can be through the input and output arguments. When the number of pieces of data (not the size) becomes large, or when the data is used in more than one function, it is not practical or convenient

---

* The two kinds of M-files in MATLAB are:

•Scripts, which do not accept input arguments or return output arguments. A script file can be given any name, as long as the name is not reserved. In the upcoming programs, in each script file we state the name of the file as a comment on the first line. When a program invokes a *script*, MATLAB executes the commands found in that file as if the content of the file had been copied into the location where it is invoked.

•Functions, which can accept input arguments and return output arguments. Internal variables are local to the function. We have already seen examples of function M-files in the previous chapters.

For more information on scripts and functions, refer to MATLAB *help*.

to pass every piece of data as input/output arguments. One possible remedy is the use of *global* statement.

If more than one function needs to share a single copy of a piece of data (e.g., an array of variables or constants), we can simply declare that piece of data as *global* in all the functions. The global declaration must occur before the data is used in a function. More than one name representing a piece of data can appear in a global statement. For example, in

```
global c   c_d   c_dd
```

arrays `c`, `c_d`, and `c_dd`, containing body coordinates, velocities, and accelerations, are declared global. The contents of these arrays can be accessed or changed in any function where this global statement appears.*

In the upcoming programs we will organize data in *cells*** as well as in arrays. As an example, consider three vector of coordinates $\mathbf{r}_1$, $\mathbf{r}_2$, and $\mathbf{r}_3$ that are arranged in an array as

```
r_array = [[0.44; -0.14] [0.68; -0.03] [0.44; 0.04]]
```

The same three vectors can also be arranged into a cell as:

```
r_cell = {[0.44; -0.14] [0.68;-0.03] [0.44; 0.04]}
```

We can access any of the vectors either from `r_array` or from `r_cell`. For example, to recall the contents of $\mathbf{r}_2$ we can use either of the followings:

```
r_array(:,2) =                  r_cell{2} =
      0.68                            0.68
     -0.03                           -0.03
```

It is our choice whether to use arrays or cells. We will use both depending on convenience in programming.

In the programs that we developed in Chapters 5 and 6, we used arrays and functions but we did not implement any scripts files or global statements. In the upcoming programs we will experiment with arrays, cells, functions, scripts, and global statements.

7.5.1 Common Scripts

The suspension examples, the filmstrip advancer, and some other examples that we consider for modeling share many common programming features. To make portions of a program of one model easily transportable to another model, we devise the

---

* There are several other ways to save and communicate data between functions. The reader is encouraged to learn about using *structures* in MATLAB. Structures are arrays with elements accessible by textual field designators.

** One way of grouping data regardless of the type or size is by *cell*. The statement `cell(m,n)` creates an $m \times n$ cell array of empty matrices. The arrays that are grouped in a cell can be of different dimensions. Furthermore, one array may contain numbers and the other may contain characters. For more detail, see `help cell` under the MATLAB prompt.

following bookkeeping strategy. This is only one of many possible ways we may organize our data.

To describe our strategy, we consider as an example the double A-arm suspension model. For this model, as we saw in Section 7.3.1, we need to describe six points on the three moving bodies: Q_1, A_1, A_2, B_2, B_3, and O_3. In addition, point E_1 will be needed for the spring-damper force in the equations of motion in the next chapter. We must define the coordinates of these points on their respective bodies (i.e., $\acute{\mathbf{s}}_i^P$) and then compute $\mathbf{s}_i^P$, $\mathbf{r}_i^P$, $\dot{\mathbf{s}}_i^P$, and $\dot{\mathbf{r}}_i^P$. To perform these computations systematically, and also allow for the number of points to vary from one model to another, we organize the local coordinates of these points into a cell as:

$$\texttt{s\_P\_p} = \{ \acute{\mathbf{s}}_1^Q \quad \acute{\mathbf{s}}_1^A \quad \acute{\mathbf{s}}_2^A \quad \acute{\mathbf{s}}_2^B \quad \acute{\mathbf{s}}_3^B \quad \acute{\mathbf{s}}_3^O \quad \acute{\mathbf{s}}_1^E \}$$

Because these points are located on different bodies, an array `iB` is constructed to contain the body indices for each vector:

```
iB =[1   1   2   2   3   3   1]
```

We also define a variable, `nsP`, representing the dimension of `s_P_p` and `iB`. Similarly, if there are any unit vectors defined on any of the moving bodies, their local components are organized in a cell as well. For example, assume two unit vectors are defined as $\acute{\mathbf{u}}_2$ and $\acute{\mathbf{u}}_3$. Then the corresponding cell and array are constructed as

$$\texttt{u\_p} = \{ \acute{\mathbf{u}}_2 \quad \acute{\mathbf{u}}_3 \}$$

```
iBu =[2   3]
```

The number of such unit vectors is denoted as `nu`. The following function M-file computes $\mathbf{s}_i^P$, $\mathbf{r}_i^P$, and $\mathbf{u}_i$. For all of the points and unit vectors that are attached to moving bodies. In this function we use the `for-end` command for the first time*.

```
function BC_vectors
% Vectors s_P, r_P and u
BC_global
     for j = 1:nsP
          i = iB(j);
               s_P{j} = A{i}*s_P_p{j}; r_P{j} = r{i} + s_P{j};
     end
     for j = 1:nu
          i = iBu(j);
               u{j} = A{i}*u_p{j};
     end
```

* The `for-end` command repeats the statements between `for` and `end`, according to the specified number.

The resultant vectors are saved in cells s_P, r_P, and u; for example,

$$\texttt{s\_P} = \{ \mathbf{s}_1^Q \quad \mathbf{s}_1^A \quad \mathbf{s}_2^A \quad \mathbf{s}_2^B \quad \mathbf{s}_3^B \quad \mathbf{s}_3^O \quad \mathbf{s}_1^E \}$$

$$\texttt{r\_P} = \{ \mathbf{r}_1^Q \quad \mathbf{r}_1^A \quad \mathbf{r}_2^A \quad \mathbf{r}_2^B \quad \mathbf{r}_3^B \quad \mathbf{r}_3^O \quad \mathbf{r}_1^E \}$$

$$\texttt{u} = \{ \mathbf{u}_2 \quad \mathbf{u}_3 \}$$

The data is communicated to and from this function through *global* statements which will be discussed shortly.

The time derivatives of the vectors are computed by the following function M-file:

```
function BC_vectors_d
% Velocity vectors: s_P_d, r_P_d and u_d
    BC_global
    for j = 1:nsP
        i = iB(j);
            s_P_d{j} = s_rot(s_P{j})*phi_d(i);
            r_P_d{j} = r_d{i} + s_P_d{j};
    end
    for j = 1:nu
        i = iBu(j);
            u_d{j} = s_rot(u{j})*phi_d(i);
    end
```

The resultant vectors are saved in; for example,

$$\texttt{s\_P\_d} = \{ \dot{\mathbf{s}}_1^Q \quad \dot{\mathbf{s}}_1^A \quad \dot{\mathbf{s}}_2^A \quad \dot{\mathbf{s}}_2^B \quad \dot{\mathbf{s}}_3^B \quad \dot{\mathbf{s}}_3^O \quad \dot{\mathbf{s}}_1^E \}$$

$$\texttt{r\_P\_d} = \{ \dot{\mathbf{r}}_1^Q \quad \dot{\mathbf{r}}_1^A \quad \dot{\mathbf{r}}_2^A \quad \dot{\mathbf{r}}_2^B \quad \dot{\mathbf{r}}_3^B \quad \dot{\mathbf{r}}_3^O \quad \dot{\mathbf{r}}_1^E \}$$

$$\texttt{u\_d} = \{ \dot{\mathbf{u}}_2 \quad \dot{\mathbf{u}}_3 \}$$

We organize points and vectors that are on the ground in cells similar to the following:

$$\texttt{s\_P0\_p} = \{ \mathbf{s}_0'^O \quad \mathbf{s}_0'^Q \quad \mathbf{s}_0'^F \},\ \texttt{nsP0}$$

$$\texttt{u0\_p} = \{ \mathbf{u}_0' \},\ \texttt{nu0}$$

For such points and vectors $\mathbf{r}_0^P = \mathbf{s}_0^P = \mathbf{s'}_0^P$, $\mathbf{u}_0 = \mathbf{u'}_0$, and their time derivatives are zeros. The following function M-file constructs the necessary vectors.

```
function BC_vectors0
% Vectors s_P_0, r_P_0, s_P_0_d, r_P_0_d, u_0 and u_0_d
      BC_global
      for j = 1:nsP0
            s_P0{j}     = s_P0_p{j}; r_P0{j}      = s_P0{j};
            s_P0_d{j}  = [0; 0];       r_P0_d{j}  = [0; 0];
      end
      for j = 1:nu0
            u0{j}      = u0_p{j}; u0_d{j} = [0; 0];
      end
```

The resultant vectors are saved in cells `s_P0`, `r_P0`, `s_P0_d`, `r_P0_d`, `u0`, and `u0_d`.

The data transfer between these function M-files and other M-files are made through the use of global statements. All the necessary variable names are organized into several global statements in a *script* M-file named `BC_global`:

```
% file name:   BC_global
global s_P_p     s_P     r_P     s_P_d     r_P_d     nsP    iB
global s_P0_p    s_P0    r_P0    s_P0_d    r_P0_d    nsP0
global u_p       u       u_d     nu        iBu
global u0_p      u0      u0_d    nu0
global phi_c     specifics
global r   phi   A   r_d   phi_d   r_dd   phi_dd
global c   c_d   c_dd   Lambda
global m   J   g   M_array   M_diag   M_inv_array   M_inv_diag
global k   L_0   d_c   f_a
global theta   theta_d   theta_dd   n_theta
global Phi   D   gamma   B   D_star   D_cut   gamma_cut   BdThetad
global Phi_driver   D_driver   rhsv_driver   rhsa_driver
global Phi_all   D_all   rhsv_all  rhsa_all
global time
global nb   ndof   nbc   njc   nv   nPhi   nPhi_cut
global alfa  betas
```

We should recognize some of the names because they have already been discussed. Some other unfamiliar names will be used in the programs in this and the upcoming chapters.

Another process that can be performed with one command is to update the rotational transformation matrices, $\mathbf{A}_i$, for all of the moving bodies. This can be performed by the following function M-file.

```
function A = BC_A_matrices(phi, nb)
% A matrices
      for i = 1:nb
            A{i} = A_matrix(phi(i));
      end
```

Another common function that is listed below transfers the translational and rotational coordinate values to a column array, c. The same function can also be used to transfer the velocity values to another column array, c_d, by using the command c_d = BC_r_to_c(r_d, phi_d, nb).

```
function c = BC_r_to_c(r, phi, nb)
% Transfer contents of r and phi to c
    for i = :nb
        j = 3*i - 2;
                    c(j:j+1) = r{i}; c(j+2) = phi(i);
    end
    c = c(:);     % Assure c is a column array
```

The scripts and function M-files that are presented in this section will be implemented in the following programs.

7.6 EXAMPLE PROGRAMS

In this section we consider the same three examples from Chapter 5 for kinematic modeling with the body coordinate method. We will develop MATLAB programs to evaluate the necessary entities for kinematic modeling. These entities will be used for various types of analyses in the upcoming chapters. In these programs we implement some of the automation and bookkeeping strategies that were discussed in Section 7.5.

We will develop a single main script (main program) to evaluate the position constraints and their corresponding Jacobian matrix and the right-hand side of the acceleration constraints. The main script must direct MATLAB to access the proper folder that contains the files for that specific example. The main script is listed as the following:

```
% file name:  BC_formulation
% Body coordinate formulation
    clear all
    addpath Basics  BC_Basics  BC_XX
    BC_global
    BC_condata            % Constant data
    BC_vardata            % Variable data
% Determine components of vectors on the ground
    BC_vectors0;
% Transfer r, phi, r_d, and phi_d to c and c_d
    c   = BC_r_to_c(r, phi, nb);
    c_d = BC_r_to_c(r_d, phi_d, nb);
% Compute A matrices, s_P and r_P vectors
    BC_A_matrices; BC_vectors;
% Evaluate position constraints and Jacobian
    Phi = BC_Phi; D = BC_jacob;
% Evaluate velocity vectors: s_P_d and r_P_d
    BC_vectors_d;
% Evaluate r-h-s array of acceleration constraints
    gamma = BC_gamma;
```

Following a `clear all` statement, an `addpath` statement directs the program to three folders: `Basics`, `BC_Basics` and `BC_XX`. The name of the folder `BC_XX` must be changed from one example to another. For example, `BC_XX` must be changed to `BC_AA` to access the double A-arm M-files. In the folder `BC_AA` (or `BC_MP_A`, etc.), MATLAB must find the following M-files:

```
BC_condata

BC_vardata

BC_Phi

BC_jacob

BC_gamma
```

The contents of these files will be discussed in Sections 7.6.1 to 7.6.3. The other M-files used by the main program are common and they have already been discussed.

7.6.1 Double A-Arm Suspension

The position constraints for the double A-arm suspension system were derived in Section 7.3.1. In this section we develop a program to evaluate the position constraints and their corresponding Jacobian matrix and the right-hand side of the acceleration constraints. For this purpose we use the main script `BC_formulation` and set the path to

```
addpath Basics  BC_Basics  BC_AA
```

Following the `clear all` and the `addpath` statements, the program invokes the script M-file `BC_global`. This is equivalent to copying all of the global statements from that script file and pasting them at the top of the program. Next the program invokes the script `BC_condata`. This file contains all of the constant geometric data for the system. Note that in this file the number of bodies, `nb`, and the number of system's degrees of freedom, `ndof`, are defined. These constant parameters can be used, wherever needed, to determine the length of some arrays, such as the array of coordinates, or the number of constraints in a system.

```
% file name:   BC_condata.m
    nb = 3; ndof = 1;
% Constant geometric data
    s_Q_0_p = [ 0.20; -0.14]; s_Q_1_p = [-0.24;  0.00];
    s_A_1_p = [ 0.18;  0.00]; s_A_2_p = [-0.07; -0.10];
    s_B_2_p = [-0.10;  0.12]; s_B_3_p = [ 0.13;  0.00];
    s_O_3_p = [-0.13;  0.00]; s_O_0_p = [ 0.32;  0.02];
    s_F_0_p = [ 0.38;  0.04]; s_E_1_p = [ 0.00;  0.00];
% Establish cells and arrays
   s_P_p = {s_Q_1_p s_A_1_p s_A_2_p s_B_2_p s_B_3_p s_O_3_p s_E_1_p};
    iB = [1 1 2 2 3 3 1]; nsP = 7;
    s_P0_p = {s_O_0_p  s_Q_0_p  s_F_0_p}; nsP0 = 3;
    u_p = {}; iBu = []; nu = 0;
    u0_p = {}; nu0 = 0;
```

TABLE 7.2

Vector	Cell Index	Vector	Cell Index
$\mathbf{s}_1^{\prime Q}$	1	$\mathbf{s}_0^{\prime O}$	1
$\mathbf{s}_1^{\prime A}$	2	$\mathbf{s}_0^{\prime Q}$	2
$\mathbf{s}_2^{\prime A}$	3	$\mathbf{s}_0^{\prime F}$	3
$\mathbf{s}_2^{\prime B}$	4		
$\mathbf{s}_3^{\prime B}$	5		
$\mathbf{s}_3^{\prime O}$	6		
$\mathbf{s}_1^{\prime E}$	7		

In this file the local coordinates of points Q, A, B, O, and E on the three moving bodies are organized into a cell, `s_P_p`, as it was discussed in Section 7.5.1. Associated with this cell the array `iB` and the variable name `nsP0` are also defined. The coordinates of point E will be used in the next chapter to determine the force of the spring-damper. The coordinates of the points on the ground are arranged in another cell named `s_P0_p` with an associated variable name `nsP0`. The vectors that are arranged in these cells and their corresponding indices (order of appearance in the cell) are listed in Table 7.2 for future reference. The x-y components (coordinates) of these vectors (points) will be computed by the function `BC_vectors`. This function also computes similar entities for any unit vector that may be present in the model. However, because there are no unit vectors for this suspension system, we must define empty cells and arrays as

```
u_p = {},   iBu = [], u0_p = {}
```

We must also define `nu = 0` and `nu0 = 0`.

The next script that the program uses is `BC_vardata`, where the coordinate and velocity values are provided.

```
% file name:  BC_vardata.m
% Body coordinates and velocities
    r = {[0.4400; -0.1408] [0.6825; -0.0364] [0.4470; 0.0480]};
    phi = [6.2800  0.0735  6.5000];
    r_d = {[0.0022; 0.6814] [-0.1191; 1.2657] [-0.1254; 0.5692]};
    phi_d = [2.8392  1.1720  4.4837];
```

Note that the translational coordinates and velocities are arranged in *cells* and the rotational entities are arranged in *arrays*.

Next, the program invokes function BC_vectors0 to compute the coordinates and velocities of points that are defined on the ground. Then the translational and rotational coordinates are transferred to the c array by invoking function BC_r_to_c. The same function then transfers the velocities to the c_d array. The function BC_A_matrices evaluates the rotational transformation matrices and the function BC_vectors computes $\mathbf{s}_i^P$ and $\mathbf{r}_i^P$ coordinates of the points that are defined on the moving bodies.

The next two functions construct the constraints and their Jacobian for the four pin joints. Note how these functions use the constructed cells to access coordinates (components) of different points (vectors).

```
function Phi = BC_Phi
% Position constraints
    BC_global
    Phi_Q = BC_Phi_rev(r_P0{2}, r_P{1});
    Phi_A = BC_Phi_rev(r_P{2} , r_P{3});
    Phi_B = BC_Phi_rev(r_P{4} , r_P{5});
    Phi_O = BC_Phi_rev(r_P{6} , r_P0{1});
        Phi = [Phi_Q; Phi_A; Phi_B; Phi_O];
```

```
function D = BC_jacob
% Jacobian matrix D
    BC_global
    [jac_Q_0, jac_Q_1] = BC_jacob_rev(s_P0{2}, s_P{1});
    [jac_A_1, jac_A_2] = BC_jacob_rev(s_P{2} , s_P{3});
    [jac_B_2, jac_B_3] = BC_jacob_rev(s_P{4} , s_P{5});
    [jac_O_3, jac_O_0] = BC_jacob_rev(s_P{6} , s_P0{1});
    z23    = zeros(2,3);
        D = [ jac_Q_1    z23        z23
              jac_A_1    jac_A_2    z23
              z23        jac_B_2    jac_B_3
              z23        z23        jac_O_3 ];
```

Next, the time derivative of $\mathbf{s}_i^P$ and $\mathbf{r}_i^P$ vectors are computed by the function BC_vectors_d. And the last function computes the right-hand-side array of the acceleration constraints.

```
function gamma = BC_gamma
% r-h-s array of acceleration constraints
    BC_global
    gam_Q = BC_gamma_rev(s_P0_d{2}, 0        , s_P_d{1} , phi_d(1));
    gam_A = BC_gamma_rev(s_P_d{2} , phi_d(1), s_P_d{3} , phi_d(2));
    gam_B = BC_gamma_rev(s_P_d{4} , phi_d(2), s_P_d{5} , phi_d(3));
    gam_O = BC_gamma_rev(s_P_d{6} , phi_d(3), s_P0_d{1}, 0);
        gamma = [gam_Q; gam_A; gam_B; gam_O];
```

Following the execution of the program, we can check for the value of any of the computed arrays and matrices. For example, we can check the values of the position constraints or the right-hand-side array of acceleration constraints by typing in the Command Window Phi or gamma to obtain the following values:

```
>> Phi
   1.0e-03 *
    0.0012
   -0.0355
    0.0333
    0.1030
   -0.0015
    0.0330
   -0.0436
   -0.0344
```

```
>> gamma

   -1.9346
    0.0062
   -1.5368
   -0.1394
    2.7014
    0.4079
    2.5523
    0.5622
```

These constraint values indicate that the coordinate values were provided with reasonable accuracy. The order of elements in these two arrays is the same as the order of constructing the constraints in the M-files. For example, the first two elements in each array correspond to the pin joint Q, the 3rd and 4th elements correspond to the pin joint A, and so on. We can also check the elements of the Jacobian matrix by typing D in the Command Window:

```
>> D
```

1.0000	0	-0.0008	0	0	0	0	0	0
0	1.0000	-0.2400	0	0	0	0	0	0
-1.0000	0	-0.0006	1.0000	0	0.1049	0	0	0
0	-1.0000	-0.1800	0	1.0000	-0.0625	0	0	0
0	0	0	-1.0000	0	0.1123	1.0000	0	-0.0280
0	0	0	0	-1.0000	0.1085	0	1.0000	0.1270
0	0	0	0	0	0	-1.0000	0	-0.0280
0	0	0	0	0	0	0	-1.0000	0.1270

For convenience, to interpret the structure of this matrix, the entries are provided here in a table. Each three columns correspond to a body and each two rows correspond to a joint. Note how the 2×3 submatrices are embedded in the larger 8×9 Jacobian.

We can also check if the given values for the velocities satisfy the constraints. In the Command Window we type:

```
>> Phi_d = D*c_d
```

This computes and reports the product $\mathbf{D}\dot{\mathbf{c}}$ as:

```
Phi_d =
    1.0e-04 *
       0.2951
      -0.0454
      -0.1978
       0.3454
      -0.3552
      -0.5432
       0.1065
       0.3437
```

Based on Eq. (7.18) we conclude that because the reported values are small, the velocities were provided with acceptable accuracy.

7.6.2 MacPherson Suspension

In this section we present three MATLAB programs to construct the necessary kinematic entities for the MacPherson suspension models that were established in Section 7.3.2. The main objective is to understand the differences between the three representations, specifically the size of each model.

Model A: For this model we use the main script `BC_formulation` and set the path to

```
addpath Basics  BC_Basics  BC_MP_A
```

The M-files associated with this model that must be saved in the folder `BC_MP_A` are listed as:

```
% file name: BC_condata.m
    nb = 3; ndof = 1;
% Constant geometric data
    s_Q_0_p = [ 0.12; -0.41]; s_Q_1_p = [-0.225; 0.00];
    s_A_1_p = [ 0.225; 0.00]; s_A_2_p = [ 0.00; -0.07];
    s_O_3_p = [-0.15;  0.00]; s_O_0_p = [ 0.41;  0.13];
    s_B_2_p = [-0.17;  0.25]; phi_c = 1.08;
    u_3_p = [1; 0];
% Establish cells and arrays
    s_P_p = {s_Q_1_p s_A_1_p s_A_2_p s_B_2_p s_O_3_p};
       iB = [1 1 2 2 3]; nsP = 5;
    s_P0_p = {s_O_0_p  s_Q_0_p}; nsP0 = 2;
    u_p = {u_3_p}; iBu = [3]; nu = 1;
    u0_p = {}; nu0 = 0;
```

```
% file name: BC_vardata.m
% Body coordinates and velocities
    r = {[0.345; -0.410] [0.5826; -0.3405] [0.4525; -0.0138]};
    phi = [0   6.08   5.000]; % angles in rad
    r_d = {[0; 0.565] [-0.0386; 1.1364] [0.082; 0.0243]};
    phi_d = [2.511   0.570   0.570];
```

```
function Phi = BC_Phi
% Position constraints
    BC_global
    Phi_Q = BC_Phi_rev(r_P0{2}, r_P{1});
    Phi_A = BC_Phi_rev(r_P{2}, r_P{3});
        d = r_P{3} - r_P{5};
    Phi_T = BC_Phi_tran(u{1}, phi(3), phi(2), d, phi_c);
    Phi_O = BC_Phi_rev(r_P0{1}, r_P{5});
      Phi = [Phi_Q; Phi_A; Phi_T; Phi_O];
```

```
function D = BC_jacob
% Jacobian matrix D
    BC_global
    [jac_Q_0 jac_Q_1] = BC_jacob_rev(s_P0{2}, s_P{1});
    [jac_A_1 jac_A_2] = BC_jacob_rev(s_P{2}, s_P{3});
         d = r_P{4} - r_P{5};
    [jac_T_3 jac_T_2] = BC_jacob_tran(u{1}, d, s_P{5}, s_P{4});
    [jac_O_0 jac_O_3] = BC_jacob_rev(s_P0{1}, s_P{5});
         z23    = zeros(2,3);
    D = [  jac_Q_1      z23          z23
           jac_A_1      jac_A_2      z23
           z23          jac_T_2      jac_T_3
           z23          z23          jac_O_3 ];
```

```
function gamma = BC_MP_A_gamma
% r-h-s array of acceleration constraints
    BC_global
    gam_Q = BC_gamma_rev(s_P0_d{2}, 0, s_P_d{1}, phi_d(1));
    gam_A = BC_gamma_rev(s_P_d{2}, phi_d(1), s_P_d{3}, phi_d(2));
           d = r_P{3} - r_P{5}; d_d = r_P_d{3} - r_P_d{5};
    gam_T = BC_gamma_tran(u_d{1}, r_d{3}, r_d{2}, d, d_d, phi_d(3));
    gam_O = BC_gamma_rev(s_P0_d{1}, 0, s_P_d{5}, phi_d(3));
           gamma = [gam_Q; gam_A; gam_T; gam_O];
```

Note that in the script `BC_condata` we have entries for unit vectors on bodies (2) and (3). The contents of the cells from this file and the corresponding cell indices are listed in Table 7.3.

If we execute the main program, we can obtain the numerical values for any of the computed arrays or matrices. For example, the position constraint values, the right-hand-side array of accelerations, or the Jacobian are reported to be:

```
>> Phi                                  >> gamma
    -0.0000                                 -1.4187
          0                                       0
    -0.0015                                 -1.4232
     0.0009                                 -0.0223
    -0.0002                                       0
    -0.0009                                 -1.2443
    -0.0000                                 -0.0138
     0.0000                                  0.0467
```

```
>> D
    1.000       0       0       0       0       0       0       0       0
        0   1.000  -0.225       0       0       0       0       0       0
   -1.000       0       0   1.000       0   0.069       0       0       0
        0  -1.000  -0.225       0   1.000  -0.014       0       0       0
        0       0       0       0       0   1.000       0       0  -1.000
        0       0       0   0.959   0.284  -0.301  -0.959  -0.284  -0.050
        0       0       0       0       0       0   1.000       0  -0.144
        0       0       0       0       0       0       0   1.000  -0.042
```

TABLE 7.3

Vector	Cell Index	Vector	Cell Index
$\mathbf{s}_1^{\prime Q}$	1	$\mathbf{s}_0^{\prime O}$	1
$\mathbf{s}_1^{\prime A}$	2	$\mathbf{s}_0^{\prime Q}$	2
$\mathbf{s}_2^{\prime A}$	3		
$\mathbf{s}_2^{\prime B}$	4	**Vector**	**Cell Index**
$\mathbf{s}_3^{\prime O}$	5	$\mathbf{u}_3^{\prime}$	1

We can also type `D*c_d` to determine whether the velocity values were provided accurately enough or not.

Model B: For this model we use the main script `BC_formulation` and set the path to

```
addpath Basics   BC_Basics   BC_MP_B
```

The M-files associated with this model that must be saved in the folder `BC_MP_B` are listed as:

```
% file name: BC_condata
    nb = 2; ndof = 1;
% Constant geometric data
    s_Q_0_p = [ 0.12; -0.41]; s_Q_1_p = [-0.225; 0.00];
    s_A_1_p = [ 0.225; 0.00]; s_A_2_p = [ 0.00; -0.07];
    s_O_0_p = [ 0.41;  0.13]; s_B_2_p = [-0.17;  0.25];
    u_2_p = [ 0.47; -0.88];
% Establish cells and arrays
    s_P_p = {s_Q_1_p  s_A_1_p  s_A_2_p  s_B_2_p};
        iB = [1 1 2 2]; nsP = 4;
    s_P0_p = {s_O_0_p  s_Q_0_p}; nsP0 = 2;
    u_p = {u_2_p}; iBu = [2]; nu = 1;
    u0_p = {}; nu0 = 0;
```

```
% file name: BC_vardata
% Body coordinates and velocities
    r = {[0.345; -0.410] [0.5826; -0.3405]}; phi = [0  6.08];
    r_d = {[0; 0.565] [-0.0386; 1.1364]}; phi_d = [2.511  0.570];
```

```
function Phi = BC_Phi
% Position constraints
    BC_global
    Phi_Q = BC_Phi_rev(r_P0{2}, r_P{1});
    Phi_A = BC_Phi_rev(r_P{2}, r_P{3});
        d = r_P{4} - r_P0{1};
    Phi_RT = BC_Phi_rev_tran(u{1}, d, 0);
        Phi = [Phi_Q; Phi_A; Phi_RT];
```

```
function D = BC_jacob
% Jacobian matrix D
    BC_global
    [jac_Q_0, jac_Q_1] = BC_jacob_rev(s_P0{2}, s_P{1});
    [jac_A_1, jac_A_2] = BC_jacob_rev(s_P{2}, s_P{3});
         d = r_P{4} - r_P0{1};
    [jac_RT_0, jac_RT_2] = BC_jacob_rev_tran(u{1}, d, s_P0{1}, s_P{4});
     z23    = zeros(2,3); z13    = zeros(1,3);
         D = [ jac_Q_1        z23
               jac_A_1      jac_A_2
               z13          jac_RT_2 ];
```

```
function gamma = BC_MP_B_gamma
% r-h-s array of acceleration constraints
    BC_global
    gam_Q = BC_gamma_rev(s_P0_d{2}, 0, s_P_d{1}, phi_d(1));
    gam_A = BC_gamma_rev(s_P_d{2}, phi_d(1), s_P_d{3}, phi_d(2));
         d = r_P{4} - r_P0{1}; d_d = r_P_d{4};
    gam_RT = BC_gamma_rev_tran(u{1}, u_d{1}, d, d_d, ...
                              s_P0_d{1}, s_P_d{4}, 0, phi_d(2));
         gamma = [gam_Q; gam_A; gam_RT];
```

Executing the program yields the following arrays and matrices:

```
>> Phi                                  >> gamma
   -0.0000                                 -1.4187
         0                                       0
   -0.0015                                 -1.4232
    0.0009                                 -0.0223
   -0.0000                                 -1.2415
```

```
>> D

   1.000         0         0         0         0         0
       0     1.000    -0.225         0         0         0
  -1.000         0         0     1.000         0     0.069
       0    -1.000    -0.225         0     1.000    -0.014
       0         0         0     0.957     0.283    -0.499
```

Model C: For this model we use the main script `BC_formulation` and set the path to

```
addpath Basics  BC_Basics  BC_MP_C
```

The M-files associated with this model that must be saved in the folder `BC_MP_C` are listed as:

```
% file name: BC_condata
    nb = 1; ndof = 1;
% Constant geometric data
    s_Q_0_p = [ 0.12; -0.41]; s_A_1_p = [ 0.00; -0.07];
    s_O_0_p = [ 0.41;  0.13]; s_B_1_p = [-0.17;  0.25];
    u_1_p = [ 0.47; -0.88]; L_AQ = 0.45;
% Establish cells and arrays
    s_P_p  = {s_A_1_p  s_B_1_p}; iB = [1 1]; nsP = 2;
    s_P0_p = {s_O_0_p  s_Q_0_p}; nsP0 = 2;
    u_p = {u_1_p}; iBu = [1]; nu = 1;
    u0_p = {}; nu0 = 0;
```

```
% file name: BC_vardata
% Body coordinates and velocities
  r   = {[0.5826; -0.3405]}; phi   = [6.08];
  r_d = {[-0.0386; 1.1364]}; phi_d = [0.570];
```

```
function Phi = BC_Phi
% Position constarints
    BC_global
        L_AQ = specifics{1}; d_AQ = r_P{1} - r_P0{2};
    Phi_AQ = BC_Phi_rev_rev(d_AQ, L_AQ);
        d = r_P{2} - r_P0{1};
    Phi_RT = BC_Phi_rev_tran(u{1}, d, 0);
        Phi = [Phi_AQ; Phi_RT];
```

```
function D = BC_jacob
% Jacobian matrix D
    BC_global
        d_AQ = r_P{1} - r_P0{2};
   [jac_AQ_0, jac_AQ_1] = BC_jacob_rev_rev(d_AQ, s_P0{2}, s_P{1});
        d = r_P{2} - r_P0{1};
   [jac_RT_0, jac_RT_1] = BC_jacob_rev_tran(u{1}, d, s_P0{1}, s_P{2});
        D = [ jac_AQ_1
              jac_RT_1 ];
```

```
function gamma = BC_gamma
% r-h-s. array of acceleration constraints
    BC_global
        d_AQ = r_P{1} - r_P0{2}; d_AQ_d = r_P_d{1};
    gam_AQ = BC_gamma_rev_rev(d_AQ, d_AQ_d, s_P0_d{2}, ...
                              s_P_d{1}, 0, phi_d(1));
        d = r_P{2} - r_P0{1}; d_d = r_P_d{2};
    gam_RT = BC_gamma_rev_tran(u{1}, u_d{1}, d, d_d, ...
                              s_P0_d{1}, s_P_d{2}, 0, phi_d(1));
        gamma = [gam_AQ; gam_RT];
```

Executing the program yields the following:

```
>> Phi                          >> gamma
   1.0e-03*
   -0.6848                         -1.2752
   -0.0037                         -1.2415
```

```
>> D
     0.4485      0.0009      0.0307
     0.9567      0.2828     -0.4989
```

At this point we can compare the size of arrays and matrices in the three models. Although the three models contain different number of variables and deal with different size arrays, the models are kinematically identical.

7.6.3 Filmstrip Advancer

The constraints for this mechanism were derived in Section 7.3.3. These constraints are identical to those for the double A-arm suspension therefore we can use the same M-files we developed for the double A-arm suspension with the exception of `BC_condata` and `BC_vardata` files. In the main script `BC_formulation`, we set the path to

```
addpath Basics   BC_Basics   BC_FS
```

The M-files associated with this model must be saved in the folder `BC_FS`. We copy the files `BC_Phi`, `BC_jacob` and `BC_gamma` from `BC_AA` folder into this folder. Then we add the following two M-files to the folder:

```
% file name:  BC_condata
    nb = 3; ndof = 1;
% Constant geometric data
    s_Q_0_p = [ 3.0; -4.5]; s_Q_1_p = [ 0.0;  0.0];
    s_A_1_p = [ 1.5;  0.0]; s_A_2_p = [ 5.0;  0.0];
    s_B_2_p = [ 0.0;  0.0]; s_B_3_p = [ 1.5;  0.0];
    s_O_3_p = [-1.5;  0.0]; s_O_0_p = [ 0.0;  0.0];
    s_C_2_p = [-5.5; -0.5];
% Establish cells and arrays
   s_P_p = {s_Q_1_p s_A_1_p s_A_2_p s_B_2_p s_B_3_p s_O_3_p s_C_2_p};
        iB = [1 1 2 2 3 3 2]; nsP = 7;
    s_P0_p = {s_O_0_p  s_Q_0_p}; nsP0 = 2;
    u_p = {}; iBu = []; nu = 0;
    u0_p = {}; nu0 = 0;
```

```
% file name:  BC_vardata
% Body coordinates and velocities
    r = {[3.0; -4.5] [2.8;  1.0] [1.4; 0.5]};
    phi = [0.5  5.0  0.3];
    r_d = {[0.0; 0.0] [-1.4; 3.9] [-0.7; 2.0]};
    phi_d = [2.8392  -0.14  1.4];
```

Executing the program provides the necessary entities for the kinematics of the system. The program also computes the coordinates and the velocity of point *C*. Typing `r_P{7}` or `r_P_d{7}` in the Command Window yields:

```
>> r_P{7}                  >> r_P_d{7}
        0.7604                    -0.6815
        6.1323                     4.1855
```

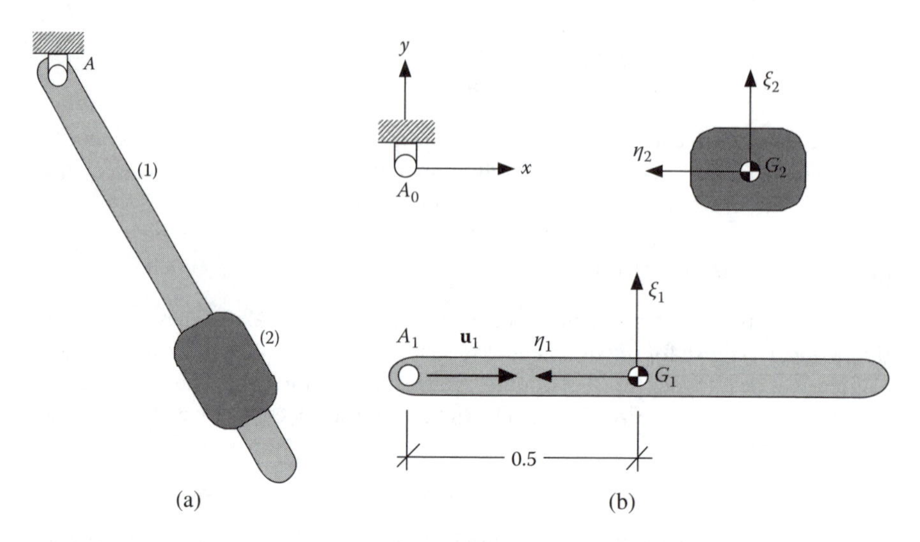

FIGURE 7.18 (a) The variable-length pendulum and (b) the necessary reference frames and vectors for kinematic modeling.

7.6.4 Variable-Length Pendulum

Although this system is much simpler than the previous examples, it is provided here because its body-coordinate formulation will be used as a starting point for the joint-coordinate formulation in Chapter 9. The description and the data for this pendulum can be found in Section 1.5.5.

The pendulum is shown in Figure 7.18(a). After defining the reference frames, as shown in Figure 7.18(b), the necessary vectors for constructing the constraints are established as

$$\mathbf{s}_1'^A = \begin{Bmatrix} 0 \\ 0.5 \end{Bmatrix}, \quad \mathbf{r}_0^A = \begin{Bmatrix} 0 \\ 0 \end{Bmatrix}, \quad \mathbf{u}_1' = \begin{Bmatrix} 0 \\ -1 \end{Bmatrix}, \quad \mathbf{d} = \mathbf{r}_2 - \mathbf{r}_1$$

Vector $\mathbf{d}$ is defined between the origins of the two body frames.

The constraints for the revolute and translational joints are written as

$$\begin{aligned} {}^{(r,2)}\mathbf{\Phi} &= \mathbf{r}_1^A = \mathbf{0} \\ {}^{(t,2)}\mathbf{\Phi} &= \begin{cases} \phi_2 - \phi_1 = 0 \\ \breve{\mathbf{u}}_1' \mathbf{d} = 0 \end{cases} \end{aligned} \tag{7.46}$$

The velocity constraints in matrix form are expressed as

$$\begin{bmatrix} \mathbf{I} & \breve{\mathbf{s}}_1^A & \mathbf{0} & \mathbf{0} \\ \hline \mathbf{0} & -1 & \mathbf{0} & 1 \\ -\breve{\mathbf{u}}_1' & -\mathbf{u}_1'\mathbf{d} & \breve{\mathbf{u}}_1' & 0 \end{bmatrix} \begin{Bmatrix} \dot{\mathbf{r}}_1 \\ \dot{\phi}_1 \\ \dot{\mathbf{r}}_2 \\ \dot{\phi}_2 \end{Bmatrix} = \mathbf{0} \tag{7.47}$$

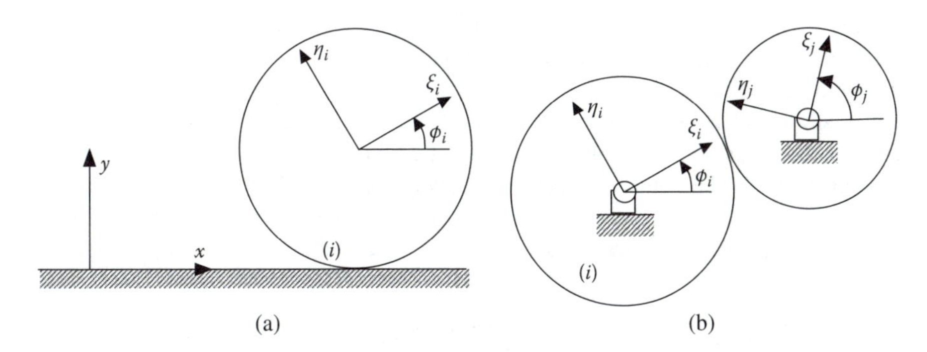

FIGURE 7.19 (a) A circular disc rolling on a flat horizontal surface; (b) two circular discs rolling against one another.

where the 4×6 coefficient matrix is the Jacobian. The acceleration constraints are expressed as

$$\begin{bmatrix} \mathbf{I} & \breve{\mathbf{s}}_1^A & \mathbf{0} & \mathbf{0} \\ \hline \mathbf{0} & -1 & \mathbf{0} & 1 \\ -\breve{\mathbf{u}}_1' & -\mathbf{u}_1'\mathbf{d} & \breve{\mathbf{u}}_1' & 0 \end{bmatrix} \begin{Bmatrix} \ddot{\mathbf{r}}_1 \\ \ddot{\phi}_1 \\ \hline \ddot{\mathbf{r}}_2 \\ \ddot{\phi}_2 \end{Bmatrix} + \begin{bmatrix} \mathbf{0} & \dot{\breve{\mathbf{s}}}_1^A & \mathbf{0} & \mathbf{0} \\ \hline \mathbf{0} & 0 & \mathbf{0} & 0 \\ -\dot{\breve{\mathbf{u}}}_1' & -\dot{\mathbf{u}}_1'\mathbf{d} - \mathbf{u}_1'\dot{\mathbf{d}} & \dot{\breve{\mathbf{u}}}_1' & 0 \end{bmatrix} \begin{Bmatrix} \dot{\mathbf{r}}_1 \\ \dot{\phi}_1 \\ \hline \dot{\mathbf{r}}_2 \\ \dot{\phi}_2 \end{Bmatrix} = 0 \quad (7.48)$$

A MATLAB program for this example will be developed in Chapter 9 when the formulation is extended from the body-coordinates to the joint-coordinates.

7.7 OTHER TYPES OF JOINTS

In addition to the kinematic joints of Section 7.2, body-coordinate formulation is suitable to describe the kinematics of some other types of joints or constraints. One example is a rigid circular disc that can roll on a flat surface as shown in Figure 7.19(a). This could represent a rigid wheel rolling on the ground. If we assume that the disc stays in contact with the ground* and there is no slipping, the rotational and translational velocity of the disc must be related as $\rho_i \dot{\phi}_i = -\dot{x}_i$, where ρ_i is the radius of the disc. Note that when the disc rolls counterclockwise (positive $\dot{\phi}_i$), the disc rolls to the left (negative $\dot{x}_i$). The integral of this relationship provides the position constraint for the *no-slip* condition as

$$\boldsymbol{\Phi}^{(n-s-1,1)} = \rho_i(\phi_i - {}^0\phi_i) + (x_i - {}^0x_i) = 0 \quad (7.49)$$

* The disc can stay in contact with the other body (e.g., the ground), either under its own weight (if it is a dynamic analysis) or through some other kinematic constraints. For example, we may set $y_i - \rho_i = 0$.

where ${}^0x_i$ and ${}^0\phi_i$ are the initial values of x and ϕ coordinates of the disc. The corresponding Jacobian matrix and the right-hand side of the acceleration constraint can be obtained easily.

A similar constraint can be derived for no-slip condition between two rigid circular discs. As shown in Figure 7.19(b), the two discs are pinned to the ground at their centers. The contact points of the two discs must have the same velocity, therefore, $\rho_i\dot{\phi}_i = -\rho_j\dot{\phi}_j$, where ρ_i and ρ_j are the radii. The integral of this relationship provides the position constraint as

$$\Phi^{(n-s-2,1)} = \rho_i\left(\phi_i - {}^0\phi_i\right) + \rho_j\left(\phi_j - {}^0\phi_j\right) = 0 \tag{7.50}$$

where ${}^0\phi_i$ and ${}^0\phi_j$ are the initial values.

The no-slip constraints of Eqs. (7.49) and (7.50) can be revised to fit other applications. For example, the flat horizontal surface of the first model could be replaced by another shape surface. This may represent a wheel going over a bump, for example. We should also note that the body-coordinate formulation is very suitable for these types of contact models, whereas formulating the same conditions with the point-coordinates would be very cumbersome.

7.8 REMARKS

The body-coordinate formulation is a powerful method to develop computer programs for general-purpose use. In this chapter we only discussed how to formulate and construct the necessary entities that will be needed for kinematic analysis of multibody systems. Although in this chapter we did not discuss anything about any types of analyses or what to do with the constructed entities. The methodology and formulations of this chapter will be extended to the dynamics in the next chapter. At this point the reader may move to the next chapter or move to Chapter 11 and learn how to perform kinematic analysis with the body-coordinate formulation.

7.9 PROBLEMS

7.1 Point A is defined on body (1) and point B is defined on body (2). The following data are provided: $\acute{\mathbf{s}}_1^A = \{0.8 \quad -0.5\}'$, $\acute{\mathbf{s}}_2^B = \{-0.3 \quad 1.2\}'$, $\mathbf{r}_1 = \{2.0 \quad 1.5\}'$, $\phi_1 = 30°$, $\mathbf{r}_2 = \{4.5 \quad 2.3\}'$, $\phi_2 = 135°$.
  (a) Determine the components of a vector $\mathbf{d}$ that connects point A to point B.
  (b) Determine the components of a unit vector along $\mathbf{d}$.

7.2 Vectors $\mathbf{a}$ and $\mathbf{b}$ are attached to bodies (1) and (2) respectively. The following data are provided: $\acute{\mathbf{a}}_1 = \{0.1 \quad -0.4\}'$, $\acute{\mathbf{b}}_2 = \{-0.5 \quad 0.2\}'$, $\mathbf{r}_1 = \{2.0 \quad 1.5\}'$, $\phi_1 = 30°$, $\mathbf{r}_2 = \{4.5 \quad 2.3\}'$, $\phi_2 = 135°$. These two vectors are supposed to be parallel. Determine the amount of constraint violation.

TABLE 7.4
Constraints, Jacobians, and Right-hand Side of the Acceleration Constraints for Four Commonly Used Joints

Revolute (Pin) joint	Translational (Sliding) joint	Revolute-Revolute joint	Revolute-Translational joint
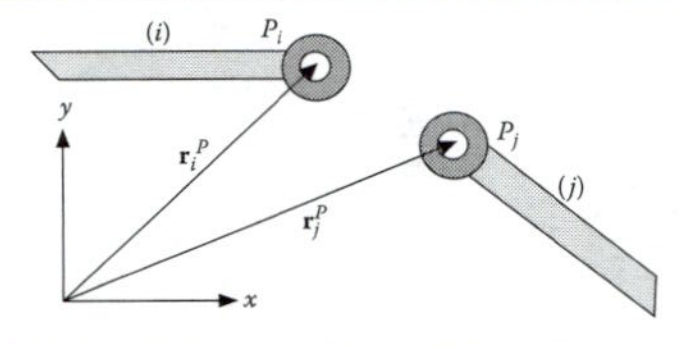	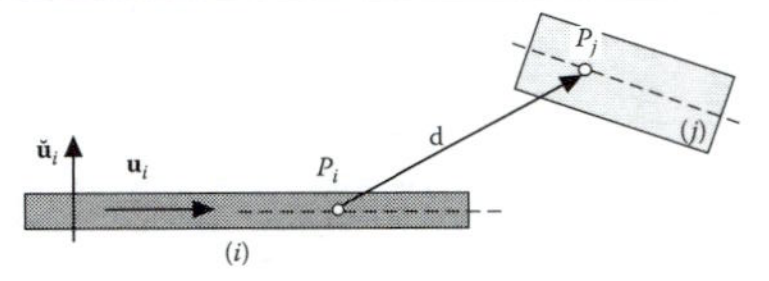	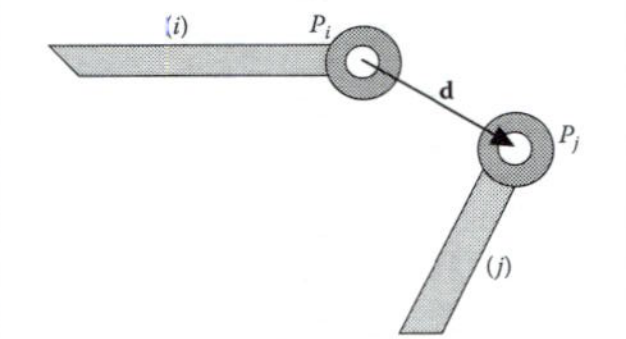	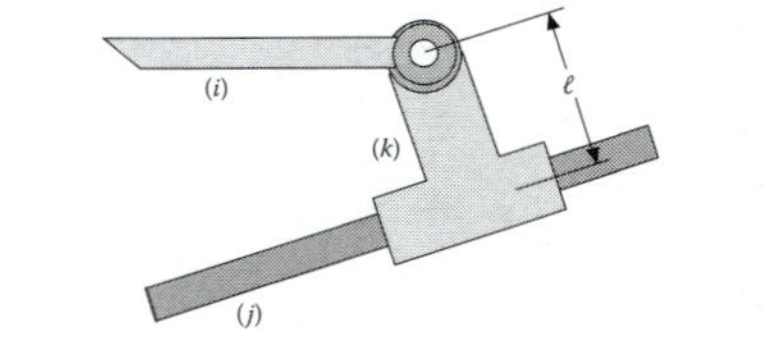
$^{(r,2)}\mathbf{\Phi} = \mathbf{r}_j^P - \mathbf{r}_i^P = \mathbf{0}$	$^{(t,2)}\mathbf{\Phi} = \begin{Bmatrix} \phi_j - \phi_i - {}^c\phi \\ \breve{\mathbf{u}}_i' \, \mathbf{d} \end{Bmatrix} = \begin{Bmatrix} 0 \\ 0 \end{Bmatrix}$	$^{(r-r,1)}\mathbf{\Phi} = \frac{1}{2}(\mathbf{d}'\mathbf{d} - \ell^2) = 0$	$^{(r-t,1)}\mathbf{\Phi} = \breve{\mathbf{u}}_j' \, \mathbf{d} - \ell = 0$
$^{(r,2)}\mathbf{D}_i = \begin{bmatrix} -\mathbf{I} & -\breve{\mathbf{s}}_i^P \end{bmatrix}$ $^{(r,2)}\mathbf{D}_j = \begin{bmatrix} \mathbf{I} & \breve{\mathbf{s}}_j^P \end{bmatrix}$	$^{(t,2)}\mathbf{D}_i = \begin{bmatrix} \mathbf{0} & -1 \\ -\breve{\mathbf{u}}_i' & -\mathbf{u}_i'(\mathbf{d} + \mathbf{s}_i^P) \end{bmatrix}$ $^{(t,2)}\mathbf{D}_j = \begin{bmatrix} \mathbf{0} & 1 \\ \breve{\mathbf{u}}_i' & \mathbf{u}_i'\mathbf{s}_j^P \end{bmatrix}$	$^{(r-r,1)}\mathbf{D}_i = \begin{bmatrix} -\mathbf{d}' & -\mathbf{d}'\breve{\mathbf{s}}_i^P \end{bmatrix}$ $^{(r-r,1)}\mathbf{D}_j = \begin{bmatrix} \mathbf{d}' & \mathbf{d}'\breve{\mathbf{s}}_j^P \end{bmatrix}$	$^{(r-t,1)}\mathbf{D}_i = \begin{bmatrix} -\breve{\mathbf{u}}_j' & -\mathbf{u}_j'\mathbf{s}_i^P \end{bmatrix}$ $^{(r-t,1)}\mathbf{D}_j = \begin{bmatrix} \breve{\mathbf{u}}_j' & \mathbf{u}_j'(\mathbf{s}_j^P - \mathbf{d}) \end{bmatrix}$
$^{(r,2)}\gamma = \dot{\breve{\mathbf{s}}}_i^P \dot{\phi}_i - \dot{\breve{\mathbf{s}}}_j^P \dot{\phi}_j$	$^{(t,2)}\gamma = \begin{Bmatrix} 0 \\ \dot{\mathbf{u}}_i'(\dot{\breve{\mathbf{d}}} - \dot{\breve{\mathbf{r}}}_i + \dot{\breve{\mathbf{r}}}_j + \mathbf{d}\dot{\phi}_i) \end{Bmatrix}$	$^{(r-r,1)}\gamma = -\dot{\mathbf{d}}'\dot{\mathbf{d}} - \mathbf{d}'(\dot{\breve{\mathbf{s}}}_j^P \dot{\phi}_j - \dot{\breve{\mathbf{s}}}_i^P \dot{\phi}_i)$	$^{(r-t,1)}\gamma = \dot{\mathbf{u}}_j'(\mathbf{d}\dot{\phi}_j + 2\dot{\breve{\mathbf{d}}}) - \breve{\mathbf{u}}_j'(\dot{\breve{\mathbf{s}}}_j^P \dot{\phi}_j - \dot{\breve{\mathbf{s}}}_i^P \dot{\phi}_i)$

7.3 Expand the following constraints and express them in terms of the coordinates of the two bodies:
(a) Translational joint; Eq. (7.4)
(b) Revolute-revolute joint; Eq. (7.5)
(c) Revolute-translational joint; Eq. (7.6)

7.4 For the translational joint, in addition to vectors $\mathbf{u}_i$ and $\mathbf{d}$, we can define vector $\mathbf{u}_j$ on body j. Analytically show that the first constraint in Eq. (7.4) (i.e., $\phi_j - \phi_i - {}^c\phi = 0$) is equivalent to enforcing $\mathbf{u}_i$ and $\mathbf{u}_j$ to be parallel; that is, $\breve{\mathbf{u}}_i' \mathbf{u}_j = 0$.

7.5 Simplify the constraints, the Jacobian, and the right-hand side of the acceleration constraints for the following joints between body i and the ground:
(a) Revolute joint
(b) Translational joint
(c) Revolute-revolute joint
(d) Revolute-translational joint

7.6 Construct the Jacobian matrix for the following constraints by taking the partial derivative of the constraints with respect to the coordinates. You may expand the constraints in terms of the coordinates before taking the partial derivatives.
(a) Revolute joint
(b) Translational joint
(c) Revolute-revolute joint
(d) Revolute-translational joint

7.7 Point P_i is defined on body i. Derive the following constraints:
(a) Keep x_i^P to be a constant c_1.
(b) Keep y_i^P to be a constant c_2.

7.8 For the constraints in Problem 7.7 derive the Jacobian and the right-hand side of the acceleration constraints.

7.9 Two axes on bodies i and j must be kept perpendicular to each other. The axis on body i is defined by a unit vector with local components $\mathbf{u}'_i = \{0.6 \quad -0.8\}'$. The axis on body j passes through two points, B and C, with local coordinates $\mathbf{s}_j'^B = \{0.5 \quad 0.4\}'$ and $\mathbf{s}_j'^C = \{1.3 \quad -0.2\}'$.
(a) Write the constraint equation(s) in term of the coordinates of the bodies.
(b) Write the velocity equation(s).
(c) Determine the entries of the Jacobian matrix.

7.10 Vectors $\mathbf{s}_1$ and $\mathbf{s}_2$, attached to bodies (1) and (2) respectively, have local components $\mathbf{s}'_1 = \{1.2 \quad -0.5\}'$ and $\mathbf{s}'_2 = \{-0.3 \quad 0.8\}'$. The array of coordinates is defined as $\mathbf{c} = \{x_1 \quad y_1 \quad \phi_1 \quad x_2 \quad y_2 \quad \phi_2\}'$.
(a) If $\phi_1 = 30°$ and $\phi_2 = 45°$, evaluate the entries of the Jacobian for the constraint $\Phi = \mathbf{s}_1' \mathbf{s}_2$.
(b) If $x_1 = 6.2$, $y_1 = 1$, $\phi_1 = 30°$, $x_2 = -1.9$, $y_2 = 2.3$, and $\phi_2 = 45°$, evaluate the entries of the Jacobian for the constraint $\Phi = \breve{\mathbf{s}}_1' \mathbf{d}$ where $\mathbf{d} = \{x_2 - x_1 \quad y_2 - y_1\}'$.

7.11 Revolute joints A and B in a four-bar mechanism are defined by the following body-fixed coordinates: $\mathbf{s}_1'^A = \{0 \quad 1.5\}'$, $\mathbf{s}_2'^A = \{-2.2 \quad 0\}'$,

$\mathbf{s}_2'^B = \{2.2 \quad 0\}'$, and $\mathbf{s}_3'^B = \{2 \quad 0\}'$. In a given configuration the coordinates of the bodies are $\mathbf{c}_1 = \{-2 \quad 2.5 \quad -12°\}'$, $\mathbf{c}_2 = \{0.5 \quad 3.6 \quad -8°\}'$, and $\mathbf{c}_3 = \{1.6 \quad 1.7 \quad 56°\}'$. Are the constraint equations for these two joints violated or not?

7.12 Two bodies, *i* and *j*, can translate and rotate in the plane. Describe the necessary constraints to keep point *P* on body *i* in contact with the flat surface of body *j* (shown as a line). Assume that this surface (line) extends without limit from either end.

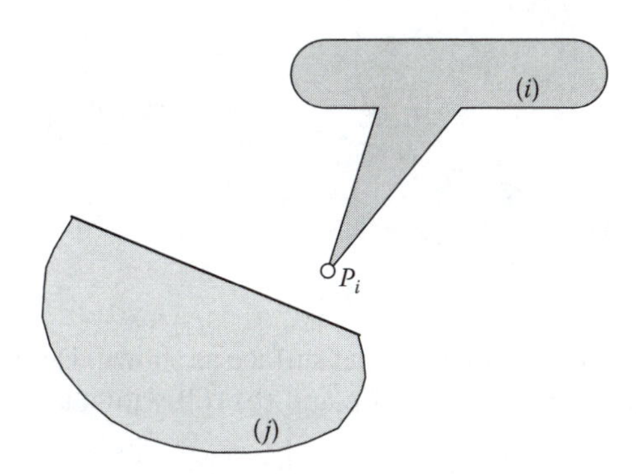

7.13 Refer to the multibody systems shown in the figure of Problem 3.6. If body-coordinates are used to model these systems, for each system determine:

(a) Number of bodies and total number of coordinates
(b) Type of joints and total number of constraints
(c) Based on the number of coordinates and constraints, determine the number of DoF

7.14 The mechanism shown has to be modeled by the body-coordinate formulation. Determine the minimum number of bodies, coordinates, and constrain equations that are needed if:

(a) Only revolute and translational joint constraints are available
(b) Revolute, translational, revolute-revolute, and revolute-translational joints are all available for formulation

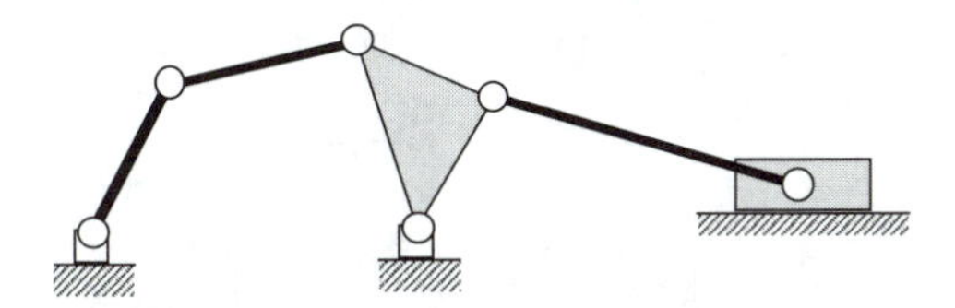

7.15 Because two bodies that are connected by a translational joint can slide relative to one another, it is possible for vector **d** (refer to Figure 7.5) to become a zero vector. Does this cause any numerical difficulty in kinematic analysis?

7.16 For the rolling discs constraint of Eq. (7.50), derive the velocity and acceleration constraints. Determine the Jacobian submatrix and the right-hand-side array of the acceleration constraint.

7.17 Revise the rolling disc constraint of Eq. (7.49) for the case where both bodies are allowed to move. Derive the Jacobian and the right-hand-side array of the acceleration constraint.

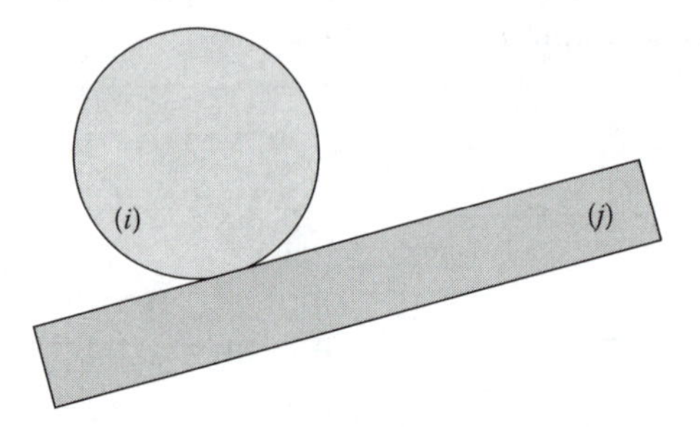

7.18 A disc can roll on an inclined flat surface as shown. Derive the necessary constraints for (a) roll with slip and (b) roll without slip.

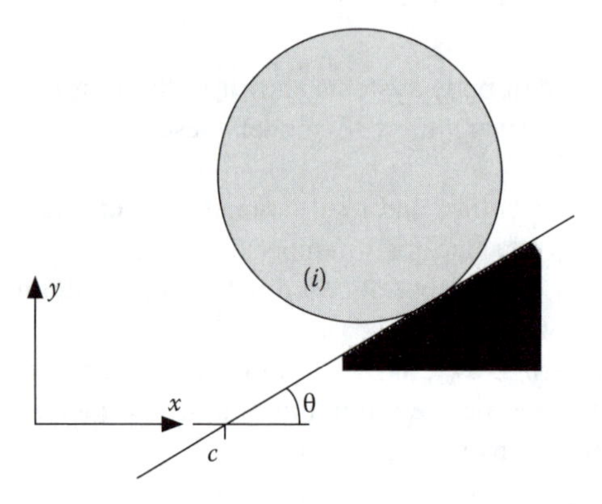

7.19 Derive the constraints for two rolling discs (no-slip) that are pinned to a third moving body. Derive the constraints for the velocities first then integrate to obtain the position constraints. Determine the Jacobian matrix and the right-hand side of the acceleration constraint.

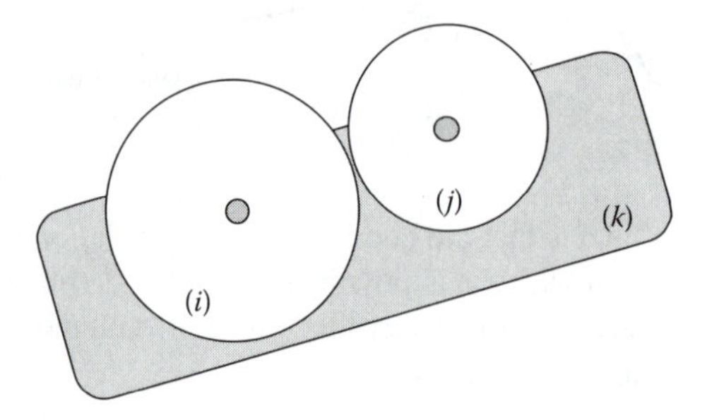

7.20 Derive the constraint equation for a belt connecting two discs that are attached to a third body by pin joints. Determine the Jacobian matrix and the right-hand side of the acceleration constraint.

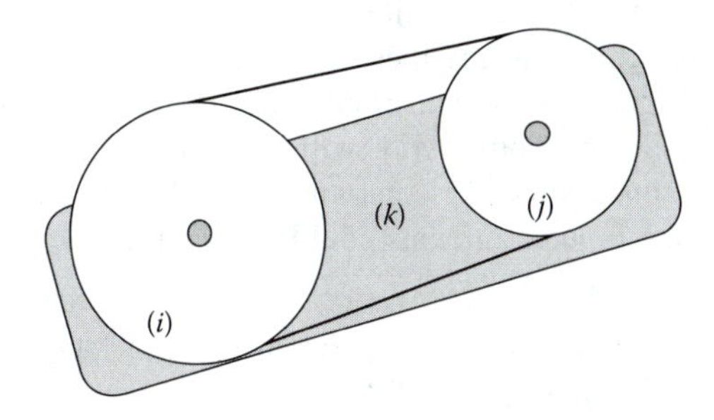

7.21 Derive the constraint and the Jacobian for a point-follower joint. Assume that the slot is described on body j by a function $\eta = f(\xi)$.

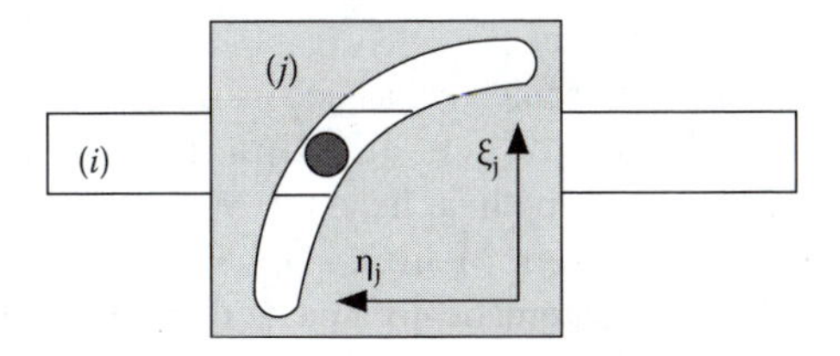

7.22 Derive velocity constraint equations to keep the translational speed of a body constant (equal to s) along a known direction denoted by a unit vector **u**. Consider two cases:
(a) Vector **u** is fixed to the ground.
(b) Vector **u** is fixed to the body.
Are the constraints holonomic or nonholonomic?

7.23 What is the size of the Jacobian matrix for a four-bar linkage modeled by three moving bodies and four revolute joints? Show the location of the nonzero entries in this matrix. What percentage of the elements of the matrix is nonzero?

7.24 What is the size of the Jacobian matrix for a slider-crank mechanism in each of the following cases? The mechanism is modeled by:
(a) Three moving bodies, three revolute joints, and one translational joint
(b) Two moving bodies, one revolute joint, one revolute-revolute joint, and one translational joint
(c) Two moving bodies, two revolute joints, and one revolute-translational joint
For each case show the location of the nonzero entries in the corresponding Jacobian. What percentage of the elements of the matrix is nonzero?

7.25 Develop three function M-files to evaluate the constraints, the Jacobian matrix, and the right-hand side of the acceleration constraints for the rigid joint. Name these functions BC_Phi_rigid, and so forth.

7.26 Develop function M-files to evaluate the constraints, the Jacobian matrix, and the right-hand side of the acceleration constraints for the simple constraints. Name these functions BC_Phi_simple, and so forth. The three simple constraints on x, y, and ϕ can be taken care of in one M-file. We could work with three flags as input arguments to direct the function to any combination of the three constraints. As an example, the input arguments could be arranged as:

```
function Phi = BC_Phi_simple( . . ., i_x, i_y, i_phi)
```

Each flag could be set to either "1" or "0" to enable or disable any of the three constraints.

7.27 Develop three function M-files to evaluate the constraints, the Jacobian matrix, and the right-hand side of the acceleration constraints for a circular disc rolling on the ground without slipping. Use a simple constraint in the y-direction and the no-slip condition of Eq. (7.49). This constraint can be used later on to approximate the wheel-ground interaction in a vehicle model.

7.28 In Section 7.5.1 we presented cells such as s_P_p and u_p to store $\mathbf{s}_i'^P$ and $\mathbf{u}_i'$ vectors. Introduce additional cells in the program BC_formulation to include $\mathbf{d}$ and $\dot{\mathbf{d}}$ vectors in cells. Then develop the necessary function M-files, similar to BC_vectors and BC_vectors_d to automatically update $\mathbf{d}$ and $\dot{\mathbf{d}}$ vectors.

7.29 In the script BC_formulation we must direct the program to different folders by manually changing the path BC_xx to, for example, BC_AA or BC_MP_A. Revise the program to prompt the user to enter the name of the desired folder from the keyboard. Then determine how to set the path dynamically in the program.

7.30 In the function M-files associated with Jacobian matrices (such as the BC_jacob in BC_AA or BC_MP_A folders), we construct small zero matrices and insert them within a larger matrix. Because these functions will be invoked numerous times in a single analysis (Chapters 11 to 14), we would like to make the insertion process computationally more efficient. Develop a script M-file to construct typical small zero matrices such as

```
z12 = zeros(1,2); z21 = z12';
z13 = zeros(1,3); z23 = zeros(2,3);
```

and so forth. Invoke this script file in BC_formulation following BC_condata and BC_vardata. In the file BC_global, insert another global statement as

```
global  z12 z21 z13 z23
```

Remove all the zero initiation of any such matrices from the BC_jacob M-files. All of the small zero arrays are now available in those M-files.

7.31 Revise the quarter-car model with double A-arm suspension from Sections 7.3.1 and 7.6.1 and allow the chassis to move along the y-axis. We may assign index (4) to the chassis and keep the indices of other bodies as in the original model, or renumber all the indices. You can put a sliding joint between the chassis and the ground or implement two simple constraints eliminating the motion of the chassis for x and ϕ. Construct all the necessary M-files.

7.32 Repeat Problem 7.31 for the quarter-car model with MacPherson suspension from Sections 7.3.2, and 7.6.2 (Model A, B or C).

8 Body-Coordinates: Dynamics

In this chapter we extend the kinematic modeling and formula derivation of previous chapters to dynamics. We learn how to construct the mass matrix, the array of applied forces, and the array of reaction forces for mechanical systems that are represented by a collection of bodies. We first consider the dynamics of systems of bodies when only force elements are present. Then, we generalize that to the system of bodies that are interconnected by force elements and kinematic joints.

Programs: In this chapter we revise some of the programs from Chapter 7 and develop several new programs. The file organization is the same as that shown in Table 7.1.

8.1 SYSTEM OF UNCONSTRAINED BODIES

A system of bodies interconnected by force elements, without the presence of kinematic joints or constraints, is referred to as a system of *unconstrained bodies*. The bodies can be free or interconnected by force elements such springs and dampers, and acted upon by gravitational force. The main objective is to develop the equations of motion for the system, and then to solve them for the acceleration of the bodies. In the process that leads to the construction of the equations of motion, it will be assumed that the coordinates and velocities of the bodies are known. Because the number of coordinates is equal to the number of DoF for any unconstrained system, the coordinates and the velocities can be assigned any arbitrary or desired values.

8.1.1 A Two-Body System

A system of two unconstrained bodies is presented in this example. As shown in Figure 8.1(a), a spring connects two bodies at A and C and a damper connects body (1) at B to the ground. The gravity also acts on the system in the negative y-direction. The following constants are provided:

$$m_1 = 0.2,\ J_1 = 0.03,\ m_2 = 0.15,\ J_2 = 0.02,\ k = 50,\ {}^0\ell = 0.2,\ d_c = 20$$

$$\mathbf{s}_1^{\prime A} = \begin{Bmatrix} 0.15 \\ 0 \end{Bmatrix},\ \mathbf{s}_1^{\prime B} = \begin{Bmatrix} -0.15 \\ 0 \end{Bmatrix},\ \mathbf{s}_2^{\prime C} = \begin{Bmatrix} 0 \\ 0.12 \end{Bmatrix},\ \mathbf{r}_0^Q = \begin{Bmatrix} 0 \\ 0 \end{Bmatrix}$$

For the instant shown, the following coordinates and velocities are given for the bodies:

$$\mathbf{r}_1 = \begin{Bmatrix} 0.02 \\ 0.2 \end{Bmatrix},\ \phi_1 = 45°,\ \dot{\mathbf{r}}_1 = \begin{Bmatrix} -0.05 \\ 0.1 \end{Bmatrix},\ \dot{\phi}_1 = -0.3 \text{ rad/sec},$$

$$\mathbf{r}_2 = \begin{Bmatrix} 0.22 \\ 0.1 \end{Bmatrix},\ \phi_2 = 15°,\ \dot{\mathbf{r}}_2 = \begin{Bmatrix} -0.09 \\ 0.13 \end{Bmatrix},\ \dot{\phi}_2 = 0.07 \text{ rad/sec}$$

To construct the equations of motion, all of the necessary arrays are computed by MATLAB and only the results are presented here*. After computing the transformation matrices $\mathbf{A}_1$ and $\mathbf{A}_2$, the **s** vectors are computed (Eq. [3.8]) to be:

$$\mathbf{s}_1^A = \begin{Bmatrix} 0.106 \\ 0.106 \end{Bmatrix},\ \mathbf{s}_1^B = \begin{Bmatrix} -0.106 \\ -0.106 \end{Bmatrix},\ \mathbf{s}_2^C = \begin{Bmatrix} -0.031 \\ 0.116 \end{Bmatrix}$$

The coordinates and velocities of points A, B, and C are determined (Eqs. [3.9] and [3.13]) to be:

$$\mathbf{r}_1^A = \begin{Bmatrix} 0.126 \\ 0.306 \end{Bmatrix},\ \mathbf{r}_1^B = \begin{Bmatrix} -0.086 \\ 0.094 \end{Bmatrix},\ \mathbf{r}_2^C = \begin{Bmatrix} 0.189 \\ 0.216 \end{Bmatrix}$$

$$\dot{\mathbf{r}}_1^A = \begin{Bmatrix} -0.018 \\ 0.068 \end{Bmatrix},\ \dot{\mathbf{r}}_1^B = \begin{Bmatrix} -0.082 \\ 0.132 \end{Bmatrix},\ \dot{\mathbf{r}}_2^C = \begin{Bmatrix} -0.098 \\ 0.128 \end{Bmatrix}$$

Having the coordinates of A and C, the pair of forces that the spring element applies on the two bodies is determined (Eqs. [4.22] to [4.25]) to be

$$^{(s)}\mathbf{f}_1^A = \begin{Bmatrix} -2.577 \\ 3.695 \end{Bmatrix},\ ^{(s)}\mathbf{f}_2^C = \begin{Bmatrix} 2.577 \\ -3.695 \end{Bmatrix}$$

For the damper between points B and Q, the pair of forces are computed to be (Eqs. [4.28] to [4.30])

$$^{(d)}\mathbf{f}_1^B = \begin{Bmatrix} 2.060 \\ -2.248 \end{Bmatrix},\ ^{(d)}\mathbf{f}_0^Q = \begin{Bmatrix} -2.060 \\ 2.248 \end{Bmatrix}$$

The force that acts at Q can be ignored because this is a nonmoving point. The weights of the bodies are determined ($g = 9.81$) as $w_1 = 1.962$ and $w_2 = 1.472$.

The spring, damper, and gravitational forces acting at their respective points on the bodies are shown in Figure 8.1(b). The spring and damper forces each incurs a moment on its corresponding body about the origin. The moments associated

* This is left as an exercise for the reader to develop a program and verify the reported results.

with ${}^{(s)}\mathbf{f}_1^A$ and ${}^{(s)}\mathbf{f}_2^C$ are computed (Eq. [4.11]) to be ${}^{(s)}n_1 = 0.665$ and ${}^{(s)}n_2 = -0.184$. The moment associated with ${}^{(d)}\mathbf{f}_1^B$ is determined to be ${}^{(d)}n_1 = 0.457$.

The array of forces for the system is constructed as

$$\mathbf{h} = \begin{Bmatrix} -2.577 + 2.060 \\ 3.695 - 2.248 - 1.962 \\ 0.665 + 0.457 \\ \hdashline 2.577 \\ -3.695 - 1.472 \\ -0.184 \end{Bmatrix} = \begin{Bmatrix} -0.517 \\ -0.516 \\ 1.122 \\ \hdashline 2.577 \\ -5.166 \\ -0.184 \end{Bmatrix}$$

Because the mass and the moment of inertia for both bodies are known, the equations of motion for the system are constructed as

$$\left[\begin{array}{ccc:ccc} 0.2 & 0 & 0 & 0 & 0 & 0 \\ 0 & 0.2 & 0 & 0 & 0 & 0 \\ 0 & 0 & 0.03 & 0 & 0 & 0 \\ \hdashline 0 & 0 & 0 & 0.15 & 0 & 0 \\ 0 & 0 & 0 & 0 & 0.15 & 0 \\ 0 & 0 & 0 & 0 & 0 & 0.02 \end{array}\right] \begin{Bmatrix} \ddot{x}_1 \\ \ddot{y}_1 \\ \ddot{\phi}_1 \\ \hdashline \ddot{x}_2 \\ \ddot{y}_2 \\ \ddot{\phi}_2 \end{Bmatrix} = \begin{Bmatrix} -0.517 \\ -0.516 \\ 1.122 \\ \hdashline 2.577 \\ -5.166 \\ -0.184 \end{Bmatrix}$$

Figure 8.1(c) shows the forces and moments that appear in the equations of motion. These forces and moments are equivalent to those shown in Figure 8.1(b). When a force is moved from its point of application to the mass center, while preserving its direction, the moment associated with that force must also be shown

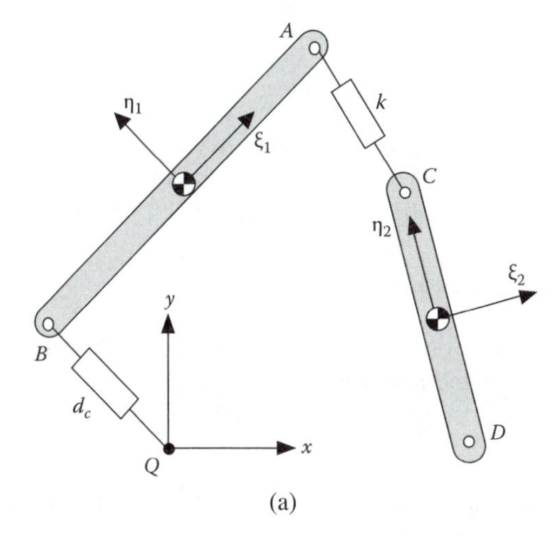

FIGURE 8.1 (a) Two unconstrained bodies connected by spring-damper elements; (b) the corresponding element forces; and (c) the equivalent forces and moments.

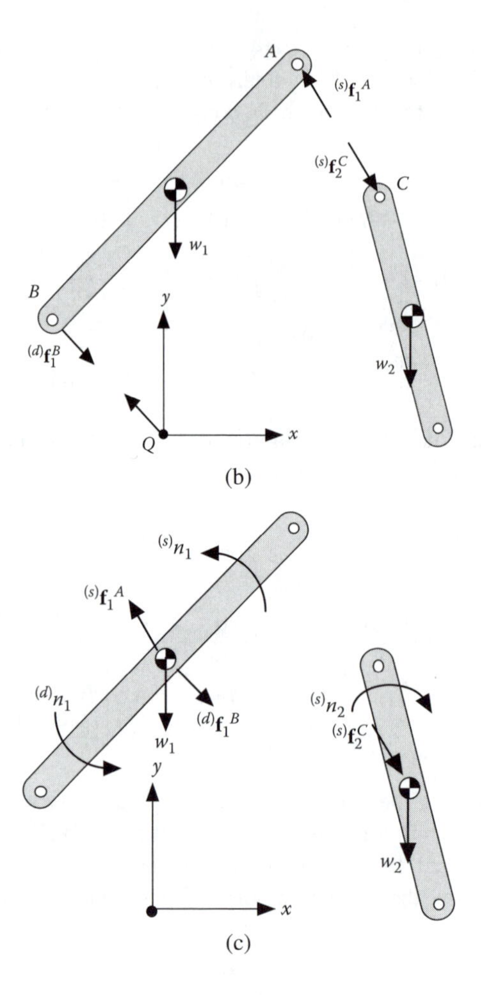

FIGURE 8.1 (Continued).

in the force-moment diagram. If we show a force at its actual point of application, as in Figure 8.1(b), we should not show the moment of that force on the diagram otherwise the moment would be counted twice which obviously is not correct.

8.1.2 Unconstrained Bodies—General

The equations of motion for a typical single body i were stated in Section 4.2.2. These equations are restated here for easy reference:

$$\begin{bmatrix} m_i\mathbf{I} & \mathbf{0} \\ \mathbf{0} & J_i \end{bmatrix} \begin{Bmatrix} \ddot{\mathbf{r}}_i \\ \ddot{\phi}_i \end{Bmatrix} = \begin{Bmatrix} \mathbf{f}_i \\ n_i \end{Bmatrix} \qquad \begin{bmatrix} m_i & 0 & 0 \\ 0 & m_i & 0 \\ 0 & 0 & J_i \end{bmatrix} \begin{Bmatrix} \ddot{x}_i \\ \ddot{y}_i \\ \ddot{\phi}_i \end{Bmatrix} = \begin{Bmatrix} f_{i(x)} \\ f_{i(y)} \\ n_i \end{Bmatrix} \tag{8.1}$$

Or,

$$\mathbf{M}_i \, \ddot{\mathbf{c}}_i = \mathbf{h}_i \tag{8.2}$$

where

$$\mathbf{M}_i = \begin{bmatrix} m_i\mathbf{I} & \mathbf{0} \\ \mathbf{0} & J_i \end{bmatrix}, \; \ddot{\mathbf{c}}_i = \begin{Bmatrix} \ddot{\mathbf{r}}_i \\ \ddot{\phi}_i \end{Bmatrix}, \; \mathbf{h}_i = \begin{Bmatrix} \mathbf{f}_i \\ n_i \end{Bmatrix}$$

For a system of n_b unconstrained bodies, the mass matrix and the arrays of accelerations and forces are constructed as

$$\mathbf{M} = \begin{bmatrix} \mathbf{M}_1 & \cdots & \mathbf{0} \\ \vdots & \ddots & \vdots \\ \mathbf{0} & \cdots & \mathbf{M}_{n_b} \end{bmatrix}, \; \mathbf{c} = \begin{Bmatrix} \ddot{\mathbf{c}}_1 \\ \vdots \\ \ddot{\mathbf{c}}_{n_b} \end{Bmatrix}, \; \mathbf{h} = \begin{Bmatrix} \mathbf{h}_1 \\ \vdots \\ \mathbf{h}_{n_b} \end{Bmatrix} \tag{8.3}$$

where $\mathbf{M}$ is a constant diagonal matrix. The equations of motion for the system are expressed as

$$\mathbf{M}\ddot{\mathbf{c}} = \mathbf{h} \tag{8.4}$$

The array of forces, $\mathbf{h}$, can have contributions from gravitational force, spring, dampers and other elements. We note that Eq. (8.4) contains as many equations as the number of accelerations. This means that these equations can be solved as a set of linear algebraic equations to determine the accelerations.

8.2 SYSTEM OF CONSTRAINED BODIES

In the presence of kinematic constraints, reaction forces between bodies must be incorporated into the equations of motion. To construct the equations of motion for a system of constrained bodies, we follow a two-step process similar to that of the system of constrained particles. We first ignore the constraints completely; that is, we assume that the system is unconstrained. For this unconstrained system, we construct the mass matrix and the array of forces as in Eq. (8.3). Then, in the second step, we return to the original constrained system and include the reaction forces. The reaction forces will be described in terms of the Jacobian matrix of the constraints and an array of Lagrange multipliers. Construction of the equations of motion will be demonstrated through several examples after generalization.

8.2.1 CONSTRAINED BODIES—GENERAL

For a system of n_b constrained bodies, the mass matrix and the array of applied forces are defined as $\mathbf{M}$ and $\mathbf{h}$, exactly as those in Eq. (8.3). For a system of constrained particles, we must revise Eq. (8.4) by including the reaction forces. The resulting equations are expressed as

$$\mathbf{M}\ddot{\mathbf{c}} = \mathbf{h} + {}^{(c)}\mathbf{h} \tag{8.5}$$

where

$${}^{(c)}\mathbf{h} = \begin{Bmatrix} {}^{(c)}\mathbf{h}_1 \\ \vdots \\ {}^{(c)}\mathbf{h}_{n_b} \end{Bmatrix} \tag{8.6}$$

represents the array of reaction (constraint) forces.

Assume that the position constraints between the n_b bodies form n_c constraint equations that are expressed in a general form as

$$\mathbf{\Phi}(\mathbf{c}) = \mathbf{0} \tag{8.7}$$

The velocity and acceleration constraints are expressed as

$$\dot{\mathbf{\Phi}} \equiv \mathbf{D}\,\dot{\mathbf{c}} = \mathbf{0} \tag{8.8}$$

$$\ddot{\mathbf{\Phi}} \equiv \mathbf{D}\,\ddot{\mathbf{c}} + \dot{\mathbf{D}}\,\dot{\mathbf{c}} = \mathbf{0} \tag{8.9}$$

where $\mathbf{D}$ is the Jacobian matrix. According to the method of Lagrange multipliers, the array of reaction forces of Eq. (8.6) can be represented as

$${}^{(c)}\mathbf{h} = \mathbf{D}'\boldsymbol{\lambda} \tag{8.10}$$

where

$$\boldsymbol{\lambda} = \begin{Bmatrix} \lambda_1 \\ \vdots \\ \lambda_{n_c} \end{Bmatrix} \tag{8.11}$$

is an array of n_c Lagrange multipliers. Then Eq. (8.5) is rewritten as:

$$\mathbf{M}\ddot{\mathbf{c}} = \mathbf{h} + \mathbf{D}'\boldsymbol{\lambda} \tag{8.12}$$

We note that Eq. (8.12) contains n_b equations and $n_b + n_c$ unknowns (accelerations and Lagrange multipliers). This means that these equations cannot be solved in their present form as a set of linear algebraic equations to determine the accelerations.

8.2.2 A Two-Body System

This system is a variation of the example in Section 8.1.1. As shown in Figure 8.2(a), a massless rod, having a length of $\ell = 0.355$, is attached between points D and B, otherwise the system is exactly the same. For the massless rod, the revolute-revolute joint constraint from Eq. (7.5) is constructed. For that, vector $\mathbf{d}$ is defined as

$$\mathbf{d} = \mathbf{r}_1^B - \mathbf{r}_2^D$$

For the equations of motion we need the Jacobian of the constraint. The Jacobian submatrices are adopted from Eq. (7.38) as

$$\mathbf{D}_1 = \begin{bmatrix} \mathbf{d}' & \mathbf{d}'\breve{\mathbf{s}}_1^B \end{bmatrix}, \mathbf{D}_2 = \begin{bmatrix} -\mathbf{d}' & -\mathbf{d}'\breve{\mathbf{s}}_2^D \end{bmatrix}$$

Substituting numerical values for the vectors, the following submatrices are obtained*:

$$\mathbf{D}_1 = \begin{bmatrix} -0.337 & 0.110 & -0.047 \end{bmatrix}, \mathbf{D}_2 = \begin{bmatrix} 0.337 & -0.110 & 0.036 \end{bmatrix}$$

Now we revise the equations of motion for the unconstrained system, from Section 8.1.1, as

$$\left[\begin{array}{ccc|ccc} 0.2 & 0 & 0 & 0 & 0 & 0 \\ 0 & 0.2 & 0 & 0 & 0 & 0 \\ 0 & 0 & 0.03 & 0 & 0 & 0 \\ \hline 0 & 0 & 0 & 0.15 & 0 & 0 \\ 0 & 0 & 0 & 0 & 0.15 & 0 \\ 0 & 0 & 0 & 0 & 0 & 0.02 \end{array}\right] \left\{\begin{array}{c} \ddot{x}_1 \\ \ddot{y}_1 \\ \ddot{\phi}_1 \\ \hline \ddot{x}_2 \\ \ddot{y}_2 \\ \ddot{\phi}_2 \end{array}\right\} = \left\{\begin{array}{c} -0.517 \\ -0.516 \\ 1.122 \\ \hline 2.577 \\ -5.166 \\ -0.184 \end{array}\right\} + \left[\begin{array}{c} -0.337 \\ 0.110 \\ -0.047 \\ \hline 0.337 \\ -0.110 \\ 0.036 \end{array}\right] \lambda$$

This represents the equations of motion for the constrained system. The Lagrange multiplier λ and the transposed Jacobian provide us the components of reaction forces and their corresponding moments. However, the value of λ is unknown at this point.

The reaction forces that act on the bodies are

$${}^{(c)}\mathbf{f}_1 = \begin{Bmatrix} -0.337 \\ 0.110 \end{Bmatrix} \lambda, \; {}^{(c)}\mathbf{f}_2 = \begin{Bmatrix} 0.337 \\ -0.110 \end{Bmatrix} \lambda$$

* Obtaining the necessary numerical values is left as an exercise for the reader.

These forces are along the axis of the rod and they are equal in magnitudes but opposite in directions. We do not know the exact directions because λ may be either positive or negative. The corresponding moments are

$$^{(c)}n_1 = -0.047\lambda,\ ^{(c)}n_2 = 0.036\lambda$$

These moments are not equal in magnitude because their moment arms are different. The reaction forces are shown in Figure 8.2(b), assuming a positive value for λ, and the equivalent force-moment representation is shown in Figure 8.2(c).

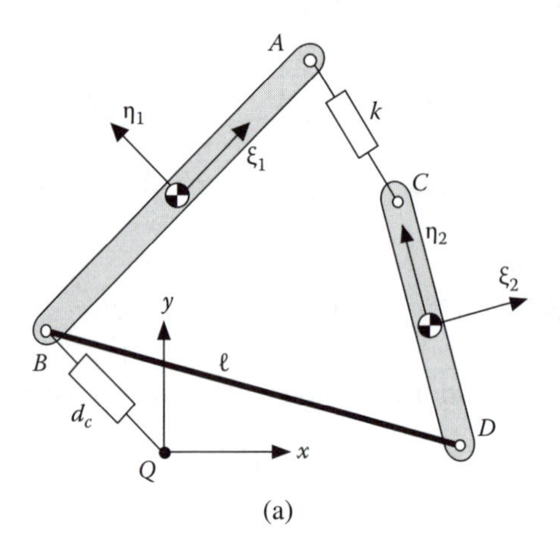

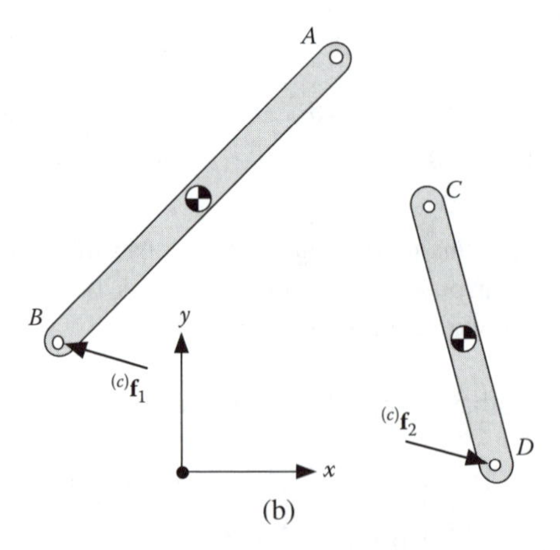

FIGURE 8.2 (a) Two constrained bodies; (b) reaction forces due to the constraint; and (c) the equivalent reaction force-moment representation.

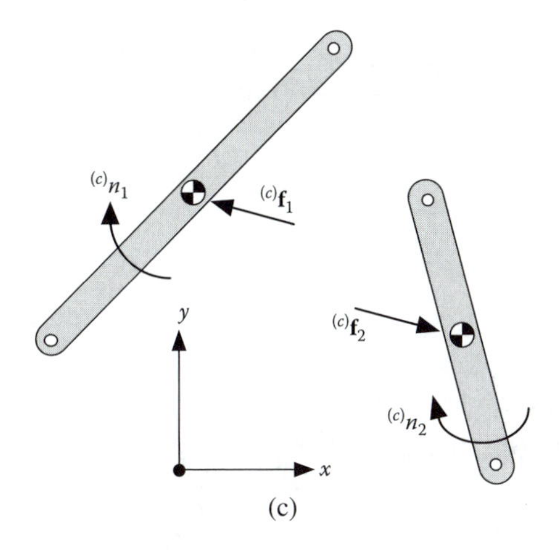

FIGURE 8.2 (Continued)

8.2.3 Reaction Forces

In the equations of motion for constrained multibody systems, we expressed the array of reaction forces in terms of the transpose of the constraint Jacobian and an array of Lagrange multipliers. In this section we verify that this presentation of reaction forces and/or moments between two bodies is valid for the commonly used joints.

For the revolute joint connecting two bodies shown in Figure 8.3(a), the reaction forces act between points P_i and P_j as shown in Figure 8.3(b). Both components of the reaction force, ${}^{(c)}\mathbf{f}$, are unknowns. Each force generates a moment about its corresponding mass center. Therefore, the array of reaction force for each body can be expressed as

$$ {}^{(c)}\mathbf{h}_i = \begin{Bmatrix} -{}^{(c)}\mathbf{f} \\ -\breve{\mathbf{s}}_i^{P\prime}\,{}^{(c)}\mathbf{f} \end{Bmatrix}, \quad {}^{(c)}\mathbf{h}_j = \begin{Bmatrix} {}^{(c)}\mathbf{f} \\ \breve{\mathbf{s}}_j^{P\prime}\,{}^{(c)}\mathbf{f} \end{Bmatrix} \tag{8.13} $$

The unknown components of the reaction force can be renamed as

$$ {}^{(c)}\mathbf{f} = \begin{Bmatrix} {}^{(c)}f_{(x)} \\ {}^{(c)}f_{(y)} \end{Bmatrix} = \begin{Bmatrix} \lambda_1 \\ \lambda_2 \end{Bmatrix} = \boldsymbol{\lambda} \tag{8.14} $$

Substituting Eq. (8.14) into Eq. (8.13) yields the array of reaction forces as

$$ {}^{(c)}\mathbf{h}_i = \begin{bmatrix} -\mathbf{I} \\ -\breve{\mathbf{s}}_i^{P\prime} \end{bmatrix} \boldsymbol{\lambda}, \quad {}^{(c)}\mathbf{h}_j = \begin{bmatrix} \mathbf{I} \\ \breve{\mathbf{s}}_j^{P\prime} \end{bmatrix} \boldsymbol{\lambda} \tag{8.15} $$

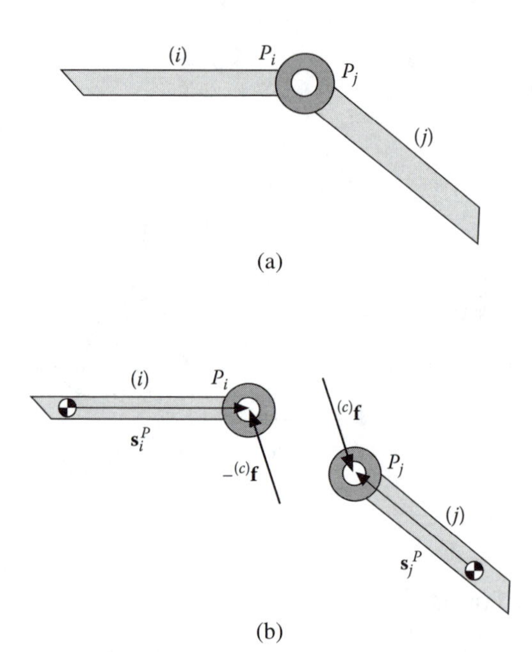

FIGURE 8.3 Reaction forces for a revolute joint.

Note that the coefficient matrices for $\boldsymbol{\lambda}$ are exactly the transpose of the sub-Jacobians for the revolute joint constraints.

For a translational joint between two bodies shown in Figure 8.4(a), we consider the reaction forces and moments shown in Figure 8.4(b). Because a translational joint does not allow relative rotation, pure reaction moments, $^{(c)}n$ and $-^{(c)}n$, are considered between the two bodies. In addition, because the two bodies cannot have relative translation along the axis perpendicular to the joint axis, reaction forces $^{(c)}\mathbf{f}$ and $-^{(c)}\mathbf{f}$ are generated perpendicular to the joint axis. These forces have their own moment arms about their corresponding body mass centers. The reaction forces, $^{(c)}\mathbf{f}$ and $-^{(c)}\mathbf{f}$, in reality are distributed forces along the contacting surfaces (contacting lines in planar description). Therefore, as concentrated forces, their exact points of application on the bodies are unknown. But, for the moment, because we have already defined point P_j on the joint axis on body j, we apply the reaction force on body j at that point. Assume that point P_i^* on body i coincides with P_j and, therefore, the reaction force on body i is applied at this point. Then, for the pure reaction moments and the reaction forces, the arrays of reaction forces are constructed as

$$^{(c)}\mathbf{h}_i = \begin{Bmatrix} -^{(c)}\mathbf{f} \\ -^{(c)}n - \breve{\mathbf{s}}_i^{P*\prime}\, {}^{(c)}\mathbf{f} \end{Bmatrix}, \quad {}^{(c)}\mathbf{h}_j = \begin{Bmatrix} ^{(c)}\mathbf{f} \\ ^{(c)}n + \breve{\mathbf{s}}_j^{P\prime}\, {}^{(c)}\mathbf{f} \end{Bmatrix} \tag{8.16}$$

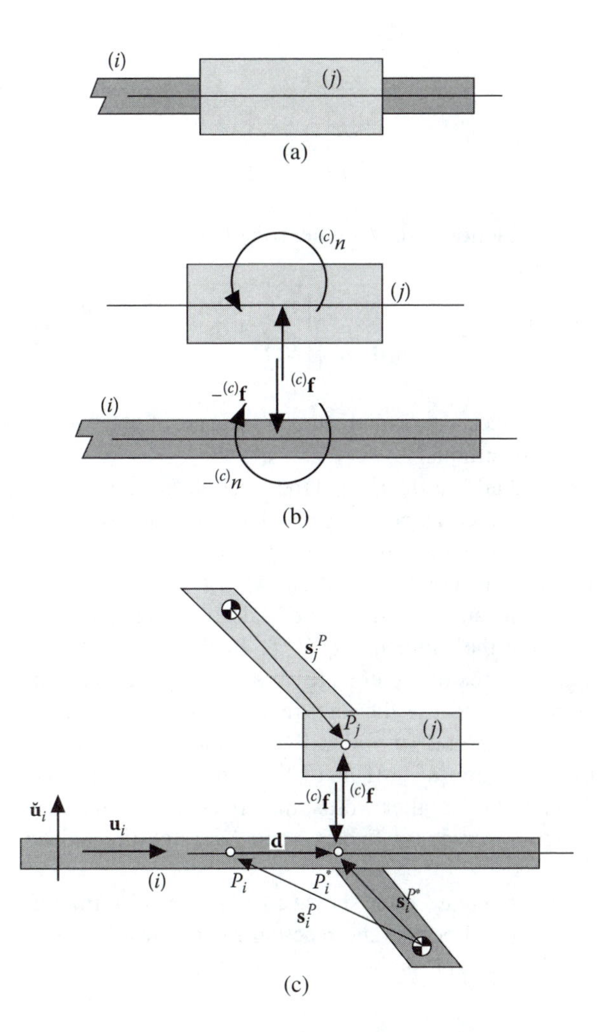

FIGURE 8.4 Reaction forces and moments for a translational joint.

Because the reaction force ${}^{(c)}\mathbf{f}$ is perpendicular to the joint axis, it can be described in terms of the unit vector $\mathbf{u}_i$ and its directional magnitude as

$$
{}^{(c)}\mathbf{f} = \breve{\mathbf{u}}_i\,{}^{(c)}f \tag{8.17}
$$

Substituting Eq. (8.17) into Eq. (8.16), using Eq. (2.16), and noting that $\mathbf{s}_i^{P*} = \mathbf{s}_i^P + \mathbf{d}$, we get

$$
{}^{(c)}\mathbf{h}_i = \begin{Bmatrix} -\breve{\mathbf{u}}_i\,{}^{(c)}f \\ -{}^{(c)}n - (\mathbf{s}_i^P + \mathbf{d})'\mathbf{u}_i\,{}^{(c)}f \end{Bmatrix}, \quad {}^{(c)}\mathbf{h}_j = \begin{Bmatrix} \breve{\mathbf{u}}_i\,{}^{(c)}f \\ {}^{(c)}n + \mathbf{s}_j^{P\prime}\mathbf{u}_i\,{}^{(c)}f \end{Bmatrix}
$$

We rename ${}^{(c)}n$ and ${}^{(c)}f$ as

$$\begin{Bmatrix} {}^{(c)}n \\ {}^{(c)}f \end{Bmatrix} \equiv \begin{Bmatrix} \lambda_1 \\ \lambda_2 \end{Bmatrix} = \boldsymbol{\lambda} \tag{8.18}$$

Then the arrays of reaction forces are described as

$$ {}^{(c)}\mathbf{h}_i = \begin{bmatrix} \mathbf{0} & -\breve{\mathbf{u}}_i \\ -1 & -(\mathbf{s}_i^P + \mathbf{d})'\mathbf{u}_i \end{bmatrix} \begin{Bmatrix} \lambda_1 \\ \lambda_2 \end{Bmatrix}, \quad {}^{(c)}\mathbf{h}_j = \begin{bmatrix} \mathbf{0} & \breve{\mathbf{u}}_i \\ 1 & \mathbf{s}_j^{P\prime}\mathbf{u}_i \end{bmatrix} \begin{Bmatrix} \lambda_1 \\ \lambda_2 \end{Bmatrix} \tag{8.19}$$

The coefficient matrices are exactly the transpose of the sub-Jacobians for the translational joint constraints.

We recall that we arbitrarily applied the reaction force on body j at point P_j, and in turn, we applied the corresponding reaction force on body i at point P_i^*. If we move point P_j to a different location on the joint axis, the moment arm for the force and, therefore, the corresponding moment, $\breve{\mathbf{s}}_j^{P\prime}\,{}^{(c)}\mathbf{f}$, will be affected. Furthermore, moving P_j to a new location causes point P_i^* to be moved as well which results in a different value for the moment $-\breve{\mathbf{s}}_i^{P^*\prime}\,{}^{(c)}\mathbf{f}$. Because both points P_j and P_i^* are moved together, the resultant change in $\breve{\mathbf{s}}_j^{P\prime}\,{}^{(c)}\mathbf{f}$ remains equal to the change in $-\breve{\mathbf{s}}_i^{P^*\prime}\,{}^{(c)}\mathbf{f}$. However, because the total reaction moments on the two bodies; that is, $-{}^{(c)}n - (\mathbf{s}_i^P + \mathbf{d})'\mathbf{u}_i\,{}^{(c)}f$ and ${}^{(c)}n + \mathbf{s}_j^{P\prime}\mathbf{u}_i\,{}^{(c)}f$, are independent of the location of these two points, any change in $-\breve{\mathbf{s}}_i^{P^*\prime}\,{}^{(c)}\mathbf{f}$ and $\breve{\mathbf{s}}_j^{P\prime}\,{}^{(c)}\mathbf{f}$ in turn causes an equal but opposite change in $-{}^{(c)}n$ and ${}^{(c)}n$. In other words, the effective total moments that appear in the arrays of reaction forces for bodies i and j are not affected by the location of point P_j.

For the revolute-revolute joint shown in Figure 8.5(a), the reaction forces are shown in Figure 8.5(b). Because the reaction forces must be along the axis of the link (i.e., vector **d**), we express the reaction force as

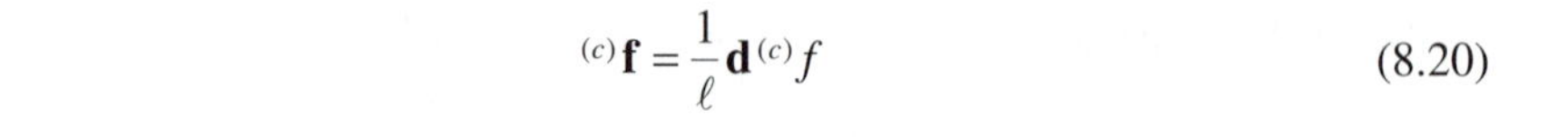

$$ {}^{(c)}\mathbf{f} = \frac{1}{\ell}\mathbf{d}\,{}^{(c)}f \tag{8.20}$$

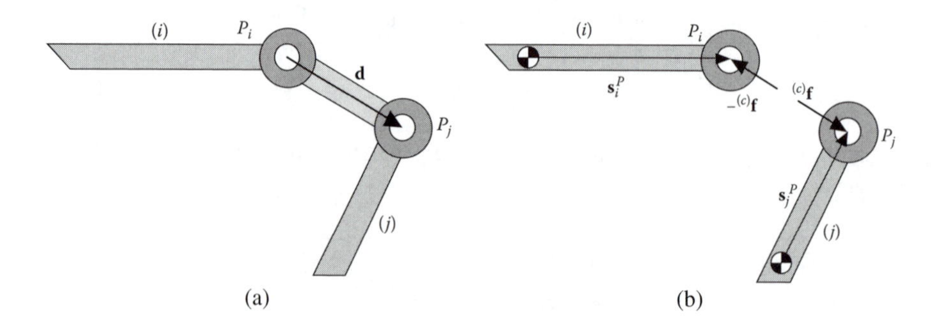

FIGURE 8.5 Reaction forces for a revolute-revolute joint.

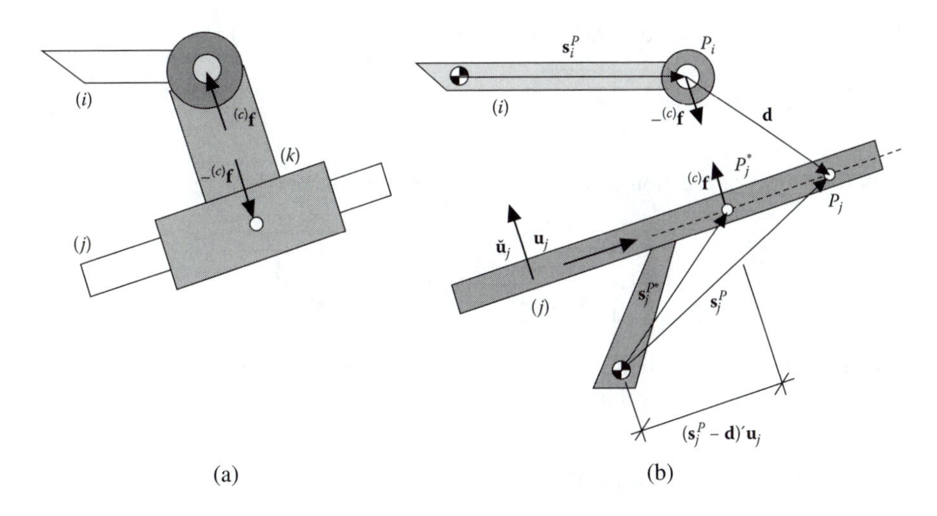

FIGURE 8.6 Reaction forces for a revolute-translational joint.

where ${}^{(c)}f$ is its unknown directional magnitude. We make the following renaming

$$\frac{1}{\ell}\,{}^{(c)}f = \lambda \tag{8.21}$$

Then the arrays of reaction forces for the two bodies are expressed as

$${}^{(c)}\mathbf{h}_i = \begin{bmatrix} -\mathbf{d} \\ -\breve{\mathbf{s}}_i^{P\prime}\mathbf{d} \end{bmatrix}\lambda,\ {}^{(c)}\mathbf{h}_j = \begin{bmatrix} \mathbf{d} \\ \breve{\mathbf{s}}_j^{P\prime}\mathbf{d} \end{bmatrix}\lambda \tag{8.22}$$

For a revolute-translational joint, as shown in Figure 8.6(a), the reaction forces between the two bodies are along an axis perpendicular to the axis of the sliding joint that passes through the center of the pin joint. This axis is established from the fact that there is a sliding joint between bodies j and k and, furthermore, there is no inertia associated with the intermediate body k. The reaction force ${}^{(c)}\mathbf{f}$ is described in terms of the unit vector $\mathbf{u}_j$ and its directional magnitude as

$${}^{(c)}\mathbf{f} = \breve{\mathbf{u}}_j\,{}^{(c)}f \tag{8.23}$$

The reaction forces are applied to points P_i and P_j^* as shown in Figure 8.6(b). The arrays of reaction forces for bodies i and j are then expressed as

$${}^{(c)}\mathbf{h}_i = \begin{Bmatrix} -\breve{\mathbf{u}}_j\,{}^{(c)}f \\ -\breve{\mathbf{s}}_i^{P\prime}\breve{\mathbf{u}}_j\,{}^{(c)}f \end{Bmatrix},\ {}^{(c)}\mathbf{h}_j = \begin{Bmatrix} \breve{\mathbf{u}}_j\,{}^{(c)}f \\ \breve{\mathbf{s}}_j^{P*\prime}\breve{\mathbf{u}}_j\,{}^{(c)}f \end{Bmatrix}$$

Noting that $\breve{\mathbf{s}}_j^{P*\prime}\breve{\mathbf{u}}_j = \mathbf{s}_j^{P*\prime}\mathbf{u}_j = (\mathbf{s}_j^P - \mathbf{d})'\mathbf{u}_j$ and renaming

$$^{(c)}f = \lambda \tag{8.24}$$

then the arrays of reaction forces for the two bodies are rewritten as

$$^{(c)}\mathbf{h}_i = \begin{bmatrix} -\breve{\mathbf{u}}_j \\ -\mathbf{s}_i^{P\prime}\mathbf{u}_j \end{bmatrix}\lambda,\ ^{(c)}\mathbf{h}_j = \begin{bmatrix} \breve{\mathbf{u}}_j \\ (\mathbf{s}_j^P - \mathbf{d})'\mathbf{u}_j \end{bmatrix}\lambda \tag{8.25}$$

We note that the coefficient matrices are the sub-Jacobians of the revolute-translational joint constraint.

8.3 EXAMPLE PROGRAMS

This section is a continuation of Section 7.6. We first construct the equations of motion for the variable-length pendulum example, and then we extend the MATLAB® programs from Chapter 7 to evaluate the array of applied forces and the mass matrix.

To construct the array of forces and the mass matrix, the following partial script must be appended to the program BC_formulation in Section 7.6:

```
% Append to the file:  BC_formulation
% ...
% Construct diagonal mass matrix and array of forces
     BC_mass_diagonal; M_diag = diag(M_array);
     h = BC_h;
```

This extension invokes a script named BC_mass_diagonal to construct the diagonal elements of two matrices—one representing the system's mass matrix and the other representing the inverse of the mass matrix—in the form of arrays. Then either of these arrays can be converted to a diagonal matrix if needed.

```
% file name: BC_mass_diagonal
% Diagonal elements of the mass matrix and its inverse
     M_array = zeros(3*nb,1); M_inv_array = zeros(3*nb,1);
     for i=1:nb
          j = 3*i - 2;
          M_array(j:j+2) = [m(i); m(i); J(i)];
          M_inv_array(j:j+2) = [1/m(i); 1/m(i); 1/J(i)];
     end
```

8.3.1 Double A-Arm Suspension

In this section we compute the necessary arrays and matrices for constructing the equations of motion for the double A-arm suspension example. In the script BC_formulation we set the path to BC_AA folder. Then, we add the inertias and the spring-damper data to the file BC_condata.

```
% Append to the file: BC_condata
% ...
% Mass, moment of inertia, and force data
    m = [2; 30; 1]; J = [0.5; 2.5; 0.5]; g = 9.81;
    k = 91600; d_c = 1433; L_0 = 0.23;
```

The array of forces is computed by the function `BC_h`. This function takes care of the spring-damper and gravitational forces.

```
function h = BC_h
% Spring-damper, gravitational, ... forces and moments
    BC_global
    d = r_P{7} - r_P0{3}; d_d = r_P_d{7};
    f_sda = pp_sda(d, d_d, k, L_0, d_c, 0);
    f_sd_1 = -f_sda;
% Gravitational force
    w = g*m;
% Force array
    h_1 = [0; -w(1); 0] + [f_sd_1; 0];
    h_2 = [0; -w(2); 0]; h_3 = [0; -w(3); 0];
    h = [h_1; h_2; h_3];
```

Executing the program provides the following values for **h** array and the diagonal elements of the mass matrix as an array:

```
>> h                                >> M_array
      1.0e+03*
       1.4313                               2.0000
      -4.3327                               2.0000
            0                               0.5000
            0                              30.0000
      -0.2943                              30.0000
            0                               2.5000
            0                               1.0000
      -0.0098                               1.0000
            0                               0.5000
```

We may also report the diagonal mass matrix `M_diag`.

8.3.2 MacPherson Suspension

This example has been modeled kinematically in Sections 7.3.2 and 7.6.2 in three forms: Models A, B, and C containing three, two, and one body respectively. In this section we construct the array of forces and the mass matrix for all three models. The structures of these programs are identical to those of the double A-arm.

Model A: For this model we revise one script files from Section 7.6.2, Model A, and develop one new function.

```
% Append to the file:  BC_condata
% ...
% Mass, moment of inertia, and force data
    m = [2; 20; 0.5]; J = [0.5; 2.5; 0.2]; g = 9.81;
    k = 22600; d_c = 1270; L_0 = 0.34;
```

```
function h = BC_h
  % Spring-damper, gravitational, ... forces and moments
    BC_global
    d = r_P{4} - r_P0{1}; d_d = r_P_d{4};
    f_sda = pp_sda(d, d_d, k, L_0, d_c, 0);
    f_sd_2 = -f_sda; n_sd_2 = s_rot(s_P{4})'*f_sd_2;
% Gravitational force
    w = g*m;
% Force array
    h_1 = [0; -w(1); 0];   h_3 = [0; -w(3); 0];
    h_2 = [0; -w(2); 0] + [f_sd_2; n_sd_2];
    h = [h_1; h_2; h_3];
```

Executing the program results into the following array of forces and the diagonal elements of the mass matrix:

```
>> h                                >> M_array
       1.0e+03*
              0                           2.0000
        -0.0196                           2.0000
              0                           0.5000
         1.2894                          20.0000
        -4.5592                          20.0000
         0.1464                           2.5000
              0                           0.5000
        -0.0049                           0.5000
              0                           0.2000
```

Model B: For this model we revise one script file from Section 7.6.2, Model B, and develop one new function.

```
% Append to the file:  BC_condata
% ...
% Mass, moment of inertia, and force data
    m = [2; 20]; J = [0.5; 2.5]; g = 9.81;
    k = 22600; d_c = 1270; L_0 = 0.34;
```

```
function h = BC_h
% Spring-damper, gravitational, ... forces and moments
    BC_global
    d = r_P{4} - r_P0{1}; d_d = r_P_d{4};
    f_sda = pp_sda(d, d_d, k, L_0, d_c, 0);
    f_sd_2 = -f_sda; n_sd_2 = s_rot(s_P{4})'*f_sd_2;
% Gravitational force
    w = g*m;
% Force array
    h_1 = [0; -w(1); 0];
    h_2 = [0; -w(2); 0] + [f_sd_2; n_sd_2];
    h = [h_1; h_2];
```

Executing the program results into the following array of forces and the diagonal elements of the mass matrix:

```
>> h                                   >> M_array
        1.0e+03*
              0                              2.0000
        -0.0196                              2.0000
              0                              0.5000
         1.2894                             20.0000
        -4.5592                             20.0000
         0.1464                              2.5000
```

Model C: For this model we revise one script file from Section 7.6.2, Model C, and develop one new function.

```
% Append to the file:   BC_condata
% ...
% Mass, moment of inertia, and force data
    m = [20]; J = [2.5]; g = 9.81;
    k = 22600; d_c = 1270; L_0 = 0.34;
```

```
function h = BC_h
% Spring-damper, gravitational, ... forces and moments
    BC_global
    d = r_P{2} - r_P0{1}; d_d = r_P_d{2};
    f_sda = pp_sda(d, d_d, k, L_0, d_c, 0);
    f_sd_1 = -f_sda; n_sd_1 = s_rot(s_P{2})'*f_sd_1;
% Gravitational force
    w = g*m;
% Force array
    h_1 = [0; -w(1); 0] + [f_sd_1; n_sd_1];
    h = [h_1];
```

Executing the program results into the following array of forces and the diagonal elements of the mass matrix:

```
>> h                                   >> M_array
    1.0e+03 *
    1.2894                                  20.0000
   -4.5592                                  20.0000
    0.1464                                   2.5000
```

8.3.3 Variable-Length Pendulum

This is a continuation of the example from Section 7.6.4. Additional constant data for this system are given as

$$m_1 = 2,\ J_1 = 0.04,\ m_2 = 1,\ J_2 = 0.1$$

$$k = 20,\ d_c = 5,\ {}^0\ell = 0.2,\ {}^{(r)}k = 60,\ {}^0\theta = 0$$

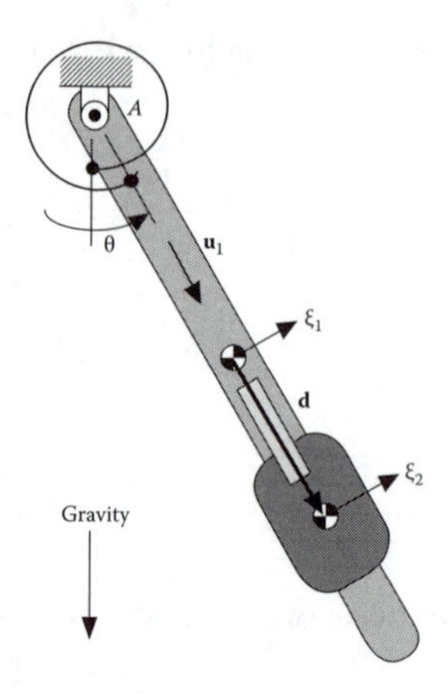

FIGURE 8.7 A variable-length pendulum with springs.

The necessary vectors are shown in Figure 8.7. We construct the equations of motion for this system in analytical and numerical forms.

Following the kinematics from Section 7.6.4, we compute $\dot{\mathbf{d}} = \dot{\mathbf{r}}_2 - \dot{\mathbf{r}}_1$ to determine the spring-damper force:

$$\ell = \left(\mathbf{d}'\mathbf{d}\right)^{\frac{1}{2}},\ \dot{\ell} = \mathbf{d}'\dot{\mathbf{d}} / \ell,\ {}^{(s,d)}f = k(\ell - {}^{0}\ell) + d_c\dot{\ell}$$

The pair of forces that the spring-damper applies on the two bodies are

$$\mathbf{f}_1 = {}^{(s,d)}f\,\mathbf{u}_1,\ \mathbf{f}_2 = -{}^{(s,d)}f\,\mathbf{u}_1$$

These forces are applied directly at the mass centers; therefore there are no associated moments. The moment of the rotational spring is determined as

$$\theta = \phi_1,\ {}^{(r-s)}n = {}^{(r)}k(\theta - {}^{0}\theta)$$

This moment is applied on body *(1)* as $n_1 = {}^{(r-s)}n$. The gravitational forces are determined as

$$\mathbf{w}_1 = \begin{Bmatrix} 0 \\ -m_1 g \end{Bmatrix},\ \mathbf{w}_2 = \begin{Bmatrix} 0 \\ -m_2 g \end{Bmatrix}$$

The array of forces for the system is constructed:

$$\mathbf{h} = \begin{Bmatrix} \mathbf{f}_1 + \mathbf{w}_1 \\ n_1 \\ \hdashline \mathbf{f}_2 + \mathbf{w}_2 \\ 0 \end{Bmatrix}$$

The equations of motion for this system are written as

$$\begin{bmatrix} m_1\mathbf{I} & \mathbf{0} & \mathbf{0} & \mathbf{0} \\ \mathbf{0} & J_1 & \mathbf{0} & \mathbf{0} \\ \mathbf{0} & \mathbf{0} & m_2\mathbf{I} & \mathbf{0} \\ \mathbf{0} & \mathbf{0} & \mathbf{0} & J_2 \end{bmatrix} \begin{Bmatrix} \ddot{\mathbf{r}}_1 \\ \ddot{\phi}_1 \\ \hline \ddot{\mathbf{r}}_2 \\ \ddot{\phi}_2 \end{Bmatrix} = \begin{Bmatrix} \mathbf{f}_1 + \mathbf{w}_1 \\ n_1 \\ \hline \mathbf{f}_2 + \mathbf{w}_2 \\ 0 \end{Bmatrix} + \left[\begin{array}{c|cc} \mathbf{I} & \mathbf{0} & -\breve{\mathbf{u}}_1 \\ \breve{\mathbf{s}}_1'^{A} & -\mathbf{u}_1'\mathbf{u}_2 & -\mathbf{d}'\mathbf{u}_1 \\ \hline \mathbf{0} & \mathbf{0} & \breve{\mathbf{u}}_1 \\ \mathbf{0} & \mathbf{u}_2'\mathbf{u}_1 & \mathbf{0} \end{array} \right] \begin{Bmatrix} \boldsymbol{\lambda}_{1,2} \\ \hline \lambda_3 \\ \lambda_4 \end{Bmatrix} \tag{8.26}$$

where $\boldsymbol{\lambda}_{1,2}$ contains two multipliers associated with the revolute joint constraints, and λ_3 and λ_4 are two multipliers associated with the translational joint constraints.

The following values are provided for the coordinates and velocities:

$$\mathbf{r}_1 = \begin{Bmatrix} 0.3 \\ -0.4 \end{Bmatrix},\ \phi_1 = 0.6435 \text{ rad},\ \dot{\mathbf{r}}_1 = \begin{Bmatrix} 0.52 \\ 0.39 \end{Bmatrix},\ \dot{\phi}_1 = 1.3 \text{ rad/sec}$$

$$\mathbf{r}_2 = \begin{Bmatrix} 0.45 \\ -0.6 \end{Bmatrix},\ \phi_2 = 0.6435 \text{ rad},\ \dot{\mathbf{r}}_2 = \begin{Bmatrix} 0.63 \\ 0.785 \end{Bmatrix},\ \dot{\phi}_2 = 1.3 \text{ rad/sec}$$

A computer program can evaluate the arrays and matrices in the equations of motion.* Then the equations of motion are numerically constructed as:

$$\left[\begin{array}{ccc|ccc} 2 & 0 & 0 & 0 & 0 & 0 \\ 0 & 2 & 0 & 0 & 0 & 0 \\ 0 & 0 & 0.04 & 0 & 0 & 0 \\ \hline 0 & 0 & 0 & 1 & 0 & 0 \\ 0 & 0 & 0 & 0 & 1 & 0 \\ 0 & 0 & 0 & 0 & 0 & 0.1 \end{array} \right] \begin{Bmatrix} \ddot{x}_1 \\ \ddot{y}_1 \\ \ddot{\phi}_1 \\ \hline \ddot{x}_2 \\ \ddot{y}_2 \\ \ddot{\phi}_2 \end{Bmatrix} = \begin{Bmatrix} 0.60 \\ -20.42 \\ -38.61 \\ \hline -0.60 \\ -9.01 \\ 0 \end{Bmatrix} + \left[\begin{array}{cc|cc} 1 & 0 & 0 & -0.8 \\ 0 & 1 & 0 & -0.6 \\ -0.4 & -0.3 & -1 & -0.25 \\ \hline 0 & 0 & 0 & -0.8 \\ 0 & 0 & 0 & 0.6 \\ 0 & 0 & 1 & 0 \end{array} \right] \begin{Bmatrix} \lambda_1 \\ \lambda_2 \\ \hline \lambda_3 \\ \lambda_4 \end{Bmatrix}$$

8.4 REMARKS

In this chapter we learned how to construct the mass matrix and the array of applied forces for a multibody system. We also discussed how to present the reaction forces in terms of the system Jacobian and an array of Lagrange multipliers. However, we did not discuss anything about how to solve the equations of motion for the unknown accelerations. At this point we may skip over the next two chapters and go directly to the analysis chapters. Or, we can continue to Chapters 9 and 10 and learn how to reduce the number of kinematic and dynamic equations by transforming them from body coordinates to joint coordinates.

* This is left as an exercise to the reader.

8.5 PROBLEMS

8.1 Two bodies are connected by a point-to-point spring as shown. The spring attachment points have the following body-fixed coordinates: $\mathbf{s}_1'^A = \{0.11 \quad 0\}'$ and $\mathbf{s}_2'^B = \{-0.15 \quad 0\}'$. In the shown configuration the two bodies have the following coordinates: $\mathbf{c}_1 = \{-0.20 \quad -0.05 \quad 90°\}'$ and $\mathbf{c}_2 = \{-0.05 \quad 0.20 \quad 0°\}'$. The inertial and spring data are given as: $m_1 = 0.2$, $J_1 = 0.03$, $m_2 = 0.15$, $J_2 = 0.02$, $k = 50$, and ${}^0\ell = 0.1$ (SI units). Construct the equations of motion.

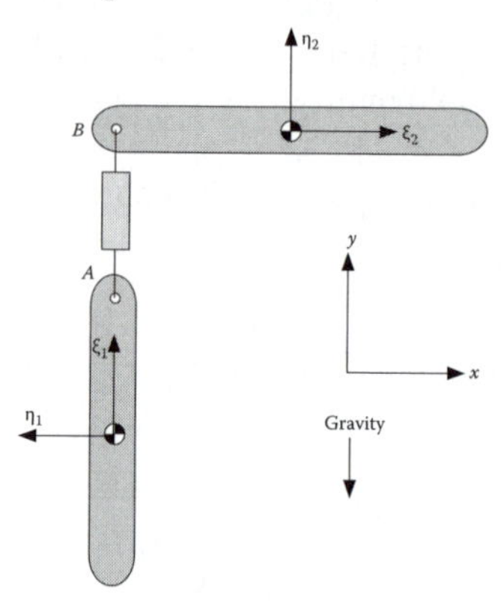

8.2 Two bodies are connected by a point-to-point spring-damper element as shown. The element attachment points have the following body-fixed coordinates: $\mathbf{s}_1'^A = \{-0.12 \quad 0\}'$ and $\mathbf{s}_2'^B = \{-0.15 \quad 0\}'$. In the shown configuration the two bodies have the following coordinates and velocities: $\mathbf{c}_1 = \{-0.16 \quad 0.06 \quad 0°\}'$, $\dot{\mathbf{c}}_1 = \{0.04 \quad -0.06 \quad 0.1\}'$, $\mathbf{c}_2 = \{-0.05 \quad 0.20 \quad 0°\}'$ and $\dot{\mathbf{c}}_2 = \{-0.01 \quad 0.03 \quad -0.2\}'$ (angular velocities are in rad/sec). The inertial and spring data are given as: $m_1 = 0.2$, $J_1 = 0.03$, $m_2 = 0.15$, $J_2 = 0.02$, $k = 50$, ${}^0\ell = 0.1$, and $d_c = 30$ (SI units). Construct the equations of motion.

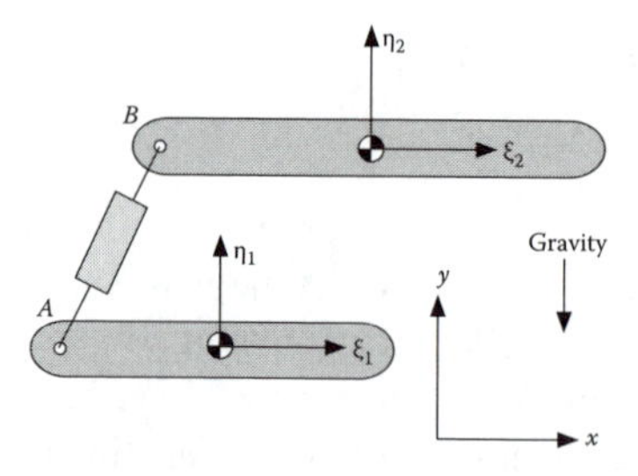

8.3 Two rods with uniform mass distribution are connected to each other and to the ground by point-to-point spring and damper elements as shown. The inertial and spring data are given as: $m_1 = 0.4$, $J_1 = 0.3$, $m_2 = 0.3$, $J_2 = 0.1$, $k_1 = 40$, ${}^0\ell_1 = 0.12$, $d_{c-1} = 12$, $k_2 = 60$, ${}^0\ell_2 = 0.2$, $k_3 = 25$, ${}^0\ell_3 = 0.13$, $d_{c-3} = 24$ (SI units). Obtain any necessary geometric data from the figure. Assume that the only nonzero velocities are $\dot{y}_1 = -0.3$ and $\dot{\phi}_2 = -0.15$. Construct the equations of motion.

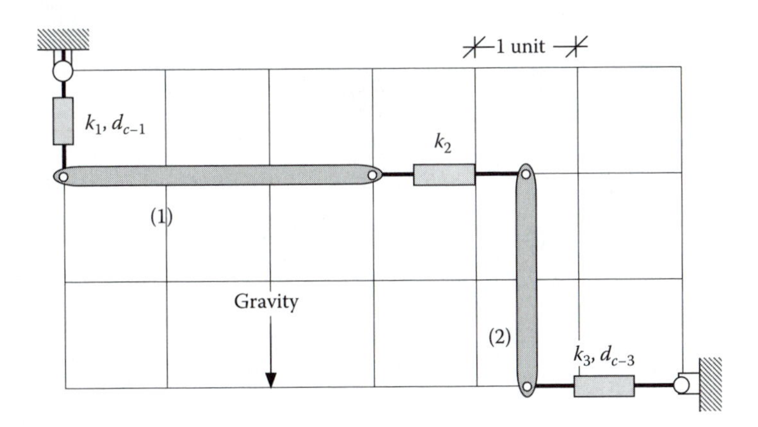

8.4 Two bodies are connected by a revolute joint as shown. The joint attachment points have the following body-fixed coordinates: $\mathbf{s}_1'^A = \{0.11 \;\; 0\}'$ and $\mathbf{s}_2'^A = \{0 \;\; 0.15\}'$. A force of 10 units acts in the negative x direction at the mass center of body (1) and another force of 15 units acts on the mass center of body (2) in the positive y-direction. In the shown configuration the two bodies have the following coordinates and velocities: $\mathbf{c}_1 = \{-0.07 \;\; 0.10 \;\; 60°\}'$, $\dot{\mathbf{c}}_1 = \{0.04 \;\; -0.06 \;\; 0.1\}'$, $\mathbf{c}_2 = \{0.10 \;\; 0.09 \;\; 45°\}'$ and $\dot{\mathbf{c}}_2 = \{0.01 \;\; -0.08 \;\; -0.2\}'$ (angular velocities are in rad/sec). The inertial data are given as: $m_1 = 0.2$, $J_1 = 0.03$, $m_2 = 0.15$, and $J_2 = 0.02$ (SI units). Construct the equations of motion (do not consider the gravitational force). Check the position and velocity constraints for any violations.

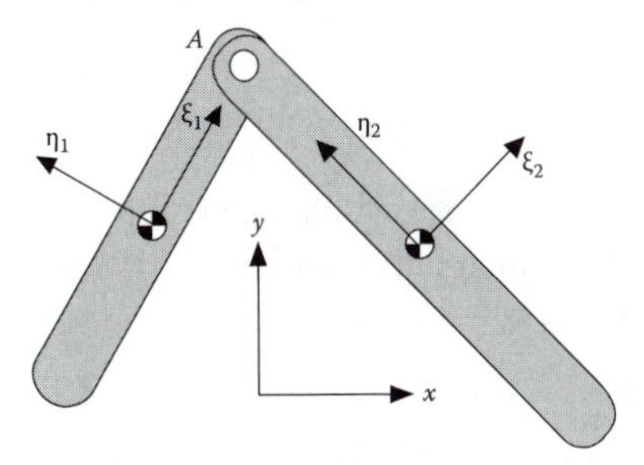

8.5 For the single pendulum shown, construct the equations of motion. The following constant data are provided: $\mathbf{r}_0^A = \{0.1 \;\; 0.3\}'$, $\mathbf{s}_1'^A = \{0.11 \;\; 0\}'$, $m_1 = 0.2$, and $J_1 = 0.03$ (SI units). Evaluate the variable elements of the equations of motion for the following coordinates and velocities: $\mathbf{c}_1 = \{0.15 \;\; 0.20 \;\; 120°\}'$ and $\dot{\mathbf{c}}_1 = \{-0.19 \;\; -0.11 \;\; -2.0\}'$.

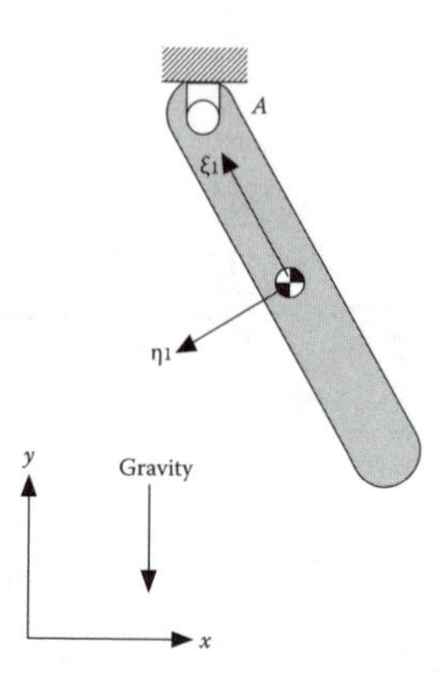

8.6 Two rods are pinned to the ground as shown. The lengths of the rods are 1.5 and 2.0 units respectively. A spring-damper-actuator element is connected between the free ends of the rods having the following characteristics: $k = 100$, ${}^0\ell = 2$, $d_c = 200$, ${}^{(a)}f = 30$. In the shown configuration, the angular velocities of the two bodies are, respectively, 1 rad/sec, CW, and 2.5 rad/sec, CCW. Construct the equations of motion for this system. Assume $m_1 = 0.8$, $J_1 = 0.15$, $m_2 = 1.2$, and $J_2 = 0.40$.

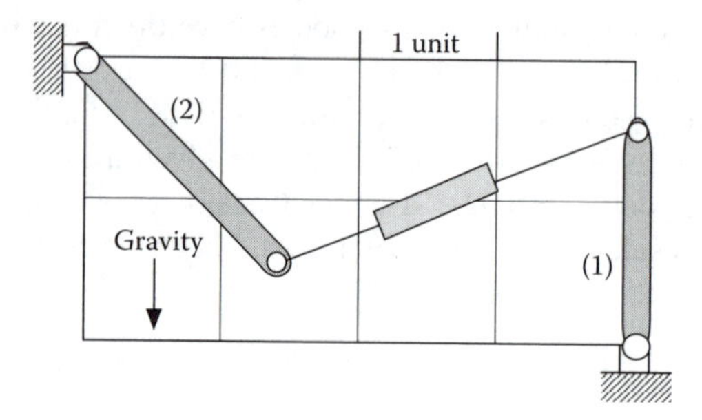

8.7 A point-to-point actuator-spring-damper is connected between two bodies at point A on link 1 and point B on link 2. The two points are positioned in their respective ξ-η frame as

$$\mathbf{s}_1'^A = \begin{Bmatrix} -0.25 \\ 0.12 \end{Bmatrix}, \; \mathbf{s}_2'^B = \begin{Bmatrix} 0.15 \\ 0.20 \end{Bmatrix}$$

The actuator pulls on the bodies with a force of 4.5 units. The spring has an undeformed length ${}^0\ell = 0.35$ and a stiffness $k = 30$. The damping coefficient is $d_c = 15$. In a given instant the two bodies have the following coordinates and velocities (angular data are in rad and rad/s):

$$\mathbf{c}_1 = \begin{Bmatrix} 0.32 \\ 0.23 \\ -0.55 \end{Bmatrix},\ \mathbf{c}_2 = \begin{Bmatrix} -0.12 \\ 0.16 \\ 0.60 \end{Bmatrix},\ \dot{\mathbf{c}}_1 = \begin{Bmatrix} -0.10 \\ 0.21 \\ 0.0 \end{Bmatrix},\ \dot{\mathbf{c}}_2 = \begin{Bmatrix} 0.18 \\ 0.09 \\ -0.20 \end{Bmatrix}$$

Assume $m_1 = 1.5$, $J_1 = 2.0$, $m_2 = 1.3$, and $J_2 = 1.8$. Construct the equations of motion.

8.8 Two rods are connected to each other by a pin join and connected to the ground by spring-damper elements as shown. The following inertial and spring-damper data are provided: $m_1 = m_2 = 6$, $J_1 = J_2 = 12.5$, $k_1 = 20$, ${}^0\ell_1 = 5$, $k_2 = 30$, ${}^0\ell_2 = 4.5$, and $d_{c-2} = 6$. Take any necessary measurements from the figure. The following velocity values are known: $\dot{x}_1 = 0.5$, $\dot{y}_1 = -0.2$, $\dot{\phi}_1 = 0.6$, and $\dot{\phi}_2 = -0.3$. Determine the remaining velocities and then construct the equations of motion.

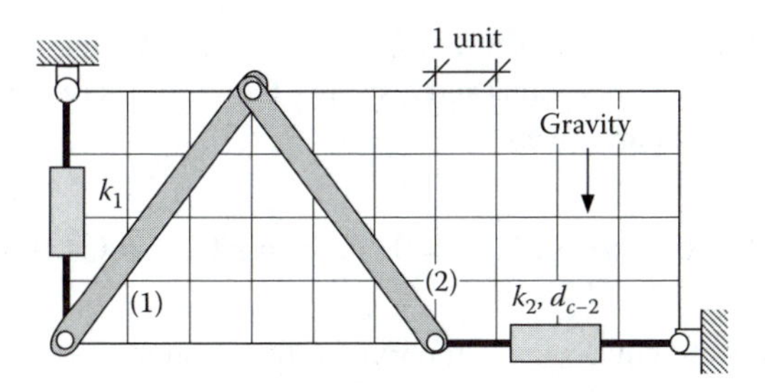

8.9 The multibody system shown consists of two moving links, a revolute joint, a sliding joint, and a spring. Gravity acts on the system. The following data are given:

$$m_1 = 5.0,\ J_1 = 0.5,\ m_2 = 3.0,\ J_2 = 0.4,\ k = 20,\ {}^0\ell = 2.0$$

Construct the equations of motion for the system.

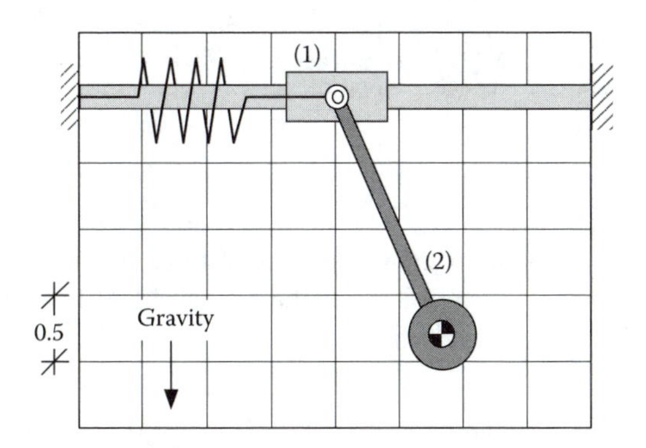

8.10 The multibody system shown is a slight variation of the system in Problem 8.9 and is represented by the same set of data. Construct the equations of motion for the system.

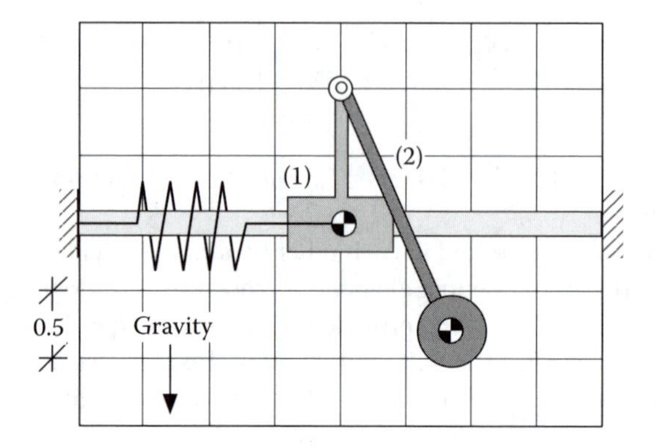

8.11 The multibody system shown consists of three moving links, two revolute joints, a sliding joint, and a spring. Gravity acts on the system. The following data are given:

$$m_1 = 5.0,\ J_1 = 0.5,\ m_2 = 1.0,\ J_2 = 0.14,\ m_3 = 3.0,\ J_3 = 0.4,\ k = 50,\ {}^0\ell = 1.0$$

Construct the equations of motion for this system.

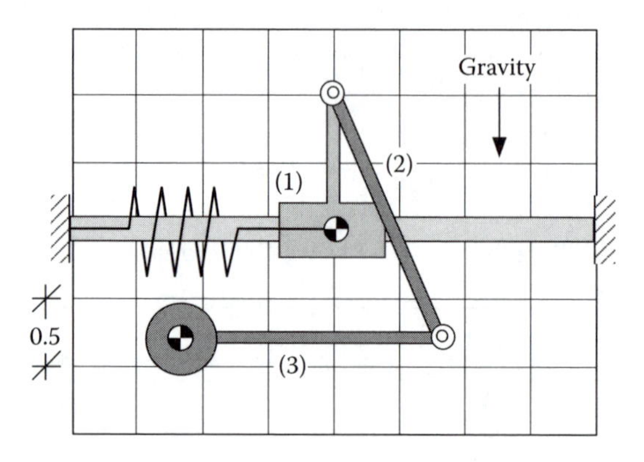

8.12 The multibody system shown consists of three moving links and three revolute joints. Gravity acts on the system. The following data are provided:

$$m_1 = 4.0,\ J_1 = 0.8,\ m_2 = m_3 = 1.5,\ J_2 = J_3 = 0.15 \text{ (SI units)}$$

Construct the equations of motion for this system.

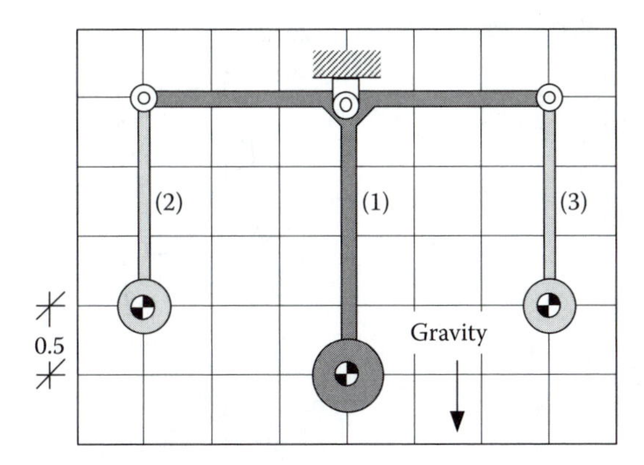

8.13 The script `BC_formulation`, in its present form, assumes any multibody system to contain constraints. Revise this script by providing logic to avoid invoking functions `BC_Phi`, `BC_jacob`, and `BC_gamma`. For the logic, we can use the data that is provided in the script `BC_condata` for the number of bodies and the number of degrees of freedom, `nb` and `ndof`. If `3*nb` is equal to `ndof`, then the system is unconstrained.

8.14 Use the revised version of the script `BC_formulation` (refer to Problem 8.13) and construct/evaluate the necessary array and matrices for the multibody systems in the following problems:
   (a) Problem 8.1
   (b) Problem 8.2
   (c) Problem 8.3
   (d) Problem 8.4
   (e) Problem 8.5
   (f) Problem 8.6
   (g) Problem 8.7
   (h) Problem 8.8
   (i) Problem 8.9
   (j) Problem 8.10
   (k) Problem 8.11
   (l) Problem 8.12

8.15 Develop a simple tire model to provide a radial force to support the quarter-car models in our example. Start with the simplest possible model and then improve on it as you are directed in the following steps:
   (a) Assume that, regardless of the camber angle, the tire model provides a force only in the y-direction that acts on the mass center of the wheel body. Consider a spring-damper element between the wheel mass center and the ground. The wheel radius could be used as the undeformed length of the spring.

(b) Revise the model by providing a logic that would allow the tire force to be computed only when the tire is in contact with the

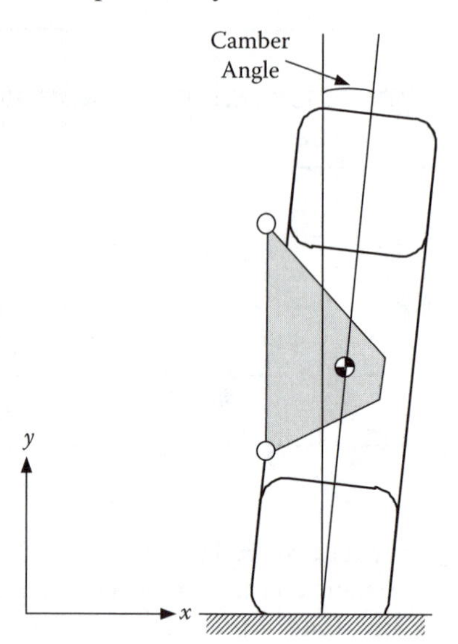

ground. In some simulations, it is possible for a tire to loose contact with the ground.

(c) Revise the model and consider the camber angle. Adjust for the contact point between the tire and the ground to move. In this case the tire force would not always be in the y-direction.

Develop an M-file for this model and save it in the folder `BC_Basics`. You may set up a flag to direct the program to either of the three models.

8.16 Implement the tire model from Problem 8.15 in the following models:

(a) The quarter-car, double A-arm suspension system, in Section 8.3.1.

(b) The quarter-car, MacPherson suspension system, in Section 8.3.2, Model A.

(c) The quarter-car, MacPherson suspension system, in Section 8.3.2, Model B.

(d) The quarter-car, MacPherson suspension system, in Section 8.3.2, Model C.

(e) The revised quarter-car double A-arm suspension system from Problem 7.31.

(f) The revised quarter-car MacPherson suspension system from Problem 7.32.

8.17 Develop an M-file for rotational spring-damper-actuator element similar to `pp_sda` M-file. Name the function and construct the input arguments as:

```
rot_sda(theta, theta_d, k, theta_0, d_c, n_a)
```

where the relative angle and its time derivative could be determined prior to invoking the new M-file as $\theta = \phi_j - \phi_i$ and $\dot{\theta} = \dot{\phi}_j - \dot{\phi}_i$.

9 Joint-Coordinates: Kinematics

The body-coordinate formulation from Chapters 7 and 8 provides a simple method to formulate the equations of motion for a multibody system and implement them in a computer program. However, the method generates a large number of equations for even simple systems. In this and the next chapter we present a method that transforms the kinematics and dynamics equations from the body-coordinates to a much smaller set. We call this the joint-coordinate method.

In the joint-coordinate method, we define a set of *joint coordinates* equal or greater than the number of system's DoF. Through a simple process, the kinematic constraints from the body-coordinate formulation are transformed into a much smaller set. For planar systems containing closed kinematic chains, the transformation results into a set of constraints identical to those obtained by the classical *vector-loop* method. As it will be seen shortly, the transformation process in the joint coordinate method can be cumbersome when compared to the vector-loop method. Therefore the joint-coordinate method is not recommended if we are only interested in the kinematics of a system. The true power of the joint coordinate method will become apparent when we extend the transformation process to the dynamic equations of motion, as will be seen in the next chapter.

Programs: Because the joint-coordinate formulation transforms the body-coordinate equations into a new form, in this chapter we use some of the programs from Chapter 7. The Current Directory in MATLAB® is a new folder called `Joint_Coord`. The file organization is shown in Table 9.1. We must provide a copy of the files `Basics` and `BC_Basics` in this directory. In addition, the M-file `BC_global` must be duplicated here as well.

9.1 VECTOR-LOOP METHOD

Vector-loop is the classical method for kinematic formulation of planar mechanisms—systems containing closed kinematic chains (loops). This methodology can be found in most textbooks on the analysis of planar mechanisms. We review this method through a simple example. We consider the planar system of Figure 7.1 that is also shown in Figure 9.1(a). For kinematic modeling, four vectors are defined forming a loop as shown in Figure 9.1(b). The vector-loop equation for this system is

$$\mathbf{d}_0 + \mathbf{d}_1 - \mathbf{d}_2 - \mathbf{d}_3 = \mathbf{0}$$

For each vector an angle is defined: θ_0, θ_1, θ_2, and θ_3, where θ_0 is a constant and $\theta_2 = \theta_3 + \frac{\pi}{2}$. For convenience, we define a unit vector $\mathbf{u}_3$ along the axis of the

TABLE 9.1

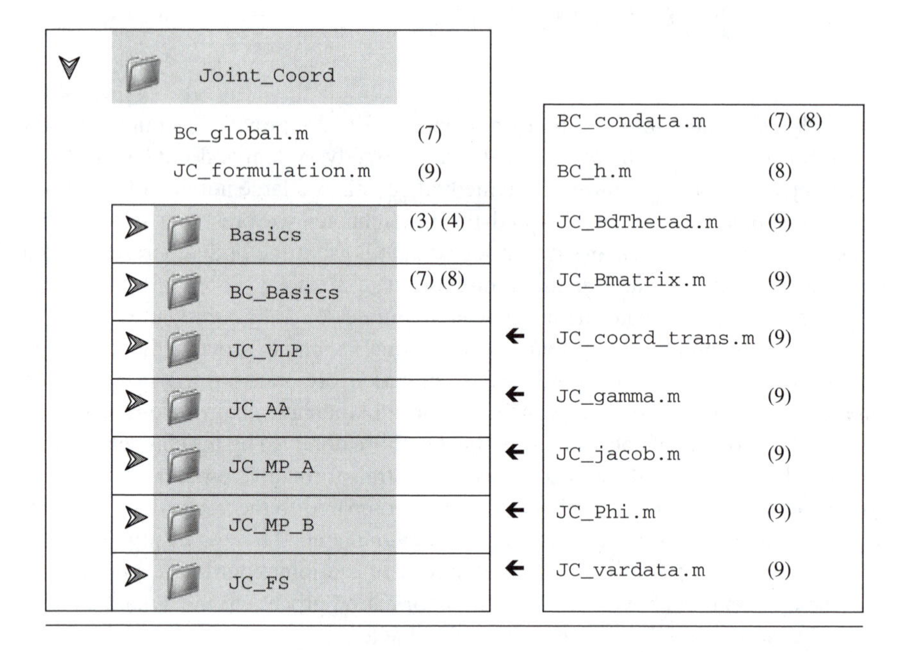

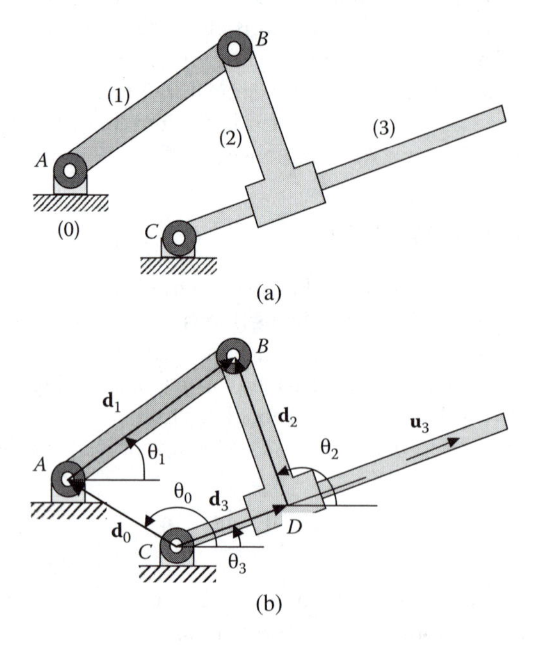

FIGURE 9.1 (a) A closed-chain mechanism and (b) its corresponding vector-loop representation.

sliding joint to express vector $\mathbf{d}_3$ as $\mathbf{d}_3 = d_3\,\mathbf{u}_3$. Then the constraint equation is rewritten as

$$\mathbf{d}_0 + \mathbf{d}_1 - \mathbf{d}_2 - d_3\,\mathbf{u}_3 = \mathbf{0} \tag{9.1}$$

This vector equation represents two algebraic equations. In these equations we have three variables: θ_1, θ_3 and d_3, therefore the mechanism has 3 – 2 = 1 degree of freedom.

The time derivative of the position constraints provides the velocity constraints as

$$\breve{\mathbf{d}}_1\dot{\theta}_1 - \breve{\mathbf{d}}_2\dot{\theta}_3 - \dot{d}_3\,\mathbf{u}_3 - d_3\,\breve{\mathbf{u}}_3\dot{\theta}_3 = \mathbf{0}$$

In matrix form the velocity constraints are expressed as

$$\begin{bmatrix} \breve{\mathbf{d}}_1 & -\mathbf{u}_3 & -(\breve{\mathbf{d}}_2 + d_3\,\breve{\mathbf{u}}_3) \end{bmatrix} \begin{Bmatrix} \dot{\theta}_1 \\ \dot{d}_3 \\ \dot{\theta}_3 \end{Bmatrix} = \mathbf{0} \tag{9.2}$$

Similarly, the time derivative of the velocity constraints yields the acceleration constraints.*

Although the vector loop method is quite simple and effective in formulating the kinematics of planar mechanisms, it is not a systematic method to implement for general-purpose formulation. The alternative is the joint coordinate method that leads to the same equations in a systematic manner.

9.2 JOINT COORDINATE METHOD

The transformation process from body coordinates to joint coordinates is well suited for computational procedures. The equations that result from this process are much smaller in number than those of the body coordinate formulation; therefore the

---

* The position and velocity constraints for this mechanism in expanded form are expressed as

$$d_0 \cos\theta_0 + d_1 \cos\theta_1 - d_2 \cos\theta_2 - d_3 \cos\theta_3 = 0$$

$$d_0 \sin\theta_0 + d_1 \sin\theta_1 - d_2 \sin\theta_2 - d_3 \sin\theta_3 = 0$$

$$-d_1 \sin\theta_1\dot{\theta}_1 + d_2 \sin\left(\theta_3 + \frac{\pi}{2}\right)\dot{\theta}_3 - \cos\theta_3\dot{d}_3 + d_3 \sin\theta_3\dot{\theta}_3 = 0$$

$$d_1 \cos\theta_1\dot{\theta}_1 - d_2 \cos\left(\theta_3 + \frac{\pi}{2}\right)\dot{\theta}_3 - \sin\theta_3\dot{d}_3 - d_3 \cos\theta_3\dot{\theta}_3 = 0$$

The expanded form of these equations is provided to remind ourselves how these constraints look like using the classical vector loop notation.

computational solution of these equations is much more efficient. In the following sections, the joint coordinate method is first presented for open-chain systems before being extended to closed-chain systems.

9.3 OPEN-CHAIN SYSTEMS

The *structure* (or the *topology*) of an open-chain system is analogous to a tree containing a root, branches, and leaves. The generic system shown in Figure 9.2(a) consists of three links, three joints, and the ground. The ground is always the *root* (or the *base*). This system contains only one *branch*. If we move from the root through the branch (i.e., move from a joint to a link to another joint and so on), we will end up at a *leaf*—the last link in a branch.

If a system is not connected to the ground by a kinematic joint, such as the system in Figure 9.2(b), it is referred to as a *floating system*. For such a system we still consider the ground as the root but we assume there is an imaginary joint that connects the ground to one of the bodies. This imaginary joint is called a *floating joint* that allows three DoF between the ground and that body; that is, this joint does not eliminate any DoF from the system.

A system may contain more than one branch. For example, the system shown in Figure 9.2(c) contains two branches. Each branch ends at a leaf; therefore there

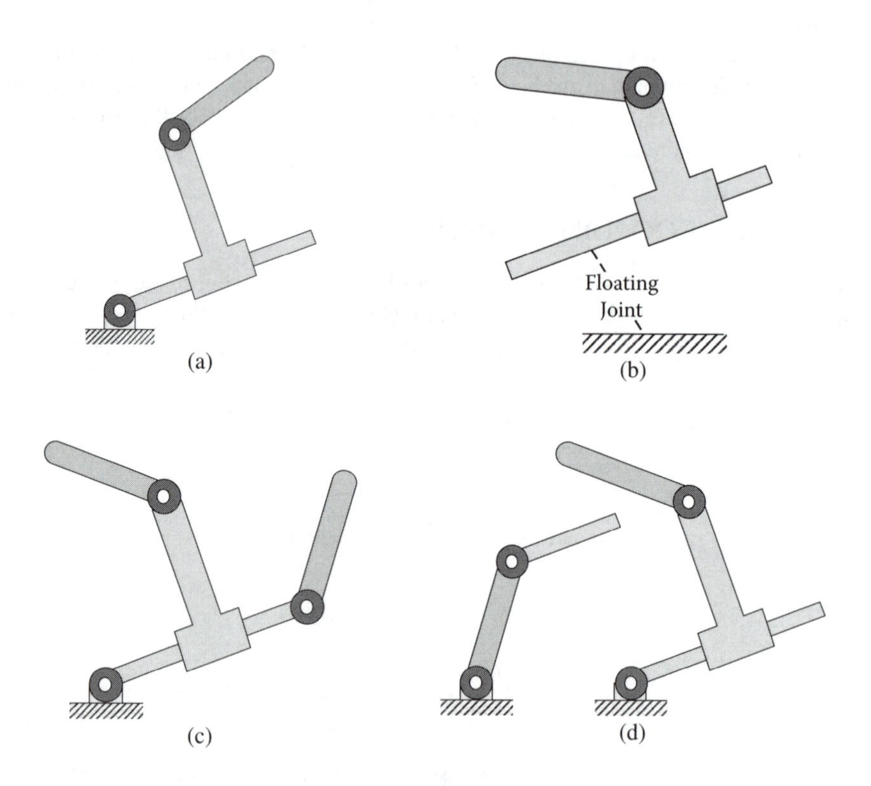

FIGURE 9.2 Various topologies of open-chain systems.

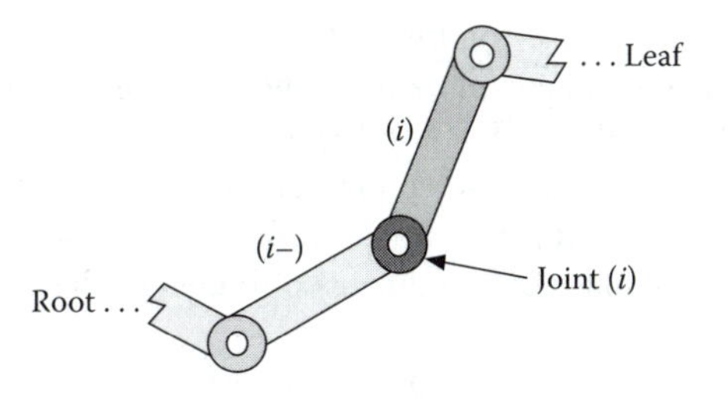

FIGURE 9.3 The joint connecting bodies (i) and $(i–)$ belongs to body (i).

are as many leaves in a system as the number of branches. A system may contain more than one tree, such as the system shown in Figure 9.2(d).

Consider two connected bodies in a branch. If the body that is closer to the leaf of that branch is referred to as body (i), then the body that is closer to the root is referred to as body $(i–)$ as shown in Figure 9.3. We say that body $(i–)$ comes *before* body (i). We also say that the joint between these two bodies belongs to body (i). Because there are as many joints in a system as the number of bodies (this includes the floating joint(s) in a system), each body owns only one joint. This description becomes important in describing the joint coordinates for a system.

The joint coordinate for a pin joint can be defined either as a relative angle or an absolute angle. As shown in Figure 9.4(a), the joint coordinate θ_i for the pin joint (i) is defined as the relative angle between bodies (i) and $(i–)$. This angle can be

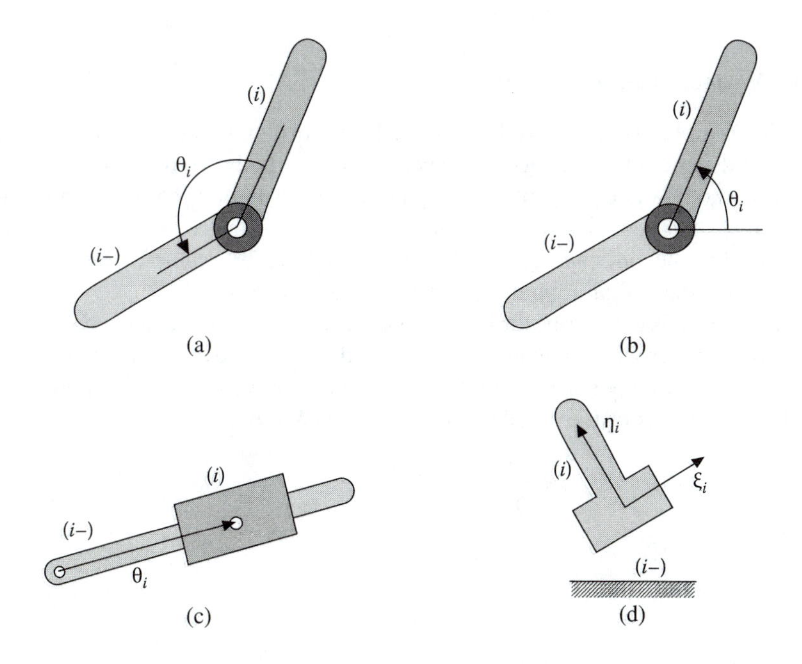

FIGURE 9.4 Joint coordinates for (a, b) a pin joint, (c) a sliding joint, and (d) a floating joint.

measured between any two axes on the bodies. For the same joint, the joint coordinate θ_i can be defined as the absolute angle between an axis on body (i) and the x-axis as shown in Figure 9.4(b). For a sliding joint, the joint coordinate θ_i is defined as the relative distance between two points on bodies (i) and $(i–)$ along the joint axis as shown in Figure 9.4(c). For a floating joint three joint coordinates must be define—they can be same as the coordinates of body (i); that is, $\boldsymbol{\theta}_i = \{x_i \quad y_i \quad \phi_i\}'$, as shown in Figure 9.4(d).

For an open-chain system, to perform the transformation, we can systematically apply the following steps:

- Assign numbers to each body (body indices). Assign index (0) to the ground. The bodies can be numbered in any desired order.
- Assign body-fixed $\xi - \eta$ frame to each body.
- Define a set of joint coordinates equal to the number of the system's DoF. This set is denoted as array $\boldsymbol{\theta}$.
- Write expressions describing the body coordinates in terms of the joint coordinates, starting from the root and moving toward the leaves.
- Obtain the velocity and acceleration expressions by taking the first and second time derivatives of the position (coordinate) expressions.
- Extract a coefficient matrix, **B**, known as the *velocity transformation matrix* from the velocity expressions.

Matrix **B** will play an important role in the transformation of the equations of motion from body coordinates to joint coordinates. The following examples show these steps in detail.

9.3.1 Variable-Length Pendulum

A simple example to consider is the variable-length pendulum of Section 7.6.4. The bodies of the pendulum, the necessary vectors, and the body-fixed frames that were used in the body-coordinate formulation are shown in Figure 9.5(a). Because the pendulum is a two DoF system, we define two joint coordinates: $\theta_1 = \phi_1$ as the joint coordinate for the pin joint A, and θ_2 (the magnitude of vector **d**) as the joint coordinate for the sliding joint as shown in Figure 9.5(b).

The next step is to describe the body coordinates in terms of the joint coordinates, starting from the ground, moving to body (1) and then to body (2). First we express the coordinates of body (1) as a function of θ_1, and then we express the coordinates of body (2) in terms of the coordinates of body (1) and θ_2:

$$\begin{aligned}
\phi_1 &= \theta_1 \\
\mathbf{r}_1 &= -\mathbf{s}_1^A \\
\phi_2 &= \phi_1 \\
\mathbf{r}_2 &= \mathbf{r}_1 + \mathbf{d} = \mathbf{r}_1 + \mathbf{u}_1\theta_2
\end{aligned} \tag{9.3}$$

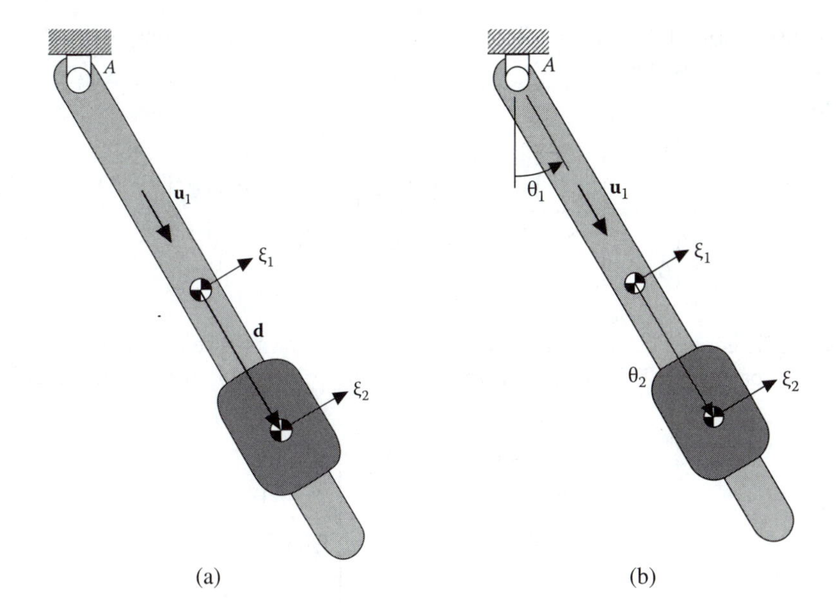

FIGURE 9.5 A variable-length pendulum.

These expressions are written in an order suitable for a recursive process; that is, if we are given values for θ_1 and θ_2, we can compute ϕ_1, $\mathbf{r}_1$, ϕ_2 and $\mathbf{r}_2$ in the order of their appearance.

The time derivative of Eq. (9.3) provides the velocity expressions as

$$\begin{aligned}
\dot{\phi}_1 &= \dot{\theta}_1 \\
\dot{\mathbf{r}}_1 &= -\breve{\mathbf{s}}_1^A \dot{\phi}_1 \\
\dot{\phi}_2 &= \dot{\phi}_1 \\
\dot{\mathbf{r}}_2 &= \dot{\mathbf{r}}_1 + \breve{\mathbf{u}}_1 \dot{\phi}_1 \theta_2 + \mathbf{u}_1 \dot{\theta}_2
\end{aligned}$$

A forward substitution yields:

$$\begin{aligned}
\dot{\phi}_1 &= \dot{\theta}_1 \\
\dot{\mathbf{r}}_1 &= -\breve{\mathbf{s}}_1^A \dot{\theta}_1 \\
\dot{\phi}_2 &= \dot{\theta}_1 \\
\dot{\mathbf{r}}_2 &= -\breve{\mathbf{s}}_1^A \dot{\theta}_1 + \breve{\mathbf{u}}_1 \theta_2 \dot{\theta}_1 + \mathbf{u}_1 \dot{\theta}_2
\end{aligned}$$

In matrix form, following a slight rearrangement of the equations, the velocity expressions are written as

$$\begin{Bmatrix} \dot{\mathbf{r}}_1 \\ \dot{\phi}_1 \\ \dot{\mathbf{r}}_2 \\ \dot{\phi}_2 \end{Bmatrix} = \left[\begin{array}{c:c} -\breve{\mathbf{s}}_1^A & \mathbf{0} \\ 1 & 0 \\ \hdashline -\breve{\mathbf{s}}_1^A + \breve{\mathbf{u}}_1\theta_2 & \mathbf{u}_1 \\ 1 & 0 \end{array}\right] \begin{Bmatrix} \dot{\theta}_1 \\ \dot{\theta}_2 \end{Bmatrix} = \left[\begin{array}{c:c} \breve{\mathbf{r}}_1 & \mathbf{0} \\ 1 & 0 \\ \hdashline \breve{\mathbf{r}}_2 & \mathbf{u}_1 \\ 1 & 0 \end{array}\right] \begin{Bmatrix} \dot{\theta}_1 \\ \dot{\theta}_2 \end{Bmatrix} \tag{9.4}$$

The coefficient matrix in this equation is shown in two forms—we can use the one that is more convenient. This equation is the *velocity transformation* matrix denoted as **B**:

$$\mathbf{B} = \left[\begin{array}{c:c} -\breve{\mathbf{s}}_1^A & \mathbf{0} \\ 1 & 0 \\ \hdashline -\breve{\mathbf{s}}_1^A + \breve{\mathbf{u}}_1\theta_2 & \mathbf{u}_1 \\ 1 & 0 \end{array}\right] = \left[\begin{array}{c:c} \breve{\mathbf{r}}_1 & \mathbf{0} \\ 1 & 0 \\ \hdashline \breve{\mathbf{r}}_2 & \mathbf{u}_1 \\ 1 & 0 \end{array}\right] \tag{9.5}$$

The time derivative of Eq. (9.4) provides the acceleration expressions as

$$\begin{Bmatrix} \ddot{\mathbf{r}}_1 \\ \ddot{\phi}_1 \\ \ddot{\mathbf{r}}_2 \\ \ddot{\phi}_2 \end{Bmatrix} = \left[\begin{array}{c:c} \breve{\mathbf{r}}_1 & \mathbf{0} \\ 1 & 0 \\ \hdashline \breve{\mathbf{r}}_2 & \mathbf{u}_1 \\ 1 & 0 \end{array}\right] \begin{Bmatrix} \ddot{\theta}_1 \\ \ddot{\theta}_2 \end{Bmatrix} + \left[\begin{array}{c:c} \dot{\breve{\mathbf{r}}}_1 & \mathbf{0} \\ 1 & 0 \\ \hdashline \dot{\breve{\mathbf{r}}}_2 & \dot{\mathbf{u}}_1 \\ 1 & 0 \end{array}\right] \begin{Bmatrix} \dot{\theta}_1 \\ \dot{\theta}_2 \end{Bmatrix} \tag{9.6}$$

where the time derivative of **B** is

$$\dot{\mathbf{B}} = \left[\begin{array}{c:c} \dot{\breve{\mathbf{r}}}_1 & \mathbf{0} \\ 1 & 0 \\ \hdashline \dot{\breve{\mathbf{r}}}_2 & \dot{\mathbf{u}}_1 \\ 1 & 0 \end{array}\right] \tag{9.7}$$

At this point we may ask: "what happened to the constraints?" The answer is that there are no constraints between the joint coordinates because the number of joint coordinates is equal to the number of system's DoF. To demonstrate what happens to the body-coordinate constraints when they undergo the transformation, we can substitute the coordinate transformation expressions of Eq. (9.3) into the position constraints of Eq. (7.46). After some manipulation of the terms, we note that all the terms on the left-hand side of the equations vanish. In other words, we end up with $\mathbf{0} = \mathbf{0}$. We can observe the same thing at the velocity and acceleration levels. This is true for any open-chain system as long as the number of defined joint coordinates is equal to the number of system's DoF.

Example 9.1

For the variable-length pendulum, show that when the velocity constraints are transformed from body coordinates to joint coordinates, they vanish.

Solution

We substitute Eq. (9.4) into the velocity constraints from Eq. (7.47):

$$\begin{bmatrix} \mathbf{I} & \breve{\mathbf{s}}_1^A & \mathbf{0} & \mathbf{0} \\ \hline \mathbf{0} & -1 & \mathbf{0} & 1 \\ -\breve{\mathbf{u}}_1' & -\mathbf{u}_1'\mathbf{d} & \breve{\mathbf{u}}_1' & 0 \end{bmatrix} \begin{bmatrix} -\breve{\mathbf{s}}_1^A & \mathbf{0} \\ 1 & 0 \\ \hline -\breve{\mathbf{s}}_1^A + \breve{\mathbf{u}}_1\theta_2 & \mathbf{u}_1 \\ 1 & 0 \end{bmatrix} \begin{Bmatrix} \dot{\theta}_1 \\ \dot{\theta}_2 \end{Bmatrix} = \mathbf{0}$$

Multiplying the two matrices yields

$$\begin{bmatrix} -\breve{\mathbf{s}}_1^A + \breve{\mathbf{s}}_1^A & \mathbf{0} \\ \hline -1+1 & 0 \\ \breve{\mathbf{u}}_1'\breve{\mathbf{s}}_1^A - \mathbf{u}_1'\mathbf{d} + \breve{\mathbf{u}}_1'(-\breve{\mathbf{s}}_1^A + \breve{\mathbf{u}}_1\theta_2) & \breve{\mathbf{u}}_1'\mathbf{u}_1 \end{bmatrix} \begin{Bmatrix} \dot{\theta}_1 \\ \dot{\theta}_2 \end{Bmatrix} \Rightarrow \begin{bmatrix} \mathbf{0} & \mathbf{0} \\ \hline 0 & 0 \\ 0 & 0 \end{bmatrix} \begin{Bmatrix} \dot{\theta}_1 \\ \dot{\theta}_2 \end{Bmatrix}$$

We note that the coefficient matrix of the joint velocities has vanished.

9.3.2 A Three-Body System

This example shows a system consisting of two trees. As shown in Figure 9.6(a) the first tree contains only one body and the second tree contains two bodies. The bodies are numbered as shown. Because this system has three DoF, three joint coordinates are defined as shown in Figure 9.6(b).

The coordinate transformation for body (1) is written as:

$$\begin{aligned} \phi_1 &= \theta_1 \\ \mathbf{r}_1 &= \mathbf{d}_0 - \mathbf{s}_1^A \end{aligned} \qquad (a.1)$$

In the second tree, if we move from the root through the first joint (pin joint C), we reach body (3). The coordinate transformation for this body is:

$$\begin{aligned} \phi_3 &= \theta_3 \\ \mathbf{r}_3 &= -\mathbf{s}_3^C \end{aligned} \qquad (a.2)$$

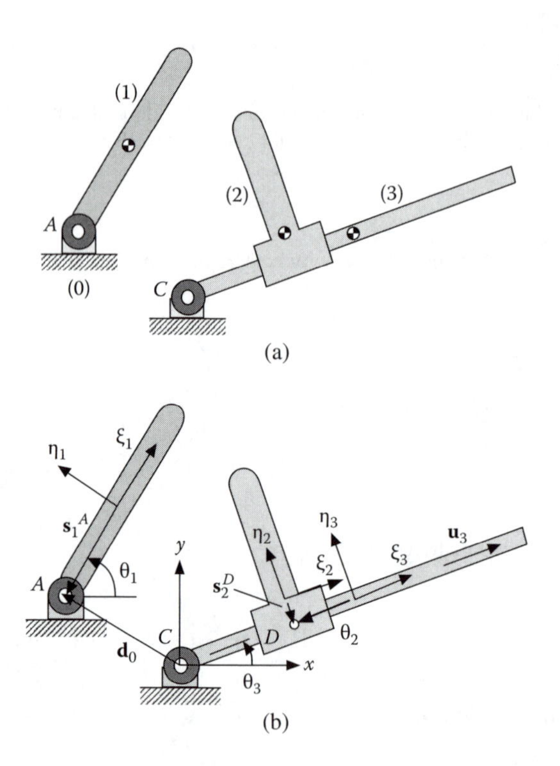

FIGURE 9.6 A system containing two trees.

In the same tree, the sliding joint takes us to body (2). The coordinates of body (2) with respect to body (3) can be written as:

$$\begin{aligned} \phi_2 &= \phi_3 \\ \mathbf{r}_2 &= \mathbf{r}_3 - \mathbf{u}_3\theta_2 - \mathbf{s}_2^D \end{aligned} \qquad (a.3)$$

Equations (*a.1*) to (*a.3*) are arranged for a recursive process to compute the body coordinates.

The time derivative of the velocity expressions provides the velocity transformation expressions as

$$\begin{aligned} \dot{\phi}_1 &= \dot{\theta}_1 \\ \dot{\mathbf{r}}_1 &= -\breve{\mathbf{s}}_1^A\dot{\phi}_1 \\ \dot{\phi}_3 &= \dot{\theta}_3 \\ \dot{\mathbf{r}}_3 &= -\breve{\mathbf{s}}_3^C\dot{\phi}_3 \\ \dot{\phi}_2 &= \dot{\phi}_3 \\ \dot{\mathbf{r}}_2 &= \dot{\mathbf{r}}_3 - \breve{\mathbf{u}}_3\dot{\phi}_3\theta_2 - \mathbf{u}_3\dot{\theta}_2 - \breve{\mathbf{s}}_2^D\dot{\phi}_2 \end{aligned} \qquad (a.4)$$

A forward substitution and rearrangement of terms yield the velocity transformation expression in matrix form as

$$\begin{Bmatrix} \dot{\mathbf{r}}_1 \\ \dot{\phi}_1 \\ \dot{\mathbf{r}}_2 \\ \dot{\phi}_2 \\ \dot{\mathbf{r}}_3 \\ \dot{\phi}_3 \end{Bmatrix} = \begin{bmatrix} -\breve{\mathbf{s}}_1^A & \mathbf{0} & \mathbf{0} \\ 1 & 0 & 0 \\ \mathbf{0} & -\mathbf{u}_3 & -\breve{\mathbf{s}}_3^C - \breve{\mathbf{u}}_3\theta_2 - \breve{\mathbf{s}}_2^D \\ 0 & 0 & 1 \\ \mathbf{0} & \mathbf{0} & -\breve{\mathbf{s}}_3^C \\ 0 & 0 & 1 \end{bmatrix} \begin{Bmatrix} \dot{\theta}_1 \\ \dot{\theta}_2 \\ \dot{\theta}_3 \end{Bmatrix} = \begin{bmatrix} \breve{\mathbf{r}}_1 - \breve{\mathbf{d}}_0 & \mathbf{0} & \mathbf{0} \\ 1 & 0 & 0 \\ \mathbf{0} & -\mathbf{u}_3 & \breve{\mathbf{r}}_2 \\ 0 & 0 & 1 \\ \mathbf{0} & \mathbf{0} & \breve{\mathbf{r}}_3 \\ 0 & 0 & 1 \end{bmatrix} \begin{Bmatrix} \dot{\theta}_1 \\ \dot{\theta}_2 \\ \dot{\theta}_3 \end{Bmatrix} \quad (9.8)$$

This yields the velocity transformation matrix as

$$\mathbf{B} = \begin{bmatrix} -\breve{\mathbf{s}}_1^A & \mathbf{0} & \mathbf{0} \\ 1 & 0 & 0 \\ \mathbf{0} & -\mathbf{u}_3 & -\breve{\mathbf{s}}_3^C - \breve{\mathbf{u}}_3\theta_2 - \breve{\mathbf{s}}_2^D \\ 0 & 0 & 1 \\ \mathbf{0} & \mathbf{0} & -\breve{\mathbf{s}}_3^C \\ 0 & 0 & 1 \end{bmatrix} = \begin{bmatrix} \breve{\mathbf{r}}_1 - \breve{\mathbf{d}}_0 & \mathbf{0} & \mathbf{0} \\ 1 & 0 & 0 \\ \mathbf{0} & -\mathbf{u}_3 & \breve{\mathbf{r}}_2 \\ 0 & 0 & 1 \\ \mathbf{0} & \mathbf{0} & \breve{\mathbf{r}}_3 \\ 0 & 0 & 1 \end{bmatrix} \quad (9.9)$$

The acceleration expressions can be found by taking the time derivative of Eq. (9.8).

9.3.3 A Floating System

An example of a floating multibody system is the robotic arm of the space shuttle. Although this is a spatial system, a simplified planar version of that is depicted in Figure 9.7. Because the orbiting shuttle is not connected to the ground, we assume

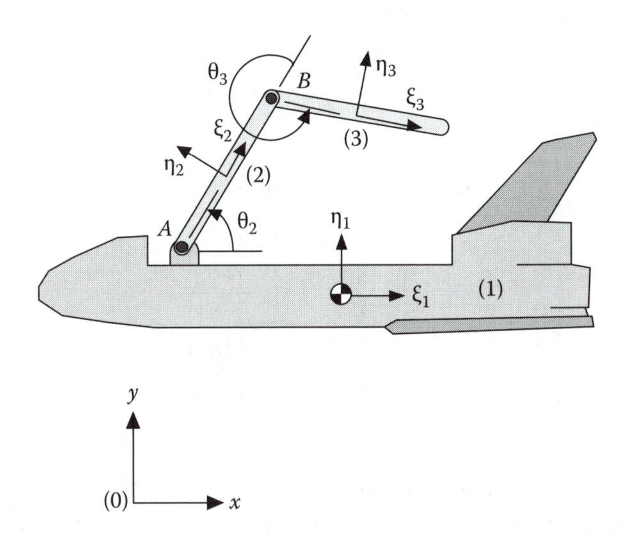

FIGURE 9.7 An example of a floating multibody system.

a floating joint between its main body, body (1), and a nonmoving reference frame x-y. The joint coordinates for this floating joint are the same as the body coordinates of body (1); that is, $\boldsymbol{\theta}_1 = \{x_1 \quad y_1 \quad \phi_1\}'$. For the pin joint at A we consider a relative angle between ξ_1 and ξ_2 axes; that is, $\theta_2 = \phi_2 - \phi_1$. Similarly for the pin joint at B we consider another relative angle $\theta_3 = \phi_3 - \phi_2$.

For the coordinate transformation we have:

$$
\begin{aligned}
\phi_1 &= \phi_1 \\
\mathbf{r}_1 &= \mathbf{r}_1 \\
\phi_2 &= \phi_1 + \theta_2 \\
\mathbf{r}_2 &= \mathbf{r}_1 + \mathbf{s}_1^A - \mathbf{s}_2^A \\
\phi_3 &= \phi_2 + \theta_3 \\
\mathbf{r}_3 &= \mathbf{r}_2 + \mathbf{s}_2^B - \mathbf{s}_3^B
\end{aligned}
\tag{a}
$$

Note that the coordinates of body (1) are part of both body coordinates and the joint coordinates.

The time derivative of the expressions in Eq. (a) after rearrangement yields:

$$
\begin{Bmatrix} \dot{\mathbf{r}}_1 \\ \dot{\phi}_1 \\ \dot{\mathbf{r}}_2 \\ \dot{\phi}_2 \\ \dot{\mathbf{r}}_3 \\ \dot{\phi}_3 \end{Bmatrix}
=
\begin{bmatrix}
\mathbf{I} & \mathbf{0} & \mathbf{0} & \mathbf{0} \\
\mathbf{0} & 1 & 0 & 0 \\
\mathbf{I} & \breve{\mathbf{s}}_1^A - \breve{\mathbf{s}}_2^A & -\breve{\mathbf{s}}_2^A & \mathbf{0} \\
0 & 1 & 1 & 0 \\
\mathbf{I} & \breve{\mathbf{s}}_1^A - \breve{\mathbf{s}}_2^A + \breve{\mathbf{s}}_2^B - \breve{\mathbf{s}}_3^B & -\breve{\mathbf{s}}_2^A + \breve{\mathbf{s}}_2^B - \breve{\mathbf{s}}_3^B & -\breve{\mathbf{s}}_3^B \\
0 & 1 & 1 & 1
\end{bmatrix}
\begin{Bmatrix} \dot{\mathbf{r}}_1 \\ \dot{\phi}_1 \\ \dot{\theta}_2 \\ \dot{\theta}_3 \end{Bmatrix}
$$

$$
=
\begin{bmatrix}
\breve{\mathbf{r}}_1 & \mathbf{0} & \mathbf{0} & \mathbf{0} \\
1 & 0 & 0 & 0 \\
\mathbf{0} & \breve{\mathbf{r}}_2 - \breve{\mathbf{r}}_1 & -\breve{\mathbf{s}}_2^A & \mathbf{0} \\
0 & 1 & 1 & 0 \\
\mathbf{0} & \breve{\mathbf{r}}_3 - \breve{\mathbf{r}}_1 & \breve{\mathbf{r}}_3 - \breve{\mathbf{r}}_2 & -\breve{\mathbf{s}}_3^B \\
0 & 1 & 1 & 1
\end{bmatrix}
\begin{Bmatrix} \dot{\mathbf{r}}_1 \\ \dot{\phi}_1 \\ \dot{\theta}_2 \\ \dot{\theta}_3 \end{Bmatrix}
\tag{9.10}
$$

The velocity transformation matrix can be extracted from this equation. The time derivative of this equation results into the acceleration transformation expressions.

9.3.4 General Formulation

For an open-chain system containing n_b bodies and having n_{dof} degrees of freedom, we define n_{dof} joint coordinates denoted as

$$\boldsymbol{\theta} = \begin{Bmatrix} \theta_1 \\ \vdots \\ \theta_{dof} \end{Bmatrix} \tag{9.11}$$

We then write expressions describing recursively the body coordinates as functions of the joint coordinates:

$$\mathbf{c} = \mathbf{c}(\boldsymbol{\theta}) \tag{9.12}$$

Most of these coordinate expressions are the same as the body-coordinate constraints arranged differently. The time derivative of the coordinate transformations yields the velocity transformation as

$$\dot{\mathbf{c}} = \mathbf{B}\dot{\boldsymbol{\theta}} \tag{9.13}$$

The acceleration transformation expressions are obtained from the time derivative of the velocity expressions as

$$\ddot{\mathbf{c}} = \mathbf{B}\ddot{\boldsymbol{\theta}} + \dot{\mathbf{B}}\dot{\boldsymbol{\theta}} \tag{9.14}$$

There are no constraints associated with the joint coordinates as long as the number of defined coordinates is equal to the number of system's DoF. We can verify this claim easily at the velocity level by substituting Eq. (9.13) into the velocity constraints of Eq. (7.18):

$$\begin{aligned} \dot{\boldsymbol{\Phi}} &= \mathbf{D}\dot{\mathbf{c}} \\ &= \mathbf{D}\mathbf{B}\dot{\boldsymbol{\theta}} = \mathbf{0} \end{aligned} \tag{9.15}$$

Because there are as many joint velocities in $\dot{\boldsymbol{\theta}}$ as the number of DoF, the elements of $\dot{\boldsymbol{\theta}}$ are independent; that is, they can be assigned arbitrary values. Hence, Eq. (9.15) can only be valid if

$$\mathbf{D}\mathbf{B} = \mathbf{0} \tag{9.16}$$

This means that the rows of **D** and the columns of **B** are orthogonal. This is a very important and useful characteristic of matrix **B**.

9.4 OPEN-CHAIN EXAMPLE PROGRAMS

In this section we develop a program to formulate the necessary arrays and matrices for the kinematics of open-chain systems. The program organization follows a strategy similar to that of the body-coordinate formulation. In fact, as shown in Table 9.1, some of the M-files that were constructed in Chapter 7 will be used here as well. The main program is named `JC_formulation`.

```
% file name:  JC_formulation
% Joint coordinate
      clear all
      addpath Basics  BC_Basics  JC_VLP
      BC_global
      BC_condata; JC_vardata;
% Determine components of vectors on the ground
      BC_vectors0;
% Coordinate transformation
      JC_coord_trans; BC_vectors;
% B matrix
      B = JC_Bmatrix;
% Velocity transformation; vectors r_P_d and s_P_d
      c_d = B*theta_d; BC_cd_to_rd; BC_vectors_d;
% B_d*theta-d array
      BdThetad = JC_BdThetad;
```

The path `JC_xx` must be set to the folder that contains the model. The new M-files will be described in the following subsection through an example.

9.4.1 Variable-Length Pendulum

The kinematic expressions for the variable-length pendulum from Section 9.3.1 are formulated in the following M-files. These files must be saved in a folder named `JC_VLP`. In that folder we must provide the constant data in a file named `BC_condata`. The values of the joint coordinates and velocities are provided in a file named `JC_vardata`:

```
% file name: BC_condata
      nb = 2; ndof = 2;
% Constant geometric data
      s_A_0_p = [ 0.0;   0.0]; s_A_1_p = [ 0.0;  0.5];
    u_1_p = [0; -1]; phi_c = 0;
% Establish cells and arrays
      s_P_p = {s_A_1_p}; iB = [1]; nsP = 1;
      s_P0_p = {s_A_0_p}; nsP0 = 1;
      u_p = {u_1_p}; iBu = [1]; nu = 1;
      u0_p = {}; nu0 = 0;
```

```
% File name:  JC_vardata
% Joint coordinates and velocities
      theta = [0.6435; 0.25]; theta_d = [-0.12; 0.2];
```

The coordinate transformations of Eq. (9.3) are formulated in the following M-file. Note that in this process the rotational transformation matrices and the coordinates of some of the points must be updated. At the end of the file, the translational and rotational coordinates are transferred into the c array for later use.

```
function JC_coord_trans
% Joint coordinate to body coordinate transformation
    BC_global
    phi(1) = theta(1); A{1} = A_matrix(phi(1));
        s_P{1} = A{1}*s_P_p{1};
    r{1} = -s_P{1};
    phi(2) = phi(1); A{2} = A_matrix(phi(2));
         u{1} = A{1}*u_p{1};
    r{2} = r{1} + u{1}*theta(2);
    c = BC_r_to_c(r, phi, nb);
```

The velocity transformation matrix from Eq. (9.5) is formulated in the following M-file:

```
function B = JC_Bmatrix
% B matrix
    BC_global
    z2 = [0 0]';
    B = [ s_rot(r{1})   z2
               1          0
          s_rot(r{2})   u{1}
               1           0 ];
```

Finally, the array $\dot{\mathbf{B}}\dot{\boldsymbol{\theta}}$, based on Eq. (9.7), is computed in the following M-file:

```
function BdThetad = JC_BdThetad
% B_d_theta_d array
    BC_global
    z2 = [0 0]';
    BdThetad = [ s_rot(r_d{1})      z2
                      0             0
                 s_rot(r_d{2})     u_d{1}
                      0             0 ]*theta_d;
```

Before we execute the program JC_formulation, we must set the path from the generic JC_xx to JC_VLP. After executing the program we can check for the computed values of the body coordinates, matrix **B**, body velocities, and the array $\dot{\mathbf{B}}\dot{\boldsymbol{\theta}}$ by typing c, B, c_d, and BdThetad in the Command Window:

```
>> c          >> B                      >> c_d        >> BdThetad
   0.3000        0.4000         0         -0.0480       -0.0043
  -0.4000        0.3000         0         -0.0360        0.0058
   0.6435        1.0000         0         -0.1200             0
   0.4500        0.6000    0.6000          0.0480       -0.0449
  -0.6000        0.4500   -0.8000         -0.2140       -0.0202
   0.6435        1.0000         0         -0.1200             0
```

Because there are no constraints associated with the joint coordinates and velocities in this example, the values of these variables in the file `JC_vardata` can be revised to any other desired values.

9.5 CLOSED-CHAIN SYSTEMS

The topology of systems containing closed chains cannot be described as a tree. In a closed chain, or a loop, if we move from one body to another body through a joint and so on, we will end up at the body that we started from. Therefore the first step in the kinematic formulation of such systems is to transform the system into an open-chain system through a process called *cut-joint*.

In the cut-joint process we remove (or cut) one joint from each loop. If the system contains only one loop, we need to remove only one joint. As an example, consider the closed-chain system of Figure 9.1(a). There are four joint in this single loop system. Therefore, there are four possible cut joints. The possible four resultant open-chain systems are shown in Figure 9.8. It is totally our choice to choose which joint to cut in a loop.

The system shown in Figure 9.9(a) contains two loops. One possible cut-joint scenario is to remove the pin joints at A and C to obtain the open-chain system shown in Figure 9.9(b). It should be obvious that many other possibilities of cut-joint exist.

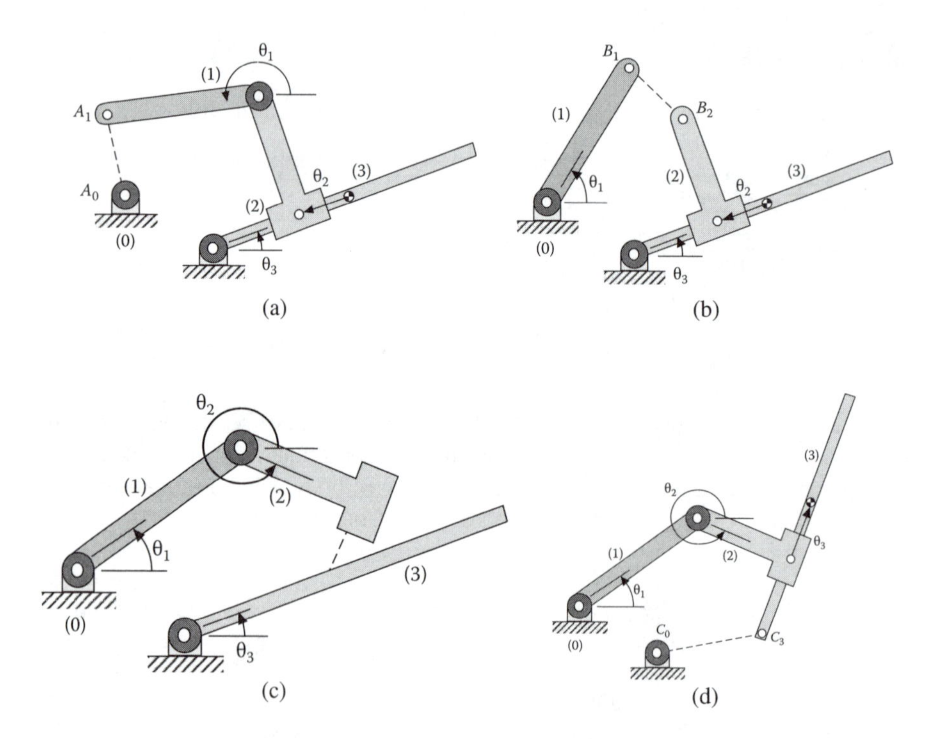

FIGURE 9.8 Four possible open-chain systems as the result of the cut-joint process.

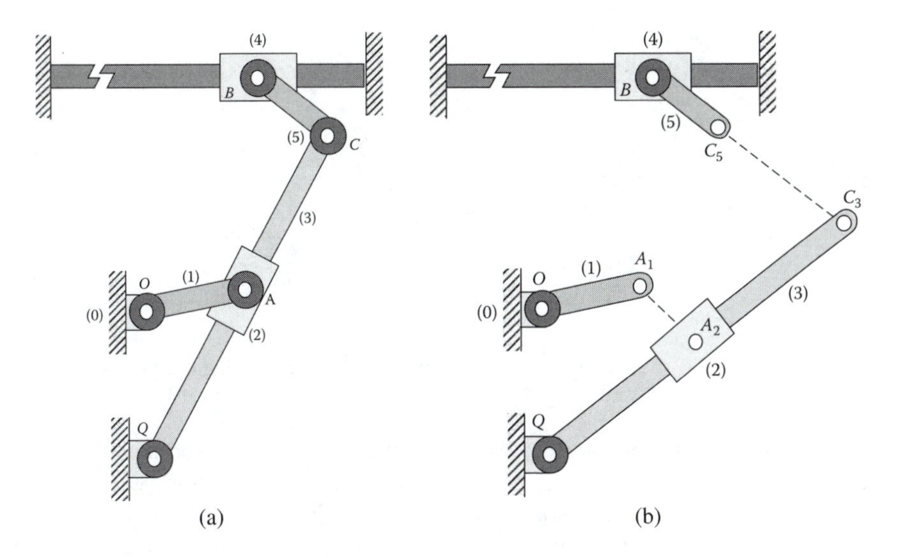

FIGURE 9.9 (a) A multiloop mechanism and (b) its corresponding open-chain systems.

Now that we have become familiar with the cut-joint concept, we can systematically apply the following steps to a closed-chain system:

- We transform the system to an open-chain system through the process of cut-joint.
- For the resultant open-chain system we assign body indices, define joint coordinates, and find the corresponding transformation expressions.
- We put the cut joints back in the system by introducing one or more constraints between the joint coordinates.

Detailed description of these steps is demonstrated through an example before generalization.

9.5.1 Slider-Crank Mechanism

The inverted slider-crank mechanism of Figure 9.1(a) is consider here as an example which is shown again in Figure 9.10(a). As we have already seen, there are four possible cut-joint representations for this system—we choose the one shown in Figure 9.10(b) because this open-chain system has already been formulated in Section 9.3.2.

In the open-chain system of Figure 9.10(b), the three joint coordinates $\boldsymbol{\theta} = \{\theta_1 \quad \theta_2 \quad \theta_3\}'$ are independent. However, when we put the pin joint back between points B_1 and B_2, the three joint coordinates become dependent. The constraint between the joint coordinates is the pin joint constraint between bodies (1) and (2) which can be obtained from Eq. (7.3) as

$$\begin{aligned} {}^{(r,2)}\boldsymbol{\Phi} &= \mathbf{r}_1^B - \mathbf{r}_2^B \\ &= \mathbf{r}_1 + \mathbf{s}_1^B - \mathbf{r}_2 + \mathbf{s}_2^B = \mathbf{0} \end{aligned} \tag{9.17}$$

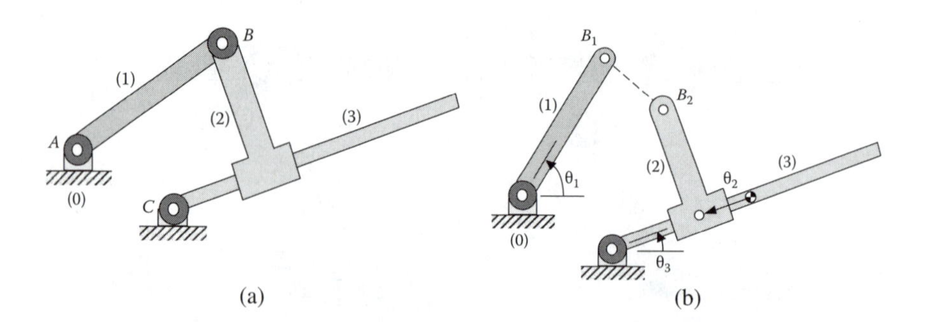

FIGURE 9.10 (a) A slider-crank mechanism and (b) its corresponding cut-joint system.

Although these equations are expressed in terms of the body coordinates, if we substitute the transformation expression (*a.1*) to (*a.3*) from Section 9.3.2 in these constraints, the constraints become functions of the joint coordinates. This substitution, however, is not performed in analytical form because the constraints and the transformation expressions are nonlinear in the coordinates.

The velocity constraint for the pin joint are obtained from Eq. (7.28) as

$$ {}^{(r,2)}\dot{\mathbf{\Phi}} = \begin{bmatrix} \mathbf{I} & \breve{\mathbf{s}}_1^B & -\mathbf{I} & -\breve{\mathbf{s}}_2^B \end{bmatrix} \begin{Bmatrix} \dot{\mathbf{r}}_1 \\ \dot{\phi}_1 \\ \dot{\mathbf{r}}_2 \\ \dot{\phi}_2 \end{Bmatrix} = \mathbf{0} \tag{9.18} $$

The portion of the velocity transformation expression of Eq. (9.8) for bodies (1) and (2) is

$$ \begin{Bmatrix} \dot{\mathbf{r}}_1 \\ \dot{\phi}_1 \\ \dot{\mathbf{r}}_2 \\ \dot{\phi}_2 \end{Bmatrix} = \begin{bmatrix} \breve{\mathbf{r}}_1 - \breve{\mathbf{d}}_0 & \mathbf{0} & \mathbf{0} \\ 1 & 0 & 0 \\ \mathbf{0} & -\mathbf{u}_3 & \breve{\mathbf{r}}_2 \\ 0 & 0 & 1 \end{bmatrix} \begin{Bmatrix} \dot{\theta}_1 \\ \dot{\theta}_2 \\ \dot{\theta}_3 \end{Bmatrix} $$

Substituting this expression into Eq. (9.18) results in

$$ {}^{(r,2)}\dot{\mathbf{\Phi}} = \begin{bmatrix} \breve{\mathbf{r}}_1^B - \breve{\mathbf{d}}_0 & \mathbf{u}_3 & -\breve{\mathbf{r}}_2^B \end{bmatrix} \begin{Bmatrix} \dot{\theta}_1 \\ \dot{\theta}_2 \\ \dot{\theta}_3 \end{Bmatrix} = \mathbf{0} \tag{9.19} $$

Now we have the velocity constraints expressed in terms of the joint velocities. We can extend this process one more step to obtain the joint acceleration constraints.

At this point we can compare the velocity constraints of Eq. (9.19) against the velocity constraints of Eq. (9.2) that were obtained by the vector-loop method. If we examine the two sets of constraints, we conclude that they are identical. The equivalent terms are:

$$\mathbf{d}_1 \equiv \mathbf{r}_1^B - \mathbf{d}_0$$

$$-\mathbf{u}_3\dot{d}_3 \equiv \mathbf{u}_3\dot{\theta}_2$$

$$\mathbf{d}_2 + d_3\,\mathbf{u}_3 \equiv \mathbf{r}_2^B$$

where the terms from Eq. (9.2) are on the left-hand side and those from Eq. (9.19) are on the right-hand side. We must note that $-\dot{d}_3$ in Eq. (9.2) is equal to $\dot{\theta}_2$ in Eq. (9.19) because vectors d_3 and θ_2 are in opposite directions and $d_3 - \theta_2$ is a constant.

Comparing the position constraints, as well as the velocity constraints, between the vector-loop method and the joint coordinate method results into the same conclusion—the two methods yield the same constraints.

9.5.2 General Formulation

For a closed-chain system assume that the corresponding open-chain system, after the cut-joints are removed, is represented by a set of joint coordinates as in Eq. (9.11). For this open-chain system the coordinate, velocity, and acceleration transformations are expressed as in Eqs. (9.12) to (9.14).

When the cut joints are put back into the system, the constraint equations for these joints from the body coordinate formulation must be considered. The position constraints can be written in a general form as

$$^*\mathbf{\Phi}(\mathbf{c}) = {}^*\mathbf{\Phi}(\mathbf{c}(\mathbf{\theta})) = \mathbf{0} \tag{9.20}$$

where the left-superscript * indicates an entity associated with the cut joints. For the velocity constraints we have

$$\begin{aligned} ^*\dot{\mathbf{\Phi}} &= {}^*\mathbf{D}\dot{\mathbf{c}} \\ &= {}^*\mathbf{D}\mathbf{B}\dot{\mathbf{\theta}} = \mathcal{D}\dot{\mathbf{\theta}} = \mathbf{0} \end{aligned} \tag{9.21}$$

where

$$\mathcal{D} = {}^*\mathbf{D}\mathbf{B} \tag{9.22}$$

is the Jacobian matrix of the cut joints transformed into the joint coordinates (velocities). Similarly the acceleration constraints for the cut joints are transformed into the joint coordinate form as

$$\begin{aligned} ^*\ddot{\mathbf{\Phi}} &= {}^*\mathbf{D}\ddot{\mathbf{c}} + {}^*\dot{\mathbf{D}}\dot{\mathbf{c}} \\ &= {}^*\mathbf{D}(\mathbf{B}\ddot{\mathbf{\theta}} + \dot{\mathbf{B}}\ddot{\mathbf{\theta}}) + {}^*\dot{\mathbf{D}}\mathbf{B}\dot{\mathbf{\theta}} \\ &= \mathcal{D}\ddot{\mathbf{\theta}} + ({}^*\mathbf{D}\dot{\mathbf{B}} + {}^*\dot{\mathbf{D}}\mathbf{B})\dot{\mathbf{\theta}} \\ &= \mathcal{D}\ddot{\mathbf{\theta}} + \dot{\mathcal{D}}\ddot{\mathbf{\theta}} = \mathbf{0} \end{aligned} \tag{9.23}$$

where

$$\dot{D} = {}^*\mathbf{D}\dot{\mathbf{B}} + {}^*\dot{\mathbf{D}}\mathbf{B} \tag{9.24}$$

The acceleration constraints can also be expressed as

$$D\dot{\theta} = \gamma \tag{9.25}$$

where

$$\begin{aligned} \gamma &= -\dot{D}\dot{\theta} \\ &= -{}^*\mathbf{D}\dot{\mathbf{B}}\dot{\theta} - {}^*\dot{\mathbf{D}}\dot{\mathbf{c}} \\ &= -{}^*\mathbf{D}\dot{\mathbf{B}}\dot{\theta} + {}^*\gamma \end{aligned} \tag{9.26}$$

The term ${}^*\gamma$ represents the right-hand side of the acceleration constraints for the cut joint in the body-coordinate formulation as described in Eq. (7.21). The right-hand side of the acceleration constraints for the cut joint in the joint coordinate formulation can be computed by either expression in Eq. (9.26).

9.6 CLOSED-CHAIN EXAMPLE PROGRAMS

In this section we extend the program from Section 9.4 to compute additional entities associated with the closed-chain systems. Here we append several statements to the file `JC_formulation` to compute the cut-joint constraints, ${}^*\mathbf{\Phi}(\mathbf{c})$, the corresponding Jacobian, D, and the right-hand-side array of the acceleration constraints, γ. However, because we would like to use the program `JC_formulation` for both open- and closed-chain systems, we need to provide some logic in the program to skip over the cut-joint constraint computations when a system is open-chain.

To determine whether a model contains constraints or not, the program can use the data from the file `BC_condata` for the number of bodies, `nb`, and the number of degrees of freedom, `ndof`. The program can calculate the number of joint coordinates and the number of cut-joint constraints as

```
njc = nb

nPhi_cut = njc - ndof
```

Then, if `nPhi_cut` is greater than zero, the program continues to compute the entities associated with the cut-joint constraints.

The following statements must be appended to the program `BC_formulation`:

```
% file name:  JC_formulation
% ...
    njc = nb; nPhi_cut = njc - ndof;
    if nPhi_cut > 0
    % Cut-joint constraints, D matrix, gamma array
        Phi_cut = JC_Phi; D_cut = JC_jacob; gamma_cut = JC_gamma;
    end
```

These statements invoke several function M-files to compute the constraints, the Jacobian, and the right-hand side of the acceleration constraints for the cut joints. Details of these functions will be demonstrated in the upcoming examples. It is important to note that some of these entities can be constructed in more than one way. For example, matrix $\boldsymbol{D}$ can be determined either numerically by the product ${}^*\mathbf{DB}$, according to Eq. (9.22), or it can first be constructed in analytical form before being evaluated numerically. For complex and large-scale systems, it is more convenient to construct and evaluate matrices ${}^*\mathbf{D}$ and $\mathbf{B}$, and then to determine $\boldsymbol{D}$ numerically. In simple problems, such as some of our examples, we determine the entries of $\boldsymbol{D}$ in closed form first. Similarly, we can evaluate the array $\boldsymbol{\gamma}$ either numerically or we can first construct it analytically before evaluating it numerically. According to Eq. (9.26), for complex problems we evaluate $\boldsymbol{\gamma}$ numerically as $-{}^*\mathbf{D}\dot{\mathbf{B}}\dot{\boldsymbol{\theta}} + {}^*\boldsymbol{\gamma}$, however in simple problems we may evaluate $\boldsymbol{\gamma}$ explicitly as $-\dot{\boldsymbol{D}}\dot{\boldsymbol{\theta}}$.

9.6.1 Double A-Arm Suspension

The double A-arm suspension system is considered here for kinematic modeling with the joint-coordinate formulation. This system was modeled kinematically by the body coordinate method in Sections 7.3.1 and 7.6.1. For the joint coordinate formulation, because this system has one closed chain, we remove the pin joint at B to create an open-chain system as shown in Figure 9.11. For the open-chain system we define three joint coordinates, θ_1, θ_2 and θ_3, that are equivalent to the absolute rotational coordinates of their corresponding bodies. We show the formulation for

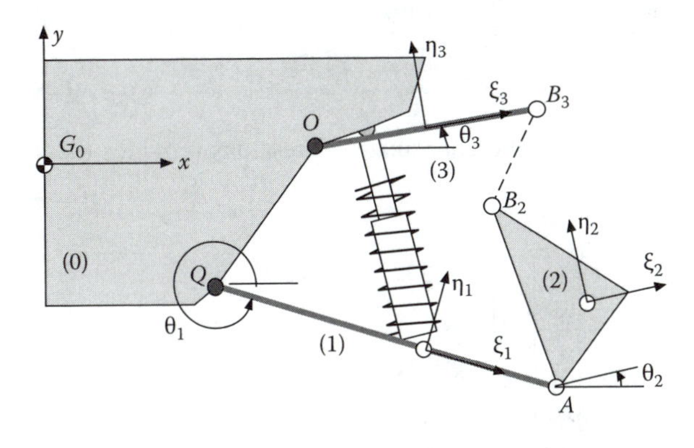

FIGURE 9.11 The double A-arm suspension system adopted for the joint-coordinate formulation.

this system as we describe the corresponding M-files. All the M-files for this model must be saved in a folder named JC_AA.

Based on Figure 9.11, we write the coordinate transformation expressions for the cut open-chain system as:

$$
\begin{aligned}
\phi_1 &= \theta_1 \\
\mathbf{r}_1 &= \mathbf{r}_0^Q - \mathbf{s}_1^Q \\
\phi_2 &= \theta_2 \\
\mathbf{r}_2 &= \mathbf{r}_1 + \mathbf{s}_1^A - \mathbf{s}_2^A \\
\phi_3 &= \theta_3 \\
\mathbf{r}_3 &= \mathbf{r}_0^O - \mathbf{s}_3^O
\end{aligned}
\tag{9.27}
$$

These expressions are evaluated in the following function M-file. Note that every time a new value for any of the rotational coordinates is determined, its corresponding **A** matrix is updated. At the end, the body coordinates are copied into the array c.

```
function JC_coord_trans
% Joint coordinate to body coordinate transformation
    BC_global
    phi(1) = theta(1); A{1} = A_matrix(phi(1));
        s_P{1} = A{1}*s_P_p{1};
    r{1} = r_P0{2} - s_P{1};
        s_P{2} = A{1}*s_P_p{2};
    phi(2) = theta(2); A{2} = A_matrix(phi(2));
        s_P{3} = A{2}*s_P_p{3};
    r{2} = r{1} + s_P{2} - s_P{3};
    phi(3) = theta(3); A{3} = A_matrix(phi(3));
        s_P{6} = A{3}*s_P_p{6};
    r{3} = r_P0{1} - s_P{6};
    c = BC_r_to_c(r, phi, nb);
```

The time derivative of the coordinate expressions provides the velocity expressions as

$$
\begin{Bmatrix} \dot{\mathbf{r}}_1 \\ \dot{\phi}_1 \\ \dot{\mathbf{r}}_2 \\ \dot{\phi}_2 \\ \dot{\mathbf{r}}_3 \\ \dot{\phi}_3 \end{Bmatrix} =
\begin{bmatrix}
-\breve{\mathbf{s}}_1^Q & \mathbf{0} & \mathbf{0} \\
1 & 0 & 0 \\
-\breve{\mathbf{s}}_1^Q + \breve{\mathbf{s}}_1^A & -\breve{\mathbf{s}}_2^A & \mathbf{0} \\
0 & 1 & 0 \\
\mathbf{0} & \mathbf{0} & -\breve{\mathbf{s}}_3^O \\
0 & 0 & 1
\end{bmatrix}
\begin{Bmatrix} \dot{\theta}_1 \\ \dot{\theta}_2 \\ \dot{\theta}_3 \end{Bmatrix}
\tag{9.28}
$$

The coefficient matrix is the **B** matrix that we evaluate by the following function:

```
function B = JC_Bmatrix
% B matrix
    BC_global
    z2 = [0 0]';
    B = [-s_rot(s_P{1})           z2                z2
           1                       0                 0
         -s_rot(s_P{1}-s_P{2})  -s_rot(s_P{3})     z2
          0                        1                 0
          z2                       z2              -s_rot(s_P{6})
          0                        0                 1 ];
```

The next function evaluates the array $\dot{\mathbf{B}}\dot{\boldsymbol{\theta}}$. This M-file is presented in a form for clarity and not for computational efficiency—it is unnecessary to carry along all the zero entries in the matrix $\dot{\mathbf{B}}$ and multiply them by the joint velocities. We should rewrite this file and construct $\dot{\mathbf{B}}\dot{\boldsymbol{\theta}}$ directly as an array and not as a product of a matrix and an array.

```
function BdThetad = JC_BdThetad
% B_d_theta_d array
    BC_global
    z2 = [0 0]';
    BdThetad = ...
         [s_P{1}*phi_d(1)              z2                  z2
              0                         0                   0
         (s_P{1}-s_P{2})*phi_d(1)   s_P{3}*phi_d(2)        z2
              0                         0                   0
              z2                        z2               s_P{6}*phi_d(3)
              0                         0                   0 ]*theta_d;
```

When the cut joint is put back in the system, the constraint equations for the pin joint at B between bodies (2) and (3) must be considered:

$$^{*}\boldsymbol{\Phi} = \mathbf{r}_3^B - \mathbf{r}_2^B = \mathbf{0} \tag{9.29}$$

The cut-joint constraints are evaluated by the following function:

```
function Phi_cut = JC_Phi
% Cut-joint constraints
    BC_global
    Phi_cut = BC_Phi_rev(r_P{4}, r_P{5});
```

The cut-joint velocity constraints are written as

$$\begin{bmatrix} -\mathbf{I} & -\breve{\mathbf{s}}_2^B & \mathbf{I} & \breve{\mathbf{s}}_3^B \end{bmatrix} \begin{Bmatrix} \dot{\mathbf{r}}_2 \\ \dot{\phi}_2 \\ \dot{\mathbf{r}}_3 \\ \dot{\phi}_3 \end{Bmatrix} = \mathbf{0}$$

The cut-joint Jacobian is ${}^*\mathbf{D} = \begin{bmatrix} -\mathbf{I} & -\breve{\mathbf{s}}_2^B & \mathbf{I} & \breve{\mathbf{s}}_3^B \end{bmatrix}$. We may write an M-file to evaluate ${}^*\mathbf{D}$, and because we already have matrix $\mathbf{B}$, we can simply evaluate the product ${}^*\mathbf{DB}$ to obtain $\boldsymbol{D}$. However, because both ${}^*\mathbf{D}$ and $\mathbf{B}$ are simple and available in closed forms, we determine the product ${}^*\mathbf{DB}$ analytically. We substitute the portion of Eq. (9.28) associated with bodies (2) and (3) in the velocity constraint, we transform the constraints to joint velocities:

$$\begin{bmatrix} -\breve{\mathbf{s}}_1^{A,Q} & -\breve{\mathbf{s}}_2^{B,A} & \breve{\mathbf{s}}_3^{B,O} \end{bmatrix} \begin{Bmatrix} \dot{\theta}_1 \\ \dot{\theta}_2 \\ \dot{\theta}_3 \end{Bmatrix} = \mathbf{0} \tag{9.30}$$

where

$$\mathbf{s}_1^{A,Q} = \mathbf{s}_1^A - \mathbf{s}_1^Q, \quad \mathbf{s}_2^{B,A} = \mathbf{s}_2^B - \mathbf{s}_2^A, \quad \mathbf{s}_3^{B,O} = \mathbf{s}_3^B - \mathbf{s}_3^O$$

and

$$\boldsymbol{D} = \begin{bmatrix} -\breve{\mathbf{s}}_1^{A,Q} & -\breve{\mathbf{s}}_2^{B,A} & \breve{\mathbf{s}}_3^{B,O} \end{bmatrix} \tag{9.31}$$

The following function evaluates the cut-joint Jacobian $\boldsymbol{D}$:

```
function D_cut = JC_jacob
% Cut-joint D matrix
    BC_global
    s_AQ = s_P{2} - s_P{1}; s_BA = s_P{4} - s_P{3};
    s_BO = s_P{5} - s_P{6};
    D_cut = [-s_rot(s_AQ)    -s_rot(s_BA)    s_rot(s_BO];
```

Because the $\boldsymbol{D}$ matrix is available in closed form, the γ array can be expressed as

$$\begin{aligned} \gamma = -\dot{\boldsymbol{D}}\dot{\theta} &= -\begin{bmatrix} -\dot{\mathbf{s}}_1^{A,Q} & -\dot{\mathbf{s}}_2^{B,A} & \dot{\mathbf{s}}_3^{B,O} \end{bmatrix} \begin{Bmatrix} \dot{\theta}_1 \\ \dot{\theta}_2 \\ \dot{\theta}_3 \end{Bmatrix} \\ &= \dot{\mathbf{s}}_1^{A,Q}\dot{\theta}_1 + \dot{\mathbf{s}}_2^{B,A}\dot{\theta}_2 - \dot{\mathbf{s}}_3^{B,O}\dot{\theta}_3 \\ &= -\mathbf{s}_1^{A,Q}(\dot{\theta}_1)^2 - \mathbf{s}_2^{B,A}(\dot{\theta}_2)^2 + \mathbf{s}_3^{B,O}(\dot{\theta}_3)^2 \end{aligned} \tag{9.32}$$

The following function evaluates this array:

```
function gamma_cut = JC_gamma
% Cut-joint -D_cut_d*theta_d
    BC_global
    s_AQ_d = s_P_d{2} - s_P_d{1}; s_BA_d = s_P_d{4} - s_P_d{3};
    s_BO_d = s_P_d{5} - s_P_d{6};
    gamma_cut = s_rot(s_AQ_d*theta_d(1) + s_BA_d*theta_d(2) ...
                - s_BO_d*theta_d(3));
```

The program `JC_formulation` accesses the constant data for this model from the file `BC_condata`. A new script named `JC_vardata` is developed that provides the values of the joint coordinates and velocities.

```
% file name:   JC_vardata
% Joint coordinates and velocities
     theta   = [6.2800;   0.0735;   6.5000];
     theta_d = [2.8392;   1.1720;   4.4837];
```

We execute the program `JC_formulation` and we check the values for the body coordinates and velocities, and the **B** matrix:

```
>> c            >> c_d          >> B
    0.4400          0.0022          0.0008         0         0
   -0.1408          0.6814          0.2400         0         0
    6.2800          2.8392          1.0000         0         0
    0.6825         -0.1191          0.0013   -0.1049         0
   -0.0365          1.2657          0.4200    0.0625         0
    0.0735          1.1720               0    1.0000         0
    0.4470         -0.1254               0         0   -0.0280
    0.0480          0.5692               0         0    0.1270
    6.5000          4.4837               0         0    1.0000
```

We can also check if the cut-joint constraints are satisfied at the coordinate and velocity levels by typing, in the Command Window, `Phi_cut` and `D_cut*theta_d`:

```
>> Phi_cut            >> D_cut*theta_d
   1.0e-04 *             1.0e-04 *
   -0.1064               -0.1514
    0.6602                0.1005
```

These values are very small. This indicates that the values of the three joint coordinates and velocities were not assigned arbitrarily. We also check for the $\boldsymbol{D}$ matrix and the γ array:

```
>> D_cut                                          >> gamma_cut
   -0.0013     0.2172    -0.0559                     1.7822
   -0.4200     0.0461     0.2539                     0.8369
```

9.6.2 MacPherson Suspension

The MacPherson suspension system was modeled kinematically with the body coordinate method in Sections 7.3.2 and 7.6.2 in three different forms. In this section we consider only two of the three forms—Models A and B—for kinematic modeling with the joint coordinate method. For these models, we save the corresponding M-files in folders named `JC_MP_A` and `JC_MP_B` respectively. A copy of the `BC_condata` for each of these models should be provided in these two folders as well.

Model A: We consider the body-coordinate description of this system from Figure 7.13. The pin joint at A is cut to create an open-chain system as shown in Figure 9.12. We define two rotational join coordinates, θ_1 and θ_3, and one sliding joint coordinate, θ_2.

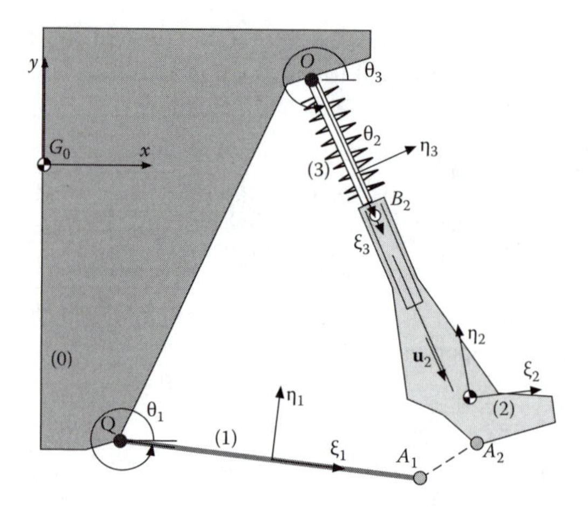

FIGURE 9.12 The MacPherson suspension modeled with three moving bodies.

The coordinate transformation expressions are written as:

$$\begin{aligned}
\phi_1 &= \theta_1 \\
\mathbf{r}_1 &= \mathbf{r}_0^Q - \mathbf{s}_1^Q \\
\phi_3 &= \theta_3 \\
\mathbf{r}_3 &= \mathbf{r}_0^O - \mathbf{s}_3^O \\
\phi_2 &= \phi_3 + {}^c\phi \\
\mathbf{r}_2 &= \mathbf{r}_0^O + \mathbf{u}_2\theta_2 - \mathbf{s}_2^B
\end{aligned} \tag{9.33}$$

where ${}^c\phi = 61.9^\circ = 1.08$ rad is the constant angle between ξ_2 and ξ_3 axes. The following function M-file performs the transformation.

```
function JC_coord_trans
% Joint coordinate to body coordinate transformation
     BC_global
     phi(1) = theta(1); A{1} = A_matrix(phi(1));
          s_P{1} = A{1}*s_P_p{1};
     r{1} = r_P0{2} - s_P{1};
     phi(3) = theta(3); A{3} = A_matrix(phi(3));
          s_P{5} = A{3}*s_P_p{5};
     r{3} = r_P0{1} - s_P{5};
     phi(2) = phi(3) + phi_c; A{2} = A_matrix(phi(2));
          u{2} = A{2}*u_p{2};
          r_P{4} = r_P0{1} + u{2}*theta(2); s_P{4} = A{2}*s_P_p{4};
     r{2} = r_P{4} - s_P{4};
     c = BC_r_to_c(r, phi, nb);
```

The time derivative of the coordinate transformations yields the **B** matrix as part of the velocity transformation expressions:

$$\begin{Bmatrix} \dot{\mathbf{r}}_1 \\ \dot{\phi}_1 \\ \dot{\mathbf{r}}_2 \\ \dot{\phi}_2 \\ \dot{\mathbf{r}}_3 \\ \dot{\phi}_3 \end{Bmatrix} = \begin{bmatrix} -\breve{\mathbf{s}}_1^Q & \mathbf{0} & \mathbf{0} \\ 1 & 0 & 0 \\ \mathbf{0} & \mathbf{u}_2 & \breve{\mathbf{u}}_2\theta_2 - \breve{\mathbf{s}}_2^B \\ 0 & 0 & 1 \\ \mathbf{0} & \mathbf{0} & -\breve{\mathbf{s}}_3^O \\ 0 & 0 & 1 \end{bmatrix} \begin{Bmatrix} \dot{\theta}_1 \\ \dot{\theta}_2 \\ \dot{\theta}_3 \end{Bmatrix} \tag{9.34}$$

The following function constructs the **B** matrix:

```
function B = JC_Bmatrix
% B matrix
     BC_global
     z2 = [0 0]';
     B = [-s_rot(s_P{1})   z2        z2
               1            0         0
               z2           u{2}      s_rot(u{2}*theta(2)-s_P{4})
               0            0         1
               z2           z2       -s_rot(s_P{5})
               0            0         1 ];
```

The array $\dot{\mathbf{B}}\dot{\boldsymbol{\theta}}$ is constructed by the following function:

```
function BdThetad = JC_BdThetad
% B_d_theta_d array
     BC_global
     BdThetad = ...
         [s_P{1}*phi_d(1)*theta_d(1)
            0
          u_d{2}*theta_d(2) + ...
          s_rot(u_d{2}*theta(2) + u{2}*theta_d(2) - s_P_d{4})*theta_d(3)
            0
          s_P{5}*phi_d(3)*theta_d(3)
            0 ];
```

The cut-joint constraints for the pin joint at A between bodies (1) and (2) are:

$$^{(*)}\boldsymbol{\Phi} = \mathbf{r}_2 + \mathbf{s}_2^A - \mathbf{r}_1 - \mathbf{s}_1^A = \mathbf{0} \tag{9.35}$$

The following function evaluates the cut-joint constraints.

```
function Phi_cut = JC_Phi
% Cut-joint constraints
     BC_global
     Phi_cut = BC_Phi_rev(r_P{2}, r_P{3});
```

The cut-joint Jacobian in body-coordinates is described as

$$^{*}\mathbf{D} = \begin{bmatrix} -\mathbf{I} & -\breve{\mathbf{s}}_1^A & \mathbf{I} & \breve{\mathbf{s}}_2^A \end{bmatrix}$$

The product $^{*}\mathbf{DB}$, after eliminating the rows of $\mathbf{B}$ associated with body (3), yields

$$\begin{aligned} \boldsymbol{D} &= \begin{bmatrix} -\breve{\mathbf{s}}_1^{A,Q} & \mathbf{u}_2 & \breve{\mathbf{u}}_2\theta_2 + \breve{\mathbf{s}}_2^{A,B} \end{bmatrix} \\ &= \begin{bmatrix} -\breve{\mathbf{s}}_1^{A,Q} & \mathbf{u}_2 & \breve{\mathbf{r}}_2^A - \breve{\mathbf{r}}_0^O \end{bmatrix} \\ &= \begin{bmatrix} \breve{\mathbf{r}}_0^Q - \breve{\mathbf{r}}_1^A & \mathbf{u}_2 & \breve{\mathbf{r}}_2^A - \breve{\mathbf{r}}_0^O \end{bmatrix} \end{aligned} \tag{9.36}$$

where

$$\mathbf{s}_1^{A,Q} = \mathbf{s}_1^A - \mathbf{s}_1^Q, \ \mathbf{s}_2^{A,B} = \mathbf{s}_2^A - \mathbf{s}_2^B$$

The following function evaluates this matrix.

```
function D_cut = JC_jacob
% Cut-joint D matrix
    BC_global
    D_cut = [s_rot(r_P0{2} - r_P{2})  u{2}  s_rot(r_P{3} - r_P0{1})];
```

The $\boldsymbol{D}$ matrix can also be obtained numerically as the product of $^{*}\mathbf{DB}$ as provided in the following function M-file. We may use either of the two M-files.

```
function D_cut = JC_jacob
% Cut-joint D matrix
    BC_global
    [jac_A_1 jac_A_2] = BC_jacob_rev(s_P{2}, s_P{3});
        z23     = zeros(2,3);
    D_star = [ jac_A_1     jac_A_2       z23 ];
    D_cut  = D_star*B;
```

The γ array is obtained as

$$\gamma = -\boldsymbol{D}\boldsymbol{\theta} = -\begin{bmatrix} -\mathbf{r}_1^A & \mathbf{u}_2 & \mathbf{r}_2^A \end{bmatrix} \begin{Bmatrix} \theta_1 \\ \theta_2 \\ \theta_3 \end{Bmatrix} \tag{9.37}$$

The following function evaluates this array.

```
function gamma_cut = JC_gamma
% Cut-joint gamma array
    BC_global
      gamma_cut = -[-s_rot(r_P_d{2})  u_d{2}  s_rot(r_P_d{3})]*theta_d;
```

A set of values for the joint coordinates and velocities are provided in the following file. These values must satisfy the cut-joint constraints at the coordinate and velocity levels.

```
% File name:   JC_vardata
% Joint coordinates and velocities
    theta = [0; 0.1995; 5.00]; theta_d = [2.511; -1.085; 0.570];
```

In the program `JC_formulation` we set the path to `JC_MP_A` folder. Based on the input values for the joint coordinates and velocities, the program computes the body coordinates, velocities, and the **B** matrix to have the following values:

```
>> c              >> c_d            >> B
     0.3450                0            0          0         0
    -0.4100           0.5650       0.2250          0         0
          0           2.5110       1.0000          0         0
     0.5825          -0.0389            0     0.2828    0.4700
    -0.3400           1.1364            0    -0.9567    0.1725
     6.0800           0.5700            0          0    1.0000
     0.4525           0.0820            0          0    0.1438
    -0.0138           0.0243            0          0    0.0425
     5.0000           0.5700            0          0    1.0000
```

The cut-joint constraints, the γ array, and the $\boldsymbol{D}$ matrix have the following values:

```
>> Phi_cut        >> gamma_cut     >> D_cut
    -0.0017          -1.6025             0     0.2828    0.5386
     0.0014           0.1748       -0.4500    -0.9567    0.1583
```

We can also evaluate the cut-joint velocity constraints, by typing in the Command Window:

```
>> D_cut*theta_d
     0.0002
    -0.0016
```

Model B: In this model we cut the revolute-translational joint between point O and body (2) as shown in Figure 9.13.* We define two rotational joint coordinates θ_1 and θ_2.

The coordinate transformation expressions are written as

$$
\begin{aligned}
\phi_1 &= \theta_1 \\
\mathbf{r}_1 &= \mathbf{r}_0^Q - \mathbf{s}_1^Q \\
\phi_2 &= \theta_2 \\
\mathbf{r}_2 &= \mathbf{r}_1 + \mathbf{s}_1^A - \mathbf{s}_2^A
\end{aligned}
\tag{9.38}
$$

* Instead of cutting the revolute-translational joint, if we cut the revolute joint as A, we will end up with three joint coordinates and two algebraic constraints.

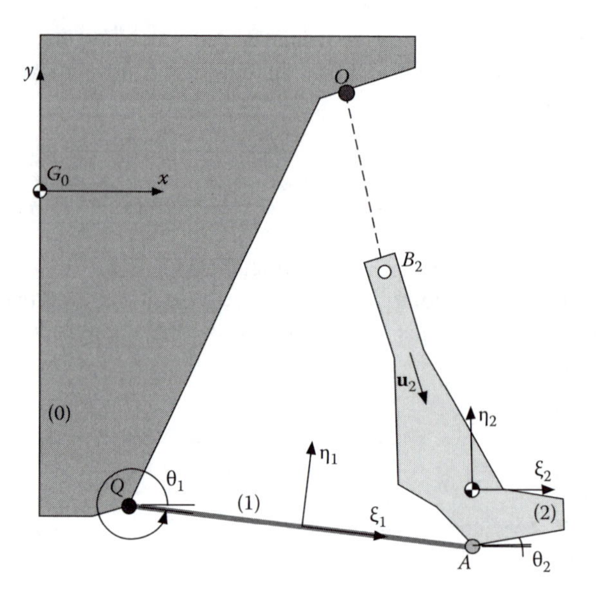

FIGURE 9.13 The MacPherson suspension modeled with two moving bodies.

The following script file performs the transformation.

```
function JC_coord_trans
% Joint coordinate to body coordinate transformation
    BC_global
    phi(1) = theta(1); A{1} = A_matrix(phi(1));
        s_P{1} = A{1}*s_P_p{1};
    r{1} = r_P0{2} - s_P{1};
        s_P{2} = A{1}*s_P_p{2};
    phi(2) = theta(2); A{2} = A_matrix(phi(2));
        s_P{3} = A{2}*s_P_p{3};
    r{2} = r{1} + s_P{2} - s_P{3};
    c = BC_r_to_c(r, phi, nb);
```

The velocity transformation expressions are

$$\begin{Bmatrix} \dot{\mathbf{r}}_1 \\ \dot{\phi}_1 \\ \dot{\mathbf{r}}_2 \\ \dot{\phi}_2 \end{Bmatrix} = \begin{bmatrix} -\breve{\mathbf{s}}_1^Q & \mathbf{0} \\ 1 & 0 \\ -\breve{\mathbf{s}}_1^Q + \breve{\mathbf{s}}_1^A & -\breve{\mathbf{s}}_2^A \\ 0 & 1 \end{bmatrix} \begin{Bmatrix} \dot{\theta}_1 \\ \dot{\theta}_2 \end{Bmatrix} \tag{9.39}$$

```
function B = JC_Bmatrix
% B matrix
    BC_global
    z2 = [0 0]';
    B = [-s_rot(s_P{1})             z2
           1                        0
           s_rot(s_P{2} - s_P{1})   -s_rot(s_P{3})
           0                        1 ];
```

The term $\dot{\mathbf{B}}\dot{\boldsymbol{\theta}}$ can be expressed as:

$$\dot{\mathbf{B}}\dot{\boldsymbol{\theta}} = \begin{bmatrix} -\dot{\breve{\mathbf{s}}}_1^Q & \mathbf{0} \\ 0 & 0 \\ -\dot{\breve{\mathbf{s}}}_1^Q + \dot{\breve{\mathbf{s}}}_1^A & -\dot{\breve{\mathbf{s}}}_2^A \\ 0 & 0 \end{bmatrix} \begin{Bmatrix} \dot{\theta}_1 \\ \dot{\theta}_2 \end{Bmatrix} = \begin{bmatrix} \mathbf{s}_1^Q \dot{\phi}_1 & \mathbf{0} \\ 0 & 0 \\ (\mathbf{s}_1^Q - \mathbf{s}_1^A)\dot{\phi}_1 & \mathbf{s}_2^A \dot{\phi}_2 \\ 0 & 0 \end{bmatrix} \begin{Bmatrix} \dot{\theta}_1 \\ \dot{\theta}_2 \end{Bmatrix} \tag{9.40}$$

This term is evaluated in the following script:

```
function BdThetad = JC_BdThetad
% B_d_theta_d array
     BC_global
     BdThetad = [s_P{1}*theta_d(1)^2
                      0
                 (s_P{1} - s_P{2})*theta_d(1)^2 + s_P{3}*theta_d(2)^2
                      0   ];
```

When the cut joint is put back in the system, the constraint equations for the revolute-translational joint between point O and body (2) is considered as:

$$^{(*)}\boldsymbol{\Phi} = \breve{\mathbf{u}}_2'\mathbf{d} = 0 \tag{9.41}$$

where $\mathbf{d} = \mathbf{r}_2^B - \mathbf{r}_0^O$. The following script files evaluate this constraint, its corresponding Jacobian matrix, and the right-hand-side array of the acceleration constraint:

```
function Phi_cut = JC_Phi
% Cut-joint constraints
     BC_global
     d = r_P{4} - r_P0{1};
     Phi_cut = BC_Phi_rev_tran(u{1}, d, 0);
```

```
function D_cut = JC_jacob
% Cut-joint D matrix
     BC_global
     d = r_P{4} - r_P0{1};
     [jac_RT_0 jac_RT_2] = BC_jacob_rev_tran(u{1}, d, s_P0{1}, s_P{4});
     D_star = [ zeros(1,3)    jac_RT_2 ];
     D_cut = D_star*B;
```

```
function gamma_cut = JC_gamma
% Cut-joint gamma array
     BC_global
     BdThetad = JC_BdThetad;
          d = r_P{4} - r_P0{1};
     gam_RT = BC_gamma_rev_tran(u{1}, u_d{1}, d, r_P_d{4}, ...
                              s_P0_d{1}, s_P_d{4}, 0, phi_d(2));
     gamma_cut = -D_star*BdThetad + gam_RT;
```

The following values for the joint coordinates and velocities are provided in the file `JC_vardata`:

```
% File name:  JC_vardata
% Joint coordinates and velocities
    theta = [0.000; -0.202]; theta_d = [2.511;  0.573];
```

We set the path in the file `JC_formulation` to `JC_MP_B` and then we execute the program. Some of the results are:

```
>> Phi_cut          >> D_cut                  >> gamma_cut
   5.2383e-04           0.1278   -0.5619           1.4750
```

9.6.3 Filmstrip Advancer

Because this system is a four-bar mechanism, we make the pin joint at B the cut-joint. Then we can use exactly the M-files that we developed for the double A-arm suspension system. We copy the folder `JC_AA` into a new folder named `JC_FS`. We replace the `BC_condata` with the one from the folder `BC_FS` and provide a set of values for the joint coordinates and velocities in the file `JC_vardata`.

```
% file name:  JC_vardata
% Joint coordinates and velocities
    theta = [0.5;  5.0;  0.3]; theta_d = [2.8392;  -0.14;  1.4];
```

9.7 REMARKS

The methodology that we discussed in this chapter allows us to transform the kinematic entities of a planar system from the body-coordinate formulation to the joint coordinate formulation. This process results in a much smaller number of variables and consequently fewer equations. For closed-chain systems, the same set of equations can be obtained with much less effort by the classical vector-loop method. Therefore, if we are only interested in the kinematics of a planar system, the joint coordinate method is not our best choice of methodologies—we could apply the vector-loop method directly. However, if our interest is in the dynamics of a system, the joint coordinate method, as it will be seen in the next chapter, will provide a powerful means to reduce the dynamic equations of motion from a large set to a much smaller set.

9.8 PROBLEMS

9.1 For the systems shown, define the necessary joint coordinates. Determine the coordinate transformation expressions, the **B** matrix, and the $\dot{\mathbf{B}}\dot{\boldsymbol{\theta}}$ array. For the joint coordinates about a pin joint, consider the global coordinate $\theta_i = \phi_i$.

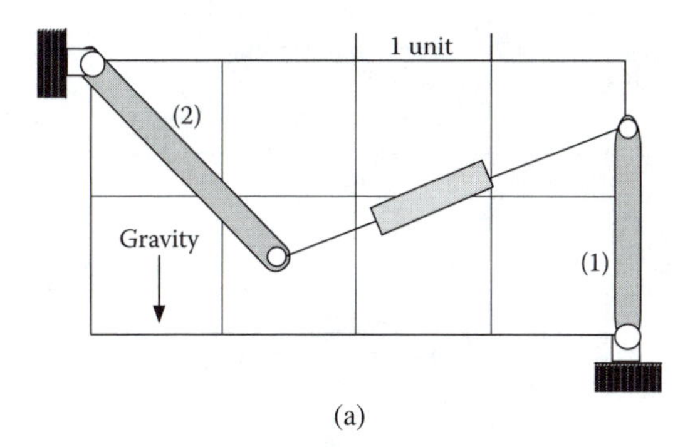

(a)

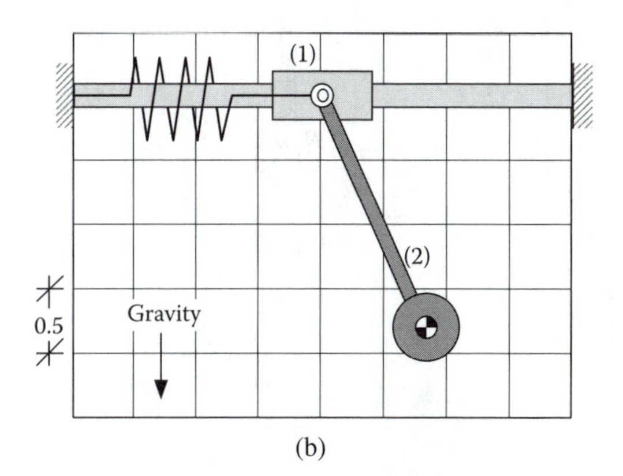

(b)

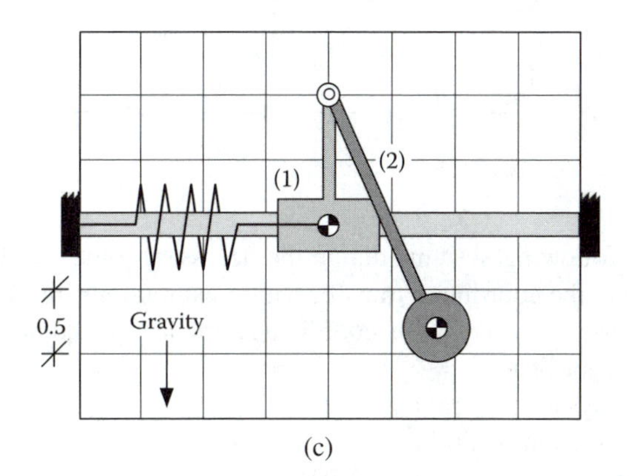

(c)

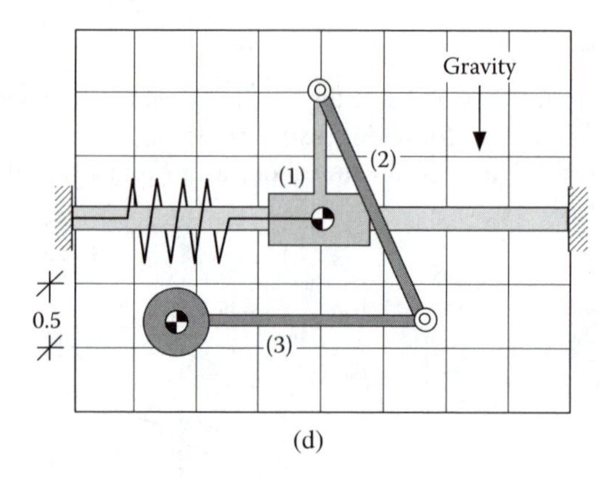

(d)

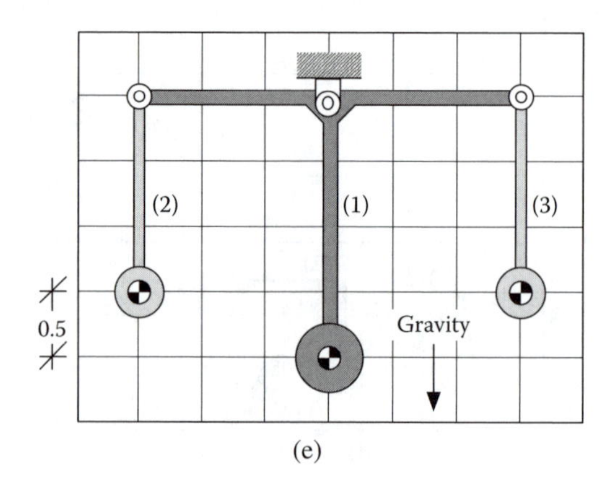

(e)

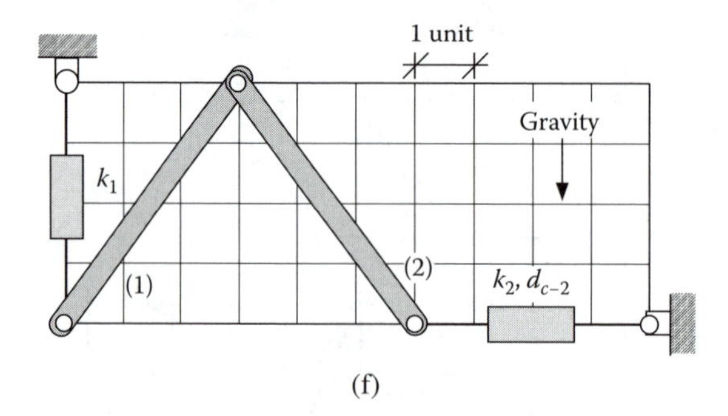

(f)

9.2 For the following systems define the necessary joint coordinates. Then determine the coordinate transformation expressions, the **B** matrix, and the $\dot{\mathbf{B}}\dot{\boldsymbol{\theta}}$ array. For the joint coordinates about a pin joint, consider the relative coordinate $\theta_i = \phi_i - \phi_{i-}$.

(a) The system of Problem 9.1(d)

(b) The system of Problem 9.1(e)

(c) The system of Problem 9.1(f)

9.3 For this four-bar mechanism, $\ell_1 = 1.0$, $\ell_2 = 1.5$, $\ell_3 = 1.2$, and $a = 1.4$. Consider the following cases for the choice of cut joint:
(a) Pin joint at A
(b) Pin joint at B
(c) Pin joint at Q
For each case define the necessary joint coordinates, assuming absolute coordinates $\theta_i = \phi_i$. Determine the coordinate transformation expressions, the **B** matrix, and the $\dot{\mathbf{B}}\dot{\boldsymbol{\theta}}$ array, the cut-joint constraints, Jacobian, and the right-hand-side array of the acceleration constraints.

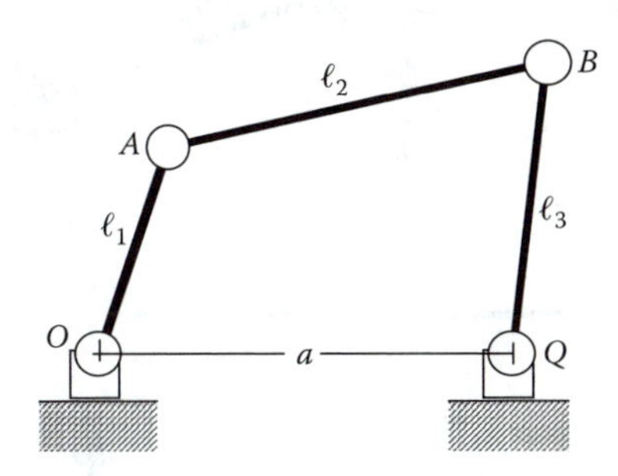

9.4 Repeat Problem 9.3, but this time consider relative joint coordinates $\theta_i = \phi_i - \phi_{i-}$.

9.5 For this slider-crank mechanism $\ell_1 = 1.0$ and $\ell_2 = 2.0$. Consider the following cases for the choice of cut joint:
(a) Pin joint at A
(b) Pin joint at B
(c) Sliding joint
(d) Revolute-revolute joint between A and B
(e) Revolute-translational joint between B and the ground (eliminate the block as a body)
For each case define the necessary joint coordinates, assuming absolute coordinates $\theta_i = \phi_i$. Determine the coordinate transformation expressions, the **B** matrix, and the $\dot{\mathbf{B}}\dot{\boldsymbol{\theta}}$ array, the cut-joint constraints, Jacobian, and the right-hand-side array of the acceleration constraints.

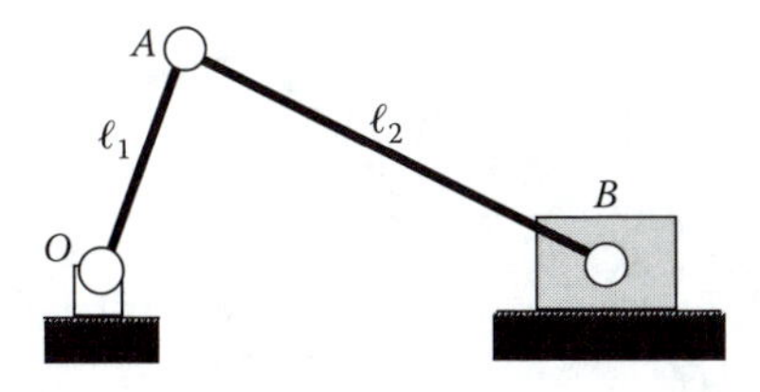

9.6 For the mechanisms shown, the following data are available:
(a) $\ell_1 = 1.0$, $\ell_2 = 1.5$, $\ell_3 = 1.2$, $\ell_4 = 1.0$, $\ell_5 = 2.0$, $a = 1.4$, $\theta = 30°$
(b) $\ell_1 = 0.12$, $\ell_2 = 0.85$, $\ell_3 = 0.74$, $\ell_4 = 0.50$, $a = 0.48$, $b = 0.75$, $c = 0.93$

(c) $\ell_1 = \ell_3 = \ell_4 = 0.23$, $\ell_2 = a = 0.43$, $\ell_5 = 0.30$, $b = 0.07$, $c = 0.22$
(d) $DA = OA = 1.25$, $DB = OE = 2.50$, $DC = 2.00$, $a = 0.64$, $b = 0.20$
For each system select a set of cut joints. Define the necessary joint coordinates. Determine the coordinate transformation expressions, the **B** matrix, and the $\dot{\mathbf{B}}\dot{\boldsymbol{\theta}}$ array, the cut-joint constraints, Jacobian, and the right-hand-side array of the acceleration constraints.

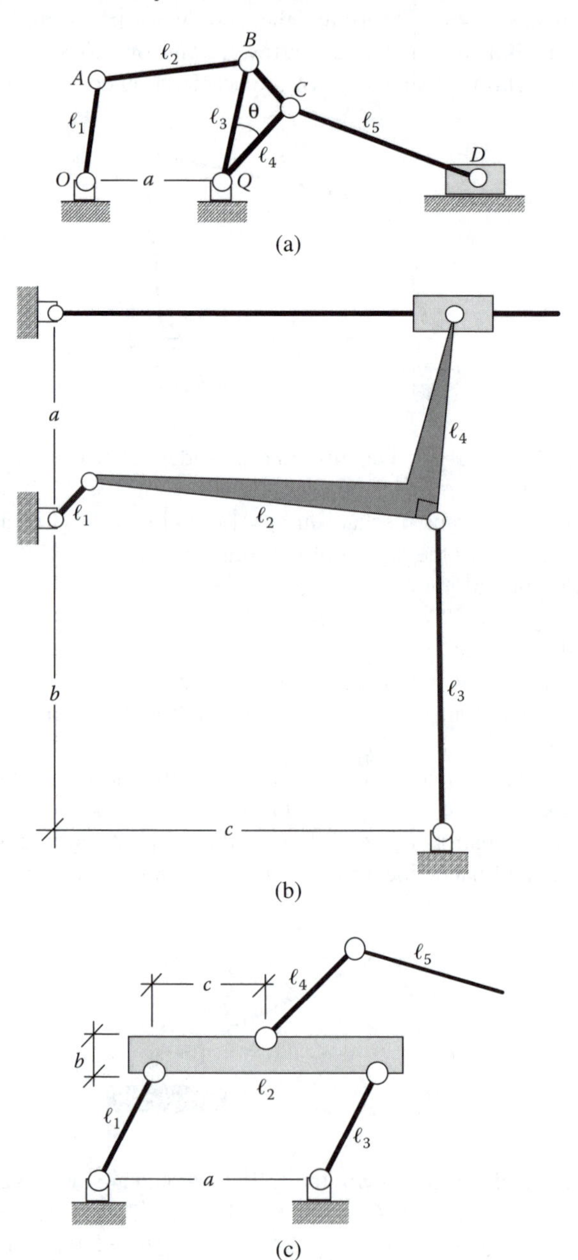

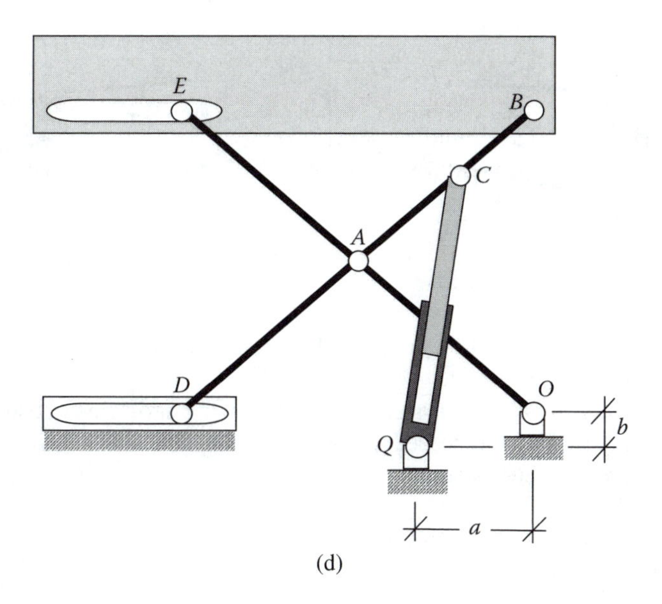

(d)

9.7 For the MacPherson suspension consider Model B for kinematic modeling with the joint coordinate method. Instead of cutting the revolute-translation joint, cut the pin joint at A. Derive the necessary kinematic entities for the system. Hint: at the revolute-sliding joint, consider two joint coordinates—one sliding and one rotational.

9.8 For the MacPherson suspension consider Model C for kinematic modeling with the joint coordinate method. Cut the revolute-revolute joint. Derive the necessary kinematic entities for the system. Compare the resultant equations against those for the same system in body-coordinates in terms of the number of variables, number of equations, and the necessary effort in deriving the equations. Is the transformation to the joint coordinate worth our effort?

9.9 Refer to Problem 7.31. Define a new set of joint coordinates for this revised model. Derive all the necessary arrays and matrices.

9.10 Refer to Problem 7.32. Define a new set of joint coordinates for this revised model (Model A or B). Derive all the necessary arrays and matrices.

9.11 Consider the frontal view of a half-car with independent double A-arm suspension systems as shown. Assume that the tire-ground interactions are modeled by force elements such as springs and dampers. Model this half-car system with the joint coordinate method. Derive the necessary kinematic entities for the system. Use the data for the quarter-car model from Section 1.5.1 for the right side and its mirror data for the left side. Hint: assume a floating joint between the ground and the chassis.

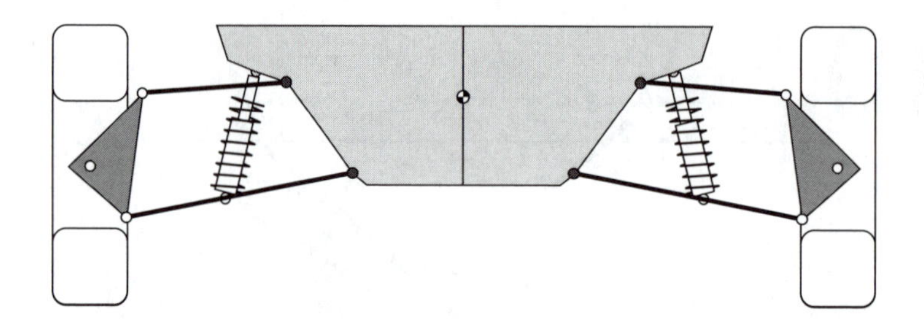

9.12 Repeat Problem 9.11 for the MacPherson suspension system from Section 1.5.2.

9.13 For each multibody system in the following problems construct the necessary M-files as needed for the `JC_formulation` program. Execute the program and observe the results.
  (a) Problem 9.1 (a), (b), (c), (d), (e), (f)
  (b) Problem 9.2 (a), (b), (c)
  (c) Problem 9.3 (a), (b), (c)
  (d) Problem 9.4 (a), (b), (c)
  (e) Problem 9.5 (a), (b), (c), (d), (e)
  (f) Problem 9.6 (a), (b), (c), (d)
  (g) Problem 9.7
  (h) Problem 9.9
  (i) Problem 9.10
  (j) Problem 9.11
  (k) Problem 9.12

9.14 Refer to Problem 7.30. Implement the same preinitialized zero arrays in the M-files of the joint coordinate formulations. The files are mainly `JC_Bmatrix`, `JC_BdThetad`, `JC_jacob`, and `JC_gamma`.

10 Joint-Coordinates: Dynamics

In this chapter we extend the kinematic modeling and formula derivation of the joint coordinate method to dynamics. This is where the effectiveness of the joint coordinate method becomes apparent. We first learn how to construct the mass matrix and the array of applied forces for open-chain mechanical systems. Then, we consider the dynamics of systems of bodies that form closed kinematic chains.

Programs: In this chapter we revise some of the programs from Chapter 9 and develop several new programs. The file organization is the same as that shown in Table 9.1.

10.1 OPEN-CHAIN SYSTEMS

The equations of motion for a multibody system containing kinematic joints were derived in Eq. (8.12). These equations are repeated here for easy reference:

$$\mathbf{M}\ddot{\mathbf{c}} = \mathbf{h} + \mathbf{D}'\boldsymbol{\lambda} \tag{10.1}$$

In Section 9.3.4, for open-chain systems, the kinematic transformation expressions were derived between the joint coordinates and the body coordinates (as well as for the velocities and accelerations). The acceleration transformations from Eq. (9.14) are repeated here as well:

$$\ddot{\mathbf{c}} = \mathbf{B}\ddot{\boldsymbol{\theta}} + \dot{\mathbf{B}}\dot{\boldsymbol{\theta}} \tag{10.2}$$

It was stated that the Jacobian of the kinematic joints, matrix $\mathbf{D}$, and the velocity transformation matrix $\mathbf{B}$ are orthogonal; that is,

$$\mathbf{D}\mathbf{B} = \mathbf{0} \tag{10.3}$$

If we substitute Eq. (10.2) into Eq. (10.1) and premultiplying the resultant equation by $\mathbf{B}'$, we obtain

$$\mathbf{B}'\mathbf{M}(\mathbf{B}\ddot{\boldsymbol{\theta}} + \dot{\mathbf{B}}\dot{\boldsymbol{\theta}}) = \mathbf{B}'\mathbf{h} + \mathbf{B}'\mathbf{D}'\boldsymbol{\lambda}$$

Because of Eq. (10.3) $\mathbf{B}'\mathbf{D}' = \mathbf{0}$, the equations of motion become

$$\boldsymbol{M}\ddot{\boldsymbol{\theta}} = \boldsymbol{h} \tag{10.4}$$

where

$$\boldsymbol{M} = \mathbf{B}'\mathbf{M}\mathbf{B} \tag{10.5}$$

$$\boldsymbol{h} = \mathbf{B}'(\mathbf{h} - \mathbf{M}\dot{\mathbf{B}}\dot{\boldsymbol{\theta}}) \tag{10.6}$$

Matrix $\boldsymbol{M}$ is the new mass matrix and vector $\boldsymbol{h}$ is the new array of forces.

We observe that the transformation of the equations of motion from the body coordinates to the joint coordinates, for open-chain systems, eliminates the reaction forces from the equations. However, if necessary, the reaction forces can be recovered with further computation. Furthermore, the number of equations in Eq. (10.4) is equal to the number of system's DoF. After evaluating the joint acceleration from Eq. (10.4), Eq. (10.2) can be evaluated recursively to determine the body accelerations. Then the body accelerations are substitute in Eq. (10.1) to obtain

$$^{(c)}\mathbf{h} = \mathbf{D}'\boldsymbol{\lambda} = \mathbf{M}\ddot{\mathbf{c}} - \mathbf{h}$$

Because $\mathbf{D}'$ is a rectangular matrix (more rows than columns), this equation cannot be solved in its present form for the Lagrange multipliers. If we premultiply the equation by $\mathbf{D}$ we obtain

$$\mathbf{D}\mathbf{D}'\boldsymbol{\lambda} = \mathbf{D}(\mathbf{M}\ddot{\mathbf{c}} - \mathbf{h}) \tag{10.7}$$

The coefficient matrix $\mathbf{D}\mathbf{D}'$ is square and, therefore, Eq. (10.7) can be solved for the Lagrange multipliers*.

10.1.1 Variable-Length Pendulum

This is a continuation of the example from Sections 7.6.4, 8.3.3, and 9.3.1. The body-coordinate equations of motion for this system were derived in Eq. (8.26). The mass matrix and the array of forces are copied from these equations as:

$$\mathbf{M} = \begin{bmatrix} m_1\mathbf{I} & \mathbf{0} & \mathbf{0} & \mathbf{0} \\ \mathbf{0} & J_1 & \mathbf{0} & \mathbf{0} \\ \mathbf{0} & \mathbf{0} & m_2\mathbf{I} & \mathbf{0} \\ \mathbf{0} & \mathbf{0} & \mathbf{0} & J_2 \end{bmatrix}, \quad \mathbf{h} = \begin{Bmatrix} \mathbf{f}_1 + \mathbf{w}_1 \\ n_1 \\ \hline \mathbf{f}_2 + \mathbf{w}_2 \\ 0 \end{Bmatrix}$$

The velocity transformation matrix and its time derivative were derived in Eqs. (9.5) and (9.6) as

$$\mathbf{B} = \left[\begin{array}{c:c} \breve{\mathbf{r}}_1 & \mathbf{0} \\ 1 & 0 \\ \hdashline \breve{\mathbf{r}}_2 & \mathbf{u}_1 \\ 1 & 0 \end{array}\right], \quad \dot{\mathbf{B}} = \left[\begin{array}{c:c} \dot{\breve{\mathbf{r}}}_1 & \mathbf{0} \\ 1 & 0 \\ \hdashline \dot{\breve{\mathbf{r}}}_2 & \dot{\mathbf{u}}_1 \\ 1 & 0 \end{array}\right]$$

---

* It is assumed that all of the kinematic constraints are independent.

Based on Eq. (10.5), the mass matrix in the joint coordinate formulation is constructed as

$$M = \mathbf{B}'\mathbf{M}\mathbf{B} = \begin{bmatrix} \breve{\mathbf{r}}_1' & 1 & \breve{\mathbf{r}}_2' & 1 \\ \mathbf{0} & 0 & \mathbf{u}_1' & 0 \end{bmatrix} \begin{bmatrix} m_1\mathbf{I} & \mathbf{0} & \mathbf{0} & \mathbf{0} \\ \mathbf{0} & J_1 & \mathbf{0} & \mathbf{0} \\ \mathbf{0} & \mathbf{0} & m_2\mathbf{I} & \mathbf{0} \\ \mathbf{0} & \mathbf{0} & \mathbf{0} & J_2 \end{bmatrix} \begin{bmatrix} \breve{\mathbf{r}}_1 & \mathbf{0} \\ 1 & 0 \\ \breve{\mathbf{r}}_2 & \mathbf{u}_1 \\ 1 & 0 \end{bmatrix}$$

$$= \begin{bmatrix} \breve{\mathbf{r}}_1' m_1 \breve{\mathbf{r}}_1 + J_1 + \breve{\mathbf{r}}_2' m_2 \breve{\mathbf{r}}_2 + J_2 & \breve{\mathbf{r}}_2' m_2 \mathbf{u}_1 \\ \mathbf{u}_1' m_2 \breve{\mathbf{r}}_2 & \mathbf{u}_1' m_2 \mathbf{u}_1 \end{bmatrix}$$

$$= \begin{bmatrix} m_1(\mathrm{r}_1)^2 + J_1 + m_2(\mathrm{r}_2)^2 + J_2 & 0 \\ 0 & m_2 \end{bmatrix} \tag{10.8}$$

where we have considered the following simplifications*:

$$\breve{\mathbf{r}}_i' m_i \breve{\mathbf{r}}_i = m_i \breve{\mathbf{r}}_i' \breve{\mathbf{r}}_i = m_i \mathbf{r}_i' \mathbf{r}_i = m_i (\mathrm{r}_i)^2 \; ; i = 1, \; 2$$

and

$$\mathbf{u}_1' \breve{\mathbf{r}}_2 = \breve{\mathbf{r}}_2' \mathbf{u}_1 = 0$$

The array of forces in the joint coordinate formulation is obtained based on Eq. (10.6) as

$$\boldsymbol{h} = \mathbf{B}'(\mathbf{h} - \mathbf{M}\dot{\mathbf{B}}\dot{\boldsymbol{\theta}})$$

$$= \begin{bmatrix} \breve{\mathbf{r}}_1' & 1 & \breve{\mathbf{r}}_2' & 1 \\ \mathbf{0} & 0 & \mathbf{u}_1' & 0 \end{bmatrix} \left\{ \begin{array}{c} \mathbf{f}_1 + \mathbf{w}_1 - m_1 \dot{\breve{\mathbf{r}}}_1 \dot{\theta}_1 \\ n_1 - J_1 \dot{\theta}_1 \\ \hline \mathbf{f}_2 + \mathbf{w}_2 - m_2(\dot{\breve{\mathbf{r}}}_2 \dot{\theta}_1 + \dot{\mathbf{u}}_1 \dot{d}_2) \\ -J_2 \dot{\theta}_1 \end{array} \right\}$$

$$= \left\{ \begin{array}{c} \breve{\mathbf{r}}_1'(\mathbf{f}_1 + \mathbf{w}_1 - m_1 \dot{\breve{\mathbf{r}}}_1 \dot{\theta}_1) + (n_1 - J_1 \dot{\theta}_1) + \breve{\mathbf{r}}_2' \left(\mathbf{f}_2 + \mathbf{w}_2 - m_2(\dot{\breve{\mathbf{r}}}_2 \dot{\theta}_1 + \dot{\mathbf{u}}_1 \dot{d}_2) \right) - J_2 \dot{\theta}_1 \\ \mathbf{u}_1' \left(\mathbf{f}_2 + \mathbf{w}_2 - m_2(\dot{\breve{\mathbf{r}}}_2 \dot{\theta}_1 + \dot{\mathbf{u}}_1 \dot{d}_2) \right) \end{array} \right\} \tag{10.9}$$

---

* In this particular example the off-diagonal elements of the mass matrix are zero. In general, this is not the case.

Although for computational purposes there is no need to simplify this equation, the array of forces can be simplified to understand how the forces and moments have been transformed.

10.2 OPEN-CHAIN EXAMPLE PROGRAM

In this section we revise the program JC_formulation from Chapter 9 to compute the mass matrix and the array of forces for the open-chain systems. Because the program already computes all the necessary kinematic entities, we first compute the array of forces and the mass matrix from the body coordinate formulation, and then we implement Eqs. (10.5) and (10.6).

```
% Append to the file:   JC_formulation
% ...
% Mass matrix and the array of forces in body coordinates
    BC_mass_diagonal; M_diag = diag(M_array); h = BC_h;
% Mass matrix and the array of forces in joint coordinates
    M_joint = B'*M_diag*B;
    h_joint = B'*(h - M_array.*BdThetad);
```

For the variable-length pendulum, we must make sure that a copy of the file BC_h is made available in the folder JC_VLP. Executing the program JC_formulation yields the following results for the variable-length pendulum example:

```
>> M_joint                          >> h_joint
    1.2025    -0.0000                   -48.8745
   -0.0000     1.0000                     6.8588
```

We note that the off-diagonal elements of the resultant mass matrix are zero and the element in second row, second column is exactly the mass of body (2) as shown analytically in Eq. (10.8).

10.3 CLOSED-CHAIN SYSTEMS

The kinematics of systems containing closed kinematic chains with the joint coordinate formulation was discussed in Section 9.5. It was shown that the process requires the system to be cut at one joint per closed chain to create an open-chain system. After deriving the necessary transformation expressions for the resultant open-chain system, the cut joints are put back and the constraints for the cut joints are considered. According to Eqs. (9.12) to (9.14), the transformation expressions are

$$\mathbf{c} = \mathbf{c}(\boldsymbol{\theta}) \tag{10.10}$$

$$\dot{\mathbf{c}} = \mathbf{B}\dot{\boldsymbol{\theta}} \tag{10.11}$$

$$\ddot{\mathbf{c}} = \mathbf{B}\ddot{\boldsymbol{\theta}} + \dot{\mathbf{B}}\dot{\boldsymbol{\theta}} \tag{10.12}$$

The kinematic constraints for the cut joints from Eqs. (9.20), (9.21), and (9.23) are repeated here for easy reference:

$$^{*}\boldsymbol{\Phi}(\mathbf{c}(\boldsymbol{\theta})) = \mathbf{0} \tag{10.13}$$

$$^{*}\dot{\boldsymbol{\Phi}} \equiv \boldsymbol{D}\dot{\boldsymbol{\theta}} = \mathbf{0} \tag{10.14}$$

$$^{*}\ddot{\boldsymbol{\Phi}} \equiv \boldsymbol{D}\ddot{\boldsymbol{\theta}} + \dot{\boldsymbol{D}}\dot{\boldsymbol{\theta}} = \mathbf{0} \tag{10.15}$$

If the joints are split into two groups—those that remain in the open chain portion and those that get cut (and put back in)—then we can express the equations of motion from Eq. (8.12) as:

$$\mathbf{M}\ddot{\mathbf{c}} = \mathbf{h} + \mathbf{D}'\boldsymbol{\lambda} + {}^{*}\mathbf{D}'\,{}^{*}\boldsymbol{\lambda} \tag{10.16}$$

where the reaction forces are split into two sets according to the grouping of the joints: $\mathbf{D}'\boldsymbol{\lambda}$ represents the reaction forces at the joints in the open chain portion and ${}^{*}\mathbf{D}'\,{}^{*}\boldsymbol{\lambda}$ represents the reaction forces at the cut joints.

To transform Eq. (10.16) to the joint coordinates, we substitute Eq. (10.12) into Eq. (10.16) and premultiply the resultant by $\mathbf{B}'$:

$$\mathbf{B}'\mathbf{M}(\mathbf{B}\ddot{\boldsymbol{\theta}} + \dot{\mathbf{B}}\dot{\boldsymbol{\theta}}) = \mathbf{B}'(\mathbf{h} + \mathbf{D}'\boldsymbol{\lambda} + {}^{*}\mathbf{D}'\,{}^{*}\boldsymbol{\lambda}) \tag{10.17}$$

Rearranging the terms and noting that $\mathbf{B}'\mathbf{D}' = \mathbf{0}$ (refer to Eq. (9.16)) result into

$$\boldsymbol{M}\ddot{\boldsymbol{\theta}} = \boldsymbol{h} + \boldsymbol{D}'\,{}^{*}\boldsymbol{\lambda} \tag{10.18}$$

where

$$\boldsymbol{M} = \mathbf{B}'\mathbf{M}\mathbf{B} \tag{10.19}$$

$$\boldsymbol{h} = \mathbf{B}'(\mathbf{h} - \mathbf{M}\dot{\mathbf{B}}\dot{\boldsymbol{\theta}}) \tag{10.20}$$

and according to Eq. (9.22),

$$\mathbf{\mathit{D}}' = \mathbf{B}'^{*}\mathbf{D}'$$

Equation (10.18) represents the equations of motion for the system, and the term $\mathbf{\mathit{D}}'^{*}\boldsymbol{\lambda}$ represents the reaction forces at the cut joints.

10.4 CLOSED-CHAIN EXAMPLE PROGRAMS

In this section we compute the necessary arrays and matrices for constructing the equations of motion for several examples. The program `JC_formulation` already contains all the necessary statements to compute the mass matrix and the array of forces for systems containing closed chains. We must make sure that a copy of the file `BC_h` is made available for each model.

10.4.1 Double A-Arm Suspension

We set the path to the folder `JC_AA` and execute the program `JC_formulation`. We obtain the following numerical values for the mass matrix and the array of forces:

```
>> M_joint                                        >> h_joint
                                                     1.0e+03 *
   5.9072      0.5713          0                  -1.1612
   0.5713      2.9470          0                  -0.0291
        0           0     0.5169                  -0.0012
```

Because there are three joint coordinates in this model, the mass matrix is 3×3 and the array of forces is 3×1. Note that the mass matrix is block-diagonal—it is composed of a 2×2 and a 1×1 submatrix placed on the diagonal. The zero elements in the third row and the third column indicate that there is no coupling between the two submatrices. This is due to the fact that the cut system is made of two separate trees. The coupling between the two trees is taken care of by the Jacobian of the cut-joint constraints.

10.4.2 MacPherson Suspension

In this section we execute the program `JC_formulation` for the MacPherson suspension, Model A and Model B. For Model A, we obtain the following values for the mass matrix and the array of forces:

```
>> M_joint                                              >> h_joint
                                                           1.0e+03 *
   0.6013           0           0                      -0.0044
        0     19.9060     -0.6420                       4.7408
        0     -0.6420      7.7247                      -0.0217
```

Similarly, for Model B we obtain the following values:

```
>> M_joint                              >> h_joint
                                          1.0e+03  *

   4.6513      0.1264                  -2.0412
   0.1264      2.5980                  -0.0146
```

Because in Model A there are three joint coordinates, the mass matrix is 3×3 and the array of forces is 3×1. Because the cut-joint representation of this model is composed of two trees, the mass matrix is composed of two uncoupled submatrices. In Model B, because there are two joint coordinates and the cut-joint system is made of one tree, the mass matrix is composed of a single 2×2 matrix.

10.5 REMARKS

In this chapter we discussed how to transform the equations of motion for a multibody system from the body coordinates to the joint coordinates. This process results in fewer number of equations—that is, smaller size mass matrix and array of forces. For open-chain systems this process eliminates the reaction forces from the equations of motion, and for closed-chain systems only the reaction forces at the cut joints remain in the equations of motion. Fewer number of equations results in more efficient analyses as far as the computational time is concerned.

10.6 PROBLEMS

10.1 Describe what the two nonzero entries of the mass matrix in Eq. (10.8) are.

10.2 Simplify the array of forces in Eq. (10.9).

10.3 For each multibody system in the following problems construct the necessary M-files as needed for the `JC_formulation` program. Execute the program and evaluate the mass matrix and the array of forces in the joint coordinate formulation for each system.
- (a) Problem 9.1 (a), (b), (c), (d), (e), (f)
- (b) Problem 9.2 (a), (b), (c)
- (c) Problem 9.7
- (d) Problem 9.9 (Observe the differences between the mass matrix and the array of forces between this model and the original model where the chassis is considered the ground.)
- (e) Problem 9.10
- (f) Problem 9.11
- (g) Problem 9.12

10.4 Add the tire from Problem 8.15 to the model from the following problems:
- (a) Problem 9.9
- (b) Problem 9.10
- (c) Problem 9.11
- (d) Problem 9.12

11 Kinematic Analysis

In Chapters 5, 7, and 9 three different formulations for constructing the necessary arrays and matrices for kinematic analysis of planar multibody systems were discussed. However, in those chapters nothing was mentioned about performing kinematic analysis. It is in this chapter that we present the fundamental tools and methodology for kinematic analysis. These fundamentals are the same regardless of the method used to formulate a problem.

Programs: In this chapter we develop programs to perform kinematic analysis with the body coordinate and the joint coordinate formulations. We leave the development of a kinematic analysis program with the point coordinate method to the reader. The M-files for the body coordinate formulation will be organized in the folder `Body_Coord` and, similarly, those for the joint coordinate formulation will be saved in the `Joint_Coord` folder. The organization for these M-files is shown in Tables 11.1 and 11.2.

11.1 UNCONSTRAINED FORMULATION

When the number of defined coordinates in any formulation is equal to the number of system's degrees of freedom, we have an unconstrained formulation. In the point-coordinate formulation this means that no constraints are defined between the points (particles). In the body-coordinate formulation, this translates to no kinematic joints in the system. For the joint-coordinate formulation an unconstrained formulation means that we have an open-chain system and no conditions are enforced between the defined joint coordinates. When there are no constraints, there are no algebraic equations to solve. Examples of such systems are the two-particle system of Section 6.1.1 and the two-body system of Section 8.1.1.

As another example, consider the variable-length pendulum from Sections 7.6.4 and 9.3.1. In the body coordinate formulation, this system is modeled with six coordinates and four position constraints as stated in Eq. (7.46). The corresponding velocity and acceleration constraints are stated in Eqs. (7.47) and (7.48). This is a *constrained* formulation and the methodology for dealing with such constraints will be discussed in the next section. However, the formulation of the same system with the joint coordinate method resulted in the coordinate transformation *expressions* (not constraints) of Eq. (9.3). Given a set of values for the joint coordinates, the body coordinates can be computed from these expressions. The expressions in Eqs. (9.4) and (9.6) provide the body velocities and accelerations in terms of the joint velocities and accelerations. Therefore, this is an *unconstrained* formulation.

TABLE 11.1

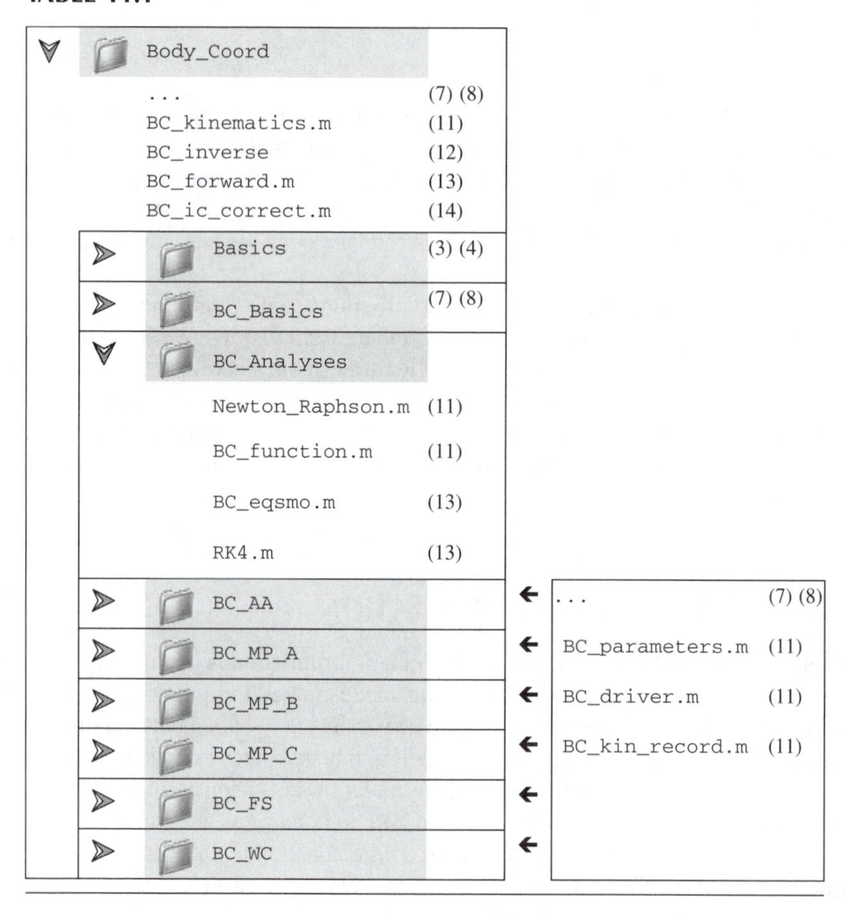

11.2 CONSTRAINED FORMULATION

When the number of defined coordinates is greater than the number of system's DoF, constraint equations are present in the kinematic formulation. Regardless of the choice of formulation, assume that a mechanical system with n_{dof} degrees of freedom is represented by n_v coordinates, $\mathbf{q}$. Then the position constraints are represented by $n_c = n_v - n_{dof}$ equations as

$$\mathbf{\Phi} \equiv \mathbf{\Phi}(\mathbf{q}) = \mathbf{0} \tag{11.1}$$

The corresponding velocity and acceleration constraints are expressed as

$$\dot{\mathbf{\Phi}} \equiv \mathbf{D}\dot{\mathbf{q}} = \mathbf{0} \tag{11.2}$$

and

$$\ddot{\mathbf{\Phi}} \equiv \mathbf{D}\ddot{\mathbf{q}} + \dot{\mathbf{D}}\dot{\mathbf{q}} = \mathbf{0} \tag{11.3}$$

TABLE 11.2

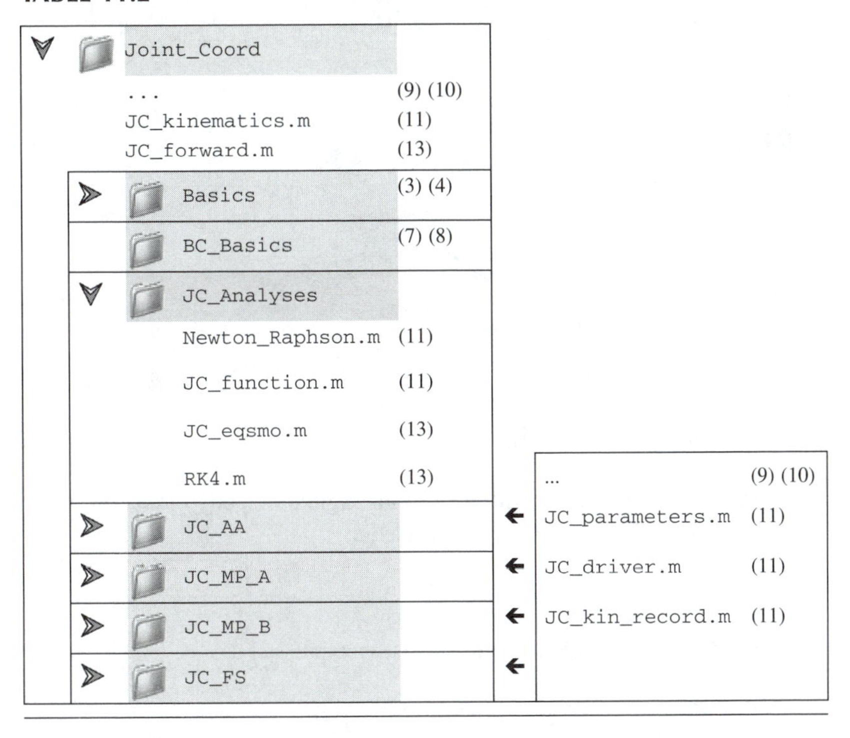

The array of the right-hand side of acceleration constraints is denoted as

$$\boldsymbol{\gamma} = -\dot{\mathbf{D}}\dot{\mathbf{q}} \tag{11.4}$$

The array of coordinates $\mathbf{q}$ could represent the array of point-coordinates, $\mathbf{r}$, the array of body-coordinates, $\mathbf{c}$, or the array of joint-coordinates, $\boldsymbol{\theta}$. Accordingly, the constraints in Eqs. (11.1) to (11.3) could represent the constraints from Chapters 5, 7, or 9.

The objective of kinematic analysis is to solve the kinematic constraints for the coordinates, velocities, and accelerations, if n_{dof} of the coordinates, velocities, and accelerations are known. This requires stating a new set of constraints, called *driver constraints*, that would provide values for n_{dof} number of coordinates (velocities and accelerations) as a function of time.

11.2.1 Driver Constraints

In kinematic analysis, the motion of one or more links must be defined using driver constraints. The number of driver constraints must be equal to the number of system's DoF.

Driver constraints, in general, are simple expressions and are normally functions of time. Examples of such drivers are shown in Figure 11.1. The rotation of link *(i)* in the system of Figure 11.1(a) is decided by a known function $f(t)$; that is,

$$^{(d-\theta,1)}\Phi \equiv \phi_i - f(t) = 0 \tag{11.5}$$

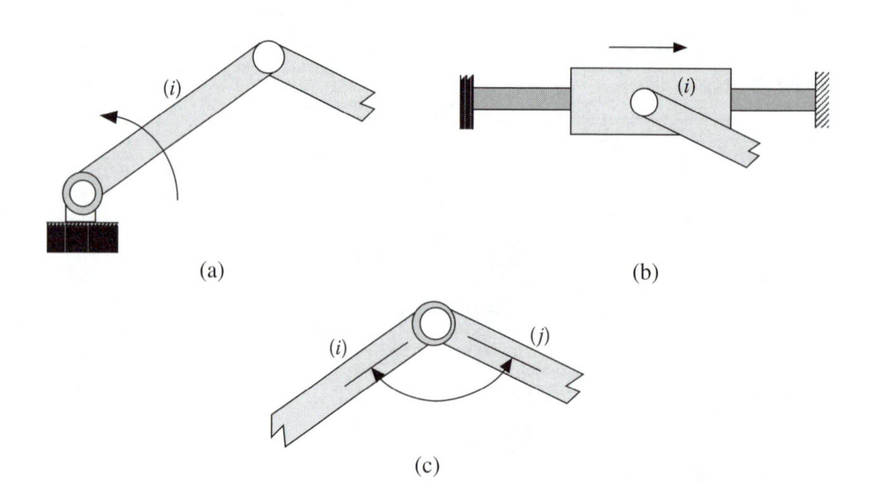

FIGURE 11.1 Various forms of drivers.

If, for example, link (i) rotates with a constant angular velocity of ω, starting from an initial angle $\phi_i = {}^0\phi_i$, then $f(t)$ is expressed as

$$f(t) = {}^0\phi_i + \omega t$$

For a constant angular acceleration of α, $f(t)$ is expressed as

$$f(t) = {}^0\phi_i + {}^0\dot{\phi}_i t + \tfrac{1}{2}\alpha t^2$$

where ${}^0\dot{\phi}_i$ is the initial angular velocity of the link.

For the slider block in Figure 11.1(b), the motion along the x-axis could be controlled by the function

$$^{(d-x,1)}\Phi \equiv x_i - f(t) = 0 \tag{11.6}$$

In this case, for example, $f(t)$ could be a sine function as

$$f(t) = {}^0x_i + a\sin(\omega t)$$

where at $t = 0$, the block is assumed to be at $x = {}^0x_i$.

Another form of driver may control the relative angle between two links such as the system shown in Figure 11.1(c). The driver constraint can be described as

$$^{(d-rot,1)}\Phi \equiv \phi_j - \phi_i - f(t) = 0 \tag{11.7}$$

or

$$^{(d-rot,1)}\Phi \equiv \theta_i - f(t) = 0 \tag{11.8}$$

Here, depending on the formulation, the driver can be expressed in terms of either the rotational coordinates of the bodies or the corresponding *relative* joint coordinate.

For a system containing n_{dof} degrees of freedom, n_{dof} driver expressions are defined as

$$^{(d)}\mathbf{\Phi} \equiv {}^{(d)}\mathbf{\Phi}(\mathbf{q}) - f(t) = \mathbf{0} \tag{11.9}$$

The first time derivative of these expressions yields the driver velocity expressions as

$$^{(d)}\dot{\mathbf{\Phi}} \equiv {}^{(d)}\mathbf{D}\dot{\mathbf{q}} - \dot{f}(t) = \mathbf{0} \tag{11.10}$$

where $^{(d)}\mathbf{D}$ is the Jacobian of the driver expressions and $\dot{f}(t)$ is the time derivative of $f(t)$. Similarly, the driver acceleration expressions are obtained from the time derivative of the velocity drivers as

$$^{(d)}\ddot{\mathbf{\Phi}} \equiv {}^{(d)}\mathbf{D}\ddot{\mathbf{q}} + {}^{(d)}\dot{\mathbf{D}}\dot{\mathbf{q}} - \ddot{f}(t) = \mathbf{0} \tag{11.11}$$

The Jacobian $^{(d)}\mathbf{D}$ for most driver expressions has a very simple form. For example, for the driver of Eq. (11.5), the Jacobian corresponding to body *(i)* is

$$^{(d-\theta,1)}\mathbf{D}_i = [0 \quad 0 \quad 1]$$

Because for most practical problems the Jacobian $^{(d)}\mathbf{D}$ is a constant matrix, $^{(d)}\dot{\mathbf{D}} = \mathbf{0}$. Then, the acceleration expressions of Eq. (11.11) is simplified to

$$^{(d)}\ddot{\mathbf{\Phi}} \equiv {}^{(d)}\mathbf{D}\ddot{\mathbf{q}} - \ddot{f}(t) = \mathbf{0} \tag{11.12}$$

11.3 SOLUTION PROCEDURES

Kinematic analysis requires that at a given time $t = {}^{i}t$, the kinematic and driver constraints to be solved for the coordinates, velocities, and accelerations. The position constraints are, in general, nonlinear algebraic equations but the velocity and acceleration constraints are linear in the velocities and accelerations respectively. Regardless of whether we solve linear or nonlinear algebraic equations, we can consider two possible arrangements for the equations.

Consider the following set of two equations in three unknowns:

$$\begin{aligned} x + 2y - z &= 2 \\ 2x - y + z &= 3 \end{aligned} \tag{a}$$

Assume that the value of one of the unknowns is given as

$$z = 3 \tag{b}$$

Equation (a) may be considered as two kinematic constraints and (b) may be viewed as a driver expression without the presence of the time variable.

In the first arrangement, the value of z is obtained from (b) and substituted in (a) to obtain two equations in two unknowns:

$$\begin{aligned} x + 2y &= 5 \\ 2x - y &= 0 \end{aligned} \qquad (c)$$

These equations are then solved to determine the unknowns: $x = 1$ and $y = 2$. This arrangement is called *coordinate partitioning* method.

In the second arrangement, (a) and (b) are considered simultaneously to form three equations in three unknowns as

$$\begin{aligned} x + 2y - z &= 2 \\ 2x - y + z &= 3 \\ z &= 3 \end{aligned} \qquad (d)$$

The solution to these simultaneous set of equations will yield: $x = 1$, $y = 2$, and $z = 3$. This arrangement is called *appended constraint* method. Although the two arrangements are not that different from each other, from the programming and computational points of view, there are advantages and disadvantages associates with each of them.

11.3.1 Coordinate Partitioning Method

In the *coordinate partitioning* (CP) method, the array of coordinates is split into two sets: the *driving coordinates* and the *driven coordinates*. The driving or the *independent* coordinates are the *known* coordinates, denoted by a left superscript (k). There are n_{dof} driving coordinates and they appear in the driver expressions. The driven or the *dependent* coordinates are the *unknown* coordinates, denoted by a left superscript (u). The number of unknowns or dependent coordinates is the same as the number of kinematic constraints, n_c.

The arrays of coordinates, velocities, and accelerations can be partitioned into *known* and *unknown* subarrays (*independent* and *dependent* subarrays) as

$$\mathbf{q} = \begin{Bmatrix} {}^{(k)}\mathbf{q} \\ {}^{(u)}\mathbf{q} \end{Bmatrix}, \quad \dot{\mathbf{q}} = \begin{Bmatrix} {}^{(k)}\dot{\mathbf{q}} \\ {}^{(u)}\dot{\mathbf{q}} \end{Bmatrix}, \quad \ddot{\mathbf{q}} = \begin{Bmatrix} {}^{(k)}\ddot{\mathbf{q}} \\ {}^{(u)}\ddot{\mathbf{q}} \end{Bmatrix} \qquad (11.13)$$

A process for kinematic analysis with the coordinate partitioning can be stated as follows:

At $t = {}^i t$:

- Determine the known coordinates, ${}^{(k)}\mathbf{q}$, from the driver expressions of Eq. (11.9).
- Solve Eq. (11.1) for the unknown coordinates, ${}^{(u)}\mathbf{q}$.
- Determine the known velocities, ${}^{(k)}\dot{\mathbf{q}}$, from the driver expressions of Eq. (11.10).
- Solve Eq. (11.2) for the unknown velocities, ${}^{(u)}\dot{\mathbf{q}}$.
- Determine the known accelerations, ${}^{(k)}\ddot{\mathbf{q}}$, from the driver expressions of Eq. (11.12).
- Solve Eq. (11.3) for the unknown accelerations, ${}^{(u)}\ddot{\mathbf{q}}$.

The process can be repeated for any other $t = {}^i t$.

EXAMPLE 11.1

Position constraints for a four-bar mechanism are given as

$$\begin{aligned} &1.0\cos\theta_1 + 3.0\cos\theta_2 - 2.2\cos\theta_3 - 2.0 = 0 \\ &1.0\sin\theta_1 + 3.0\sin\theta_2 - 2.2\sin\theta_3 - 0.5 = 0 \end{aligned} \qquad (a)$$

The driver expression is given as

$$\theta_1 - \frac{\pi}{2} - 2\pi t = 0 \qquad (b)$$

Solve these equations and their corresponding velocity equations at $t = 0$ using the CP method.

Solution:

At $t = 0$ the value of the independent coordinate θ_1 is found from (b) to be $\theta_1 = \frac{\pi}{2}$. Then (a) is written as

$$\begin{aligned} &\Phi_1 = 3.0\cos\theta_2 - 2.2\cos\theta_3 + 1.0\cos\frac{\pi}{2} - 2.0 \\ &\Phi_2 = 3.0\sin\theta_2 - 2.2\sin\theta_3 + 1.0\sin\frac{\pi}{2} - 0.5 \end{aligned} \qquad (c)$$

Solving these nonlinear equations, by a method that will be discussed later, yields $\theta_2 = 0.58$ rad and $\theta_3 = 1.34$ rad. The time derivative of (b) yields the velocity driver expression as

$$\dot{\theta}_1 - 2\pi = 0 \qquad (d)$$

The velocity constraints are obtained from the time derivative of the position constraints:

$$\begin{aligned}
&-\sin\theta_1\,\dot{\theta}_1 - 3.0\sin\theta_2\,\dot{\theta}_2 + 2.2\sin\theta_3\,\dot{\theta}_3 = 0\\
&\cos\theta_1\,\dot{\theta}_1 + 3.0\cos\theta_2\,\dot{\theta}_2 - 2.2\cos\theta_3\,\dot{\theta}_3 = 0
\end{aligned} \qquad (e)$$

Substituting the known values of the coordinates and $\dot{\theta}_1 = 2\pi$ in (e) yields

$$\begin{bmatrix} -1.64 & 2.14 \\ 2.51 & -0.51 \end{bmatrix} \begin{Bmatrix} \dot{\theta}_2 \\ \dot{\theta}_3 \end{Bmatrix} = \begin{Bmatrix} 6.28 \\ 0 \end{Bmatrix} \qquad (f)$$

Solving these linear algebraic equations yields $\dot{\theta}_2 = 0.71$ and $\dot{\theta}_3 = 3.50$ rad/sec.

11.3.2 Appended Constraint Method

In the *appended constraint* (AC) method, the n_{dof} driver expressions are appended to the n_c kinematic constraints to form a total of n_v constraints. For position analysis, Eqs. (11.1) and (11.9) form a set of constraints as

$$\mathbf{\Phi}(\mathbf{q}, t) = \begin{Bmatrix} \mathbf{\Phi}(\mathbf{q}) \\ {}^{(d)}\mathbf{\Phi}(\mathbf{q}) - \boldsymbol{f}(t) \end{Bmatrix} = \begin{Bmatrix} \mathbf{0} \\ \mathbf{0} \end{Bmatrix} \qquad (11.14)$$

Similarly, Eqs. (11.2) and (11.10) form the velocity constraints as

$$\dot{\mathbf{\Phi}}(\mathbf{q}, t) = \begin{Bmatrix} \mathbf{D}\dot{\mathbf{q}} \\ {}^{(d)}\mathbf{D}\dot{\mathbf{q}} - \dot{\boldsymbol{f}}(t) \end{Bmatrix} = \begin{Bmatrix} \mathbf{0} \\ \mathbf{0} \end{Bmatrix}$$

or

$$\begin{bmatrix} \mathbf{D} \\ {}^{(d)}\mathbf{D} \end{bmatrix} \dot{\mathbf{q}} = \begin{Bmatrix} \mathbf{0} \\ \dot{\boldsymbol{f}}(t) \end{Bmatrix} \qquad (11.15)$$

The complete set of acceleration constraints is obtained from Eqs. (11.3) and (11.12) as

$$\ddot{\mathbf{\Phi}}(\mathbf{q}, t) = \begin{Bmatrix} \mathbf{D}\ddot{\mathbf{q}} + \dot{\mathbf{D}}\dot{\mathbf{q}} \\ {}^{(d)}\mathbf{D}\ddot{\mathbf{q}} - \ddot{\boldsymbol{f}}(t) \end{Bmatrix} = \begin{Bmatrix} \mathbf{0} \\ \mathbf{0} \end{Bmatrix}$$

or

$$\begin{bmatrix} \mathbf{D} \\ {}^{(d)}\mathbf{D} \end{bmatrix} \ddot{\mathbf{q}} = \begin{Bmatrix} \boldsymbol{\gamma} \\ \ddot{\boldsymbol{f}}(t) \end{Bmatrix} \qquad (11.16)$$

At a given time $t = {}^{i}t$, Eqs. (11.14) to (11.16) can be solved to obtain the coordinates, velocities, and accelerations.

A process for kinematic analysis with the appended constraint method can be stated as follows:

At $t = {}^i t$:

- Solve Eq. (11.14) for the unknown coordinates, $\mathbf{q}$.
- Solve Eq. (11.15) for the unknown velocities, $\dot{\mathbf{q}}$.
- Solve Eq. (11.16) for the unknown accelerations, $\ddot{\mathbf{q}}$.

The process can be repeated for any other $t = {}^i t$.

EXAMPLE 11.2

Arrange the position and velocity constraints from Example 11.1 for the AC method and then solve them for the unknowns.

Solution:

The driver and kinematic constraints are appended together as

$$\begin{aligned} \Phi_1 &= \cos\theta_1 + 3\cos\theta_2 - 2.2\cos\theta_3 - 2 = 0 \\ \Phi_2 &= \sin\theta_1 + 3\sin\theta_2 - 2.2\sin\theta_3 - 0.5 = 0 \\ \Phi_3 &= \theta_1 - \frac{\pi}{2} - 2\pi t = 0 \end{aligned} \tag{a}$$

At $t = 0$, solving these equations yields $\theta_1 = 1.57$, $\theta_2 = 0.58$, and $\theta_3 = 1.34$. The velocity equations are arranged as

$$\begin{bmatrix} -\sin\theta_1 & -3\sin\theta_2 & 2.2\sin\theta_3 \\ \cos\theta_1 & 3\cos\theta_2 & -2.2\cos\theta_3 \\ 1 & 0 & 0 \end{bmatrix} \begin{Bmatrix} \dot{\theta}_1 \\ \dot{\theta}_2 \\ \dot{\theta}_3 \end{Bmatrix} = \begin{Bmatrix} 0 \\ 0 \\ 2\pi \end{Bmatrix} \tag{b}$$

For the known values of the position coordinates at $t = 0$, the velocity equations become:

$$\begin{bmatrix} -1 & -1.64 & 2.14 \\ 0 & 2.51 & -0.51 \\ 1 & 0 & 0 \end{bmatrix} \begin{Bmatrix} \dot{\theta}_1 \\ \dot{\theta}_2 \\ \dot{\theta}_3 \end{Bmatrix} = \begin{Bmatrix} 0 \\ 0 \\ 2\pi \end{Bmatrix} \tag{c}$$

These equations are solved to find $\dot{\theta}_1 = 6.28$, $\dot{\theta}_2 = 0.71$, and $\dot{\theta}_3 = 3.50$.

At this point it may not be clear why we would want to use the AC method instead of the CP method. It is obvious that in the CP method the number of equations, and hence the number of unknowns, is smaller than that of the AC method. Therefore, the computational effort with the CP method is much less than the AC method. However, for developing a general-purpose kinematic analysis program, the *programming effort* that the AC method requires is much less than that of the CP method. This is due to the fact that in the AC method, the constraint equations keep their original forms—we do not need to split a set of equations into the known and unknown parts and move the known part to the right-hand side. Another important point is that in problems with a large number of equations and coordinates, appending one or two driving equations to the original set will not make much difference in the computation time.

11.4 LINEAR ALGEBRAIC EQUATIONS

The velocity and acceleration constraints are linear in the velocities and accelerations respectively. Solving these equations for the velocities and accelerations requires solving simultaneous linear algebraic equations. The position constraints, in contrast, are nonlinear algebraic equations. As it will be seen in the next section, nonlinear algebraic equations can be approximated as linear algebraic equations and solved iteratively for the unknowns.

Linear algebraic equations can be expressed in a general form as

$$a_{11}x_1 + a_{12}x_2 + \ldots + a_{1n}x_n = c_1$$

$$a_{21}x_1 + a_{22}x_2 + \ldots + a_{2n}x_n = c_2$$

$$\ldots$$

$$a_{n1}x_1 + a_{n2}x_2 + \ldots + a_{nn}x_n = c_n$$

In compact matrix form, these equations are expressed as

$$\mathbf{A}\,\mathbf{x} = \mathbf{c} \tag{11.17}$$

Our objective is to solve such a set of equations for the unknown array $\mathbf{x}$, assuming that the matrix of coefficients, $\mathbf{A}$, and the array $\mathbf{c}$ are known.

One obvious method of solution is to write Eq. (11.17) as

$$\mathbf{x} = \mathbf{A}^{-1}\,\mathbf{c} \tag{11.18}$$

This process requires inverting matrix $\mathbf{A}$.

Example 11.3

Solve the velocity constraints from Example 11.2 by inverting the coefficient matrix.

Solution:

We develop a MATLAB® program for Eq. (*c*) of Example 11.2.

```
% Example 11.3
% Define the coefficient matrix and the rhs array
    D = [-1    -1.64     2.14
          0     2.51    -0.51
          1     0        0    ];
    rhsv = [0   0   2*pi]';
    theta_d = inv(D)*rhsv
```

Executing this program yields:

```
theta_d =
     6.2832
     0.7066
     3.4776
```

Although matrix inversion provides us with the answer, computing the inverse of the coefficient matrix, for large-scale problems, is computationally inefficient. Our objective is to find the unknowns and not necessarily the inverse of the coefficient matrix.

There are other computational methods that can solve Eq. (11.17) for the unknowns more efficiently. Gaussian elimination, Gauss-Jordan reduction, and L-U factorization are examples of such methods. For detailed description of these methods, interested reader should refer to any textbook on numerical methods.

The Gaussian elimination method operates on matrix **A** and array **c** simultaneously while transforming **A** to a triangular matrix. Then a back (or forward) substitution on the triangular matrix provides the solution.

In Example 11.3, instead of inverting the coefficient matrix, we can solve for the unknowns using the *backslash* operator (also known as *matrix left division*) in MATLAB. The backslash operator invokes some form of Gaussian elimination on the matrix.

```
% ...
    theta_d = D\rhsv
```

For any nonsingular matrix **A**, it can be proven that there exists an *upper triangular matrix* **U** with nonzero diagonal elements and a *lower triangular matrix* **L** with unit diagonal elements such that

$$\mathbf{A} = \mathbf{L}\,\mathbf{U} \tag{11.19}$$

The process of factorizing **A** into the product **LU** is called L-U factorization. Once the **L** and **U** matrices are obtained, by whatever method, the operation

$$\mathbf{A}\,\mathbf{x} = \mathbf{L}\,\mathbf{U}\,\mathbf{x} = \mathbf{c}$$

is performed in two steps:

$$\mathbf{L}\,\mathbf{y} = \mathbf{c} \tag{11.20}$$

and

$$\mathbf{U}\,\mathbf{x} = \mathbf{y} \tag{11.21}$$

Equation (11.20) is first solved for **y** and Eq. (11.21) is then solved for **x**.

In Example 11.3 we can solve for the unknowns using the following statements:

```
% ...
    [L, U, P] = lu(D)
            y = L\(P*rhsv)
    theta_d = U\y
```

The first statement transforms the coefficient matrix D into matrices L, U, and P:

```
L =                              U =                        P =
    1.000        0        0   -1.000   -1.640    2.140    1  0  0
        0    1.000        0        0    2.510   -0.510    0  1  0
   -1.000   -0.653    1.000        0        0    1.807    0  0  1
```

We note that the diagonal elements of L are ones. Matrix P, called the *permutation* matrix, keeps track of any row and/or column interchange in the process. Interchanging rows and columns is based on a concept known as *pivoting*. Pivoting reduces the amount of truncation error when some elements of the matrix become very small in magnitude during the process of factorization. In our example the diagonal elements of P are all ones. This indicates that no rows or columns have been interchanged.

The two-step process yields the solution to the problem:

```
y =                  theta_d =

        0               6.2832
        0               0.7066
   6.2832               3.4776
```

To have a better understanding of the permutation matrix P, we revise the coefficient matrix from Example 11.3 to have a redundant row.

EXAMPLE 11.4

```
% Coefficient matrix and the rhs array
    D = [-1    -1.64    2.14
          1     1.64   -2.14
          1     0       0    ];
    rhsv = [0   0   2*pi]';
% L-U factorization and solution
    [L, U, P] = lu(D)
           y = L\(P*rhsv)
    theta_d = U\y
```

Executing the program yields the following:

```
L =             U =                          P =
   1  0  0     -1.000   -1.640   2.140       1  0  0
  -1  1  0          0   -1.640   2.140       0  0  1
  -1  0  1          0        0       0       0  1  0
```

The program also displays the following warning:

"Warning: Matrix is singular to working precision"

The elements of matrix P tell us that the operation moved the second row to the third row and vice-versa. We also note that the third diagonal element of U is a zero. This indicates that the row which ended up in the third row (the original second row) is redundant.

11.5 NONLINEAR ALGEBRAIC EQUATIONS

One of the most frequently occurring problems in scientific work is to find the roots of one or a set of nonlinear algebraic equations of the form

$$\mathbf{\Phi}(\mathbf{q}) = \mathbf{0} \tag{11.22}$$

Kinematic analysis of mechanical systems is one example for which the solution of position constraints in the form of Eq. (11.22) is required. The most common and frequently used method for finding the zeros of nonlinear algebraic equations in the form of Eq. (11.22) is known as the Newton-Raphson method which is an iterative approach.

11.5.1 NEWTON-RAPHSON METHOD FOR ONE EQUATION IN ONE UNKNOWN

Consider the following nonlinear equation in one unknown q:

$$\Phi(q) = 0 \tag{11.23}$$

The Newton-Raphson iteration for solving this single equation is stated as follows (left superscript i represents the iteration number):

Assume a solution $q = {}^iq$:

- Evaluate the function ${}^i\Phi = \Phi({}^iq)$;
- If $|{}^i\Phi| \leq \varepsilon$, then ${}^iq$ is the solution to $\Phi(q) = 0$, stop iteration;
- Otherwise, evaluate ${}^iD = \frac{d\Phi(q)}{dq}$ for $q = {}^iq$;
- Compute ${}^i\Delta q = -\frac{{}^i\Phi}{{}^iD}$;
- Improve the estimated solution as ${}^{i+1}q = {}^iq + {}^i\Delta q$;

Repeat the process.

In this process ε is a small number that we accept as "zero".

Example 11.5

Use the Newton-Raphson method to find the root(s) of the equation

$$y = x^3 - 3x^2 - 10x + 24$$

Solution:

The derivative of y with respect to x is

$$dy/dx = 3x^2 - 6x - 10$$

The calculations can be done with the following MATLAB program:

```
% Newton-Raphson process
    x = input(' estimate for x = ?  ')
for n = 1:10
    % Constraints and Jacobian
        y = x^3 - 3*x^2 - 10*x + 24;
        dy_dx = 3*x^2 - 6*x - 10;
    % Are the constraints violated?
    if abs(y) < 0.0001
        x
        break
    end
    % Solve for corrections
        d_x = -y/dy_dx;
    % Correct estimates
        x = x + d_x;
end
```

The program asks us for an estimate for the unknown x. Entering a value starts the Newton-Raphson process; for example, if we enter 10, the iteration converges to x = 4. We can also print intermediate values of x and y during the process to have a better feel for the convergence to the solution. An estimate of 10 for x takes eight iterations to find the solution:

n	x	y
1.0000	10.0000	624.0000
2.0000	7.2870	178.7667
3.0000	5.5937	49.2200
4.0000	4.6153	12.2555
5.0000	4.1478	2.2688
6.0000	4.0121	0.1713
7.0000	4.0001	0.0013
8.0000	4.0000	0.0000

Because there are three roots for this function, we can try different estimates to find the other roots.

The Newton-Raphson method, when it works, is very efficient but restrictions on the method are seldom discussed. The method may converge to an unwanted solution if the initial estimate is not close to the desired solution. Because the method could diverge, it is essential to terminate the process after a finite number of iterations.

11.5.2 Newton-Raphson Method for n Equations in n Unknowns

Consider n nonlinear algebraic equations in n unknowns,

$$\mathbf{\Phi}(\mathbf{q}) = \mathbf{0} \tag{11.24}$$

The Newton-Raphson algorithm for n equations is the n-dimensional version of Eq. (11.23). In the algorithm, we need the Jacobian matrix

$$\mathbf{D} = \frac{\partial \mathbf{\Phi}}{\partial \mathbf{q}}$$

The algorithm is stated as follows:

Assume a solution $\mathbf{q} = {}^i\mathbf{q}$:

- Evaluate the function ${}^i\mathbf{\Phi} = \mathbf{\Phi}({}^i\mathbf{q})$;
- If $\sqrt{{}^i\mathbf{\Phi}'\,{}^i\mathbf{\Phi}} \le \varepsilon$, then ${}^i\mathbf{q}$ is the solution to $\mathbf{\Phi}(\mathbf{q}) = \mathbf{0}$, stop iteration;
- Otherwise, evaluate ${}^i\mathbf{D}$ for $\mathbf{q} = {}^i\mathbf{q}$;
- Solve ${}^i\mathbf{D}\,{}^i\Delta\mathbf{q} = -{}^i\mathbf{\Phi}$ for ${}^i\Delta\mathbf{q}$;
- Improve the estimated solution as ${}^{i+1}\mathbf{q} = {}^i\mathbf{q} + {}^i\Delta\mathbf{q}$.

Repeat the process.

The array ${}^{i}\boldsymbol{\Phi}$ is known as the array of *residuals*, which corresponds to the *violation* in the equations or the constraint, and the term $\sqrt{{}^{i}\boldsymbol{\Phi}'\,{}^{i}\boldsymbol{\Phi}}$ is the *norm* of the residuals. If this term is smaller than the specified tolerance ε, then every residual is small enough to stop the iteration.

EXAMPLE 11.6

From Example 11.2, consider Eq. (*a*) and its Jacobian from Eq. (*b*). Develop a general-use Newton-Raphson program to find the solution to this set of equations. Consider starting estimates $\theta_1 = 1.57$, $\theta_2 = 0.5$, and $\theta_3 = 1.2$.

Solution

We develop the following scripts:

```
    clear all;          % Example 11.6
% Starting estimates
    theta = [1.57; 0.5; 1.2];
% Compute the roots of the function Ex_11_6
      coords = Newton_Raphson(theta, 3, 0.0001, 10, @Ex_11_6)
```

The main program provides the initial values for the unknowns and it invokes the function Newton-Raphson. The initial estimates, the number of coordinates, the desired tolerance, the allowed number of iterations, and a function name are sent to Newton-Raphson. The function whose name is passed on will be invoked by Newton-Raphson to obtain the residuals and the Jacobian matrix. In MATLAB, @ is called a *function handle.*

```
function coords = Newton_Raphson(q, n, tol, iter_max, fun)
% Newton-Raphson process
     coords = zeros(iter_max, n+1); % space to save iteration results
    flag=0;
for i = 1:iter_max
    % Constraints and Jacobian
        [Phi, D] = fun(q);
    % Are the constraints violated?
        ff=sqrt(Phi'*Phi);
        coords(i,:) = [ff q']; % save violation and coordinates
    if ff < tol
        flag=1;
        break
    end
    % Solve for corrections
        delta_q = -D\Phi;
    % Correct estimates
        q = q + delta_q;
end
if flag == 0
    'Convergence failed in Newton-Raphson'
    return
end
```

The function Newton_Raphson is written in a form that can be used in this and in other programs. In this program an array named coords is constructed for saving the results of the iterations. In this array, in each iteration, the norm of the constraint violations are saved in the first column, and the coordinates are saved in the remaining columns.

```
function [Phi, D] = Ex_11_6(theta)
% Constraints and Jacobian
    Phi = [(cos(theta(1)) + 3*cos(theta(2)) - 2.2*cos(theta(3)) - 2)
           (sin(theta(1)) + 3*sin(theta(2)) - 2.2*sin(theta(3)) - 0.5)
           (theta(1) - pi/2)];
    D   = [-sin(theta(1))  -3*sin(theta(2))   2.2*sin(theta(3))
            cos(theta(1))   3*cos(theta(2))  -2.2*cos(theta(3))
            1               0                 0                  ];
```

Executing the program yields the following results in the coords array:

```
    0.1984    1.5700    0.5000    1.2000
    0.0151    1.5708    0.5849    1.3398
    0.0001    1.5708    0.5782    1.3357
         0         0         0         0
      ...
```

The norm of the constraint violations is reduced from 0.1984 to 0.0001 in three iterations. The solution is:

```
theta =

    1.5708
    0.5782
    1.3357
```

11.6 BODY COORDINATE FORMULATION

A program to perform kinematic analysis for systems that are modeled by the body coordinate formulation is presented in this section. The program is based on the appended constraint method of Section 11.3.2. At any given time step, the program uses the Newton-Raphson iteration to correct the coordinate values, and then it solves for the velocities and accelerations at that instant. The main script for this program is listed as follows:

```
% file name:  BC_kinematics
% Kinematic analysis with body coordinate formulation
    clear all
    addpath Basics  BC_Basics  BC_Analyses
    folder = input(' which folder ? ', 's'); addpath (folder)
    BC_global
    BC_condata; BC_vardata; nbc = 3*nb;
% Determine components of vectors on the ground
    BC_vectors0;
% Transfer r, phi, r_d, and phi_d to c and c_d
    c    = BC_r_to_c(r, phi, nb);
```

```
% Parameters for kinematic analysis
     BC_parameters;
% Arrays for the r-h-s of velocity constraints and saving results
     rhsv_all = zeros(nbc, 1); record = zeros(n_steps, 20);
% Kinematic loop
     for n = 1:n_steps
     % Compute (correct) coordinates
         coords = Newton_Raphson(c, nbc, NR_tol, NR_iter,
@BC_function);
     % Compute velocities
         rhsv_all(nbc) = rhsv_driver; c_d = D_all\rhsv_all;
     % Update velocity vectors r_d, phi_d, s_P_d and r_P_d
         BC_cd_to_rd; BC_vectors_d;
     % Construct gamma array then compute accelerations
         rhsa_all = [BC_gamma; rhsa_driver]; c_dd = D_all\rhsa_all;
     % Save results
         BC_kin_record;
     % Increment time
         time = time + delta_t;
     end
```

Many of the statements in this script have been borrowed from the script `BC_formulation`. This program, in addition to the standard `addpath` statement, prompts the user for the name of the folder that contains the M-files for the model. For example, in response to the prompt "`which folder?`" we could respond: `BC_FS`.

The necessary parameters for kinematic analysis and Newton-Raphson algorithm are provided in a script named `BC_parameters`. This script must be supplied for each problem separately. One of the parameters in this script is named `n_step` which is the number of time steps for that particular kinematic analysis. This parameter is used to initialize a storage array as `record(n_steps, 20)`. Arbitrarily, we have allowed room for 20 values to be saved in each time step. The user can change this number if necessary.

The functions `Newton_Raphson` and `BC_function` are saved in the folder `BC_Analyses`. The input arguments in the `Newton_Raphson` function are:

`(c, nbc, NR_tol, NR_iter, @BC_function)`

These are the array of coordinates, the number of coordinates, the acceptable tolerance for the residuals, the allowed maximum number of iterations, and the name of the function M-file in which the residuals and the Jacobian can be found. The function `BC_function` is constructed in a general form as follows:

```
function [Phi_all, D_all] = BC_function(q)
% This function constructs array of residuals and  Jacobian matrix
%     for the kinematic and driver constraints
     BC_global
     c = q;
% Compute A matrices, s_P and r_P vectors
     BC_c_to_r; BC_A_matrices; BC_vectors;
% Evaluate position constraints and Jacobian
     BC_driver;
     Phi_all = [BC_Phi; Phi_driver]; D_all = [BC_jacob; D_driver];
```

This function accesses the functions containing the kinematic constraints and the Jacobian, BC_Phi and BC_jacob, for the model. In addition, it accesses a new function, BC_driver, to obtain the driver constraint and its Jacobian. This function will be discussed further for individual examples.

The number of kinematic steps is defined by the parameter n_steps. Depending on the time step delta_t, the user must set this parameter accordingly. Because after each time interval the results are overwritten, the program at the end of each time step saves the computed values for those variables that we are interested in the array record. The user must specify the contents of this array in the script BC_kin_record for each model.

11.6.1 Filmstrip Advancer

The necessary kinematic entities for this system were constructed in several M-files in Section 7.6.3. In this section we use these files and perform a kinematic analysis for one complete revolution of the crank, body (1). According to the data from Section 1.5.3, the crank must rotate at a constant speed of 60π rad/sec, CW. We assume a starting angle of ${}^0\phi_1 = 0.5$ rad for the crank. The driver expression and its time derivatives are defined as

$$^{(d)}\Phi \equiv \phi_1 - 0.5 + 60\pi t = 0 \ , \ ^{(d)}\dot{\Phi} \equiv \dot{\phi}_1 + 60\pi = 0 \ , \ ^{(d)}\ddot{\Phi} \equiv \ddot{\phi}_1 = 0$$

For this driver the following M-file is constructed:

```
function BC_driver
% Driver constraint, Jacobian, r-h-s of vel. and acc. constraints
    BC_global
    Phi_driver  = phi(1) - 0.5 + 60*pi*time; % constraint
    D_driver = zeros(1, nbc); D_driver(3) = 1; % Jacobian
    rhsv_driver = -60*pi; % r-h-s of velocity constraint
    rhsa_driver = 0.0; % r-h-s of acceleration constraint
```

The necessary parameters for kinematic analysis of this mechanism are defined in the following script:

```
% File name:  BC_parameters
% Parameters for kinematic analysis
    time = 0.0; delta_t = 1/(30*100);;
    NR_tol = 0.0001; NR_iter = 10; n_steps = 101;
```

The time step and the number of steps are set according to the angular velocity of the crank, to allow the crank to rotate one complete revolution.

Because we are interested in the path of point C and its velocity, we save the corresponding computed values at every time step in the array record. In addition, we save the angle of the crank at every step. The corresponding script file is:

```
% File name:  BC_kin_record
    record(n, 1:6) = [time  phi(1)  r_P{7}'  r_P_d{7}'];
```

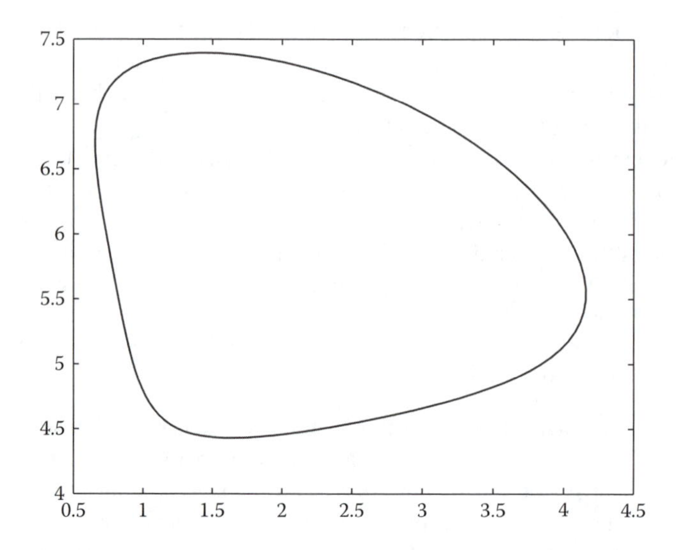

FIGURE 11.2 Path of point *C*.

We execute the program `BC_kinematics` and direct it to use the folder `BC_FS`. We may look at the results for the first five steps by typing in the Command Window:

```
>> record(1:5,:)

        0    0.5000    0.7192    6.0964    45.1222   -281.0649
   0.0003    0.4372    0.7346    6.0024    47.0002   -282.8583
   0.0007    0.3743    0.7505    5.9079    48.1980   -283.5702
   0.0010    0.3115    0.7666    5.8135    48.8879   -283.1696
   0.0013    0.2487    0.7830    5.7193    49.2414   -281.6288
```

We can also plot the path of point *C* by typing in the Command Window:

```
>> plot(record(:,3), record(:,4))
```

This yields the plot shown in Figure 11.2.

11.6.2 Web-Cutter

The web-cutter mechanism from Section 1.5.4 is considered here for kinematic analysis. Because this system is a four-bar mechanism and the joints are labeled the same as those in the double A-arm and the filmstrip four-bars, we can use the M-files from one of those models. We duplicate the folder `BC_FS`, name the new folder `BC_WC`, and revise the `BC_condata` and `BC_vardata` files according to the data given in Figure 1.12. Note that in the file `BC_vardata`, the coordinates are only estimates—Newton-Raphson iteration will correct them. The velocities are all set to zeros—correct values for the velocities will be determined according to the parameters of the driver expression. The crank is rotated with a constant angular velocity of 2π rad/sec, *CCW.* The revised M-files for this system are the following:

```
% file name:  BC_condata
% Web Cutter
    nb = 3; ndof = 1;
% Constant geometric data
    s_Q_0_p = [ 0.0;  0.0]; s_Q_1_p = [ 0.0;   0.0];
    s_A_1_p = [ 0.1;  0.0]; s_A_2_p = [ 0.0;  -0.7];
    s_B_2_p = [ 0.0;  0.0]; s_B_3_p = [ 0.0;   0.0];
    s_O_3_p = [ 1.0;  0.0]; s_O_0_p = [ 0.65; -0.1];
    s_C_2_p = [ 0.6;  0.1]; s_D_3_p = [ 0.35;  0.5];
% Establish cells and arrays
    s_P_p = {s_Q_1_p  s_A_1_p  s_A_2_p  s_B_2_p ...
             s_B_3_p  s_O_3_p  s_C_2_p  s_D_3_p};
       iB = [1 1 2 2 3 3 2 3]; nsP = 8;
    s_P0_p = {s_O_0_p  s_Q_0_p}; nsP0 = 2;
    u_p = {}; iBu = []; nu = 0;
    u0_p = {}; nu0 = 0;
% Mass, moment of inertia, and force data
    m = [1; 10; 10]; J = [0.1; 5.0; 4.0]; g = 9.81;
    k = 0; d_c = 0;
```

```
% file name:  BC_vardata
% Body coordinates and velocities
      r = {[0.0; 0.0] [0.05; 0.7] [0.65; -0.1]};
    phi = [20*pi/180   0   300*pi/180];
    r_d = {[0.0; 0.0] [0.0; 0.0] [0.0; 0.0]}; phi_d = [0  0  0];
```

```
% File name:  BC_parameters
% Parameters for kinematic analysis
    time = 0.0; delta_t = 1/60;
    NR_tol = 0.0001; NR_iter = 10; n_steps = 61;
```

```
function BC_driver
% Driver constraint, Jacobian, r-h-s of vel. and acc. constraints
    BC_global
    Phi_driver  = phi(1) - 20*pi/180 - 2*pi*time; % constraint
    D_driver = zeros(1, nbc); D_driver(3) = 1; % Jacobian
    rhsv_driver = 2*pi; % r-h-s of velocity constraint
    rhsa_driver = 0.0; % r-h-s of acceleration constraint
```

```
% File name:  BC_kin_record
  record(n,1:10) = [time phi(1) r_P{7}' r_P{8}' r_P_d{7}' r_P_d{8}'];
```

Note that in this program we keep track of two points, C and D.

We execute the program `BC_kinematics` and direct it to the folder `BC_WC`. We then construct the following script to create two plots:

```
% File name:  BC_plot
% Plot paths of C and D
    figure(1)
     plot(record(:,3), record(:,4), 'r--') % response in red/dashed
     hold on
     plot(record(:,5), record(:,6), 'g') % response in green
```

```
% Plot y_C and y_D versus phi(1)
    figure(2)
        plot(record(:,2), record(:,4), 'r--')
        hold on
        plot(record(:,2), record(:,6), 'g')
```

Executing this script results into the plots shown in Figure 11.3. The paths of points C and D intersect at two points: the first intersect cuts the web and the second intersect is when the blades separate.

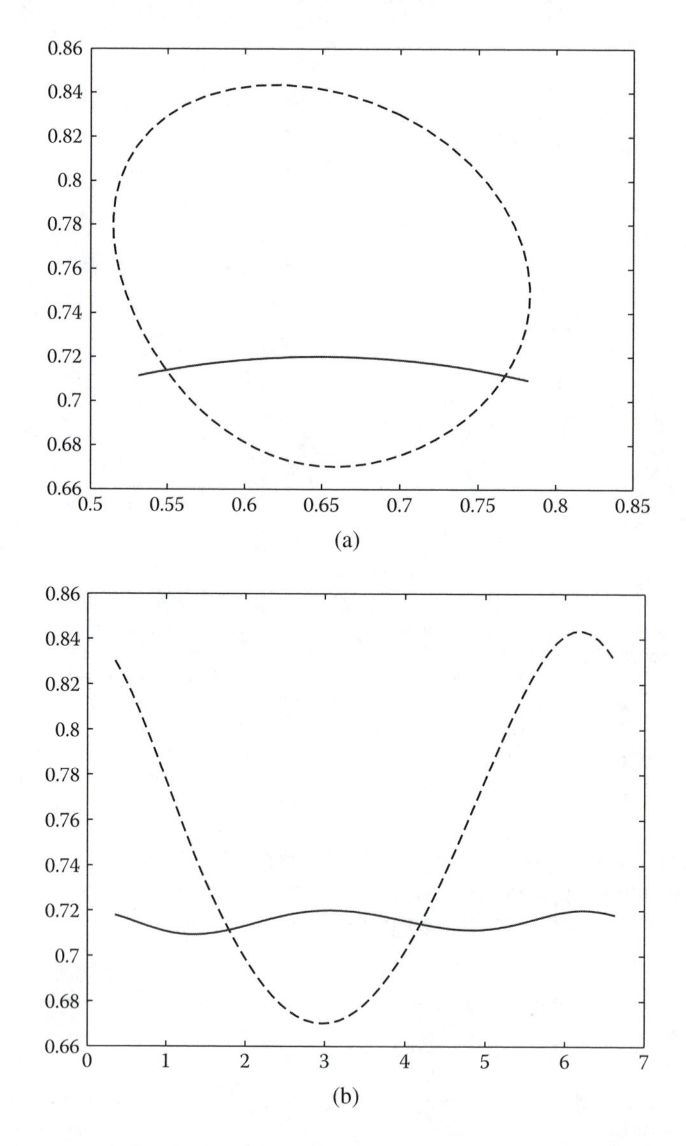

FIGURE 11.3 (a) Paths of points C and D; (b) y-coordinates of points C and D versus crank angle.

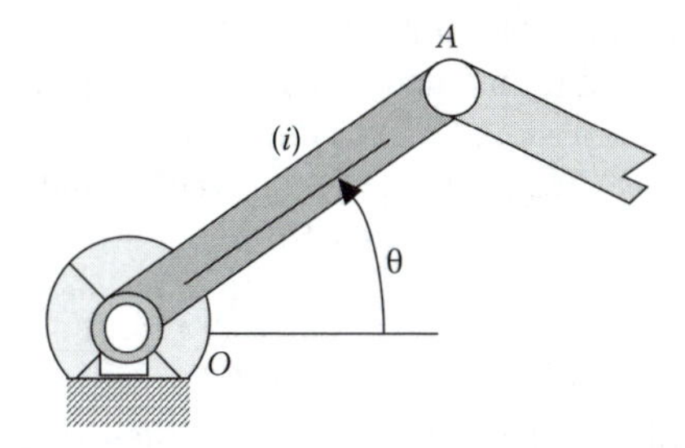

FIGURE 11.4 A driver link.

11.7 POINT COORDINATE FORMULATION

In most applications of kinematic analysis, a rotating motor controls the input link. In the point coordinate formulation describing a driver expression is as simple as in the body coordinate formulation but it may not appear to be that obvious at first.

Assume that link i of a typical system, as shown in Figure 11.4, is controlled by a rotating motor. The motion of the link, assuming a constant angular velocity ω, can be described as

$$\theta(t) = {}^{0}\theta + \omega t \tag{11.25}$$

where ${}^{0}\theta$ is the initial angle of the link. The coordinates of the primary point A can be described as a function of θ as

$$^{(d-\theta,2)}\boldsymbol{\Phi} \equiv \mathbf{r}^{A} - \ell^{A,O}\mathbf{u}(\theta) = \mathbf{0} \tag{11.26}$$

where $\mathbf{u}(\theta)$ is defined as

$$\mathbf{u}(\theta) = \begin{Bmatrix} \cos\theta \\ \sin\theta \end{Bmatrix} \tag{11.27}$$

Equation (11.26) represents two driver constraints, because both x^{A} and y^{A} can be defined as a function of the angle of the link. The first and second time derivatives of Eqs. (11.25) and (11.26) provide the driver expressions for the velocity and acceleration analyses.

Developing a MATLAB program to perform kinematic analysis with the point coordinate formulation is similar to that of the body coordinate formulation. However, because the M-files in Chapter 5 do not use the same global statements and data structures as in the body coordinate formulation, revising the program `BC_kinematics` for the point coordinate formulation requires more than a few minor changes. We leave the task of developing a kinematic analysis program for the point coordinate formulation (call it PC_kinematics) to the reader. We recommend that the programs of Chapter 5 (and Chapter 6) be reconstructed and reorganized in a form similar to those of Chapter 7 (and Chapter 8).

11.7.1 Filmstrip Advancer

In this section we consider the point coordinate formulation of the filmstrip advancer from Section 5.1.3, which is shown again in Figure 11.5. We reformulate the system with two primary points at A and B, and we consider point C as a secondary point.

We write the following kinematic constraints and driver expressions:

$$\frac{1}{2}\left(\mathbf{d}^{B,A'}\mathbf{d}^{B,A} - (\ell^{B,A})^2\right) = 0$$

$$\frac{1}{2}\left(\mathbf{d}^{B,O'}\mathbf{d}^{B,O} - (\ell^{B,O})^2\right) = 0$$

$$\mathbf{d}^{A,Q} - \ell^{A,O}\mathbf{u}(\theta) = \mathbf{0}$$

where $\mathbf{d}^{B,A} = \mathbf{r}^B - \mathbf{r}^A$, $\mathbf{d}^{A,Q} = \mathbf{r}^A - \mathbf{r}^Q$, and $\mathbf{d}^{B,O} = \mathbf{r}^B - \mathbf{r}^O$. Note that the two driver expressions have replaced the length constraint between points A and Q. For a known value of θ, these equations can be solved by the Newton-Raphson iteration for the four primary coordinates.

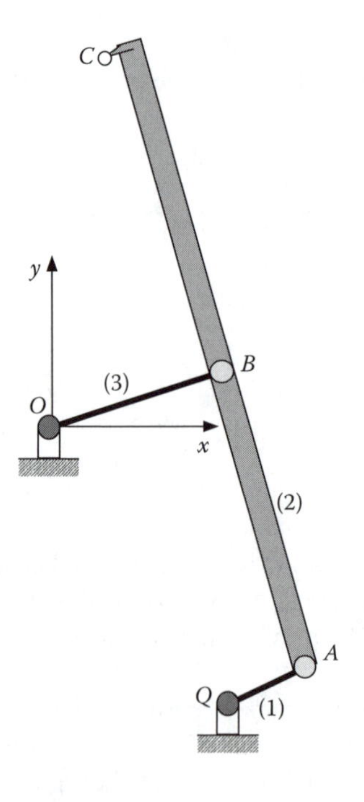

FIGURE 11.5 Filmstrip advancer mechanism.

The velocity and acceleration constraints are expressed as

$$\mathbf{d}^{B,A\prime}(\dot{\mathbf{r}}^B - \dot{\mathbf{r}}^A) = 0 \qquad \mathbf{d}^{B,A\prime}(\ddot{\mathbf{r}}^B - \ddot{\mathbf{r}}^A) = -\dot{\mathbf{d}}^{B,A\prime}\dot{\mathbf{d}}^{B,A}$$

$$\mathbf{d}^{B,O\prime}\dot{\mathbf{r}}^B = 0 \qquad \mathbf{d}^{B,O\prime}\ddot{\mathbf{r}}^B = -\dot{\mathbf{d}}^{B,O\prime}\dot{\mathbf{d}}^{B,O}$$

$$\dot{\mathbf{r}}^A = \ell^{A,O}\omega\breve{\mathbf{u}}(\theta) \qquad \ddot{\mathbf{r}}^A = -\ell^{A,O}\omega^2\mathbf{u}(\theta)$$

The Jacobian matrix for the system is derived from the velocity constraints to be

$$\mathbf{D} = \begin{bmatrix} -\mathbf{d}^{B,A\prime} & \mathbf{d}^{B,A\prime} \\ \mathbf{0} & \mathbf{d}^{B,O\prime} \\ \mathbf{I} & \mathbf{0} \end{bmatrix} \tag{11.28}$$

The velocity and acceleration equations are solved as linear algebraic equation. The kinematics of point C can be determined as a secondary point.

11.8 JOINT COORDINATE FORMULATION

In this section we develop a kinematic analysis program, similar to that in Section 11.6, for systems that are modeled by the joint coordinate formulation. The main script for this program is listed as follows:

```
% file name:  JC_kinematics
% Kinematic analysis with joint coordinate formulation
    clear all
    addpath Basics  BC_Basics  JC_Analyses
    folder = input(' which folder ? ', 's'); addpath (folder)
    BC_global
    BC_condata; JC_vardata;
    nbc = 3*nb; njc = nb; nPhi_cut = njc - ndof;
% Determine components of vectors on the ground
    BC_vectors0;
% Parameters for kinematic analysis
    BC_parameters;
% Arrays for the r-h-s of velocity constraints and to save results
    rhsv_all = zeros(njc, 1); record = zeros(n_steps, 20);
% Kinematic loop
    for n = 1:n_steps
    % Compute (correct) coordinates
      coords = Newton_Raphson(theta, njc, NR_tol, NR_iter, @JC_function);
    % Compute velocities
      rhsv_all(njc) = rhsv_driver; theta_d = D_all\rhsv_all;
    % Update velocity vectors r_d, phi_d, s_P_d and r_P_d
      B = JC_Bmatrix; c_d = B*theta_d; BC_cd_to_rd; BC_vectors_d;
    % Construct gamma array the compute accelerations
      rhsa_all = [JC_gamma; rhsa_driver]; theta_dd = D_all\rhsa_all;
    % Save the results
      JC_kin_record;
    % Increment time
      time = time + delta_t;
    end
```

A copy of the file `BC_parameters` for each model must be provided in the corresponding folder.

The functions `Newton_Raphson` and `JC_function` are saved in the folder `JC_Analyses`. The function `JC_function` is provided in a general form as follows:

```
function [Phi_all, D_all] = JC_function(q)
% This function constructs array of residuals and  Jacobian matrix
%      for the kinematic and driver constraints
     BC_global
     c = q;
% Compute A matrices, s_P and r_P vectors
     BC_c_to_r; BC_A_matrices; BC_vectors;
% Evaluate position constraints and Jacobian
     BC_driver;
     Phi_all = [BC_Phi; Phi_driver]; D_all = [BC_jacob; D_driver];
```

11.8.1 Filmstrip Advancer

For this system, we construct a driver function and a script file to save some of the results as follows:

```
function JC_driver
% Driver constraint
     BC_global
     Phi_driver  = theta(1) - 0.5 + 60*pi*time; % constraint
     D_driver = [1   0   0]; % Jacobian
     rhsv_driver = -60*pi; % r-h-s of velocity constraint
     rhsa_driver = 0.0; % r-h-s of acceleration constraint
```

```
% File name:   JC_kin_record
     record(n, 1:9) = [time  phi(1)  r_P{7}'  r_P_d{7}' theta_dd'];
```

We execute the program `JC_kinematics` and direct the program to the folder `JC_FS`. We may plot the path of point *C* which yields a plot identical to that of Figure 11.2. Because we have saved the three joint accelerations, we may plot $\ddot{\theta}_2$ versus time using the commands

```
>> plot(record(:,1), record(:,8))
>> xlabel('Time')
>> ylabel('theta-dd-2')
```

This yields the plot shown in Figure 11.6.

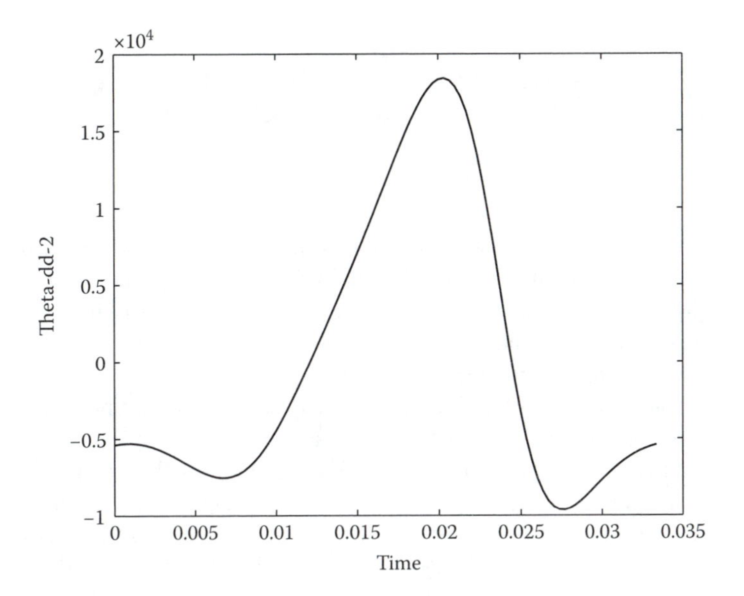

FIGURE 11.6 Joint acceleration $\ddot{\theta}_2$ versus time.

11.9 REMARKS

The fundamentals of kinematic analysis remain the same regardless of the choice of the coordinates used in a formulation. Kinematic analysis relies on the solution of linear and nonlinear algebraic equations. Based on this simple process several computer programs were presented to determine the kinematics of a variety of planar multibody systems. These programs have been developed for easy adoption from one example to another, from one formulation to a different one, and from one form of analysis to another form. Therefore, to keep this flexibility in the programs, computational efficiency has been overlooked—one can find many unnecessary steps in these programs for any particular application. These programs can be revised to improve their computational efficiency.

11.10 PROBLEMS

11.1 Consider the slider-crank mechanism shown. Assume $\ell_1 = 0.12$ and $\ell_2 = 0.26$. Apply the classical vector-loop method and write the constraint equations in terms of the coordinates θ_1, θ_2, and d. Derive the corresponding velocity and acceleration constraints. If the crank is the driver, assume a driver expression as $\theta_1 = 0.8 + 0.1t$. Formulate the position, velocity, and acceleration constraints and the Jacobian matrix for:

(a) The coordinate partitioning method

(b) The appended constraint method

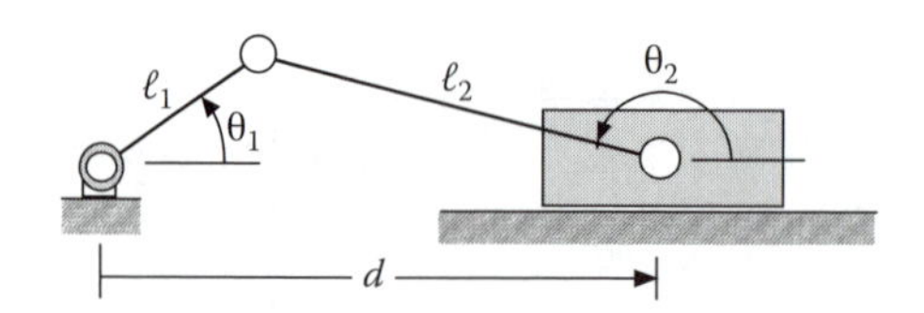

11.2 In Problem 11.1, assume that the slider block is the driver following the expression $d = 0.25 - 0.2t - 0.03t^2$. Formulate the position, velocity, and acceleration constraints and the Jacobian matrix for:
(a) The coordinate partitioning method
(b) The appended constraint method

11.3 Points P and Q are defined on the coupler of a four-bar mechanism as shown. Assume $\ell_1 = 0.05$, $\ell_2 = 0.12$, $\ell_3 = 0.08$, $a = 0.10$, $\ell_4 = 0.015$, and $\ell_5 = 0.02$. Apply the classical vector-loop method and write the constraint equations in terms of the coordinates θ_1, θ_2, and θ_3. Write expressions for x^P, y^P, x^Q, and y^Q in terms of the coordinates. Derive the corresponding velocity and acceleration constraints and expressions. Assume a driver expression link *(1)* as $\theta_1 = 0.75 + 0.1t$. Formulate the problem for:
(a) The coordinate partitioning method
(b) The appended constraint method

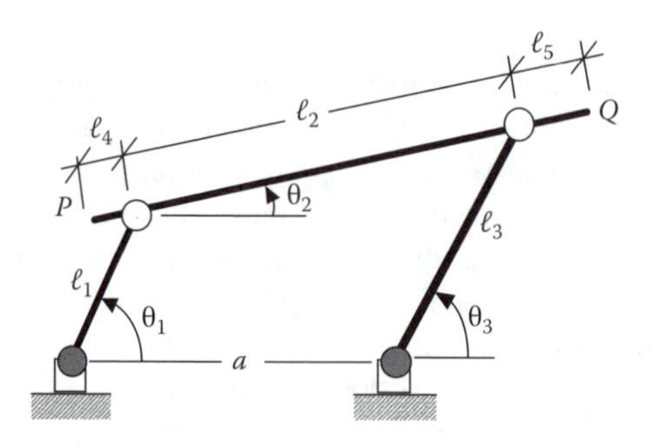

11.4 For each of the following set of linear algebraic equations, find the unknowns:
(a) Use the backslash operator.
(b) Use matrix inverse command.
(c) Use LU factorization command.

Set 1:
$$\begin{bmatrix} 2 & -1 & 0 & 1 \\ -2 & 2 & 0 & -3 \\ 4 & 1 & 3 & -1 \\ 0 & -1 & -2 & 5 \end{bmatrix} \begin{Bmatrix} x_1 \\ x_2 \\ x_3 \\ x_4 \end{Bmatrix} = \begin{Bmatrix} 2 \\ 8 \\ 1 \\ 2 \end{Bmatrix}$$

Set 2:

$$\begin{bmatrix} 3 & 1 & -1 & 2 \\ -6 & -2 & 4 & 3 \\ 0 & 3 & 2 & -2 \\ 1 & 1 & -5 & 6 \end{bmatrix} \begin{Bmatrix} x_1 \\ x_2 \\ x_3 \\ x_4 \end{Bmatrix} = \begin{Bmatrix} 0 \\ 4 \\ 1 \\ -10 \end{Bmatrix}$$

Set 3:

$$\begin{bmatrix} 2 & -1 & 0 & 1 \\ 2 & 2 & 3 & -2 \\ 4 & 1 & 3 & -1 \\ 0 & 1 & 1 & -1 \end{bmatrix} \begin{Bmatrix} x_1 \\ x_2 \\ x_3 \\ x_4 \end{Bmatrix} = \begin{Bmatrix} 2 \\ -1 \\ 1 \\ -1 \end{Bmatrix}$$

11.5 Refer to Problem 11.1. Develop a MATLAB program and use the `Newton-Raphson` function to solve the position constraints at $t = 0$:
(a) Apply the coordinate partitioning method.
(b) Apply the appended constraint method.

11.6 Refer to Problem 11.2. Develop a MATLAB program to solve the position constraints at $t = 0$:
(a) Apply the coordinate partitioning method.
(b) Apply thc appended constraint method.

11.7 Refer to Problem 11.3. Develop a MATLAB program to solve the position constraints at $t = 0$:
(a) Apply the coordinate partitioning method.
(b) Apply the appended constraint method.
After the coordinates are determined, compute the coordinates of points P and Q.

11.8 For the system shown assume no slipping between the disc and the ground. The constant lengths are $OA = 0.15$, $AB = 0.2$, $BC = 0.1$, and $r = 0.12$. Use the classical vector-loop method and write the constraints for this system in terms of θ_1, θ_2, θ_3, and d. Note that two equations can be written for the loop closure and one for the no-slip condition. The system was initially assembled for $d = 0.3$ and $\theta_3 = 120°$. For the crank angle $\theta_1 = 60°$, apply the Newton-Raphson algorithm to solve the constraint equations for the unknown coordinates. Show the intermediate results in a table.

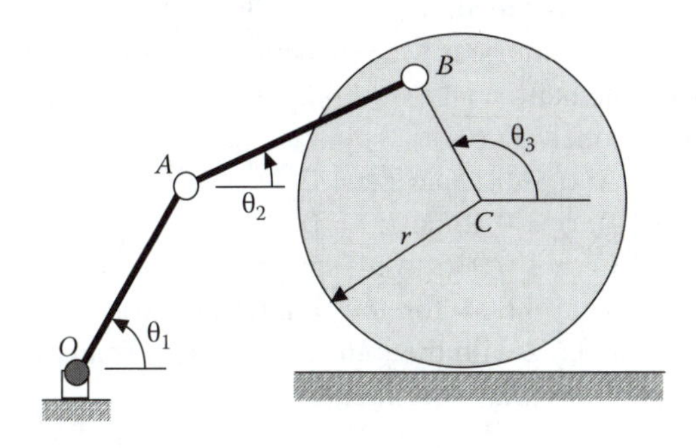

11.9 For the mechanism shown use the vector-loop method to derive the position, velocity, and acceleration constraints. Assume the following lengths: $OB = 0.03$, $CD = 0.18$, and $OE = 0.09$. Develop a MATLAB program to solve for the kinematics of the system for a complete revolution of the crank. Assume the crank OB rotates with a constant angular velocity of 0.2 rad/sec, CCW.

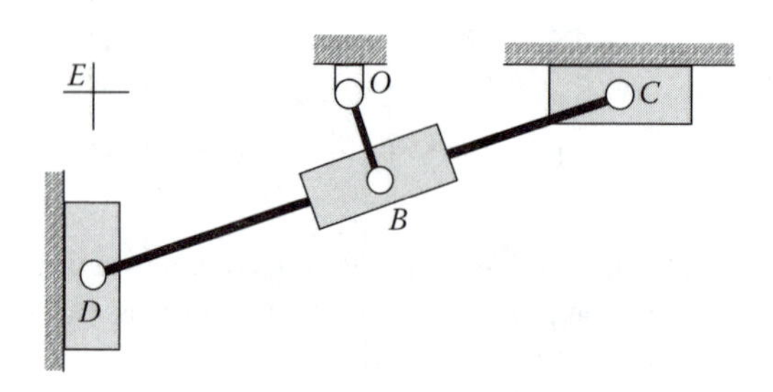

11.10 Formulate Problem 11.9 with the point-coordinate method and then develop a MATLAB program.

11.11 Consider the filmstrip advancer model in Section 11.6.1 or 11.8.1. Run the simulation and answer the following questions:
- (a) How far apart should the sprocket holes be placed on the filmstrip?
- (b) What is the mean speed of the filmstrip during the phase that is being pulled by the mechanism?
- (c) What is the mean speed of the filmstrip during one revolution of the crank?

11.12 Consider the web-cutter model in Section 11.6.2. Run the simulation and answer the following questions:
- (a) What is the angle of the crank at the time of cutting?
- (b) What should be the velocity of the web at the time of cutting?

11.13 Develop the necessary M-files and simulate the kinematics of the web-cutter mechanism with the joint coordinate formulation using the script `JC_kinematics`.

11.14 Consider the quarter-car A-arm suspension model from Section 7.6.1 for kinematic analysis. Apply a driver expression on the motion of the knuckle in the y-direction as $y_2 + 0.036 - 0.07\sin 2\pi t = 0$.

11.15 Consider the quarter-car MacPherson suspension model from Section 7.6.2 for kinematic analysis. Apply a driver expression in one of the following models:
- (a) Model A: $\phi_1 - 0.15\sin 2\pi t = 0$
- (b) Model B: $\phi_1 - 0.15\sin 2\pi t = 0$
- (c) Model C: $y_1 + 0.036 - 0.07\sin 2\pi t = 0$

11.16 Repeat Problem 11.14 for the quarter-car A-arm suspension model from Section 9.6.2 with the joint-coordinate formulation. Note that y_2 is not one of the joint coordinates.

11.17 Develop a MATLAB program to perform kinematic analysis with the point coordinate formulation similar to that of the body coordinate formulation. Name the script `PC_kinematics`.

11.18 Consider the six-bar mechanism shown. The following data are provided: $\ell_1 = 0.12$, $\ell_2 = 0.85$, $\ell_3 = 0.74$, $\ell_4 = 0.50$, $a = 0.48$, $b = 0.75$, and $c = 0.93$. Perform a kinematic analysis of this system by rotating the crank with a constant angular velocity of 1 rad/sec. Show that the link QC dwells.

(a) Apply the body coordinate formulation; `BC_kinematics`
(b) Apply the joint coordinate formulation; `JC_kinematics`
(a) Apply the body coordinate formulation; `PC_kinematics` (refer to Problem 11.17)

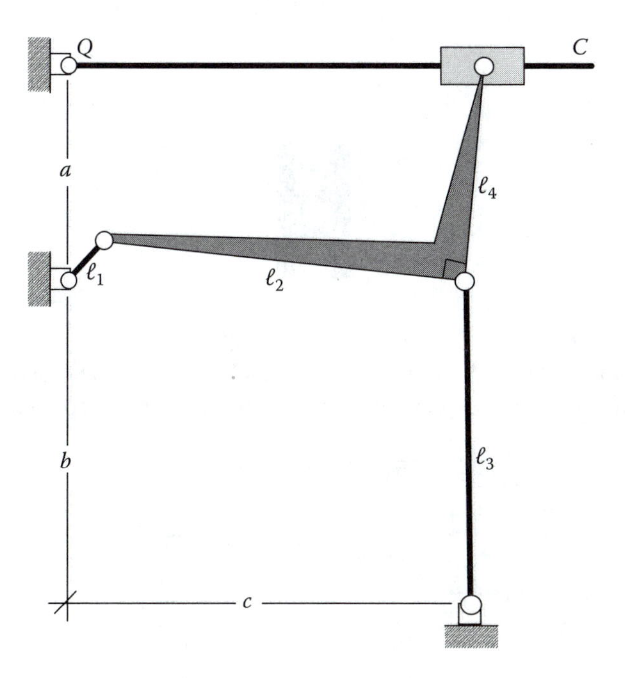

11.19 The disc of the mechanism shown rotates with a constant angular velocity of π rad/sec, CW. Perform a kinematic analysis of this system for a complete revolution of the disc. Consider the following dimensions: $\ell_1 = 0.31$, $\ell_2 = 0.75$, $\ell_3 = 0.93$, $a = 0.20$ and $b = 0.075$.

(a) Model the system using five bodies, four pin joints, and three sliding joints.
(b) Model the system using two bodies, one pin joint, and three revolute-translational joints.

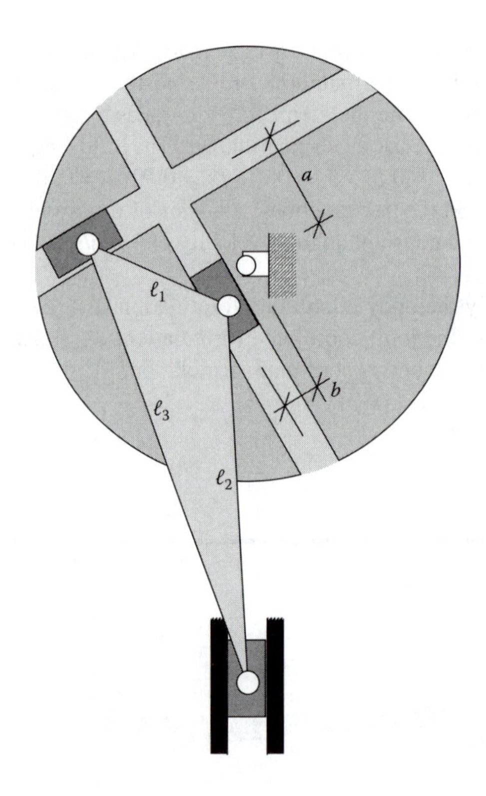

11.20 For this lift mechanism the following data are provided: $DA = OA = 1.25$, $DB = OE = 2.50$, $DC = 2.00$, $a = 0.64$ and $b = 0.20$. Apply a driver expression between points Q and C representing

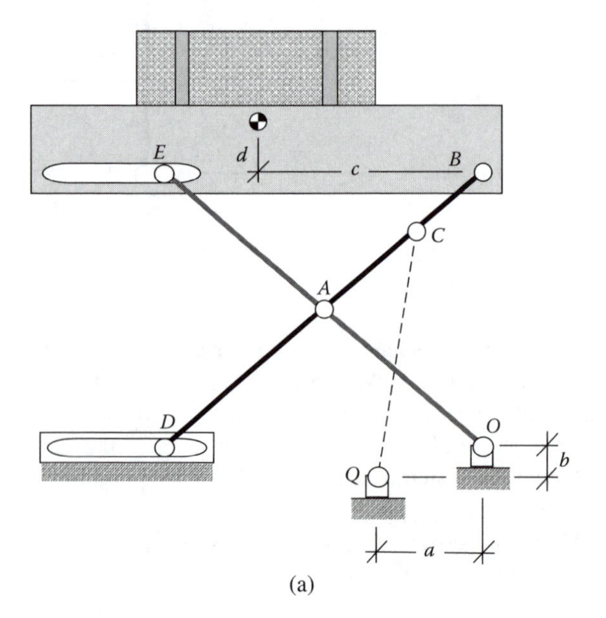

(a)

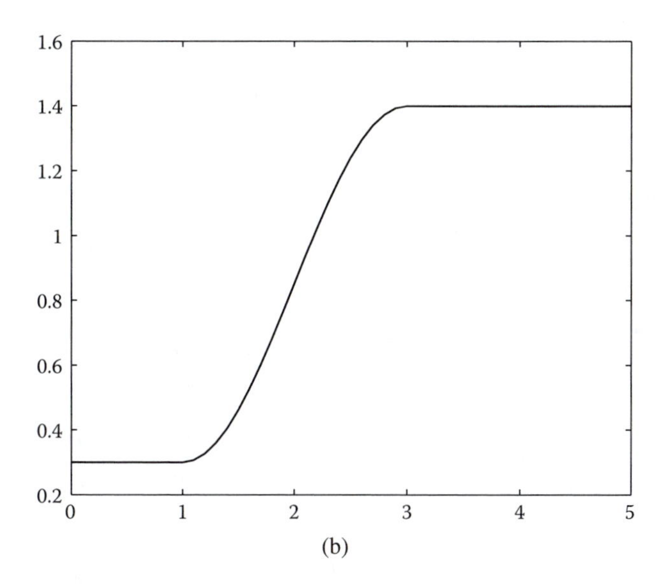

(b)

a hydraulic actuator. The driver must control the length of the vector connecting the two points according to the function shown, going from a minimum length of 0.3 to a maximum length of 1.4. The transition between the two limits can be achieved by a cosine function during a period of, for example, 2 seconds. Perform a kinematic analysis of the mechanism.

11.21 A hydraulic actuator controls the unloading process of a dump truck. The dimensions are provided as $AC = EF = 0.4$, $CD = ED = 1.3$, $BE = CK = 2.0$, $FK = 1.6$, $a = 1.6$ and $b = 0.2$. Apply a driver expression similar to the function in Problem 11.19. Determine the minimum and maximum lengths of the actuator. Allow 5 seconds for the unloading process to be completed.

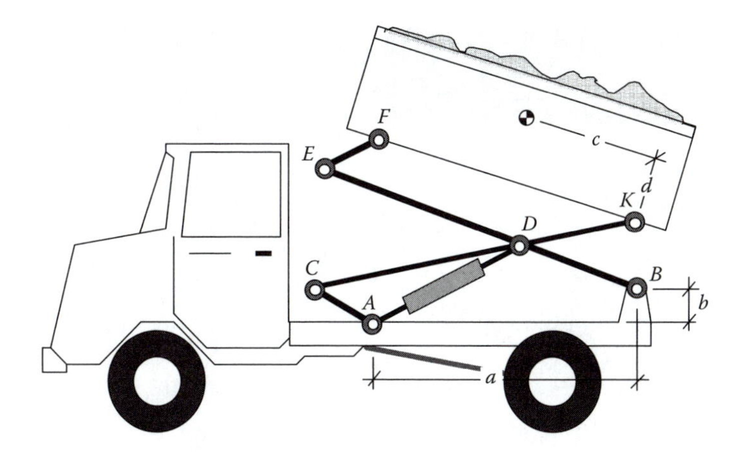

12 Inverse Dynamic Analysis

The equations of motion for a mechanical system represent the relationship between the forces that act on the system and its motion. If a specified motion of a mechanical system is sought and some of the applied forces and moments are known, it could be possible to determine the remaining forces and moments that cause that particular motion. When the desired motion is known and the objective is to find the unknown forces, the process is referred to as *inverse dynamic analysis*. In a particular form of inverse dynamic analysis that we consider in this chapter, we only need to solve algebraic equations. Inverse dynamics is an extension of kinematic analysis in which the reaction forces and some of the unknown applied forces and moments are computed. The fundamentals of inverse dynamic analysis are the same regardless of the method used to formulate a problem.

Programs: The new M-files that are developed in this chapter must be organized according to the file structure shown in Table 11.1.

12.1 UNCONSTRAINED FORMULATION

Consider the robotic arm shown in Figure 12.1. Assume that the system is modeled with the joint coordinate formulation where the defined coordinates are θ_1, θ_2 and θ_3. Two rotational motors at A and B, and a linear actuator along the sliding joint T control the motion of this system. The motors and the actuator must keep the *tool* that is being held by the end-effector in the right orientation and move it to the correct position. Assume that three driver expressions provide the desired motion:

$$\begin{aligned} {}^{(d)}\Phi_1 &= \theta_1 - f_1(t) = 0 \\ {}^{(d)}\Phi_2 &= \theta_2 - f_2(t) = 0 \\ {}^{(d)}\Phi_3 &= \theta_3 - f_3(t) = 0 \end{aligned} \tag{12.1}$$

The corresponding velocity and acceleration drivers are

$$\begin{bmatrix} 1 & 0 & 0 \\ 0 & 1 & 0 \\ 0 & 0 & 1 \end{bmatrix} \begin{Bmatrix} \dot{\theta}_1 \\ \dot{\theta}_2 \\ \dot{\theta}_3 \end{Bmatrix} = \begin{Bmatrix} \dot{f}_1(t) \\ \dot{f}_2(t) \\ \dot{f}_3(t) \end{Bmatrix} \tag{12.2}$$

and

$$\begin{bmatrix} 1 & 0 & 0 \\ 0 & 1 & 0 \\ 0 & 0 & 1 \end{bmatrix} \begin{Bmatrix} \ddot{\theta}_1 \\ \ddot{\theta}_2 \\ \ddot{\theta}_3 \end{Bmatrix} = \begin{Bmatrix} \ddot{f}_1(t) \\ \ddot{f}_2(t) \\ \ddot{f}_3(t) \end{Bmatrix} \tag{12.3}$$

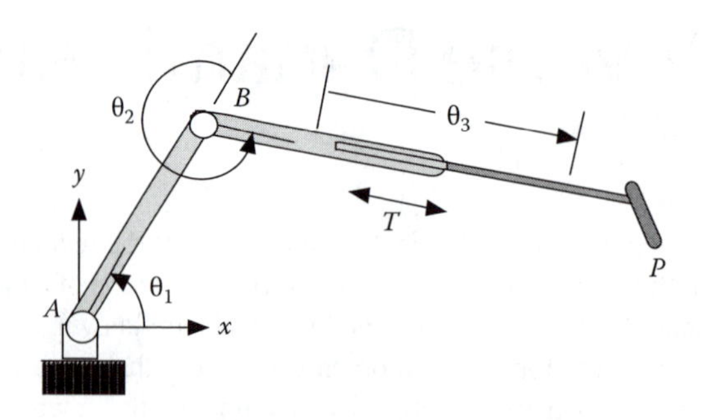

FIGURE 12.1 A robotic arm formulated with the joint coordinates.

The Jacobian of these driver constraints is ${}^{(d)}\boldsymbol{D} = \mathbf{I}$. The equations of motion for this system must contain the reaction forces/moments associated with these driver constraints:

$$\boldsymbol{M}\ddot{\boldsymbol{\theta}} = \boldsymbol{f} + {}^{(d)}\boldsymbol{D}'\,{}^{(d)}\boldsymbol{\lambda} \tag{12.4}$$

In these equations ${}^{(d)}\boldsymbol{\lambda}$ represents three Lagrange multipliers associated with the three driver constraints.

At any given time t^i, Eqs. (12.1) to (12.3) provide the values of $\boldsymbol{\theta}$, $\dot{\boldsymbol{\theta}}$, and $\ddot{\boldsymbol{\theta}}$. As in any other open-chain system, the mass matrix $\boldsymbol{M}$ and the array of forces $\boldsymbol{f}$ are evaluated. Because $\ddot{\boldsymbol{\theta}}$ is a known array, Eq. (12.4) is rearranged as

$${}^{(d)}\boldsymbol{D}'\,{}^{(d)}\boldsymbol{\lambda} = \boldsymbol{M}\ddot{\boldsymbol{\theta}} - \boldsymbol{f} \tag{12.5}$$

Furthermore, because in this example ${}^{(d)}\boldsymbol{D} = \mathbf{I}$, we have

$${}^{(d)}\boldsymbol{\lambda} = \boldsymbol{M}\ddot{\boldsymbol{\theta}} - \boldsymbol{f} \tag{12.6}$$

The three computed Lagrange multipliers represent the torques that the two rotational motors and the force that the linear actuator must supply at time t^i to maintain the motion that is described by the driver expressions.

12.1.1 General Procedure

Assume that a multibody system is modeled without any constraints; that is, the number of coordinates, n_v, is equal to the number of system's degrees of freedom, n_{dof}. Assume that the motion of the system is described by n_{dof} drivers and their first and second time derivatives as

$${}^{(d)}\boldsymbol{\Phi} = \boldsymbol{\Phi}(\mathbf{q}) - \boldsymbol{f}(t) = \mathbf{0} \tag{12.7}$$

$${}^{(d)}\dot{\boldsymbol{\Phi}} = {}^{(d)}\mathbf{D}\dot{\mathbf{q}} - \dot{\boldsymbol{f}}(t) = \mathbf{0} \tag{12.8}$$

$${}^{(d)}\ddot{\boldsymbol{\Phi}} = {}^{(d)}\mathbf{D}\ddot{\mathbf{q}} + {}^{(d)}\dot{\mathbf{D}}\dot{\mathbf{q}} - \ddot{\boldsymbol{f}}(t) = \mathbf{0} \tag{12.9}$$

The equations of motion for this system are expressed as

$$\mathbf{M}\ddot{\mathbf{q}} = \mathbf{h} + {}^{(d)}\mathbf{D}'\,{}^{(d)}\boldsymbol{\lambda}$$

or

$${}^{(d)}\mathbf{D}'\,{}^{(d)}\boldsymbol{\lambda} = \mathbf{M}\ddot{\mathbf{q}} - \mathbf{h} \tag{12.10}$$

The Jacobian ${}^{(d)}\mathbf{D}$ is a square matrix therefore Eq. (12.10) can be solved for the Lagrange multipliers ${}^{(d)}\boldsymbol{\lambda}$. Proper interpretation of these multipliers yields the forces and moments that the driver motors/actuators must provide to generate the desired motion.

A process for the inverse dynamic analysis can be stated as:

> At $t = {}^{i}t$:
>
> - Solve (or evaluate) Eqs. (12.7) to (12.9) for $\mathbf{q}$, $\dot{\mathbf{q}}$, and $\ddot{\mathbf{q}}$.
> - Solve Eq. (12.10) for the Lagrange multipliers, ${}^{(d)}\boldsymbol{\lambda}$.
>
> The process can be repeated for any other $t = {}^{i}t$.

12.2 CONSTRAINED FORMULATION

The multibody system of Figure 12.1 was modeled in Section 12.1 by the joint coordinate formulation. This system can also be modeled by the point-coordinate or the body-coordinate formulations. Assume that the system is described in body coordinates, as shown in Figure 12.2, by a 9-array of coordinate $\mathbf{c}$ and its derivatives $\dot{\mathbf{c}}$ and $\ddot{\mathbf{c}}$. Six kinematic constraints—four for the two pin joints and two for the sliding joint—must be written. These constraints yield a 6×9 Jacobian, $\mathbf{D}$.

The driver expressions of Eqs. (12.1) must be written in terms of the body coordinates as

$$\begin{aligned}
{}^{(d)}\Phi_1 &= \phi_1 - f_1(t) = 0 \\
{}^{(d)}\Phi_2 &= \phi_2 - \phi_1 - f_2(t) = 0 \\
{}^{(d)}\Phi_3 &= \tfrac{1}{2}\left(\mathbf{d}'\mathbf{d} - (f_3(t))^2\right) = 0
\end{aligned} \tag{12.11}$$

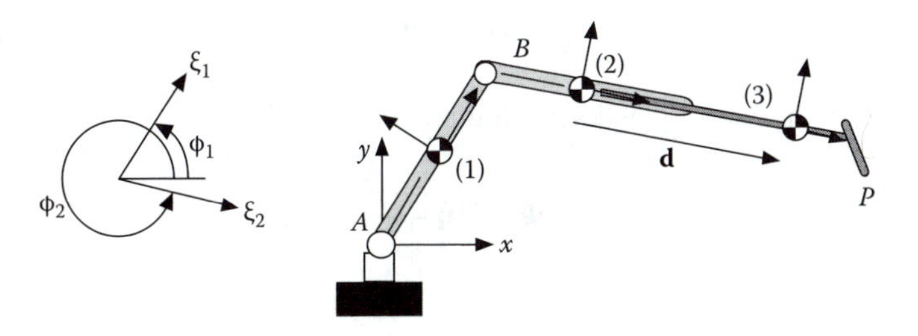

FIGURE 12.2 A robotic arm described for body coordinates formulation.

where $\mathbf{d} = \mathbf{r}_3 - \mathbf{r}_2$. The corresponding velocity drivers are

$$^{(d)}\mathbf{D}\,\dot{\mathbf{c}} = \begin{Bmatrix} \dot{f}_1(t) \\ \dot{f}_2(t) \\ f_3(t)\dot{f}_3(t) \end{Bmatrix} \tag{12.12}$$

where

$$^{(d)}\mathbf{D} = \left[\begin{array}{cc|cc|cc} \mathbf{0} & 1 & \mathbf{0} & 0 & \mathbf{0} & 0 \\ \mathbf{0} & -1 & \mathbf{0} & 1 & \mathbf{0} & 0 \\ \mathbf{0} & 0 & -\mathbf{d}' & 0 & \mathbf{d}' & 0 \end{array}\right] \tag{12.13}$$

The acceleration drivers are written as

$$^{(d)}\mathbf{D}\,\ddot{\mathbf{c}} = \begin{Bmatrix} \ddot{f}_1(t) \\ \ddot{f}_2(t) \\ -\dot{\mathbf{d}}'\dot{\mathbf{d}} + f_3(t)\ddot{f}_3(t) + (\dot{f}_3(t))^2 \end{Bmatrix} \tag{12.14}$$

The equations of motion for this system must contain the reaction forces/moments associated with these driver constraints:

$$\mathbf{M}\,\ddot{\mathbf{c}} = \mathbf{h} + \mathbf{D}'\boldsymbol{\lambda} \; + {}^{(d)}\mathbf{D}'\,{}^{(d)}\boldsymbol{\lambda} \tag{12.15}$$

In these equations $\mathbf{D}'\boldsymbol{\lambda}$ represents the reaction forces at the joints and ${}^{(d)}\mathbf{D}'\,{}^{(d)}\boldsymbol{\lambda}$ provides the reaction force/moments associated with the three driver constraints. The later set represents exactly the forces and the moment that the actuators must provide to maintain the desired motion.

12.2.1 General Procedure

Assume that a multibody system that has n_{dof} degrees of freedom is represented by n_v coordinates, $\mathbf{q}$, and $n_c = n_v - n_{dof}$ kinematic constraints as

$$\boldsymbol{\Phi}(\mathbf{q}) = \mathbf{0} \tag{12.16}$$

The corresponding velocity and acceleration constraints are:

$$\dot{\boldsymbol{\Phi}} = \mathbf{D}\,\dot{\mathbf{q}} = \mathbf{0} \tag{12.17}$$

and

$$\ddot{\boldsymbol{\Phi}} = \mathbf{D}\,\ddot{\mathbf{q}} + \dot{\mathbf{D}}\,\dot{\mathbf{q}} = \mathbf{0} \tag{12.18}$$

For the system to undergo a desired motion, we must describe n_{dof} drivers and their first and second time derivatives as

$$^{(d)}\mathbf{\Phi} = \mathbf{\Phi}(\mathbf{q}, t) = \mathbf{0} \tag{12.19}$$

$$^{(d)}\dot{\mathbf{\Phi}} = \dot{\mathbf{\Phi}}(\mathbf{q}, t) = \mathbf{0} \tag{12.20}$$

$$^{(d)}\ddot{\mathbf{\Phi}} = \ddot{\mathbf{\Phi}}(\mathbf{q}, t) = \mathbf{0} \tag{12.21}$$

Because the driver expressions could be either linear or nonlinear in terms of the coordinates and the given function $f(t)$, Eqs. (12.19) to (12.21) have been expressed in a general form. If the Jacobian of the driver constraints is denoted as $^{(d)}\mathbf{D}$, then the equations of motion can be expressed as

$$\mathbf{M}\ddot{\mathbf{q}} = \mathbf{h} + \mathbf{D}'\boldsymbol{\lambda} + {}^{(d)}\mathbf{D}'\,{}^{(d)}\boldsymbol{\lambda} \tag{12.22}$$

or

$$[\mathbf{D}' \quad {}^{(d)}\mathbf{D}']\begin{Bmatrix} \boldsymbol{\lambda} \\ {}^{(d)}\boldsymbol{\lambda} \end{Bmatrix} = \mathbf{M}\ddot{\mathbf{q}} - \mathbf{h} \tag{12.23}$$

Because $[\mathbf{D} \quad {}^{(d)}\mathbf{D}]$ is a square matrix, Eq. (12.23) can be solved for the Lagrange multipliers $\boldsymbol{\lambda}$ and $^{(d)}\boldsymbol{\lambda}$. Proper interpretation of the elements of $^{(d)}\boldsymbol{\lambda}$ yields the forces and moments that the driver motors/actuators must provide to have the desired motion.

For systems having only one degree of freedom, Eq. (12.23) can be simplified because there is only one driver equation, and therefore, we have only one multiplier in $^{(d)}\boldsymbol{\lambda}$. We premultiply Eq. (12.22) by $\dot{\mathbf{q}}'$ to get:

$$\dot{\mathbf{q}}'\mathbf{M}\ddot{\mathbf{q}} = \dot{\mathbf{q}}'\mathbf{h} + \dot{\mathbf{q}}'\mathbf{D}'\boldsymbol{\lambda} + \dot{\mathbf{q}}'\,{}^{(d)}\mathbf{D}'\,{}^{(d)}\boldsymbol{\lambda} \tag{12.24}$$

Based on Eq. (12.17), the term $\dot{\mathbf{q}}'\mathbf{D}'\boldsymbol{\lambda}$ vanishes. Therefore, Eq. (12.24) is rearranged as

$$^{(d)}\lambda = \dot{\mathbf{q}}'(\mathbf{M}\ddot{\mathbf{q}} - \mathbf{h}) \,/\, {}^{(d)}\mathbf{D}\dot{\mathbf{q}} \tag{12.25}$$

Note that because $^{(d)}\lambda$ is a single multiplier, $^{(d)}\mathbf{D}'$ is a 9×1 matrix and, therefore, $\dot{\mathbf{q}}'\,{}^{(d)}\mathbf{D}' = {}^{(d)}\mathbf{D}\dot{\mathbf{q}}$ is a scalar. Furthermore, note that the array $\dot{\mathbf{q}}$ does not have to contain the actual velocity components of the system. As long as $\dot{\mathbf{q}}$ contains a feasible set of velocities—that is, it satisfies the velocity constraints of Eq. (12.17)—it can be used in our formula without changing the results.*

* In some textbooks, Eq. (12.25) is called the *virtual work equation*. Because the terms of this equation have the unit of power, we will refer to it as the *power equation*.

A process for the inverse dynamic analysis of constrained systems can be stated as:

At $t = {}^i t$:

- Solve Eqs. (12.16) and (12.9) for $\mathbf{q}$.
- Solve Eqs. (12.17) and (12.20) for $\dot{\mathbf{q}}$.
- Solve Eqs. (12.18) and (12.21) for $\ddot{\mathbf{q}}$.
- Solve Eq. (12.23) for the Lagrange multipliers, $\boldsymbol{\lambda}$ and ${}^{(d)}\boldsymbol{\lambda}$, or if a single DoF system, you may solve Eq. (12.25) instead for ${}^{(d)}\boldsymbol{\lambda}$.

The process can be repeated for any other $t = {}^i t$.

This algorithm will be implemented in the upcoming MATLAB® program.

EXAMPLE 12.1

The slider-crank shown in Figure 12.3 represents a single-cylinder engine. In a configuration where the crank angle is 15° and its angular velocity is 6π rad/sec, a force of –25 units acts on the slider-block. If the angular velocity of the crank is a constant, what is the load on the crank? Consider the following inertial values:

$$m_1 = 1.0, J_1 = 0.1, m_2 = 2.0, J_2 = 0.6, m_3 = 3.0, J_3 = 0.5:$$

Solution:

We model this system with the joint coordinate formulation. For $\theta_1 = 0.2618$ rad, $\dot{\theta}_1 = 6\pi$, and $\ddot{\theta}_1 = 0$, we determine the coordinates, velocities, and accelerations based on the procedure described in Chapter 9:

$$\boldsymbol{\theta} = \begin{Bmatrix} 0.2618 \\ 6.1959 \\ 0.2080 \end{Bmatrix}, \dot{\boldsymbol{\theta}} = \begin{Bmatrix} 18.8496 \\ -5.7736 \\ -0.3204 \end{Bmatrix}, \ddot{\boldsymbol{\theta}} = \begin{Bmatrix} 0 \\ 25.9308 \\ -22.1116 \end{Bmatrix}$$

Continuing with the procedures of Chapters 9 and 10, we determine the Jacobian matrix associated with the cut-joint at B, the mass matrix, and the array of forces:

$$\boldsymbol{D} = \begin{bmatrix} 0.0129 & -0.0139 & 1.0 \\ -0.0483 & -0.1594 & 0 \end{bmatrix}, \boldsymbol{M} = \begin{bmatrix} 0.1056 & 0.0038 & 0 \\ 0.0038 & 0.6032 & 0 \\ 0 & 0 & 3.0 \end{bmatrix}, \boldsymbol{f} = \begin{Bmatrix} -0.0456 \\ 0.4861 \\ 16.0221 \end{Bmatrix}$$

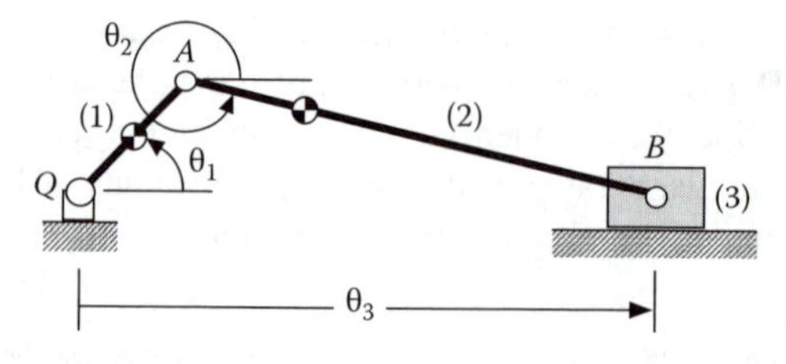

FIGURE 12.3 A slider-crank mechanism.

Because the unknown torque is associated with the joint coordinate θ_1, we add a third row to the Jacobian matrix and rewrite Eq. (12.23) as

$$\begin{bmatrix} 0.0129 & -0.0483 & 1.0 \\ -0.0139 & -0.1594 & 0 \\ 1.0 & 0 & 0 \end{bmatrix} \begin{Bmatrix} \lambda_1 \\ \lambda_2 \\ {}^{(d)}\lambda \end{Bmatrix} = \begin{bmatrix} 0.1056 & 0.0038 & 0 \\ 0.0038 & 0.6032 & 0 \\ 0 & 0 & 3.0 \end{bmatrix} \begin{Bmatrix} 0 \\ 25.9308 \\ -22.1116 \end{Bmatrix} - \begin{Bmatrix} -0.0456 \\ 0.4861 \\ 16.0221 \end{Bmatrix}$$

We solve these equations for the three Lagrange multiplier to obtain:

$$\begin{Bmatrix} \lambda_1 \\ \lambda_2 \\ {}^{(d)}\lambda \end{Bmatrix} = \begin{Bmatrix} -82.3569 \\ -87.8777 \\ -3.0353 \end{Bmatrix}$$

The resistive torque about the pin joint Q is 3.0353 units, CW.

We can also solve this problem using Eq. (12.25). Based on this equation we have

$$ {}^{(d)}\lambda = \begin{Bmatrix} 18.8496 & -5.7736 & -0.3204 \end{Bmatrix} \left(\begin{bmatrix} 0.1056 & 0.0038 & 0 \\ 0.0038 & 0.6032 & 0 \\ 0 & 0 & 3.0 \end{bmatrix} \begin{Bmatrix} 0 \\ 25.9308 \\ -22.1116 \end{Bmatrix} - \begin{Bmatrix} -0.0456 \\ 0.4861 \\ 16.0221 \end{Bmatrix} \right) / \left(\begin{bmatrix} 1.0 & 0 & 0 \end{bmatrix} \right) \begin{Bmatrix} 18.8496 \\ -5.7736 \\ -0.3204 \end{Bmatrix}$$

This single equation yields ${}^{(d)}\lambda = -3.0353$ which represents the load on the crank.

12.3 DIFFERENT JACOBIAN MATRICES

In some applications of inverse dynamics it becomes necessary to use two different Jacobian matrices in kinematic analysis and in solving the equations of motion. This type of applications is demonstrated here with one example.

We consider the robotic device from the preceding sections. Assume that the end effector of this robot has to follow a specified path; in this case a straight line as shown in Figure 12.4. The end-effector sprays paint on an object along the path and, therefore, it must keep a defined constant angle with the path; for example, it must remain perpendicular to the path. The objective of this inverse dynamic analysis is to determine the required torques and the force in the actuators that provide such motion for the end-effector in the operating range of the robotic device.

Assume that this system is modeled in body coordinate formulation. The conditions for point P on the end-effector can be described as

$$\mathbf{r}_3^P = \mathbf{r}^C + \breve{\mathbf{u}}(a + bt) \tag{12.26}$$

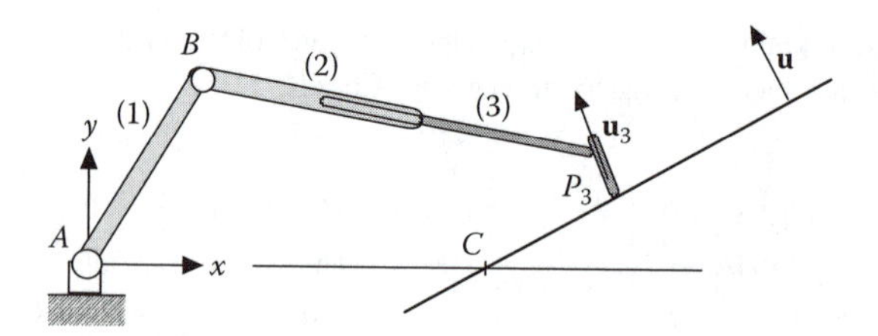

FIGURE 12.4 The end-effector of this robot must follow a straight line.

where a is the initial distance of point P from C, and b is the speed of point P along the path. Vector $\mathbf{u}$ represents a unit vector perpendicular to the path and $\mathbf{r}^C$ is a fixed point on the ground where the path intersects the x-axis. In order for the end-effector to remain perpendicular to the path, we define a unit vector $\mathbf{u}_3$ on body (3) as shown. Then we enforce the condition

$$\breve{\mathbf{u}}'\mathbf{u}_3 = 0 \tag{12.27}$$

The velocity and acceleration constraints associated with Eqs. (12.26) and (12.27) are:

$$\begin{bmatrix} \mathbf{I} & \breve{\mathbf{s}}_3^P \\ \mathbf{0} & 1 \end{bmatrix} \begin{Bmatrix} \dot{\mathbf{r}}_3 \\ \dot{\phi}_3 \end{Bmatrix} = \begin{Bmatrix} b\breve{\mathbf{u}} \\ 0 \end{Bmatrix} \tag{12.28}$$

and

$$\begin{bmatrix} \mathbf{I} & \breve{\mathbf{s}}_3^P \\ \mathbf{0} & 1 \end{bmatrix} \begin{Bmatrix} \ddot{\mathbf{r}}_3 \\ \ddot{\phi}_3 \end{Bmatrix} = \begin{Bmatrix} -\dot{\breve{\mathbf{s}}}_3^P \dot{\phi}_3 \\ 0 \end{Bmatrix} \tag{12.29}$$

The constraints in Eqs. (12.26) and (12.27), and the corresponding velocity and acceleration constraints, must be used in the kinematic analysis as drivers to determine the kinematics of the system.

Following the kinematic analysis, in the inverse dynamic equations, instead of the Jacobian from Eq. (12.28), we use the Jacobian from Eq. (12.13). The Lagrange multipliers associated with this Jacobian yield the required actuator torques and forces.

12.4 BODY COORDINATE FORMULATION

Program development for inverse dynamics with the body coordinate formulation requires a simple extension to the program we developed in Chapter 11 for kinematic analysis. For inverse dynamics we add the formulas from Eqs. (12.23) and/or (12.25) to the time-loop of kinematic analysis in the program `BC_kinematics`:

```
% file name:  BC_inverse
% Body coordinate formulation
% ...
    BC_parameters;
% Ask user for the type of solution?
    sol = input(' enter 1 if power equation, 2 otherwise: ');
% Kinematic loop
    for n = 1:n_steps
    % ...
  % Construct gamma array then compute accelerations
        rhsa_all = [BC_gamma; rhsa_driver]; c_dd = D_all\rhsa_all;
    % Evaluate array of forces
        h = BC_h;
    % Construct diagonal mass matrix
        BC_mass_diagonal; % M_diag = diag(M_array);
    % Method of solution?
        if sol == 1
        % Power equation
          Lambda_dr = (c_d'*(M_array.*c_dd - h))/(D_driver*c_d);
        else
        % Standard method
            Lambda = D_all'\(M_array.*c_dd - h);
            Lambda_dr = Lambda(nbc);
        end
      % Save the results
          BC_inv_record;
      % Increment time
          time = time + delta_t;
    end
```

In the time loop, after evaluating the accelerations, the program evaluates the array of forces and the mass matrix. Then, based on the user's response to an earlier prompt, the program employs either Eq. (12.23) or Eq. (12.25) to solve the problem. Note how the program takes advantage of the fact that the mass matrix is available as a column array when it computes the accelerations. In the script `BC_inv_record` we can specify and save any of the variables for post-processing.

12.4.1 Filmstrip Advancer

For this system we continue with the kinematic analysis scenario from Section 11.6.1. Assume that when point C of link (2) engages with the filmstrip, a constant resistive force of 600 units acts on point C. Our objective is to determine the torque that the motor has to supply during one cycle of the crank rotation. For post-processing purpose, we save some of the computed variables through the following script:

```
% File name:  BC_inv_record
    record(n, 1:6) = [time  phi(1)  r_P{7}'  c_dd(6)  Lambda_dr];
```

We first execute the program BC_inverse without any resistive force, using the *power method*. After the simulation is completed, we plot the Lagrange multiplier associated with the driver constraint versus the crank angle by typing:

```
>> plot(record(:,2), record(:,6))
```

In a second simulation, we include the resistive force. For this purpose, we monitor the x-coordinate of point C to determine whether it is engaged with the filmstrip or not. If we have engagement, then we apply the resistive forces on C. We revise the M-file BC_h as:

```
function h = BC_h
% Force array
    BC_global
    h = zeros(9,1);
% Force acting at C during engagement
    r_C = r_P{7}; f = [0; 600];
    if r_C(1) < 1.0
        h(4:5) = f; h(6) = s_rot(s_P{7})'*f;
    end
```

We execute the program BC_inverse again. After the simulation is completed, we type the following commands:

```
>> hold on
>> plot(record(:,2), record(:,6), 'r--')
```

These commands yield two plots as shown in Figure 12.5. Note that during engagement, the motor provides a larger torque.

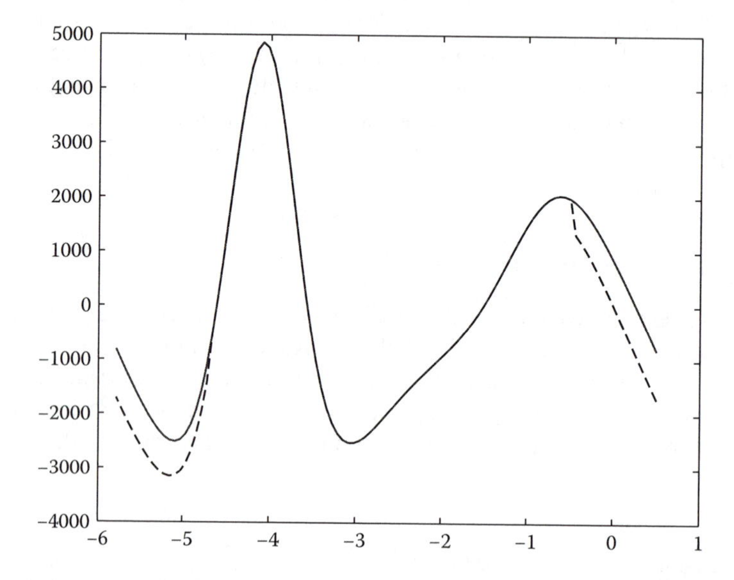

FIGURE 12.5 Torque of the motor versus crank angle (no load in solid; with load in dashed).

12.4.2 Web-Cutter

For the web-cutter mechanism, we continue with the kinematic analysis scenario of Section 11.6.2. Assume that the thickness of the web is 0.02 units and during the cut the web applies resistive forces of 2,000 units on points *C* and *D*. Our objective is to determine the torque of the motor during one cycle of operation. For this purpose, we monitor the distance between points *C* and *D* by constructing a vector $\mathbf{d} = \mathbf{r}_2^C - \mathbf{r}_3^D$. We also monitor the velocity of point *C*. When the magnitude of $\mathbf{d}$ becomes smaller than the thickness of the web, and the velocity of point *C* is downward, then the resistive forces are applied on *C* and *D*. We revise the M-file `BC_h` for this model and construct a new `BC_inv_record` as:

```
function h = BC_h
    BC_global
% Force array
    h = zeros(9,1); thickness = 0.02; f = [0; 2000];
% Include resistive force from the web only during cutting
    d = r_P{7} - r_P{8}; d_mag = sqrt(d'*d); r_C_d = r_P_d{7};
    if d_mag < thickness
        if r_C_d(2) < 0
            h(4:5) =  f; h(6) =  s_rot(s_P{7})'*f;
            h(7:8) = -f; h(9) = -s_rot(s_P{8})'*f;
        end
    end
```

```
% File name:  BC_inv_record
    record(n, 1:7) = [time  phi(1)  r_P{7}'  r_P{8}'  Lambda_dr];
```

We execute the program twice—with and without the resistive forces. If we plot the values of the Lagrange multiplier associated with the driver constraint from both simulations versus the crank angle, we obtain the plots shown in Figure 12.6.

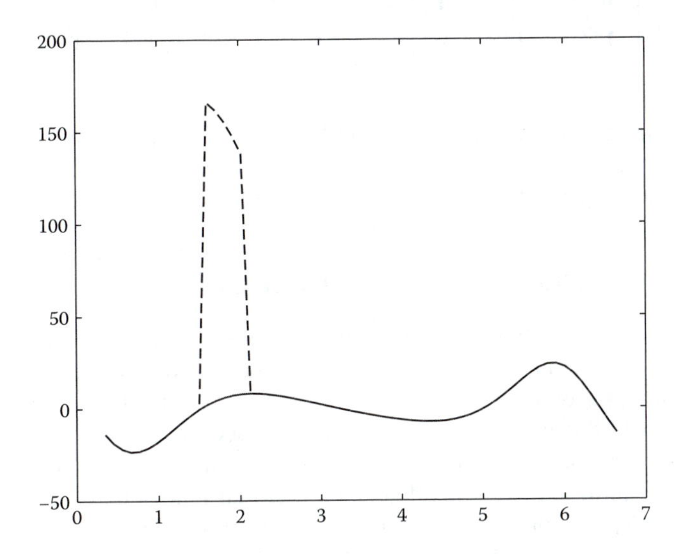

FIGURE 12.6 Torque of the motor versus crank angle (no load in solid; with load in dashed).

12.5 POINT COORDINATE FORMULATION

Inverse dynamics with the point coordinate formulation follows a process identical to that described in Section 12.2.1 with few minor differences. One difference is the way the driver force/torque appears in the equations of motion. If a rotational motor is acting as a driver, its applied torque is replaced by an equivalent set of forces in the equations of motion. Therefore, after we determine the driver force from inverse dynamic analysis, these forces must be converted to an equivalent driver torque. The other minor difference is that a driver force normally is represented by two unknown components in two separate equations. That means there must be two unknown Lagrange multipliers associated with such a force. Therefore, we cannot apply the power formula of Eq. (12.25)—we can only use Eq. (12.23). These points will be clarified with the filmstrip advancer mechanism in the following subsection.

Developing a program for inverse dynamics with the point coordinate formulation is left to the reader. A process similar to that of the `BC_inverse` can be followed in developing this new program.

12.5.1 FILMSTRIP ADVANCER

The kinematic analysis of this mechanism with the point coordinate formulation was discussed in Section 11.7.1. Using the Jacobian matrix from Eq. (11.28) and assuming an approximated mass matrix (we can also use an exact mass matrix), the inverse dynamic equations for this system based on Eq. (12.23) can be written as:

$$\begin{bmatrix} -\mathbf{d}^{B,A} & \mathbf{0} & \vdots & \mathbf{I} \\ \mathbf{d}^{B,A} & \mathbf{d}^{B,O} & \vdots & \mathbf{0} \end{bmatrix} \begin{Bmatrix} \lambda_1 \\ \lambda_2 \\ {}^{(d)}\boldsymbol{\lambda} \end{Bmatrix} = \begin{bmatrix} m^A\mathbf{I} & \mathbf{0} \\ \mathbf{0} & m^A\mathbf{I} \end{bmatrix} \begin{Bmatrix} \ddot{\mathbf{r}}^A \\ \ddot{\mathbf{r}}^B \end{Bmatrix} - \begin{Bmatrix} \mathbf{f}^A \\ \mathbf{f}^B \end{Bmatrix} \tag{12.30}$$

The Lagrange multipliers associated with the two driver expressions are ${}^{(d)}\boldsymbol{\lambda} = \{{}^{(d)}\lambda_1 \quad {}^{(d)}\lambda_2\}'$ that represent the components of a driver force acting on A. The driver torque can be computed as ${}^{(d)}n = \breve{\mathbf{d}}^{B,O'}\,{}^{(d)}\boldsymbol{\lambda}$.

12.6 JOINT COORDINATE FORMULATION

When open-chain multibody systems are modeled by the joint coordinate formulation, the resultant equations of motion are unconstrained. Therefore inverse dynamic analysis of such systems follows the procedure that we discussed in Section 12.1.1. The example of robot device in Section 12.1 represents exactly the type of formulation for open-chain systems with the joint coordinates. In contrast, for multibody systems that contain closed kinematic chains, the equations of motion with the joint coordinates are constrained; that is, reaction forces at the cut joints appear in the equations. Inverse dynamic analysis of such systems follows the procedure of Section 12.2.1. The slider-crank formulation in Example 12.1

represents exactly this type of process. Developing a MATLAB program for inverse dynamics with the joint coordinate formulation is left as an exercise for the reader.

12.7 REMARKS

In this chapter we learned how to extend a kinematic analysis program to perform inverse dynamics. For systems with one degree of freedom, the force analysis step can be transformed into a single equation in one unknown. For systems containing more that one degree of freedom, such as a robotic device, we are required to solve a set of simultaneous algebraic equations. These methodologies were applied to develop a program for the body coordinate formulation. Developing similar programs for other formulations is left to the reader.

12.8 PROBLEMS

12.1 Refer to Problem 11.13. Determine the force associated with the driver expression. Plot the magnitude of the force versus y_2.

12.2 Refer to Problem 11.14. Determine the torque or the force associated with the driver expression in each model.

12.3 Refer to Problem 11.19. Assume that the mass centers of the two rods are at *A* and each rod has a mass of 25 and a moment of inertia of 8. The container and the load have a total mass of 450 and a moment of inertia of 150, and the combined mass center is positioned at $c = 1.34$ and $d = 0.38$. Determine the actuator force for moving the load according to the defined driver.

12.4 Refer to Problem 11.20. The mass center of each link is at its geometric center and the mass and moment of inertia data are: $m_{E,F} = m_{A,c} = 0.4$, $J_{E,F} = J_{A,c} = 0.005$, $m_{C,K} = m_{B,E} = 2.0$, and $J_{C,K} = J_{B,E} = 0.7$. The container and the load have a total mass of 1,000 and a moment of inertia of 27. The combined mass center is positioned at $c = 0.8$ and $d = 0.4$. Determine the necessary actuator force for moving the load according to the defined driver.

12.5 A typical dough-kneader mechanism is shown.* The crank *OB* goes through a complete rotation in 3 seconds. Consider the following dimensions: $OB = AC = BD = 0.06$, $AB = CD = 0.13$, $QA = 0.15$, $BP = 0.26$, $a = 0.155$ and $b = 0.09$. Assume a resistive force of 100 acting on point *P* that is always in the opposite direction of the velocity of *P*. Because the rotational velocity of the crank is relatively slow, the inertia terms are insignificant. Therefore assume each moving link has a mass of 0.2 and a moment of inertia of 0.1. Determine the required torque from the motor.

* This mechanism has been adopted for simulation from the following reference:
Artobolevsky, I. I., *Mechanisms in Modern Engineering Design*, Vol. I, MIR Publishers, Moscow, 1975.

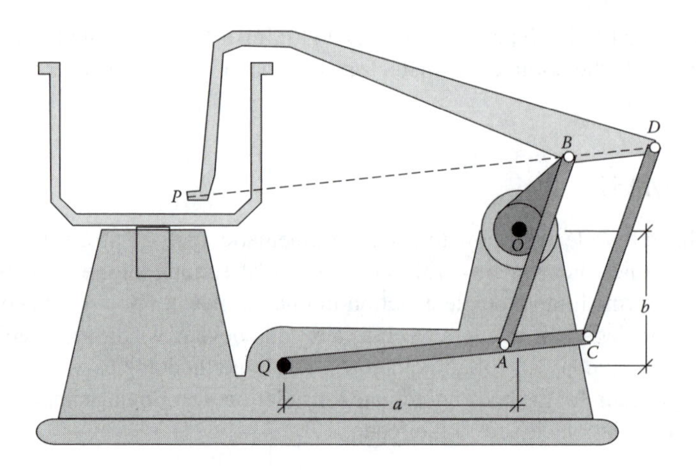

12.6 The simple robotic device shown contains three moving links, two rotational actuators about the pin joints at O and A, and a translational actuator between points A and B. Link OA has a length of 1 unit. For each link assume a mass of 1 and a moment of inertia of 0.1. Construct three driver expressions as:

$$^{(d)}\Phi_1 = \phi_1 - f_1(t) = 0$$

$$^{(d)}\Phi_2 = \phi_2 - \phi_1 - f_2(t) = 0$$

$$^{(d)}\Phi_3 = \tfrac{1}{2}\left(\mathbf{d}'\mathbf{d} - (f_3(t))^2\right) = 0$$

where the three time functions must be described according to the curves shown:

$f_1(t)$ is a cosine function that varies ϕ_1 from $\pi / 4$ to $\pi / 2$ in 4 seconds.
$f_2(t)$ is a cosine function that varies $\phi_2 - \phi_1$ from $3\pi / 4$ to $\pi / 4$ in 4 seconds.
$f_3(t)$ is a cosine function that varies the length of vector $\mathbf{d}$, connecting point A to point B, from 0.9 to 0.3 units in 4 seconds.

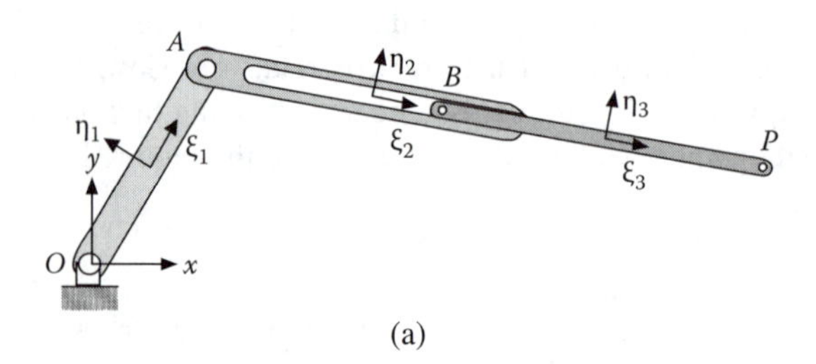

(a)

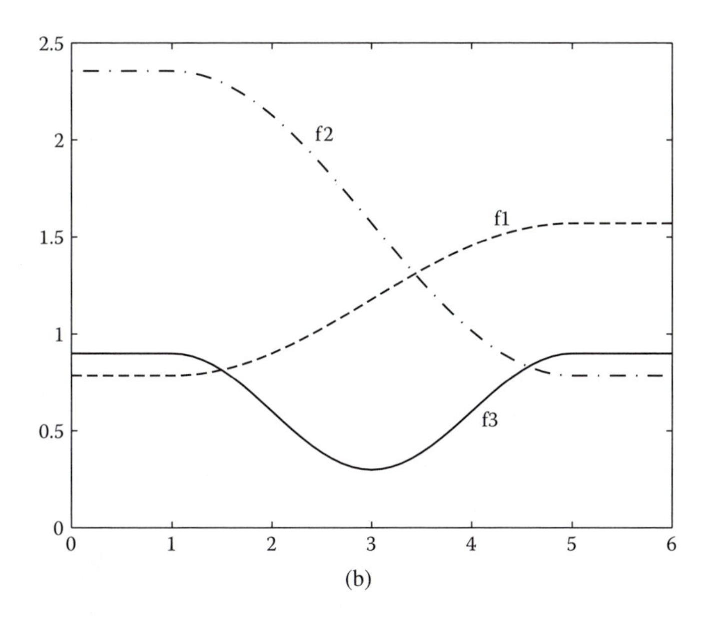

(b)

Perform an inverse dynamic analysis and determine the required torques/force at the actuators.

12.7 Point P of the robotic device in Problem 12.6 must follow a circular path centered at the coordinates $x = 2.5$ and $y = 1.0$ having a radius of $R = 0.5$. While point P follows this circular path, link (3) must keep a constant angle of $\phi_3 = 30°$. Furthermore, point P must negotiate the path in 6 seconds with a constant speed. Determine the required torques/force of the actuators. You need to revise the program `BC_inverse` for this problem.

12.8 Revise the program `BC_inverse` such that it would automatically choose the solution method: if a system has only 1 DoF, the program would apply method 1 (power formula), otherwise the program would apply method 2 (simultaneous equations).

12.9 Develop a MATLAB program to perform inverse dynamics with the joint coordinate formulation. Small portions from the program `JC_formulation` can be combined with the program `JC_kinematics` to create this new program; name it `JC_inverse`. You may develop two different versions of the program: one for open-chain systems and one for closed-chain systems. You can also develop a single program with logical statements to handle both open- and closed-chain formulations.

12.10 Develop a MATLAB program to perform inverse dynamics with the point coordinate formulation.

13 Forward Dynamic Analysis

In forward dynamic analysis the objective is to predict the motion of a multibody system under given applied loads starting from a set of known initial conditions for the coordinates and velocities. Unlike kinematic analysis that the prescribed motion of the driving coordinates determines the motion of the system, in forward dynamic analysis it is the applied forces and moments that are used to determine the motion. For this purpose, the equations of motion are treated as second-order differential equations. In this chapter several procedures for numerical solution of the equations of motion are presented. These procedures utilize numerical methods for solving first-order differential equations.

If the equations of motion do not contain any algebraic constraints, then a numerical integration algorithm can be applied to them directly. However, when the equations of motion contain algebraic constraints, in addition to the standard numerical integration processes, some other issues must be considered. The techniques and algorithms that are discussed in this chapter can be used to solve the equations of motion based on any formulation. We first consider the process of forward dynamics for unconstrained systems and then we discuss constrained formulations.

Programs: The new M-files that are developed in this chapter must be organized according to the file structure shown in Table 11.1 and Table 11.2.

13.1 UNCONSTRAINED FORMULATION

Unconstrained equations of motion for a system of particles or bodies have the following general form:

$$\mathbf{M}\,\ddot{\mathbf{q}} = \mathbf{h} \tag{13.1}$$

These equations may represent the equations of motion for a system of unconstrained particles, a set of unconstrained bodies, or an open-chain system represented by joint-coordinates. When the unknowns are the accelerations, $\ddot{\mathbf{q}}$, these equations are treated as linear algebraic equations. When the unknowns are the coordinates and velocities, the same equations are treated as second-order ordinary differential equations (ODE).

To treat the equations of motion as second-order differential equations and integrate them numerically, two integration arrays are defined as

$$\mathbf{z} = \begin{Bmatrix} \mathbf{q} \\ \dot{\mathbf{q}} \end{Bmatrix} \qquad \dot{\mathbf{z}} = \begin{Bmatrix} \dot{\mathbf{q}} \\ \ddot{\mathbf{q}} \end{Bmatrix} \tag{13.2}$$

If we integrate the array $\dot{\mathbf{z}}$, containing the velocities and accelerations, we obtain the array $\mathbf{z}$, containing the coordinates and velocities for the next time step. Obviously, this requires a set of initial values for the coordinates and velocities to start the integration process.

EXAMPLE 13.1

For the system of two unconstrained particles in Section 6.1.1, define the integration arrays.

Solution:

For the two particles, the coordinate and velocity vectors were defined as: $\mathbf{r}^A$, $\mathbf{r}^B$, $\dot{\mathbf{r}}^A$, and $\dot{\mathbf{r}}^B$. Therefore, the integration arrays are constructed as:

$$\mathbf{z} = \begin{Bmatrix} \mathbf{r}^A \\ \mathbf{r}^B \\ \dot{\mathbf{r}}^A \\ \dot{\mathbf{r}}^B \end{Bmatrix} \qquad \dot{\mathbf{z}} = \begin{Bmatrix} \dot{\mathbf{r}}^A \\ \dot{\mathbf{r}}^B \\ \ddot{\mathbf{r}}^A \\ \ddot{\mathbf{r}}^B \end{Bmatrix}$$

EXAMPLE 13.2

For the variable-length pendulum of Sections 9.3.1 and 10.1.1, define the integration arrays.

Solution:

For this system, two joint coordinates were defined as θ_1 and θ_2. Therefore, the integration arrays are constructed as:

$$\mathbf{z} = \begin{Bmatrix} \theta_1 \\ \theta_2 \\ \dot{\theta}_1 \\ \dot{\theta}_2 \end{Bmatrix} \qquad \dot{\mathbf{z}} = \begin{Bmatrix} \dot{\theta}_1 \\ \dot{\theta}_2 \\ \ddot{\theta}_1 \\ \ddot{\theta}_2 \end{Bmatrix}$$

13.1.1 Initial-Value Problems

Most general-purpose numerical integration algorithms are developed to solve first-order differential equations. A first-order differential equation is written as

$$\dot{z} = f(z, t) \tag{13.3}$$

where t is referred to as the *independent* variable and $z = z(t)$ is considered a *dependent* variable. Equation (13.3) has a family of solutions z(t). The choice of an initial value ${}^0z$ serves to determine one of the solutions of the family. The initial value ${}^0z$ could be defined for any value of ${}^0t$, although it is often assumed that a transformation has been made so that ${}^0t = 0$. This does not affect the solution or methods used to approximate the solution.

In general, if n dependent variables are involved and n first-order differential equations are available, Eq. (13.3) is written in vector form as

$$\dot{\mathbf{z}} = \boldsymbol{f}(\mathbf{z}, t) \tag{13.4}$$

with a set of initial conditions ${}^0\mathbf{z}$. In dynamics problems, t represents the time and it may not appear explicitly in the equations.

Solving initial-value problems involves a step-by-step process in which a sequence of discrete points ${}^0t$, ${}^1t$, ${}^2t$, ... is generated. The discrete points may have either constant or variable spacing ${}^i\Delta t = {}^{i+1}t - {}^it$, where ${}^i\Delta t$ is the step size for any discrete point ${}^it$. At each point ${}^it$, the *solution* z(${}^it$) is approximated by a *number* ${}^iz$.

Because no numerical algorithm is capable of finding z(${}^it$) exactly, the quantity ${}^i\varepsilon = \left| z({}^it) - {}^iz \right|$ is defined to represent the *total error* at $t = {}^it$. The total error consists of two components: *a truncation error* and a *round-off error*. The truncation error depends on the nature of the numerical algorithm used in computing ${}^iz$. The round-off error is due to the finite word length in a computer. In this textbook, when the term *numerical error* is used, we mean truncation error. The accuracy of an algorithm is directly proportional to its *order*. In general, higher the order, more accurate the algorithm can predict a solution. Although there exist many algorithms for solving initial-value problems, most of them are based on two basic approaches: *Taylor series expansion* and *polynomial approximation*.

Algorithms that are based on Taylor series expansion require the derivatives of the function with respect to t. A k-th order algorithm needs the derivatives up to k–1. These derivatives, in general, are not readily available. Therefore, the higher-order algorithms that are accurate are not computationally efficient.

A classical theory asserts that any continuous function can be approximated arbitrarily within any closed interval by a polynomial of sufficiently high degree, even if the solution is not a polynomial. In view of this theory, a polynomial of sufficiently high order can, in principle, be used to calculate z(${}^{i+1}t$) to any desired accuracy. A k-th-degree polynomial is referred to as an algorithm of order k. In practice, the amount of computation increases with the order of polynomial. In contrast to the procedure in the Taylor algorithms, information from previous time steps is utilized in most polynomial approximation algorithms to compute ${}^{i+1}z$. An algorithm that utilizes information from time steps prior to ${}^it$ is called a *multistep* algorithm. The polynomial approximation algorithms can be categorized as *explicit algorithms* or *implicit algorithms*. The implicit algorithms are iterative. In these algorithms, an estimate of ${}^{i+1}z$ is required to start the iteration of the formula. To obtain a relatively good estimate for ${}^{i+1}z$, an explicit formula can be used. This step is known as a *predictor* step. Then, the implicit formula is used to correct the predicted value of ${}^{i+1}z$. This step is known as a *corrector* step.

Normally, an algorithm iterates on the corrector step. Taylor algorithms are single-step, explicit algorithms.

Another commonly used family of explicit algorithms is the Runge-Kutta which can be considered as a revised form of Taylor algorithms. In the following section we review the fourth order Runge-Kutta algorithm.

13.1.2 Runge-Kutta Algorithms

For a relatively large step size and reasonable accuracy, the *fourth-order Runge-Kutta* is a widely used algorithm. This algorithm is stated as:

$$^{i+1}z = {}^{i}z + \Delta t\, f \tag{13.5}$$

where

$$f = \tfrac{1}{6}(f_1 + 2f_2 + 2f_3 + f_4) \tag{13.6}$$

and

$$\begin{aligned} f_1 &= f({}^{i}z,\ {}^{i}t) \\ f_2 &= f\left({}^{i}z + \frac{\Delta t}{2} f_1,\ {}^{i}t + \frac{\Delta t}{2}\right) \\ f_3 &= f\left({}^{i}z + \frac{\Delta t}{2} f_2,\ {}^{i}t + \frac{\Delta t}{2}\right) \\ f_4 &= f\left({}^{i}z + \Delta t\, f_3,\ {}^{i}t + \Delta t\right) \end{aligned} \tag{13.7}$$

Because the algorithm is a fourth-order one, the truncation error remains relatively small even for a relatively large step size. The major disadvantage of this algorithm is that the function $f(z, t)$ must be evaluated four times at each time step. In addition, the values of the function are not used in any subsequent computations. Hence, this algorithm is not computationally as efficient as some of the multistep algorithms.

Example 13.3

Refer to Example 13.1. Use the fourth-order Runge-Kutta algorithm to find the response of the system of two unconstrained particles for 10 seconds.

Solution:

The array of forces and the mass matrix for this system were formulated in a program in Section 6.1.1. We borrow the statements from that program and rearrange them in a main script and a function called `Ex_13_3`. Then we formulate the Runge-Kutta algorithm into another function, named `RK4`, as listed in the following:

```
% Example 13.3
    clear all
    addpath Basics  PC_Basics
    global k_1  L_1_0  d_c_1  k_2  L_2_0  r_O  M_inv_array
% Define particle masses and spring-damper characteristics
    m_A = 3;  m_B = 4;
    k_1 = 50; L_1_0 = 1.1; d_c_1 = 20; k_2 = 100; L_2_0 = 1.2;
% Initial values for coordinates and velocities
    r_A = [0.5; -1]; r_A_d = [-0.2; 0.4]; r_O = [0; 0];
    r_B = [2; -1];    r_B_d = [-0.1; 0.3];
% Construct the inverse of mass matrix (array)
    M_inv_array = [1/m_A; 1/m_A; 1/m_B; 1/m_B];
% Initialize z array
    z = [r_A; r_B; r_A_d; r_B_d];
% Set time parameters
    Tspan = [0.0:0.05:10.0];
% Integrate
    [T, zT] = RK4(@Ex_13_3, Tspan, z);
```

```
function [T, zT] = RK4(fun, Tspan, z)
% 4th order Runge-Kutta algorithm
    T = Tspan'; zT = z'; nT = size(Tspan);
for i=1:(nT(2) - 1)
    t  = Tspan(i); dt = Tspan(i + 1) - Tspan(i);
    f1 = fun(t, z);
    f2 = fun((t + dt/2), (z + dt*f1/2));
    f3 = fun((t + dt/2), (z + dt*f2/2));
    f4 = fun((t + dt  ), (z + dt*f3   ));
    z  = z + dt*(f1 + 2*(f2 + f3) + f4)/6;
    zT = [zT; z'];
end
```

```
function z_d = Ex_14_3(t, z)
   global k_1  L_1_0  d_c_1  k_2  L_2_0  r_O  M_inv_array
% Unpack the contents of z array
   r_A = z(1:2); r_B = z(3:4); r_A_d = z(5:6); r_B_d = z(7:8);
% compute d1, d1_d, d2 and d2_d vectors
   d1 = r_A - r_O; d1_d = r_A_d;
   d2 = r_B - r_A; d2_d = r_B_d - r_A_d;
% Compute forces for element 1
   f_sda_1 = pp_sda(d1, d1_d, k_1, L_1_0, d_c_1, 0);
   f_O_1 =  f_sda_1; f_A_1 = -f_sda_1;
% Compute forces for element 2
   f_sda_2 = pp_sda(d2, d2_d, k_2, L_2_0, 0, 0);
   f_A_2 =  f_sda_2; f_B_2 = -f_sda_2;
% Compute total force acting on A and B
   f_A = f_A_1 + f_A_2; f_B = f_B_2;
% Construct the 4x1 array of forces
   h = [f_A; f_B];
% Compute accelerations
   r_dd = M_inv_array.*h;
% pack velocities and accelerations into z_d array
   z_d = [z(5:8); r_dd];
```

The function RK4 is written in a general form that can be used in the future programs as well. The input-output arguments of this function are organized to be the same as a built-in MATLAB® function, ode45, which will be used later on. The output of RK4 contains two arrays, T and zT. The column array T contains the time marks of the integration. The array zT contains the time record of z' (coordinates and velocities). The accelerations are not returned from RK4—if the accelerations are needed, they must be recomputed (this is done to be consistent with the output of ode45).

Executing the program results into the coordinates and velocities of the two particles during 10 seconds. We may plot, for example, y^B versus x^B by typing in the Command Window the following:

```
>> plot(zT(:,3), zT(:,4))
```

13.1.3 VARIABLE STEP SIZE

One of the most important problems in using a constant-step numerical integration algorithm, such as the fourth-order Runge-Kutta, is the selection of a proper step size. A large step size may cause erroneous results. A step size too small may yield accurate results while increasing the computation time unreasonably. Therefore, it is important to choose a reasonably small step size to obtain accurate results without unnecessarily increasing the computation time. A thorough discussion on the subject of time-step selection is outside the scope of this textbook. The interested reader may refer to textbooks on the subject of numerical solution of differential equations. In this section, two highly simple examples are provided to familiarize the reader with this important point. These examples deal with vibratory motion which has not been covered in this textbook either.

The simplest form of vibratory motion is a *simple harmonic motion* which is described by the second-order differential equation

$$\ddot{q} + \omega^2 q = 0 \tag{13.8}$$

where ω is a *real* number. The frequency and the period of oscillation of this single-degree-of-freedom system, respectively, are

$$f = \frac{\omega}{2\pi}, \quad \tau = \frac{2\pi}{\omega} \tag{13.9}$$

If Eq. (13.8) is solved numerically, the step size Δt must be much smaller than the period τ.

As the first example, consider the one-dimensional motion of the mass-spring system shown in Figure 13.1(a). The only external force acting on the system is the gravity. The equations of motion for this system, in the y-direction, is given by

$$m\ddot{y} = -mg - k(y - {}^0\ell)$$

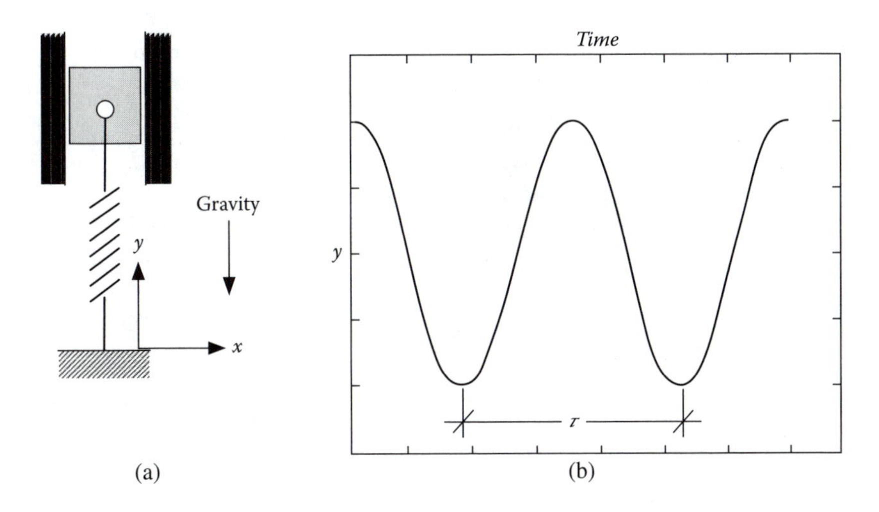

FIGURE 13.1 (a) A one-dimensional oscillating mass-spring system, and (b) a full cycle of its response denoted by τ.

or

$$\ddot{y}+\frac{k}{m}y=-g+\frac{k}{m}{}^{0}\ell \tag{13.10}$$

For a system in free vibration, if the left sides of Eqs. (13.8) and (13.10) are compared, it is found that

$$\omega^2=\frac{k}{m},\quad \tau=2\pi\sqrt{\frac{m}{k}} \tag{13.11}$$

Given the values of m and k, the time period τ shown in Figure 13.1(b) can be calculated. Assuming that $m=1$ and $k=100$, the time period is found to be $\tau=0.628$ seconds.

As another example, consider the single pendulum shown in Figure 13.2. Assuming that the length of the rod is ℓ, its mass and moment of inertia are m and $m\ell^2/12$, and the only external force acting on the system is the gravity, the equation of motion is found to be

$$\tfrac{1}{3}m\ell^2\ddot{\theta}+\tfrac{1}{2}mg\ell\sin\theta=0 \tag{13.12}$$

For oscillations of small amplitudes, $\sin\theta$ can be replace by θ, and Eq. (13.12) is linearized as

$$\tfrac{1}{3}m\ell^2\ddot{\theta}+\tfrac{1}{2}mg\ell\theta=0 \tag{13.13}$$

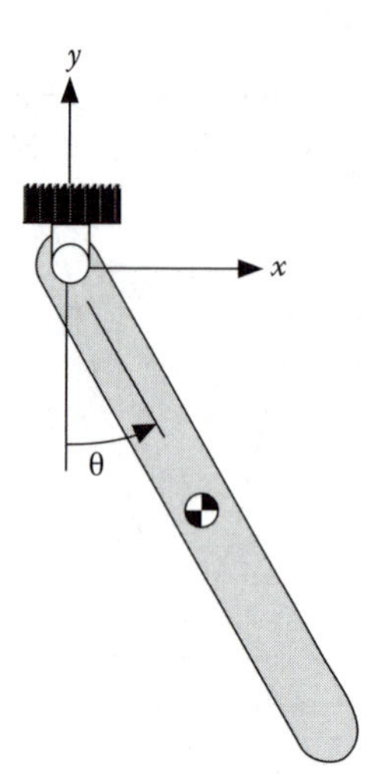

FIGURE 13.2 A single pendulum.

Comparing Eqs. (13.8) and (13.13) yields

$$\omega^2 = \frac{g}{6\ell}, \quad \tau = 2\pi\sqrt{\frac{6\ell}{g}} \tag{13.14}$$

Assuming numerical values of $\ell = 1$ and $g = 9.81$, the time period is found to be $\tau = 4.9$.

For numerical integration, a reasonable value for Δt could be:

$$\Delta t \le \tau / 10 \tag{13.15}$$

For the mass-spring example, the integration time-step should be $\Delta t \le 0.06$, and for the pendulum example, the time-step should be $\Delta t \le 0.4$. These examples show how the time period of the oscillation can be calculated. For more complex systems, the calculation of natural frequencies will not be that simple. For multi-body systems, the linearization of the equations of motion in explicit form can be rather cumbersome. For systems with more than 1 DoF, there will be more than one natural frequency. For such systems, the highest frequency must be found, and a step size much smaller than the period of the highest frequency must be selected. Because the equations of motion are generally nonlinear in terms of the coordinates, linearization of these equations yields a time step which is valid only in the configuration around which the system is linearized. In a different configuration, the linearization process may yield a different time period, and consequently a different time step size.

To avoid the difficulties associated with the selection of a proper size for the time step, it is strongly recommended that a variable-step size algorithm be used for dynamic analysis. Many well-developed algorithms of this sort are available which determine a proper time step and will adjust the time step size automatically during simulation. One such algorithm is the `ode45` in MATLAB.* The function `ode45` can be invoked by the following statement:

```
[T, zT] = ode45(@Example, Tspan, z)
```

The input-output arguments for this function are the same as those for function `RK4`. The main difference is that the value we specify for Δt (in `Tspan`) is actually the reporting time interval and not the time-step for integration. For more detail on this function refer to the MATLAB help.

13.1.4 General Procedure

Assume that a multibody system is modeled without any constraints and, therefore, the number of defined coordinates, n_v, is equal to the number of system's degrees of freedom, n_{dof}. A process for forward dynamic analysis can be stated as:

- Define initial conditions for the coordinates and velocities; that is, ${}^0\mathbf{q}$ and ${}^0\dot{\mathbf{q}}$.
- Initialize an integration array as $\mathbf{z} = \begin{Bmatrix} {}^0\mathbf{q} \\ {}^0\dot{\mathbf{q}} \end{Bmatrix}$.
- Establish time parameters for integration.
- Invoke a numerical integrator such as `ode45`. The integrator sends $\mathbf{z}$, for $t = {}^i t$, to a function and asks for $\dot{\mathbf{z}}$:
  - Transfer the contents of $\mathbf{z}$ to $\mathbf{q}$, and $\dot{\mathbf{q}}$.
  - Compute accelerations $\ddot{\mathbf{q}}$.
  - Transfer the contents of $\dot{\mathbf{q}}$, and $\ddot{\mathbf{q}}$ to a $\dot{\mathbf{z}}$ array as $\dot{\mathbf{z}} = \begin{Bmatrix} \dot{\mathbf{q}} \\ \ddot{\mathbf{q}} \end{Bmatrix}$.
  - Return the $\dot{\mathbf{z}}$ array to the integrator.
  - This function is invoked several times by the integrator for every time step.
- Recover $\ddot{\mathbf{q}}$ if necessary.

Most numerical integrators, such as `ode45` or `RK4`, do not return the computed accelerations to the main script, so the accelerations are lost. To recover the accelerations, after the integration process is completed, knowing the coordinates and

---

* The function `ode45` uses two error tolerances: a relative error tolerance, `RelTol`, and an absolute error tolerance, `AbsTol`. The default setting for these tolerances are `RelTol = 1e-3` and `AbsTol = 1e-6`. These tolerances can be reset by the user as shown in the following statements:

```
options = odeset('RelTol', 1e-6, 'AbsTol', 1e-9);
[T, zT] = ode45(@BC_eqsmo, Tspan, z, options);
```

If we observe significant numerical error when we simulate the response of a multibody system, tightening the error tolerances may remedy the problem. This in turn significantly increases the computational time.

velocities at every reported time step, we recomputed the accelerations by reconstructing the equations of motion.

13.2 CONSTRAINED FORMULATION

The kinematic constraints and the equations of motion for a system of constrained particles or bodies are expressed as

$$\mathbf{\Phi}(\mathbf{q}) = \mathbf{0} \tag{13.16}$$

$$\dot{\mathbf{\Phi}} = \mathbf{D}\dot{\mathbf{q}} = \mathbf{0} \tag{13.17}$$

$$\ddot{\mathbf{\Phi}} = \mathbf{D}\ddot{\mathbf{q}} - \boldsymbol{\gamma} = \mathbf{0} \tag{13.18}$$

$$\mathbf{M}\,\ddot{\mathbf{q}} = \mathbf{h} + \mathbf{D}'\boldsymbol{\lambda} \tag{13.19}$$

These equations of motion could represent a system of constrained particles, a set of bodies formulated in body coordinates, or a closed-chain system represented by joint-coordinates. If the unknowns are the accelerations, $\ddot{\mathbf{q}}$, and Lagrange multipliers, $\boldsymbol{\lambda}$, these equations are treated as linear algebraic equations. However, when the unknowns are the coordinates and velocities, the same equations must be treated as second-order differential-algebraic equations (DAE).

As algebraic equations, Eq. (13.19) must be solved for the accelerations. However, there are more unknowns in these equations than the number of equations. To make the number of equations equal to the number of unknowns, we append the acceleration constraints to Eq. (13.19) and rearrange them in matrix form as

$$\begin{bmatrix} \mathbf{M} & -\mathbf{D}' \\ \mathbf{D} & \mathbf{0} \end{bmatrix} \begin{Bmatrix} \ddot{\mathbf{q}} \\ \boldsymbol{\lambda} \end{Bmatrix} = \begin{Bmatrix} \mathbf{h} \\ \boldsymbol{\gamma} \end{Bmatrix} \tag{13.20}$$

These equations can be solved for the unknowns $\ddot{\mathbf{q}}$ and $\boldsymbol{\lambda}$.

As differential equations, Eq. (13.19) must be integrated numerically. For that purpose, similar to the unconstrained formulations, two integration arrays are defined as

$$\mathbf{z} = \begin{Bmatrix} \mathbf{q} \\ \dot{\mathbf{q}} \end{Bmatrix} \quad \dot{\mathbf{z}} = \begin{Bmatrix} \dot{\mathbf{q}} \\ \ddot{\mathbf{q}} \end{Bmatrix} \tag{13.21}$$

Then, if we integrate $\dot{\mathbf{z}}$, which contains velocities and accelerations, we obtain $\mathbf{z}$—that is, coordinates and velocities for the next time step. This requires a set of initial values for the coordinates and velocities to start the integration process. The initial values cannot be provided arbitrarily—the coordinates and velocities must satisfy the position and velocity constraints respectively.

EXAMPLE 13.4

For the system of two-particle system of Section 6.2.1, show the necessary equations to solve for the accelerations and Lagrange multipliers. Then define the integration arrays.

Solution:

The equations of motion for this system were derived in Section 6.2.1. We append the acceleration constraint for the rod to the equations of motion to get:

$$\left[\begin{array}{cc|c} 3\mathbf{I} & \mathbf{0} & -\mathbf{d}^{B,A} \\ \mathbf{0} & 4\mathbf{I} & \mathbf{d}^{B,A} \\ \hline -\mathbf{d}^{B,A\prime} & \mathbf{d}^{B,A\prime} & \mathbf{0} \end{array}\right] \left\{\begin{array}{c} \ddot{\mathbf{r}}^A \\ \ddot{\mathbf{r}}^B \\ \hline \lambda \end{array}\right\} = \left\{\begin{array}{c} \mathbf{f}^A \\ \mathbf{f}^B \\ \hline -\dot{\mathbf{d}}^{B,A\prime}\dot{\mathbf{d}}^{B,A} \end{array}\right\}$$

These five algebraic equations can be solved for four accelerations and one Lagrange multiplier. The integration arrays for this system are the same as those for the unconstrained system of two particles in Example 13.1.

EXAMPLE 13.5

For the single pendulum shown in Figure 13.3, write the necessary equations for computing the accelerations. Assume the only force acting on the pendulum is the gravity, and the mass and moment of inertia are m_1 and J_1. Also, define the integration arrays for this system.

Solution:

The force vector acting on the body is $\mathbf{f}_1 = \left\{\begin{array}{c} 0 \\ -m_1 g \end{array}\right\}$. Considering the pin joint at A, the equations of motion and the acceleration constraints are expressed as

$$\left[\begin{array}{cc|c} m_1\mathbf{I} & \mathbf{0} & \mathbf{I} \\ \mathbf{0} & J_1 & \breve{\mathbf{s}}_1^{A\prime} \\ \hline \mathbf{I} & \breve{\mathbf{s}}_1^A & \mathbf{0} \end{array}\right] \left\{\begin{array}{c} \ddot{\mathbf{r}}_1 \\ \ddot{\phi}_1 \\ \hline \lambda \end{array}\right\} = \left\{\begin{array}{c} \mathbf{f}_1 \\ 0 \\ \hline \mathbf{s}_1^A\dot{\phi}_1^2 \end{array}\right\}$$

Two 6×1 integration arrays are defined as:

$$\mathbf{z} = \left\{\begin{array}{c} \mathbf{r}_1 \\ \phi_1 \\ \dot{\mathbf{r}}_1 \\ \dot{\phi}_1 \end{array}\right\} \qquad \dot{\mathbf{z}} = \left\{\begin{array}{c} \dot{\mathbf{r}}_1 \\ \dot{\phi}_1 \\ \ddot{\mathbf{r}}_1 \\ \ddot{\phi}_1 \end{array}\right\}$$

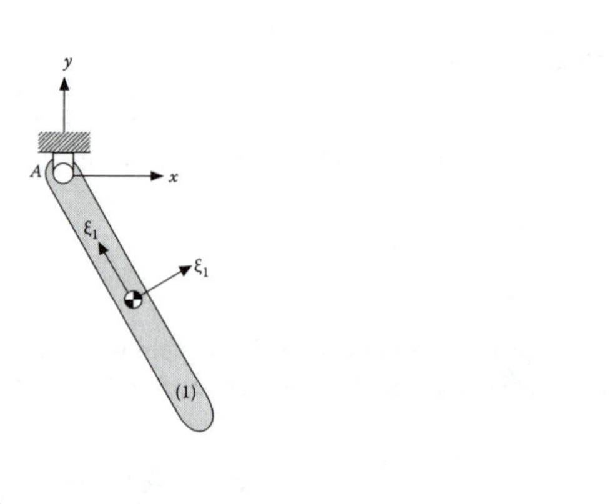

FIGURE 13.3 A single pendulum.

To determine the accelerations at a given time step, we must solve Eq. (13.20). These equations can be constructed exactly as they appear in Eq. (13.20)—that is, constructing a large coefficient matrix and the right-hand-side array. The coefficient matrix contains many zero entries—**M** is diagonal, **D** contains numerous zero entries, and we also have a zero matrix at the lower right corner. Any operation on this large coefficient matrix is computationally inefficient. To improve the computational efficiency, we find the accelerations and the Lagrange multipliers in a two-step process. We rewrite Eq. (13.19) as

$$\ddot{\mathbf{q}} = \mathbf{M}^{-1}(\mathbf{h} + \mathbf{D}'\boldsymbol{\lambda}) \tag{13.22}$$

Then we substitute this equation into Eq. (13.18) to get

$$\mathbf{D}\mathbf{M}^{-1}(\mathbf{h} + \mathbf{D}'\boldsymbol{\lambda}) = \boldsymbol{\gamma}$$

or

$$\mathbf{D}\mathbf{M}^{-1}\mathbf{D}'\boldsymbol{\lambda} = \boldsymbol{\gamma} - \mathbf{D}\mathbf{M}^{-1}\mathbf{h} \tag{13.23}$$

Because $\mathbf{D}\mathbf{M}^{-1}\mathbf{D}'$ is a square matrix, we first solve Eq. (13.23) for $\boldsymbol{\lambda}$, then we solve Eq. (13.22) for $\ddot{\mathbf{q}}$. Also note that **M** is a constant diagonal matrix, therefore $\mathbf{M}^{-1}$ can be computed easily and only once. If these steps were performed properly, the computational effort would be much less than that of solving Eq. (13.20).

13.2.1 Initial Conditions

Solving second-order differential equations requires initial conditions on the coordinates and velocities. Although these initial conditions must represent the state of the system at the initial time, in general, they can be assigned arbitrary values if the system is unconstrained. However, when the system is constrained, the initial conditions must satisfy the constraints at position and velocity levels; that is,

$$\mathbf{\Phi}(^0\mathbf{q}) = \mathbf{0} \tag{13.24}$$

$$\dot{\mathbf{\Phi}} = {}^0\mathbf{D}\,{}^0\dot{\mathbf{q}} = \mathbf{0} \tag{13.25}$$

If the integration starts with a set of initial conditions not satisfying the constraints, the integrated response will be, most likely, erroneous. In Chapter 14 we will discuss several methods for correcting the initial conditions.

13.2.2 General Procedure

Assume that a multibody system with n_{dof} degrees of freedom is represented by an n_v coordinates and $n_c = n_v - n_{dof}$ kinematic constraints. A process for forward dynamic analysis can be stated as:

- Define initial conditions for the coordinates and velocities—that is, ${}^0\mathbf{q}$ and ${}^0\dot{\mathbf{q}}$. The initial conditions must satisfy the constraints at the coordinate and velocity levels.
- Initialize an integration array as $\mathbf{z} = \begin{Bmatrix} {}^0\mathbf{q} \\ {}^0\dot{\mathbf{q}} \end{Bmatrix}$.
- Establish time parameters for integration.
- Invoke a numerical integrator such as `ode45`. The integrator sends $\mathbf{z}$, for $t = {}^it$, to a function and asks for $\dot{\mathbf{z}}$:
  - Transfer the contents of $\mathbf{z}$ to $\mathbf{q}$, and $\dot{\mathbf{q}}$.
  - Compute accelerations, $\ddot{\mathbf{q}}$, and Lagrange multipliers, $\boldsymbol{\lambda}$.
  - Transfer the contents of $\dot{\mathbf{q}}$, and $\ddot{\mathbf{q}}$ to an array $\dot{\mathbf{z}}$ as $\dot{\mathbf{z}} = \begin{Bmatrix} \dot{\mathbf{q}} \\ \ddot{\mathbf{q}} \end{Bmatrix}$.
  - Return $\dot{\mathbf{z}}$ to the integrator.
  - This function is invoked several times by the integrator for every time step.
- Recover $\ddot{\mathbf{q}}$ and $\boldsymbol{\lambda}$ if necessary.

Example 13.6

Refer to Example 13.4. Use `ode45` to find the response of the system of two constrained particles for 1 second.

Solution:

The equations of motion from Section 6.2.1 are formulated in the following program:

```
clear all
   addpath Basics  PC_Basics
   global k_1  L_1_0  d_c_1  m_A  m_B  g  r_O  M  M_inv_array
% Define particle masses
   M_A = 3;   M_B = 3; G = 9.81;
```

```
% Define particle masses and spring-damper characteristics
   m_A = 3;   m_B = 4;
   k_1 = 50; L_1_0 = 1.1; d_c_1 = 20;
% Initial values for coordinates and velocities
   r_A = [0.5; -1]; r_A_d = [0.0; 0.0]; r_O = [0; 0];
   r_B = [2; -1];    r_B_d = [0.0; 0.0];
% Construct the inverse of mass matrix (array)
   M_inv_array = [1/m_A; 1/m_A; 1/m_B; 1/m_B];
   M = diag(M_inv_array);
% Initialize z array
   z = [r_A; r_B; r_A_d; r_B_d];
```

```
% Set time parameters
    Tspan = [0.0:0.02:1.0];
% Integrate
    [T, zT] = ode45(@Ex_13_6, Tspan, z);
% Animation
    nT = size(T);
    plot_trace(nT, r_O, zT)
```

```
function z_d = Ex_13_6(t, z)
    global k_1  L_1_0  d_c_1  m_A  m_B  g  r_O  M  M_inv_array
% Unpack the contents of z array
    r_A = z(1:2); r_B = z(3:4); r_A_d = z(5:6); r_B_d = z(7:8);
% compute d1, d1_d, d2 and d2_d vectors
    d1 = r_A - r_O; d1_d = r_A_d;
    d2 = r_B - r_A; d2_d = r_B_d - r_A_d;
% Jacobian and r-h-s of acceleration constraints
    [jac_A2, jac_B2] = PC_jacob_L(d2);
          D = [jac_A2  jac_B2];
    gamma = PC_gamma_L(d2_d);
% Compute the force array
    f_sda_1 = pp_sda(d1, d1_d, k_1, L_1_0, d_c_1, 0);
    f_A = -f_sda_1 + [0; -m_A*g]; f_B = [0; -m_B*g];
          h = [f_A; f_B];
% Compute accelerations and Lagrange multipliers
    sol = [M -D'; D 0]\[h; gamma];
% pack velocities and accelerations into z_d array
    z_d = [z(5:8); sol(1:4)];
```

```
function plot_trace(nT, r_O, zT)
% Plot the trace of the response
    plot(r_O(1), r_O(2), 'ko'), hold on
    for i=1:nT(1)
          plot(zT(i,1), zT(i,2), 'ro')
          plot(zT(i,3), zT(i,4), 'ro')
          axis ([-2.5 2.5 -4.5 0.5])
          line([zT(i,1), zT(i,3)], [zT(i,2), zT(i,4)])
    end
```

```
function plot_anim(nT, r_O, zT)
% Animate the response
    for i=1:nT(1)
        drawnow
        plot(r_O(1), r_O(2), 'ko'), hold on
        plot(zT(i,1), zT(i,2), 'ro')
```

```
        plot(zT(i,3), zT(i,4), 'ro')
        axis ([-2.5 2.5 -4.5 0.5])
        line([zT(i,1), r_O(1)], [zT(i,2), r_O(2)], 'color', 'g')
        line([zT(i,1), zT(i,3)], [zT(i,2), zT(i,4)])
        hold off
    end
```

Executing the program results into the coordinates and velocities of the two particles during 1 second. The function `plot_trace` provides the trace of the positions of the rod at every reporting time interval as shown in Figure 13.4. We can obtain an animation of the positions of the rod by typing in the Command Window

```
>> plot_anim(nT, r_O, zT)
```

In this animation, in addition to the rod, we will see the spring as a green line.

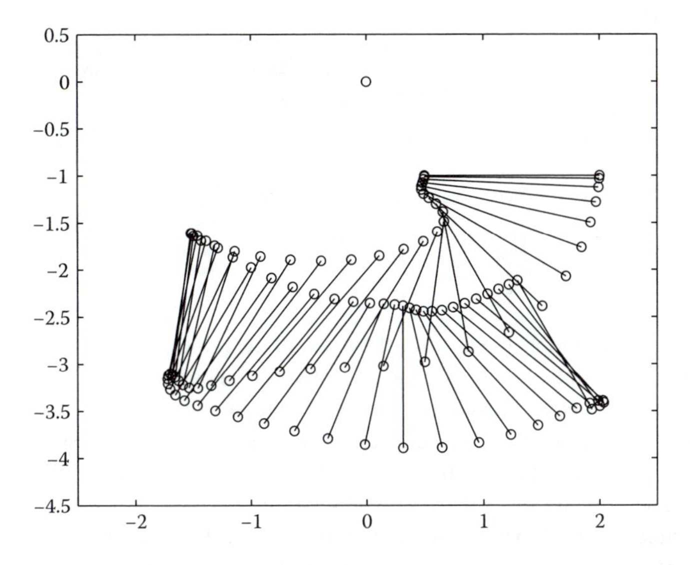

FIGURE 13.4 Trace of the rod in the *x*-*y* reference frame.

13.3 BODY COORDINATE FORMULATION

In this section we present a program to perform forward dynamic analysis for systems that are modeled by the body coordinate formulation. The program uses the integration function `ode45` to predict the response of the system during the specified time period. The main script named `BC_forward` is listed as the following:

```
% file name:  BC_forward
% Forward dynamic analysis: Body coordinate formulation
     clear all
     addpath Basics  BC_Basics  BC_Analyses
     folder = input(' which folder ? ', 's'); addpath (folder)
     BC_global
     BC_condata; BC_vardata; nbc = 3*nb;
% Determine components of vectors on the ground
     BC_vectors0;
% Transfer r, phi, r_d, and phi_d to c and c_d
     c   = BC_r_to_c(r, phi, nb);
     c_d = BC_r_to_c(r_d, phi_d, nb);
% Construct diagonal mass matrix
     BC_mass_diagonal; M_inv_diag = diag(M_inv_array);
% Initialize z array
     z = [c; c_d];
% Establish time parameters
     time = 0.0;
     period = input(' time period ? '); time_end = time + period;
     delta_t = input(' delta_t ? ');
     Tspan = [time:delta_t:time_end];
% Begin integration
     [T, zT] = ode45(@BC_eqsmo, Tspan, z);
% Re-compute and save accelerations, Lagrange multipliers, etc.
     nT = size(T); zT_d = zeros(nT(1),(2*nbc-ndof));
     for i=1:nT(1)
     % Unpack zT into c and c_d arrays
          z_d = BC_eqsmo(T(i), zT(i,:)');
     % Pack c_dd and Lagrange multipliers into zT_d
          zT_d(i,:) = [z_d(nbc+1:2*nbc)' Lambda'];
     end
```

The program prompts the user for the name of the folder containing the model, the time period for the integration, and the time step for reporting. Note that the time step for integration is decided by ode45 independent of the reporting time step. The integrator returns the arrays of coordinates and velocities at every time step (in the array zT). The program recomputes the accelerations and Lagrange multipliers for every reporting time step and saves the results in the array zT_d.

The integration algorithm sends the array z to the function BC_eqsmo. In this function the contents of z_d are retrieved and passed on to the arrays c and c_d.

```
function z_d = BC_eqsmo(t, z)
    BC_global
    time = t;
% Unpack z into coordinate and velocity sub-arrays
    c = z(1:nbc); c_d = z(nbc+1:2*nbc);
% Transfer c and c_d to r, phi, r_d, and phi_d
    BC_c_to_r; BC_cd_to_rd;
% Compute A matrices, s_P, r_P, s_P_d and r_P_d vectors
    BC_A_matrices; BC_vectors; BC_vectors_d;
```

```
% Evaluate Jacobian and r-h-s array of accelerations
    D = BC_jacob; gamma = BC_gamma;
% Evaluate array of forces
    h = BC_h;
% Solve for accelerations and Lagrange multipliers
    DMinv = D*M_inv_diag;
    Lambda = (DMinv*D')\(gamma - DMinv*h);
    c_dd = M_inv_array.*(h + D'*Lambda);
% Pack velocities and accelerations into z_d
    z_d = [c_d; c_dd];
```

This function, after evaluating the array of forces, determines the Lagrange multipliers and the accelerations based on the two-step process of Eqs. (13.23) and (13.22). The computed accelerations and the velocities are saved in the array `z_d` where it is returned to the integration function.

13.3.1 Double A-Arm Suspension

In this section we perform a simulation with the double A-arm suspension by dropping it from its initial position until it comes to rest. For this purpose, we first construct a very simple tire model.

In this model we assume that when the tire is in contact with the ground it acts as a spring-damper element. Furthermore, it we assume that the tire force always acts in the vertical direction at a defined point on the wheel axis. The graphical presentation of such a tire model is shown in Figure 13.5. The characteristics of the spring-damper must represent the radial characteristics of the tire. The undeformed length of the spring is the undeformed radius of the tire. Knowing the coordinates and the velocity of a point, such as G_2 or C, on the wheel axis, and the y-coordinate of the ground, we can compute the tire radial force.

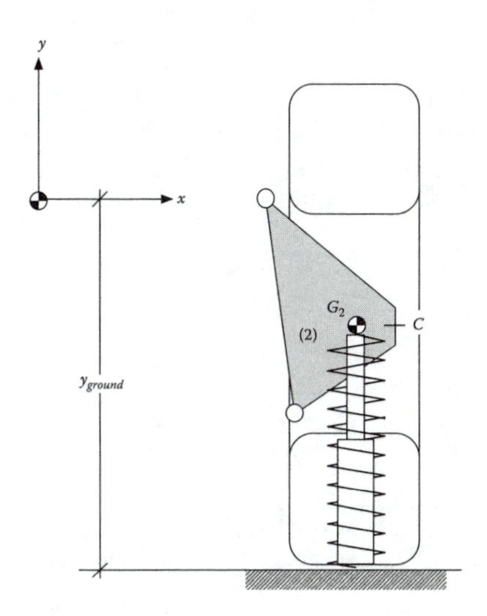

FIGURE 13.5 Graphical presentation of a simple tire model.

We revise two of the files in the BC_AA folder, BC_condata and BC_h, as listed in the followings. We set all the velocities to zero in the file BC_vardata to represent releasing the system from rest.

```
% file name:   BC_condata
% ...
      s_C_2_p = [ 0.03;   0.00]; s_H_0_p = [ 0.00; -0.38];
% Establish cells and arrays
      s_P_p = {s_Q_1_p s_A_1_p s_A_2_p s_B_2_p ...
               s_B_3_p s_O_3_p s_E_1_p s_C_2_p};
          iB = [1 1 2 2 3 3 1 2]; nsP = 8;
      s_P0_p = {s_O_0_p s_Q_0_p s_F_0_p s_H_0_p}; nsP0 = 4;
      u_p = {}; iBu = []; nu = 0;
      u0_p = {}; nu0 = 0;
% Mass, moment of inertia, and force data
     m = [2; 30; 1]; J = [0.5; 2.5; 0.5]; g = 9.81;
      k = [91600 58770]; d_c = [1433 2500]; L_0 = [0.23 0.35];
```

In the revised BC_condata file, we have added two more points: C_2 on the wheel axis and H_0 on the ground. The x-coordinate of point H_0 will not be used in the model, therefore H_0 could be any point on the ground. We also add the tire stiffness, damping, and radius to this file. For this purpose, we define k, d_c, and L_0 as arrays where the first element in each array belongs to the suspension and the second element belongs to the tire.

The revised version of the file BC_h is listed in the following. The tire force is computed by invoking the function BC_radial.

```
function h = BC_h
% Suspension Spring-damper
      BC_global
      d = r_P{7} - r_P0{3}; d_d = r_P_d{7};
      f_sda = pp_sda(d, d_d, k(1), L_0(1), d_c(1), 0);
      f_sd_1 = -f_sda;
% Tire spring damper
      r_C = r_P{8}; r_C_d = r_P_d{8}; r_H = r_P0{4};
      f_tire = BC_radial(r_C(2), r_C_d(2), r_H(2), k(2), L_0(2), d_c(2));
      f_sd_2 = f_tire; n_sd_2 = s_rot(s_P{8})'*f_sd_2;
% Gravitational force
      w = g*m;
% Force array
      h_1 = [0; -w(1); 0] + [f_sd_1; 0];
      h_2 = [0; -w(2); 0] + [f_sd_2; n_sd_2]; h_3 = [0; -w(3); 0];
      h = [h_1; h_2; h_3];
```

```
function f = BC_radial(yT, y_d, yH, k, R_0, d_c)
% Tire radial force (uni-axial)
      f = [0; 0]; R = yT - yH; dR = R_0 - R;
      if dR > 0
          f = [0; (k*dR - d_c*y_d)];
      end
```

We use the same initial conditions as listed in the file BC_vardata in Section 7.6.1. We execute the program BC_forward and set the simulation

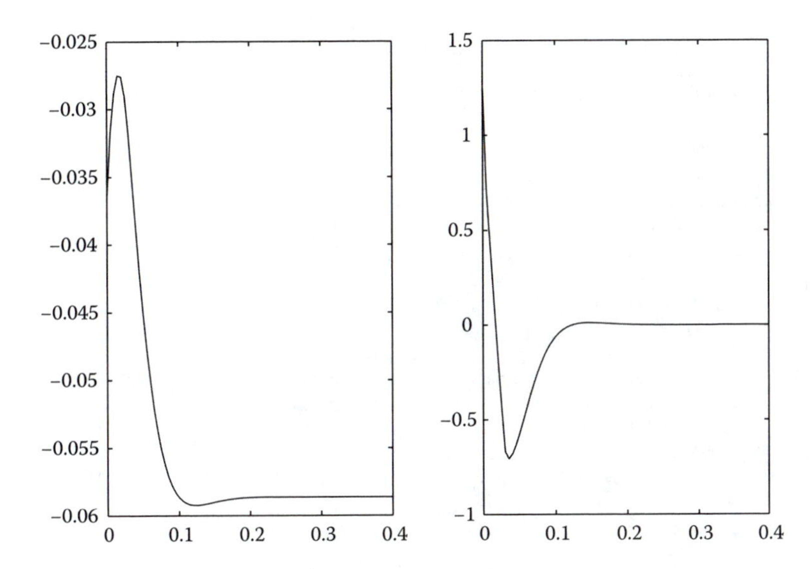

FIGURE 13.6 Vertical position and velocity of the mass center of body (2) as a function of time.

period and reporting time interval to 0.4 and 0.005 seconds respectively. Then in the Command Window we type the following:

```
>> subplot(1,2,1)
>> plot(T,zT(:,5))
>> subplot(1,2,2)
>> plot(T,zT(:,14))
```

These commands yield two plots as shown in Figure 13.6.

13.4 JOINT COORDINATE FORMULATION

Developing a program for forward dynamics with the joint coordinate formulation is very similar to that of the body coordinate formulation. One difference is that in the joint coordinate formulation we need to know whether a system is closed chain or open—that is, whether the formulation is constrained or not. The following is the main script for the program `JC_forward`.

```
% file name:  JC_forward
% Forward dynamic analysis: Joint coordinate formulation
    clear all
    addpath Basics  BC_Basics  JC_Analyses
    folder = input(' which folder ? ', 's'); addpath (folder)
    BC_global
    BC_condata; JC_vardata;
    nbc = 3*nb; njc = nb; nPhi_cut = njc - ndof;
```

```
% Determine components of vectors on the ground
      BC_vectors0;
% Initialize z array
      z = [theta; theta_d];
% Construct inverse of diagonal mass matrix
      BC_mass_diagonal; M_inv_diag = diag(M_inv_array);
% Establish time parameters
      time = 0.0;
      period = input(' time period ? '); time_end = time + period;
      delta_t = input(' delta_t ? ');
      Tspan = [time:delta_t:time_end];
% Begin integration
      [T, zT] = ode45(@JC_eqsmo, Tspan, z);
% Re-compute and save accelerations, Lagrange multipliers, etc.
      nT = size(T); zT_d = zeros(nT(1),(njc+nPhi_cut));
      for i=1:nT(1)
      % Unpack zT into theta and theta_d arrays
          theta = zT(i,1:njc)'; theta_d = zT(i,njc+1:2*njc)';
          z_d = JC_eqsmo(T(i), zT(i,:)');
      % Pack theta_dd and Lagrange multipliers into zT_d
          if nPhi_cut > 0
              zT_d(i,:) = [z_d(njc+1:2*njc)' Lambda'];
          else
              zT_d(i,:) = [z_d(njc+1:2*njc)'];
          end
      end
```

As we can see, the structure of this program is the same as that of the program `BC_forward`.

The integration algorithm sends the array `z` to the function `JC_eqsmo`. In this function the contents of `z` are retrieved and passed on to the arrays `theta` and `theta_d`. After determining the values of the joint accelerations, the contents of arrays `theta_d` and `theta_dd`. are saved in the array `z_d` and returned to the integrator. With this program we can simulate the response of both open and closed chain systems.

```
function z_d = JC_eqsmo(t, z)
% Coordinate transformation and B matrix
    BC_global
    time = t;
% Unpack z into joint coordinate and velocity sub-arrays
    theta = z(1:njc); theta_d = z(njc+1:2*njc);
% Coordinates, vectors, and B matrix
    JC_coord_trans; BC_vectors; B = JC_Bmatrix;
% Velocity transformation; vectors r_P_d and s_P_d; B_d*theta_d array
   c_d = B*theta_d; BC_cd_to_rd; BC_vectors_d; BdThetad = JC_BdThetad;
% Mass matrix and  array of forces in body coordinates
   BC_mass_diagonal; M_diag = diag(M_array); h = BC_h;
% Mass matrix and array of forces in joint coordinates
   M_joint = B'*M_diag*B; h_joint = B'*(h - M_array.*BdThetad);
% Solve for accelerations and Lagrange multipliers
if nPhi_cut > 0
```

```
% Closed-chain
    D_cut = JC_jacob; gamma_cut = JC_gamma; % Cut-joints
    % Stabilization feedback terms
    if alfa > 0.01
           Phi_cut = JC_Phi; Phi_cut_d = D_cut*theta_d;
           gamma_cut = gamma_cut - 2*alfa*Phi_cut_d - betas*Phi_cut;
    end
    Minv = inv(M_joint); DMinv = D_cut*Minv;
    Lambda = (DMinv*D_cut')\(gamma_cut - DMinv*h_joint);
    theta_dd = Minv*(h_joint + D_cut'*Lambda);
else
% Open-chain
    theta_dd = M_joint\h_joint;
end
% Pack velocities and accelerations into z_d
    z_d = [theta_d; theta_dd];
```

13.4.1 Variable-Length Pendulum

This open-chain system was discussed in Sections 9.4.1 and 10.1.1. In this section we perform a forward dynamic analysis starting with the following initial conditions:

```
% File name: JC_vardata
% Joint coordinates and velocities
    theta = [0.6435; 0.25]; theta_d = [0; 0];
```

We execute the program `JC_forward` for 5 seconds. Then we plot the two joint coordinates versus time and obtain the results shown in Figure 13.7. From the plots we observe two natural frequencies due to the oscillations of the pendulum and of the slider block.

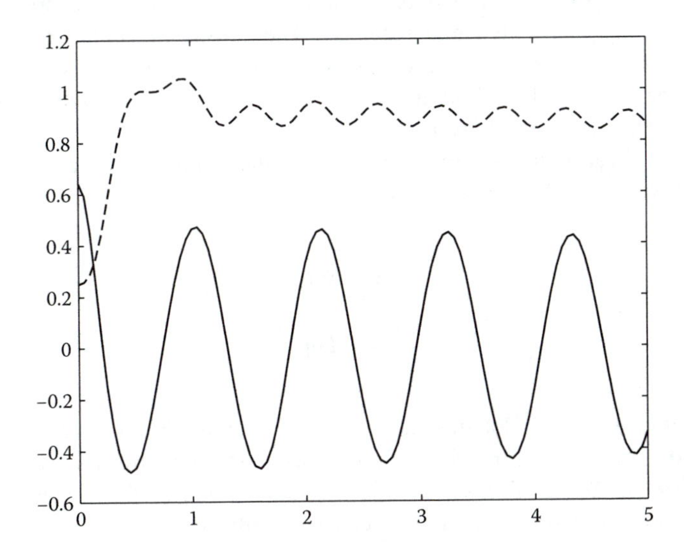

FIGURE 13.7 Angular position of body (1) (solid line) and relative position of body (2) from the origin (dashed line) as a function of time.

13.5 POINT COORDINATE FORMULATION

Forward dynamics with the point coordinate formulation follows a process similar to that of the body or the coordinate formulations. Developing a computer program for forward dynamics with the point coordinate formulation is left to the reader. A structure simpler than that of the `BC_forward` can be followed in developing the new program.

13.6 CONSTRAINT VIOLATION

The procedure for forward dynamic analysis that we presented in Section 13.2.2 does not guarantee that the coordinate and velocity constraints are not violated as the integration process progresses. The violation is due to the presence of numerical error that is inherent to any numerical integration routine. In that procedure only the constraints at the acceleration level are used—the constraints at the coordinate and velocity levels have not been used anywhere in the process. Therefore, when the integration algorithm predicts values for the coordinate and velocities at the next time step, the predicted (approximated) values may not satisfy their corresponding constraints.

Violation of constraints may be negligible in some problems but significant in other problems. In the upcoming sections we discuss several techniques for controlling the growth of the constraint violation or eliminating it completely.

13.6.1 Constraint Violation Stabilization Method

In this section we discuss a simple method to control the constraint violation. This method *may* control the growth of constraint violation, in some problems, and provide stability in the dynamic response. The *constraint violation stabilization method* is an extension of the feedback control theory applied to the equation of motion. One of the goals in designing a feedback controller is to suppress the growth of error and achieve a stable response.

This stabilization method requires the coordinate and velocity constraints to be evaluated before we determine the accelerations:

$$\mathbf{\Phi} = \mathbf{\Phi}(\mathbf{q}) \tag{13.26}$$

$$\dot{\mathbf{\Phi}} = \mathbf{D}\dot{\mathbf{q}} \tag{13.27}$$

If the coordinates and velocities satisfy the constraints, then we should get $\mathbf{\Phi} = \mathbf{0}$ and $\dot{\mathbf{\Phi}} = \mathbf{0}$. However, numerical error causes the constraints evaluation to yield $\mathbf{\Phi} \neq \mathbf{0}$ and $\dot{\mathbf{\Phi}} \neq \mathbf{0}$. The stabilization method modifies the acceleration constraints of Eq. (13.18) by adding two feedback terms as

$$\ddot{\mathbf{\Phi}} \equiv \mathbf{D}\ddot{\mathbf{q}} - \boldsymbol{\gamma} + 2\alpha\dot{\mathbf{\Phi}} + \beta^2\mathbf{\Phi} = \mathbf{0} \tag{13.28}$$

where α and β are two positive parameters. Hence, Eq. (13.20) becomes:

$$\begin{bmatrix} \mathbf{M} & -\mathbf{D}^T \\ \mathbf{D} & \mathbf{0} \end{bmatrix} \begin{Bmatrix} \ddot{\mathbf{q}} \\ \boldsymbol{\lambda} \end{Bmatrix} = \begin{Bmatrix} \mathbf{h} \\ \boldsymbol{\gamma} - 2\alpha\dot{\boldsymbol{\Phi}} - \beta^2\boldsymbol{\Phi} \end{Bmatrix} \tag{13.29}$$

Then, in the procedure of Section 13.2.2, instead of solving Eq. (13.20) for the accelerations and Lagrange multipliers, we solve Eq. (13.29).

We should be reminded that the feedback terms stabilize a linear system of second-order differential equations, whereas the equations of motion for multibody systems are nonlinear. Therefore, the feedback terms may not always be effective in controlling the error. In some problems the feedback terms may actually cause the constraint violations to grow more rapidly. In general, for nonzero values of α and β, the solution oscillates about the "correct" solution. The amplitude and frequency of the oscillation, caused by the stabilization terms, depend on the values of α and β. In most practical problems, a range of values between 1 and 10 for these parameters might be adequate. When $\alpha = \beta$, critical damping is achieved which usually stabilizes the response more quickly.

The following are the revisions to some of the files for forward dynamics with either the body or the joint-coordinate formulation to include the feedback stabilization terms:

```
% File name:  BC_forward (or JC_forward)
% ...
% Prompt user to define values for alpha and beta
    alfa = input(' alpha ? '); betas = alfa^2;
% Begin integration
% ...
```

```
function c_dd = BC_eqsmo
% ...
% Evaluate Jacobian and r-h-s array of accelerations
    D = BC_jacob; gamma = BC_gamma;
    Phi = BC_Phi; Phi_d = D*c_d;
    gamma = gamma - 2*alfa*Phi_d - betas*Phi;

% ...
```

```
function z_d = JC_eqsmo(t, z)
% ...
if nPhi_cut > 0
% Closed-chain
    D_cut = JC_jacob; gamma_cut = JC_gamma; % Cut-joints
    % Stabilization feedback terms
    if alfa > 0.01
        Phi_cut = JC_Phi; Phi_cut_d = D_cut*theta_d;
        gamma_cut = gamma_cut - 2*alfa*Phi_cut_d - betas*Phi_cut;
    end
    Minv = inv(M_joint); DMinv = D_cut*Minv;
% ...
```

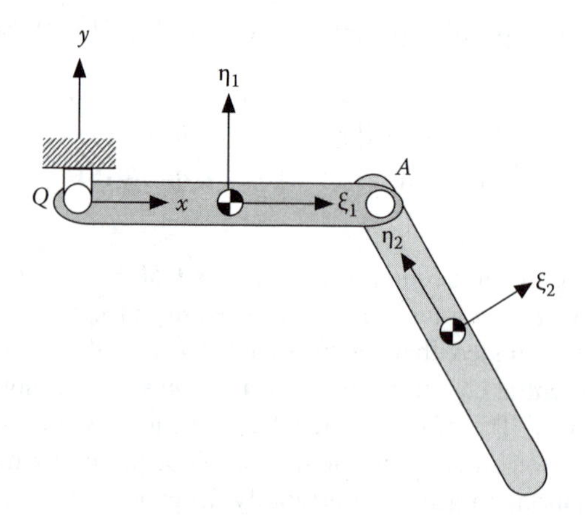

FIGURE 13.8 A double pendulum.

In the script `BC_forward` or `JC_forward` the user is prompted to enter a value for the parameter *alpha*. The parameter *beta* is set to be equal to *alpha*. These parameters are communicated to the function `BC_eqsmo` or `JC_eqsmo` via a global statement. In the function `BC_eqsmo` and `JC_eqsmo` the coordinate and velocity constraints are evaluated, and the results are included in the right-hand-side array of the acceleration constraints.

To observe how the feedback terms control the accumulation of constraint error, we consider a simple example. The example is a double pendulum shown in Figure 13.8. We release the pendulum from an initial state and use `ode45` to integrate the equations of motion. We simulate the response of the system for three different values of α; 0, 1, and 5 using the program `BC_forward`. After each simulation we run the following script to compute and save the square root of the sum of squares of the constraint violations at the coordinate and velocity levels.

```
% File name:  BC_violations
% Compute and save constraint violations
    nT = size(T);
    record = zeros(nT(1),2);
    for i=1:nT(1)
    % Unpack zT into c and c_d arrays
        c = zT(i,1:nbc)'; c_d = zT(i,nbc+1:2*nbc)';
        time = T(i);
    % Evaluate constraints
        BC_c_to_r; BC_A_matrices; BC_vectors;
        Phi = BC_Phi; D = BC_jacob; Phi_d = D*c_d;
        P2 = sqrt(Phi'*Phi); Pd2 = sqrt(Phi_d'*Phi_d);
        record(i,:) = [P2  Pd2];
    end
```

The plots of the square root of the sum of squares of the constraint violations at the coordinate level versus time are shown in Figure 13.9. We note that for $\alpha = 0$,

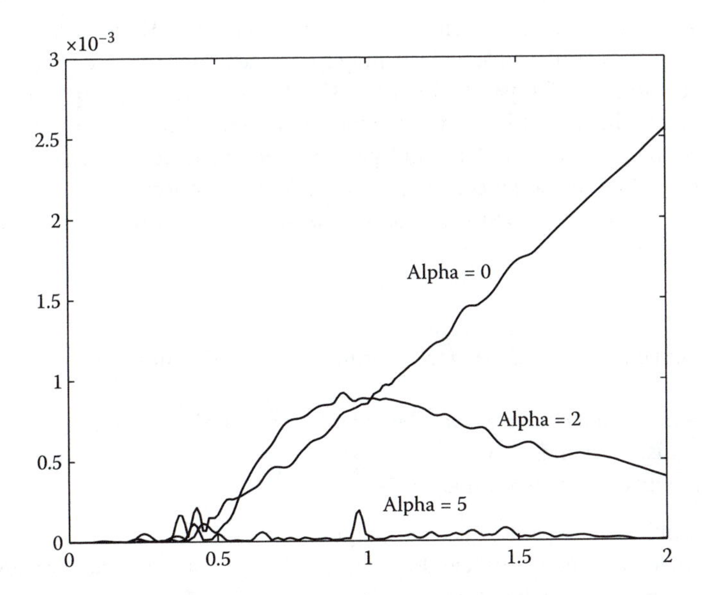

FIGURE 13.9 Constraint violation versus time for three different values of parameter α.

the integration causes the largest violation and as the value of α is increased, the stabilization terms provide some degree of control over the violations. However, based on these results we should not get the impression that by increasing the value of α there will be more control over the constraint violations.

13.6.2 Coordinate Partitioning Method

The coordinate partitioning method controls the accumulation of the numerical error quite differently from the constraint violation stabilization method. This method is based on the same concept as for kinematic analysis in Section 11.3.1. The method makes use of the fact that, when constraints are present, the coordinates are not independent. The coordinates, and their first and second time derivatives, are partitioned into *independent* and *dependent* sets as:

$$\mathbf{q} = \begin{Bmatrix} {}^{(k)}\mathbf{q} \\ {}^{(u)}\mathbf{q} \end{Bmatrix}, \quad \dot{\mathbf{q}} = \begin{Bmatrix} {}^{(k)}\dot{\mathbf{q}} \\ {}^{(u)}\dot{\mathbf{q}} \end{Bmatrix}, \quad \ddot{\mathbf{q}} = \begin{Bmatrix} {}^{(k)}\ddot{\mathbf{q}} \\ {}^{(u)}\ddot{\mathbf{q}} \end{Bmatrix} \tag{13.30}$$

We are reminded that there are as many independent coordinates as the number of degrees of freedom and as many dependent coordinates as the number of constraints. Based on this partitioning, the integration arrays of Eq. (13.21) are redefined in smaller arrays as

$$\mathbf{z} = \begin{Bmatrix} {}^{(k)}\mathbf{q} \\ {}^{(k)}\dot{\mathbf{q}} \end{Bmatrix}, \quad \dot{\mathbf{z}} = \begin{Bmatrix} {}^{(k)}\dot{\mathbf{q}} \\ {}^{(k)}\ddot{\mathbf{q}} \end{Bmatrix} \tag{13.31}$$

Note that the arrays $\mathbf{z}$ and $\dot{\mathbf{z}}$ each contain $2 \times n_{dof}$ variables. Because the integration algorithm predicts the contents of the $\mathbf{z}$ array for the next time step (i.e., ${}^{(k)}\mathbf{q}$ and ${}^{(k)}\dot{\mathbf{q}}$), then the constraints of Eqs. (13.16) and (13.17) are solved for ${}^{(u)}\mathbf{q}$ and ${}^{(u)}\dot{\mathbf{q}}$. The constraints can be solved by either method of Section 11.3.1 or Section 11.3.2. In these processes, the values of the independent coordinates or velocities, instead of being specified by the driver constraints, come from the $\mathbf{z}$ array.

A process for forward dynamic analysis with the coordinate partitioning method can be stated as:

Determine a set of independent coordinates and velocities, ${}^{(k)}\mathbf{q}$ and ${}^{(k)}\dot{\mathbf{q}}$.

- Define initial conditions for the independent coordinates and velocities—that is, ${}^{0(k)}\mathbf{q}$ and ${}^{0(k)}\dot{\mathbf{q}}$.
- Initialize an integration array as $\mathbf{z} = \begin{Bmatrix} {}^{0(k)}\mathbf{q} \\ {}^{0(u)}\dot{\mathbf{q}} \end{Bmatrix}$.
- Establish time parameters for integration.
- Invoke a numerical integrator such as `ode45`. The integrator sends $\mathbf{z}$, for $t = {}^{i}t$, to a function and asks for $\dot{\mathbf{z}}$:
  - Transfer the contents of $\mathbf{z}$ to ${}^{(k)}\mathbf{q}$ and ${}^{(k)}\dot{\mathbf{q}}$.
  - Solve the constraints of Eqs. (13.16) and (13.17) for ${}^{(u)}\mathbf{q}$ and ${}^{(u)}\dot{\mathbf{q}}$. Now all of the coordinates and velocities are known—that is, $\mathbf{q}$ and $\dot{\mathbf{q}}$.
  - Solve Eq. (13.20) or Eqs. (13.23) and (13.22) for accelerations, $\ddot{\mathbf{q}}$, and Lagrange multipliers, $\boldsymbol{\lambda}$.
  - Partition the acceleration array into independent and dependent sets—that is, ${}^{(k)}\ddot{\mathbf{q}}$ and ${}^{(u)}\ddot{\mathbf{q}}$.
  - Transfer the contents of ${}^{(k)}\dot{\mathbf{q}}$ and ${}^{(k)}\ddot{\mathbf{q}}$ to $\dot{\mathbf{z}}$ array as $\dot{\mathbf{z}} = \begin{Bmatrix} {}^{(k)}\dot{\mathbf{q}} \\ {}^{(k)}\ddot{\mathbf{q}} \end{Bmatrix}$.
  - Return the $\dot{\mathbf{z}}$ array to the integrator.
  - This process is repeated several times for every time step.
- Recover $\ddot{\mathbf{q}}$ and $\boldsymbol{\lambda}$ if necessary.

This process ensures that all of the constraints are satisfied at every solution time step.

Although the coordinate partitioning method in principle can eliminate the constraint violation completely, the actual process may run into some difficulties. One troublesome part of this process is the step where the independent coordinates are known and the position constraints are to be solved for the dependent coordinates. Because the position constraints are solved by iterative methods such as Newton-Raphson, the process requires relatively accurate estimates for the dependent coordinates in every time step. Finding a good estimate for the coordinates is not a problem because the values of ${}^{(u)}\mathbf{q}$ at ${}^{i-1}t$ can be used as estimates for ${}^{(u)}\mathbf{q}$ at ${}^{i}t$. This estimate will yield a solution in most cases however, in cases where the independent set of coordinates has been poorly selected, the iterative process may fail to converge to a solution.

Proper partitioning of the coordinates into independent and dependent sets is critical in controlling the accumulation of numerical error. To clarify this

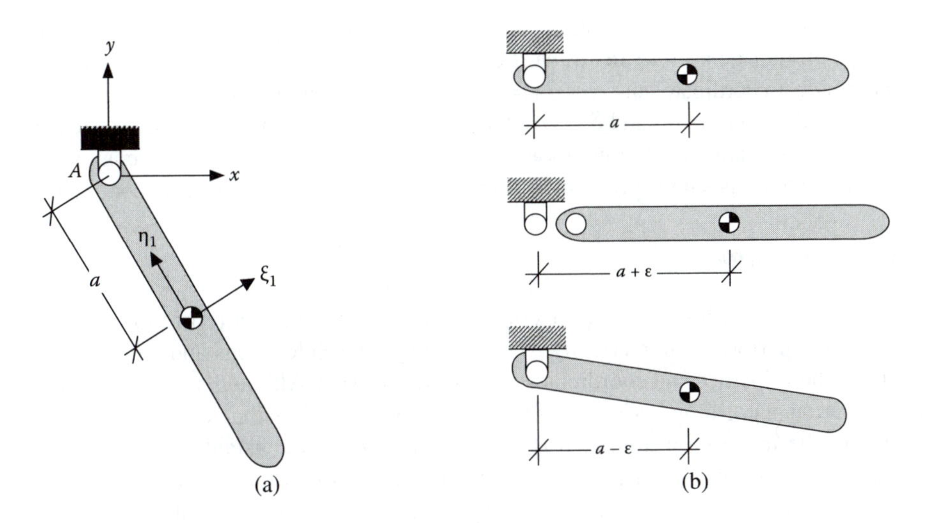

FIGURE 13.10 (a) A single pendulum and (b) three possible scenarios due to integration error.

point, we provide a simple example. The example is the single pendulum shown in Figure 13.10(a). This is a 1 DoF system and, therefore, there is only one independent coordinate. The body coordinates of this pendulum are x, y and ϕ. The constraint equations for the revolute joint at A are:

$$\begin{aligned} x - a\cos\phi &= 0 \\ y + a\sin\phi &= 0 \end{aligned} \tag{a}$$

Assume that the values of these coordinates are the result of an integration process and the corresponding numerical error are δx, δy, and $\delta\phi$, which are related through the constraints as

$$\begin{aligned} \delta x + a\sin\phi\,\delta\phi &= 0 \\ \delta y - a\cos\phi\,\delta\phi &= 0 \end{aligned} \tag{b}$$

We first assume that out of these three coordinates, x is selected to be the independent coordinate. An integration error $\delta x = \varepsilon$ in predicting the value of x yields errors in the values of ϕ and y as $\delta\phi = -(1/a\sin\phi)\varepsilon$ and $\delta y = -(\cos\phi/\sin\phi)\varepsilon$. As long as the rotational angle ϕ is not close to 0 or π, we will not have any problems in the process. But when ϕ is close to 0 or π, a small error in predicting x yields large errors in the computed values of ϕ and y. This situation is demonstrated graphically in Figure 13.10(b). When ϕ is exactly 0, the predicted value of x should be exactly a—that is, $\delta x = \varepsilon = 0$. But if the predicted value is $x = a + \varepsilon$, and $\varepsilon > 0$, then the constraints can never be satisfied—Newton-Raphson iteration will fail to provide a solution for ϕ and y. If $\varepsilon < 0$, solution to ϕ and y will contain large errors as illustrated in the figure.

Similar phenomenon exists if y is selected as the independent coordinate and the pendulum gets close to the vertical position. However, if we select ϕ as the independent coordinate, an error $\delta\phi = \varepsilon$ will result in errors in the dependent coordinates as $\delta x = -a \sin\phi\varepsilon$ and $\delta y = a \cos\phi\varepsilon$. These errors will not cause any particular difficulties, for any orientation of the pendulum, and the integration can be continued. Therefore, for this simple problem, ϕ is the best choice for the independent coordinate.

13.7 REMARKS

Forward dynamic analysis requires the integration of the equations of motion. This process is performed numerically by integrating the accelerations and velocities to obtain the velocities and coordinates at the next time step. Although many numerical integration algorithms exit for our purpose, an algorithm that can dynamically control the integration time step is recommended over a constant step algorithm. Regardless of what algorithm we select for integration, we must not forget that numerical integration only provides an approximation to the exact solution; that is, numerical error is always present in a solution. As long as the equations of motion are unconstrained, the error caused by numerical integration does not cause any further problems. When constraints are present, however, numerical integration error may cause violation in the constraints at the coordinate and velocity levels. If these violations are not properly dealt with, the simulated response could be erroneous. In general, if we can reduce the number of constraints—which can be the result of reducing the number of coordinates—we reduce the chance of violating the constraints.

13.8 PROBLEMS

13.1 Develop a MATLAB program to perform forward dynamic analysis with the point coordinate formulation similar to that of the body coordinate formulation. Name the script `PC_dynamics`.

13.2 For the following simulations either use the program `PC_dynamics` from Problem 13.1 or develop individual programs to perform forward dynamics for each system:
   (a) Refer to Problem 6.1
   (b) Refer to Problem 6.2
   (c) Refer to Problem 6.4
   (d) Refer to Problem 6.5(a)
   (e) Refer to Problem 6.5(b)
   (f) Refer to Problem 6.6
   (g) Refer to Problems 6.12(a) and 6.13
   (h) Refer to Problems 6.12(b) and 6.13
   (i) Refer to Problems 6.12(c) and 6.13
   (j) Refer to Problem 6.16
   (k) Refer to Problem 6.17
   (l) Refer to Problem 6.18
   (m) Refer to Problem 6.19

(n) MacPherson suspension from Section 6.2.2
(n.1) Approximate mass distribution
(n.2) Exact mass distribution (refer to Problem 6.15)
(o) Double A-arm suspension from Section 6.3.4
(n.1) Approximate mass distribution
(n.2) Exact mass distribution

The response of each simulation could be reported numerically or it could be viewed in the form of plots. However, simple animation routine could be developed to view the motion from each simulation.

13.3 In this exercise we experiment with the size of the time step when we integrate with constant time step algorithms. Use the forward dynamics program and the variable-length pendulum model in the joint coordinate formulation, `JC_forward` and `JC_VLP`. Perform the following simulations and plot the responses. For example, plot θ_1 versus time from each simulation, all on one plot.

(a) Integrate by `ode45` and choose a time interval of 0.05 (this is the reporting time interval and not the integration time interval). Plot the response in "blue".
(b) Integrate by `RK4` and choose a time interval of 0.05 (this is the integration and the reporting time interval). Plot the response in "green".
(c) Integrate by `RK4` and choose a time interval of 0.1. Plot the response in "yellow".
(d) Integrate by `RK4` and choose a time interval of 0.2. Plot the response in "black".
(e) Integrate by `RK4` and choose a time interval of 0.3. Plot the response in "red".
(f) Integrate by `RK4` and choose a time interval of 0.5.

What do you conclude from the results?

There are two natural frequencies for this system because it has two degrees of freedom. The response from the `ode45` simulation clearly shows the higher of the two natural frequencies. Roughly determine the time period, τ, from this response and show that if $\Delta t \leq \tau/10$, `RK4` can produce reasonably acceptable results.

13.4 In this exercise we compare results from simulating a multibody system that is modeled by two different formulations, and the influence of the size of the integration time step on the computed response. For this purpose we compare the models of the variable-length pendulum with the body coordinate and the joint coordinate formulations. In the following simulations, execute `BC_forward` with `BC_VLP` and `JC_forward` with `JC_VLP`:. In each exercise, plot ϕ_1 versus time from `BC_forward` and θ_1 versus time from `JC_forward` on one plot.

(a) Use `ode45` (use this response as reference)
(b) Use `RK4` with $\Delta t = 0.1$
(c) Use `RK4` with $\Delta t = 0.2$

As the integration time step increases, which formulation yields a response that is further away from the correct response?

13.5 In this exercise we experiment with different values of the feedback parameter α in controlling the error due to constraint violation. Use the script `BC_forward` and the model `BC_VLP`. Execute the script with the following integrator and feedback parameter values:
(a) Use `ode45` and $\alpha = 0$ (use this response as reference)
(b) Use `RK4`, $\Delta t = 0.2$, and $\alpha = 0$
(c) Use `RK4`, $\Delta t = 0.2$, and $\alpha = 5$
What do you observe?

13.6 The speed governor shown consists of a plunger that moves outward as the angular speed increases until it contacts switch S. When the angular velocity is zero, the plunger is in the position shown and the spring is undeformed. The stiffness of the spring is 800. The spring in the model could be attached between any two points along the axis of the translational joint—for example, between points O and G. For the disk and the plunger assume the following inertia data: $m_d = 1.5$, $J_d = 0.5$, $m_p = 0.25$, and $J_p = 0.1$. A torque of 20 acts on the disk in the CCW direction. Perform a forward dynamic analysis and determine the angular velocity at which the plunger contacts the switch.

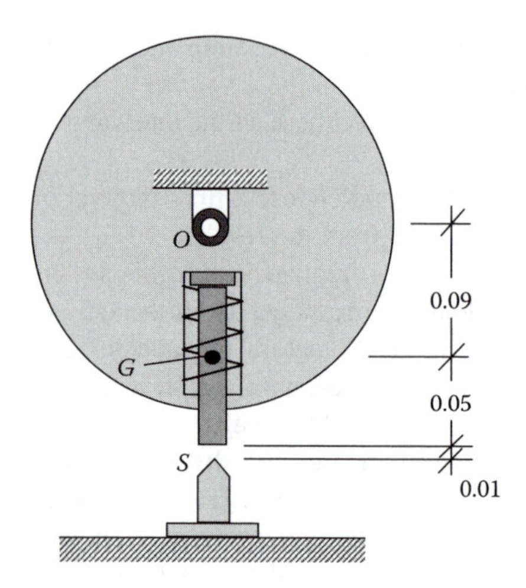

13.7 Simulate the bouncing of a rubber ball against the ground. Assume that the ground surface is flat and is positioned at $y = 0$. Develop a *unilateral* spring-damper element (function M-file). When the distance of the ball center from the ground becomes smaller than its radius, then the spring-damper element becomes active. Otherwise, the element should not contribute any forces. Test the new M-file by assuming the following data for the simulation: $k = 100$, $d_c = 10$, $m = 1$, $J = 0.1$, and $R = 0.1$. Release the ball from a height of 2.0.

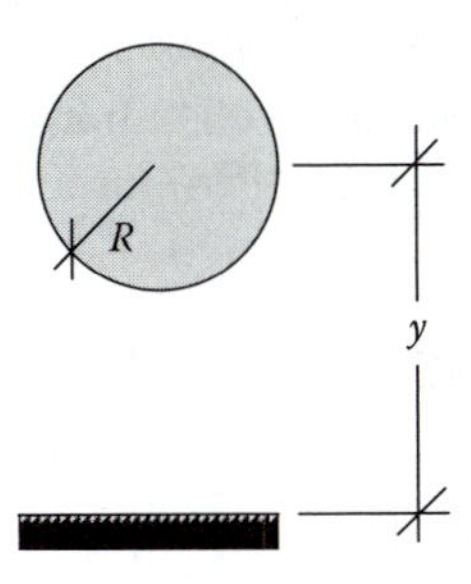

13.8 Repeat Problem 13.7 and assume that the ground surface makes an angle $\theta = 15°$ with the horizontal direction.

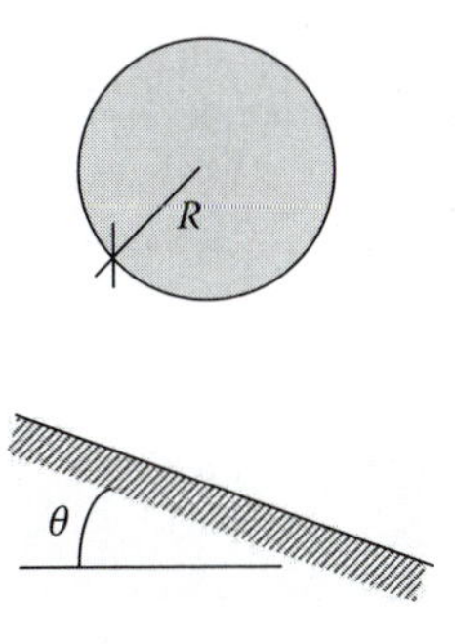

13.9 Two rubber balls are constrained to move in the vertical direction inside a frictionless cylinder as shown. Consider unilateral spring-damper elements between the balls and between the lower ball and the ground. Use the same data as in Problem 13.7. Release the balls from any desired heights and simulate the response of the system.

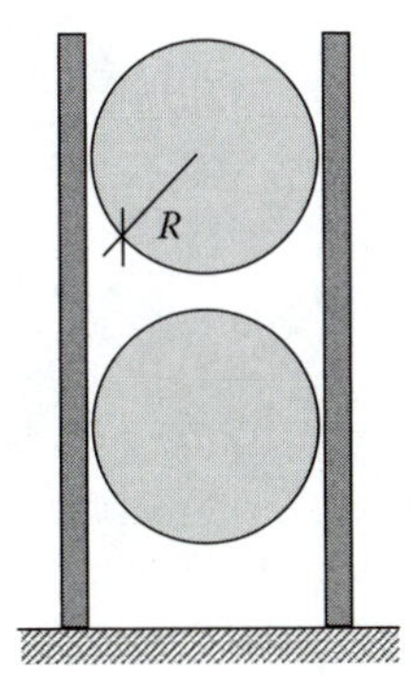

13.10 The apparatus shown consists of three pendulums terminating in rubber balls which can be modeled as three bodies and three pin joints. The interaction between the balls can be modeled by unilateral spring-damper elements. Assume $\ell = 0.3$, $R = 0.05$, $k = 5{,}000$, $d_c = 10$, and for each ball assume $m = 0.5$ and $J = 0.002$. Initially move one pendulum to a 45° orientation and release. Perform a forward dynamic analysis of this scenario.

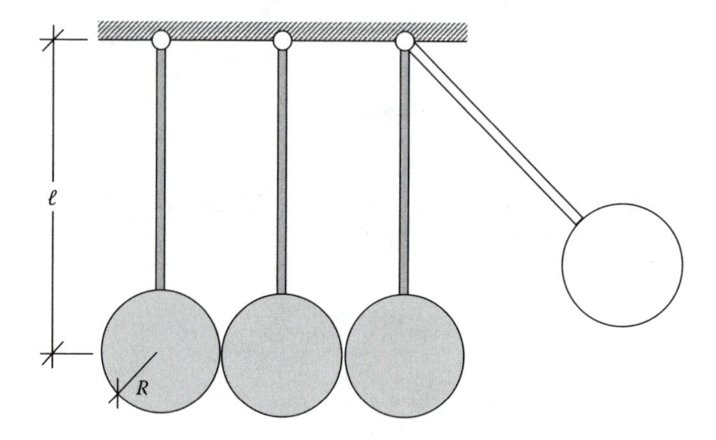

14 Complementary Analyses

Besides the standard kinematic, inverse dynamic, and forward dynamic analyses, there are other types of analyses that could be performed on multibody systems. Several of such complementary analyses methods will be discussed in this chapter. Specifically, we will discuss static and static equilibrium analyses, and several methods for correcting initial conditions that are required prior to performing a forward dynamic analysis. We will also discuss a methodology to perform kinematics, inverse dynamics, and forward dynamics through integration, all in one program. We will discuss issues associated with redundant constraints, we will look more closely at inclusion of friction in multibody models and, finally, we will show how to include structural deformation in a body following a simple procedure.

Programs: The new M-files that are developed in this chapter must be organized according to the file structure shown in Table 11.1 and Table 11.2.

14.1 STATIC ANALYSIS

A mechanical system becomes a structure (a nonmoving system) when the number of DoF is zero. In this case the number of independent constraint equations is equal to the number of defined coordinates; that is, $n_c = n_v$. As an example, consider the structure shown in Figure 14.1. Assume that this system is modeled with the body-coordinate formulation. Because there are 8 links and 12 pin joints (when n links come together we must consider $n - 1$ joints), we have $n_v = 3 \times 8 = 24$ coordinates and $n_c = 2 \times 12 = 24$ constraints.

Assume that the constraint equations for a structure are expressed as n_c equations:

$$\mathbf{\Phi}(\mathbf{q}) = \mathbf{0} \tag{14.1}$$

where $\mathbf{q}$ contains $n_v = n_c$ coordinates. Because there are as many coordinates as constraint equations, the equations can be solved to find the values of the coordinates.

The equations of motion for any constrained mechanical system can be written as

$$\mathbf{M}\ddot{\mathbf{q}} - \mathbf{D}'\boldsymbol{\lambda} = \mathbf{h}$$

Because for a structure $\dot{\mathbf{q}} = \ddot{\mathbf{q}} = \mathbf{0}$, the equations of motion are simplified to

$$\mathbf{D}'\boldsymbol{\lambda} = -\mathbf{h} \tag{14.2}$$

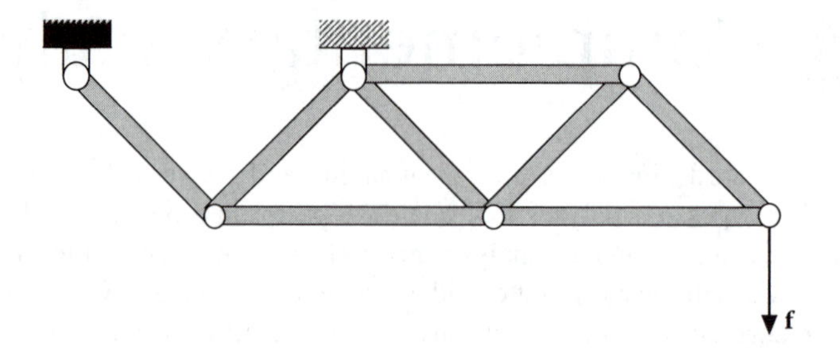

FIGURE 14.1 A planar structure subject to external forces.

where **h** contains the known applied loads. Because $n_c = n_v$, the Jacobian is a square matrix and hence Eq. (14.2) can be solved for the Lagrange multipliers. After determining the Lagrange multipliers, the constraint reaction forces can be computed at each joint.

Example 14.1

Consider the structure shown in Figure 14.2(a). The constant lengths are given as $\ell_1 = 5$, $\ell_2 = \sqrt{13}$, and $a = 2$. The links have negligible weight. An external load of 100 units acts at point A in the negative x-direction. Formulate the static equations for this structure with the point-coordinates and determine the reaction forces in the links.

Solution

The coordinates of the primary point A are determined to be $x^A = 4$ and $y^A = 3$. Two length constraints yield a Jacobian as $\mathbf{D} = \begin{bmatrix} x^A & y^A \\ x^A - a & y^A \end{bmatrix} = \begin{bmatrix} 4 & 3 \\ 2 & 3 \end{bmatrix}$. The applied force acting at A is described as $\mathbf{f} = \{100 \quad 0\}'$. Then, Eq. (14.2) is constructed as

$$\begin{bmatrix} 4 & 2 \\ 3 & 3 \end{bmatrix} \begin{Bmatrix} \lambda_1 \\ \lambda_2 \end{Bmatrix} = -\begin{Bmatrix} -100 \\ 0 \end{Bmatrix}$$

Solving these equations yields $\lambda_1 = 50$ and $\lambda_2 = -50$. The reaction forces that the links apply on point A are

$${}^{(c)}\mathbf{f}_1 = \begin{Bmatrix} 4 \\ 3 \end{Bmatrix} \lambda_1, \quad {}^{(c)}\mathbf{f}_2 = \begin{Bmatrix} 2 \\ 3 \end{Bmatrix} \lambda_2$$

Free-body diagrams for point A and the two links are shown in Figure 14.2(b), indicating that link (1) is in compression and link (2) is in tension.

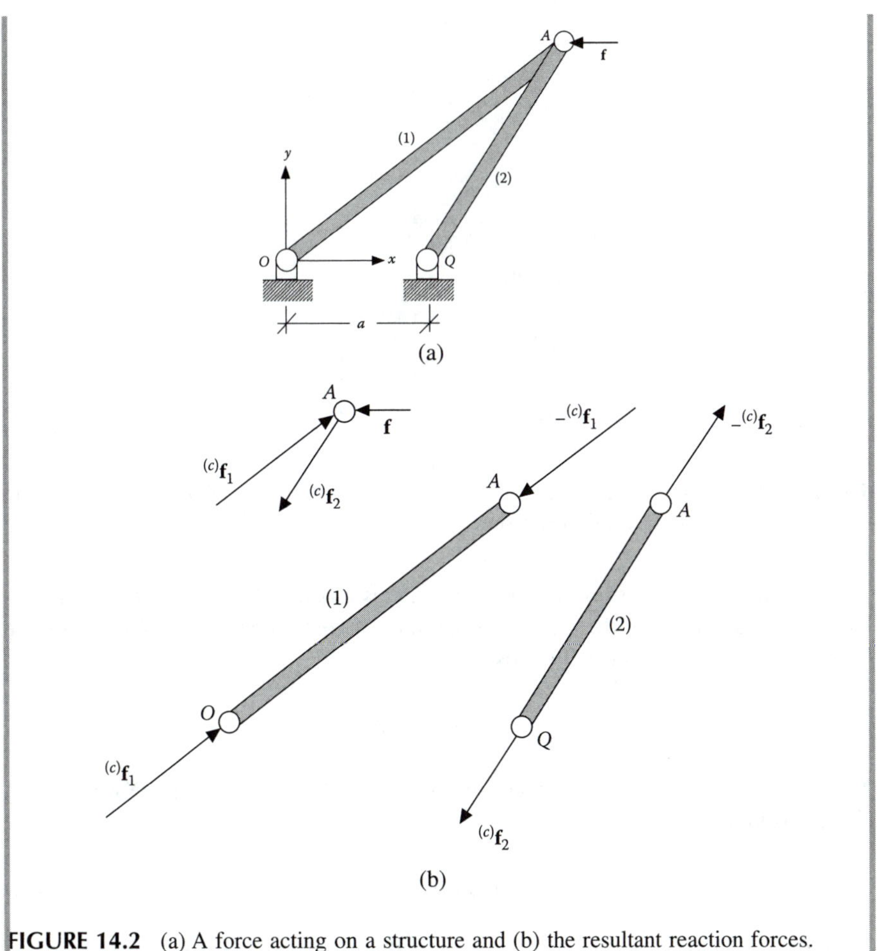

FIGURE 14.2 (a) A force acting on a structure and (b) the resultant reaction forces.

14.2 STATIC EQUILIBRIUM

A multibody system is said to be in state of static equilibrium when all the accelerations are zeros. The velocities could be zeros or constants. In such a state, the applied and the internal forces are in balance. In some applications of multibody systems, it may be required to start a forward dynamic analysis from the state of static equilibrium. As an example consider the single pendulum shown in Figure 14.3, orientation (a). There are two states of static equilibrium for this pendulum; one stable, orientation (b), and one unstable, orientation (c). In the stable state, the potential energy of the system is at its minimum and in the unstable state the potential energy is at its maximum. There are a variety of methods to determine the state of static equilibrium of a system.

One method for determining the state of static equilibrium is to construct a *potential energy function* for the system and minimize it. This approach is outside the scope of methods that have been discussed in this textbook.

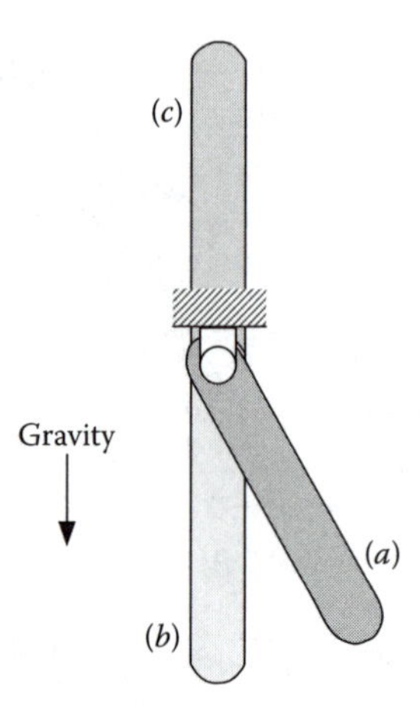

FIGURE 14.3 A single pendulum in nonstatic equilibrium (a), stable static equilibrium (b), and unstable static equilibrium states (c).

Another method is to set all of the velocities and accelerations in the equations of motion to zero. If the equations of motion are unconstrained (i.e., in the form $\mathbf{M}\ddot{\mathbf{q}} = \mathbf{h}$), then $\dot{\mathbf{q}} = \ddot{\mathbf{q}} = \mathbf{0}$ yields

$$\mathbf{h} = \mathbf{h}(\mathbf{q}) = \mathbf{0} \tag{14.3}$$

If the equations of motion are constrained and of the form $\mathbf{M}\ddot{\mathbf{q}} - \mathbf{D}'\boldsymbol{\lambda} = \mathbf{h}$, then $\dot{\mathbf{q}} = \ddot{\mathbf{q}} = \mathbf{0}$ yields

$$\mathbf{D}'\boldsymbol{\lambda} + \mathbf{h}(\mathbf{q}) = \mathbf{0} \tag{14.4}$$

With these equations we must also consider the coordinate constraints:

$$\boldsymbol{\Phi}(\mathbf{q}) = \mathbf{0} \tag{14.5}$$

If the equations are unconstrained, we must solve Eq. (14.3) for $\mathbf{q}$. If the equations are constrained, then we must solve Eqs. (14.4) and (14.5) for $\mathbf{q}$ and $\boldsymbol{\lambda}$. These are algebraic equation nonlinear in $\mathbf{q}$, therefore, they must be solved by iterative methods such as Newton-Raphson. Newton-Raphson method requires partial derivative of the equations with respect to the unknowns. Because computing these derivatives is not a simple task, we consider a third method for finding the static equilibrium state.

The third method is called the *fictitious damping* method. This method is based on the fact that a mechanical system with no damping elements can oscillate about its static equilibrium state. If dampers, that are energy-dissipating elements, are added to the system, the total energy of the system will decrease as time passes. The oscillation will be slowed, and finally the system will reach its static equilibrium state. If a system does not have adequate damping, we can add *fictitious* damping in the equations of motion. Then we perform a forward dynamic analysis until the system reaches its static equilibrium state.

The fictitious damping term that we add to the system does not have to represent any actual damping elements. The fictitious damping can be in analytical form added to the equations of motion. For unconstrained equations of motion, a damping array is included with the array of forces as

$$\mathbf{M}\ddot{\mathbf{q}} = \mathbf{h} + {}^{(f-d)}\mathbf{h} \tag{14.6}$$

For constrained equations of motion, the equations are modified as

$$\mathbf{M}\ddot{\mathbf{q}} - \mathbf{D}'\boldsymbol{\lambda} = \mathbf{h} + {}^{(f-d)}\mathbf{h} \tag{14.7}$$

The fictitious damping array can be defined as

$${}^{(f-d)}\mathbf{h} = -d_c\dot{\mathbf{q}} \tag{14.8}$$

or

$${}^{(f-d)}\mathbf{h} = -{}^{(d_c)}\mathbf{D}\dot{\mathbf{q}} \tag{14.9}$$

In these definitions d_c is a single damping coefficient and ${}^{(d_c)}\mathbf{D}$ is a diagonal damping matrix. The diagonal elements of ${}^{(d_c)}\mathbf{D}$ must be positive values representing the fictitious damping coefficients. The values of these coefficients do not change the static equilibrium state, but influence the speed of reaching that state. Implementation of this method in any of the forward dynamic analysis programs is a simple task.

EXAMPLE 14.2

Find the static equilibrium state for the double-pendulum system of Figure 13.8 in Section 13.5.1 using the fictitious damping method.

Solution

We refer to the model of this system in the body-coordinate formulation and use the main script `BC_forward`. In the file `BC_h` for the double pendulum, we add one statement at the end as

```
h = h - 10*c_d
```

where we assume a damping coefficient $d_c = 10$. We simulate the response twice, with and without damping. The *y*-coordinate of links (1) and (2) are plotted as shown in Figure 14.4. We note that with damping, $y_1 = -1$ and $y_2 = -3$ are the results at the static equilibrium.

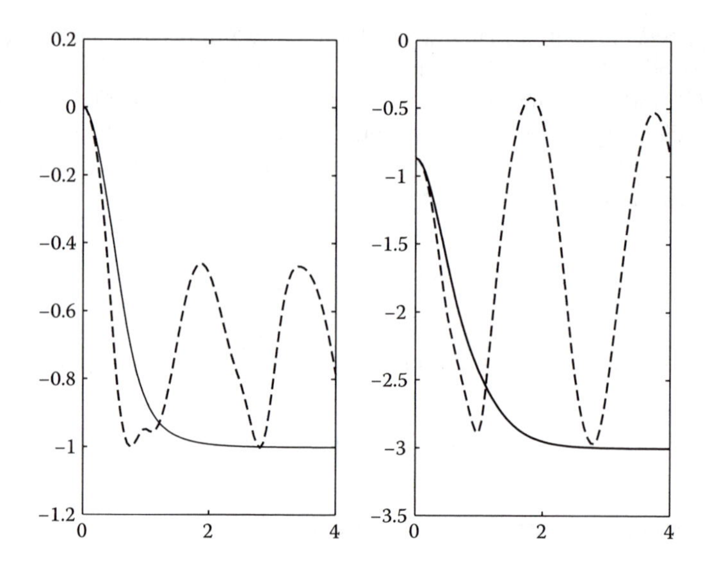

FIGURE 14.4 The y-coordinate of links (1) and (2) versus time without (dashed lines) and with (solid lines) fictitious damping.

14.3 INITIAL CONDITION CORRECTION

A forward dynamic analysis of constrained equations of motion requires a set of initial conditions on coordinates and velocities that satisfy the corresponding constraints. Failure in providing such initial conditions, most likely, results into erroneous responses. For highly simple multibody systems, we could come up with correct initial conditions at both the coordinate and velocity levels by either inspection or hand calculation. However, for more complex systems, determining a correct set of initial conditions may not be a simple task. In this section we discuss several computational procedures for correcting the initial conditions.

To determine a set of initial conditions that satisfy the constraints, we must solve the corresponding constraints at the coordinate or at the velocity level. The solution methods that are presented in this section can be illustrated through a simple example, using a set of linear algebraic equations. Assume that a set of initial conditions (estimates) on three variables are given as $x^e = 0.9$, $y^e = 2.1$ and $z^e = 3.2$. These estimates must satisfy the equations

$$\begin{aligned}\Phi_1 &= x + 2y - z - 2 = 0 \\ \Phi_2 &= 2x - y + z - 3 = 0\end{aligned} \tag{a}$$

Evaluating the constraints reveals that $\Phi_1 = 3.9$ and $\Phi_2 = -0.5$; that is, the constraints are not satisfied and therefore the initial conditions must be corrected.

We can correct the initial conditions by keeping the value of one of the variables unchanged and solve for the other two variables. For example, we can keep $x = x^e = 0.9$ and solve the equations for the other variables; that is, we solve the following equations:

$$\begin{bmatrix} 1 & 2 & -1 \\ 2 & -1 & 1 \\ 1 & 0 & 0 \end{bmatrix} \begin{Bmatrix} x \\ y \\ z \end{Bmatrix} = \begin{Bmatrix} 2 \\ 3 \\ 0.9 \end{Bmatrix} \tag{b}$$

The solution yields $x = 0.9$, $y = 1.1333$, and $z = 1.1667$. Obviously, if we select y or z to keep their given values, the results would be different.

We can solve the equations in a different manner. Because the estimated values yield violations Φ_1 and Φ_2, *(a)* could be expressed as

$$\begin{bmatrix} 1 & 2 & -1 \\ 2 & -1 & 1 \end{bmatrix} \begin{Bmatrix} \Delta x \\ \Delta y \\ \Delta z \end{Bmatrix} = -\begin{Bmatrix} \Phi_1 \\ \Phi_2 \end{Bmatrix} \tag{c}$$

where

$$\Delta x = x - x^e, \quad \Delta y = y - y^e, \quad \Delta z = z - z^e$$

We solve *(c)* in MATLAB® using the *backslash* operator:

```
D = [1  2 -1;2 -1  2]; rhs = [2; 3];
sol_e = [0.9; 2.1; 3.2]; % estimated values (to be corrected)
Phi = D*sol_e - rhs;   % violations
d_sol = -D\Phi;         % corrections
sol = sol_e + d_sol    % corrected solution
```

Executing this script yields $x = -0.32$, $y = 2.76$ and $z = 3.2$. We note that the process selected z to be the *best* variable to keep its value, and it corrected the values of x and y. The *best* variable could be picked by performing some form of matrix operation, such as Gaussian elimination or LU factorization with pivoting, on the coefficient matrix of the equations. Through this process one of the columns of the coefficient matrix, which is associated with one of the variables, will be picked to be the independent variable. Note that the *backslash* operator works on square and nonsquare matrices.

A third way to solve the equations is by a method based on minimizing the sum of squares of the corrections, as it will be explained shortly. We replace the last two lines in the preceding script by the following lines:

```
d_sol = -D'*((D*D')\Phi);    % corrections
sol = sol_e + d_sol          % corrected solution
```

Executing this script yields $x = -0.058$, $y = 2.256$, and $z = 2.570$. In this solution, all three estimates have been adjusted. We can check the sum of squares of corrections for all three solutions:

Solution 1: sum of squares of corrections = 5.0689
Solution 2: sum of squares of corrections = 1.9240
Solution 3: sum of squares of corrections = 1.1302

It is obvious that the third solution has the smallest corrections.

The preceding three solution schemes can be generalized for a multibody system having n_{dof} degrees of freedom. The coordinate and velocity constraints are expressed as

$$\mathbf{\Phi}(\mathbf{q}) = \mathbf{0} \tag{14.10}$$

$$\mathbf{D}\dot{\mathbf{q}} = \mathbf{0} \tag{14.11}$$

The first method is fundamentally a kinematic analysis at the coordinate and velocity levels. Any of the methods that we discussed in Chapter 11 could be applied here. For example, we can apply the appended constraint method and express Eqs. (14.10) and (14.11) as

$$\begin{aligned} &\mathbf{\Phi}({}^{(k)}\mathbf{q},\ {}^{(u)}\mathbf{q}) = \mathbf{0} \\ &{}^{(k)}\mathbf{q} = {}^{(k)}\mathbf{q}^{e} \end{aligned} \tag{14.12}$$

$$\begin{bmatrix} {}^{(k)}\mathbf{D} & {}^{(u)}\mathbf{D} \\ {}^{(k)}\mathbf{I} & \mathbf{0} \end{bmatrix} \begin{Bmatrix} {}^{(k)}\dot{\mathbf{q}} \\ {}^{(u)}\dot{\mathbf{q}} \end{Bmatrix} = \begin{Bmatrix} \mathbf{0} \\ {}^{(k)}\dot{\mathbf{q}}^{e} \end{Bmatrix} \tag{14.13}$$

where ${}^{(k)}\mathbf{q}^{e}$ and ${}^{(k)}\dot{\mathbf{q}}^{e}$ are the estimated initial conditions. Equation (14.12) is solved iteratively by the Newton-Raphson method, and Eq. (14.13) is solved as a set of linear algebraic equations. In this process we decide what variables should keep their estimated values.

In the second method, we let the backslash operator in MATLAB select the variables that should keep their values. For this purpose, we operate directly on the rectangular Jacobian matrix $\mathbf{D}$ and solve Eqs. (14.10) and (14.11). We can easily implement this process in any of the analysis programs.

The third process could be considered the best solution method if we have no preference on which variable to keep its value. The constraints of Eq. (14.10) are solved by the Newton-Raphson method using the correction steps as

$$
\begin{aligned}
{}^{i}\Delta\mathbf{q} &= -{}^{i}\mathbf{D}'({}^{i}\mathbf{D}\,{}^{i}\mathbf{D}')^{-1}\,{}^{i}\mathbf{\Phi} \\
{}^{i+1}\mathbf{q} &= {}^{i}\mathbf{q} + {}^{i}\Delta\mathbf{q}
\end{aligned}
\tag{14.14}
$$

After the coordinates are corrected, the velocities can be corrected in one step as

$$
\begin{aligned}
\Delta\dot{\mathbf{q}} &= -\mathbf{D}'(\mathbf{D}\,\mathbf{D}')^{-1}\mathbf{D}\dot{\mathbf{q}}^{e} \\
\dot{\mathbf{q}} &= \dot{\mathbf{q}}^{e} + \Delta\dot{\mathbf{q}}
\end{aligned}
\tag{14.15}
$$

In problems where all the initial velocities are zero, regardless of the complexity of the system, the velocity constraints are automatically satisfied.

Proof 14.1

To minimize the sum of square of corrections in the array of velocity, $\mathbf{v}$, that satisfies Eq. $\mathbf{D}\mathbf{v} = \mathbf{0}$, we follow the following process. Assume that an estimate for the array of velocity is given as $\mathbf{v}^{e}$. This estimated array results into $\mathbf{D}\mathbf{v}^{e} = \boldsymbol{\varepsilon}$. The correction in the velocity array is denoted as $\mathbf{v}^{e} - \mathbf{v} = -\Delta\mathbf{v}$. A minimization problem is stated as:

$$\text{Minimize } f = \Delta\mathbf{v}'\Delta\mathbf{v} \text{ subject to constraints } \mathbf{D}\mathbf{v} = \mathbf{0}$$

This constrained minimization problem can be restated as

$$\text{Minimize } f = \Delta\mathbf{v}'\Delta\mathbf{v} - \Delta\mathbf{v}'\mathbf{D}'\boldsymbol{\lambda} \tag{a}$$

where $\boldsymbol{\lambda}$ is an array of Lagrange multipliers.[a] The function in (a) is minimized when

$$\frac{\partial f}{\partial \Delta\mathbf{v}} = 2\Delta\mathbf{v} - \mathbf{D}'\boldsymbol{\lambda} = \mathbf{0} \tag{b}$$

Premultiplying this equation by $\mathbf{D}$ results into $2\mathbf{D}\Delta\mathbf{v} = \mathbf{D}\mathbf{D}'\boldsymbol{\lambda}$, or $\boldsymbol{\lambda} = -2(\mathbf{D}\mathbf{D}')^{-1}\boldsymbol{\varepsilon}$. Substituting this result into (*b*) yields

$$\Delta\mathbf{v} = -\mathbf{D}'(\mathbf{D}\mathbf{D}')^{-1}\boldsymbol{\varepsilon} \tag{c}$$

Then $\mathbf{v} = \mathbf{v}^{e} + \Delta\mathbf{v}$. If $\mathbf{v} \equiv \dot{\mathbf{q}}$, then (*c*) represents Eq. (14.15). If $\mathbf{v} \equiv \mathbf{q}$, then (*c*) represents Eq. (14.14) and must be solved iteratively because the constraints are nonlinear algebraic equations.

---

[a] Interested reader should refer to any textbook on the subject of optimization.

14.3.1 Body-Coordinate Formulation

In this section we implement the initial condition correction methods of Section 14.3 in a MATLAB program for body-coordinate formulation. For this purpose we revise the `BC_kinematics` program by performing the Newton-Raphson iteration in the main script.

```
% file name:  BC_ic_correct
% Initial conditions correction: Body coordinate formulation
     clear all
     addpath Basics  BC_Basics  BC_Analyses
     folder = input(' which folder ? ', 's'); addpath (folder)
     BC_global
     BC_condata; BC_vardata;
     nbc = 3*nb;
% Determine components of vectors on the ground
     BC_vectors0;
% Parameters for kinematic analysis
     BC_parameters; n_steps = 1;
% Transfer r and phi to c; r_d and phi_d to c_d
     c = BC_r_to_c(r, phi, nb); c_d = BC_r_to_c(r_d, phi_d, nb);
% Ask for the method
     i_method = input(' Method 1, 2, or 3? ');
     if i_method == 1
        i_coord = input(' Coordinate number to remain unchanged? ');
           D_driver = zeros(1, nbc); D_driver(i_coord) = 1;
           rhsv = zeros(nbc,1); rhsv(i_coord) = c_d(i_coord);
     end
% Coordinate correction
     flag=0;
for n = 1:NR_iter
     % Compute A matrices, vectors, constraints and Jacobian
           BC_A_matrices; BC_vectors; Phi = BC_Phi; D = BC_jacob;
     % Are the constraints violated?
           ff=sqrt(Phi'*Phi);
     if ff < NR_tol
           flag=1; break
     end
% Solve for corrections
     if i_method == 1
           D_all = [D; D_driver]; Phi_all = [Phi; 0];
           delta_c = -D_all\Phi_all;
     elseif i_method == 2
           delta_c = -D\Phi;
     else
           delta_c = -D'*((D*D')\Phi);
     end
     % Correct estimates
           c = c + delta_c; BC_c_to_r;
end
if flag == 0
     error('Convergence failed in Newton-Raphson');
end
```

```
% Velocity correction
    if i_method == 1
        c_d = D_all\rhsv;
    elseif i_method == 2
        Phi_d = D*c_d; c_d = c_d - D\Phi_d;
    else
        c_d = c_d - D'*((D*D')\(D*c_d));;
    end
```

The program prompts the user for the *type of method* to be used. User must respond by typing 1, 2, or 3. If the requested method is 1, then the program asks for the *index of the coordinate* (and the corresponding velocity) that should remain unchanged. The user must know the number of coordinates in the model and their order of appearance in the c (or c_d) array. For example, if we wish to keep the value of y_2 unchanged, we must respond 5. This portion of the program could be revised for user friendliness.

As an example, we try to correct the estimates for the coordinates and velocities of the double A-arm suspension system. We refer to Section 7.3.1 and find a set of coordinates that led to the constraint violations shown in Figure 7.11. These coordinates and a set of estimates for the velocities are arranged in the file BC_vardata as:

```
% file name:  BC_vardata.m
% Body coordinates and velocities
    r = {[0.54; -0.20] [0.75; -0.07] [0.49; 0.02]};
    phi = [(340*pi/180)  0  (350*pi/180)];
    r_d = {[0.1; 0.8] [-0.1; 1.5] [-0.2; 0.6]};
    phi_d = [2.8  1.5  4.5];
```

We execute the program BC_ic_correct three times, trying all the three methods. In the first try, we ask for method 1 and coordinate number 3 (i.e., ϕ_1). Then we print the corrected results for c and c_d. Then we try for method 2, and then for method 3. The results are summarized below:

Method 1:

```
c':    0.426 -0.222  5.934  0.665 -0.184  -0.007  0.443 -0.022  5.956
c_d':  0      0      0     -1.477 -0.940 -13.296  0.064  0.189  1.534
```

Method 2:

```
c':    0.432 -0.200  6.030  0.676 -0.145  0.003  0.448 -0.003  6.109
c_d':  0.172  0.666  2.867  0.257  1.197  0.446  0.102  0.576  4.500
```

Method 3:

```
c':    0.432 -0.203  6.018  0.675 -0.150  0.001  0.448 -0.005  6.089
c_d':  0.184  0.676  2.917  0.280  1.211  0.414  0.115  0.585  4.586
```

Any of these sets of results can be used as initial conditions for this model to start a forward dynamic analysis.

14.4 THREE COMBINED ANALYSES BY INTEGRATION

In kinematic analysis of Chapter 11 we solved the position constraints as nonlinear algebraic equations, and velocity and acceleration constraints as linear algebraic equations. In Chapter 12 we extended the kinematic analysis to inverse dynamics by adding another step of solving linear algebraic equations to obtain the Lagrange multipliers leading to the reaction forces. In this section we show how to perform kinematic and inverse dynamic analyses by integration. In other words, we can make both kinematic and inverse dynamic analyses *to appear* as a form of forward dynamic analysis.

For kinematic analysis by integration we consider the acceleration constraints from Eq. (11.16), containing the kinematic and the driver constraints:

$$\begin{bmatrix} \mathbf{D} \\ {}^{(d)}\mathbf{D} \end{bmatrix} \ddot{\mathbf{q}} = \begin{Bmatrix} \boldsymbol{\gamma} \\ \ddot{f}(t) \end{Bmatrix} \tag{14.16}$$

The overall Jacobian is a square matrix. Therefore, at any given instant, if $\mathbf{q}$ and $\dot{\mathbf{q}}$ are available, Eq. (14.16) can be solved for the accelerations, $\ddot{\mathbf{q}}$. The arrays $\mathbf{q}$ and $\dot{\mathbf{q}}$ can be determined from integrating $\dot{\mathbf{q}}$ and $\ddot{\mathbf{q}}$. In the kinematic algorithms of Chapter 11, we determined $\mathbf{q}$ and $\dot{\mathbf{q}}$ from the solution of the coordinate and velocity constraints—that is, by solving nonlinear and linear algebraic equations. But in the process that is presented in this section, we determine $\mathbf{q}$ and $\dot{\mathbf{q}}$ by integration.

Following the determinations of the accelerations, we can extend the process to inverse dynamics by considering Eq. (12.23); that is,

$$[\mathbf{D}' \quad {}^{(d)}\mathbf{D}'] \begin{Bmatrix} \boldsymbol{\lambda} \\ {}^{(d)}\boldsymbol{\lambda} \end{Bmatrix} = \mathbf{M}\ddot{\mathbf{q}} - \mathbf{h} \tag{14.17}$$

The solution of this equation yields the Lagrange multipliers $\boldsymbol{\lambda}$ and ${}^{(d)}\boldsymbol{\lambda}$.

Because the kinematic and inverse dynamic analyses can be performed by integration, we can combine this process with that of the forward dynamic analysis to have a single formulation, and hence a single program, to perform all the three analyses. Assume that a system is represented by n_v coordinates, n_c kinematic constraints, and n_d driver constraints. Appending the acceleration equations for both the kinematic and driver constraints to the equations of motion results into the following set of equations:

$$\begin{bmatrix} \mathbf{M} & -\mathbf{D}' & -{}^{(d)}\mathbf{D}' \\ \mathbf{D} & \mathbf{0} & \mathbf{0} \\ {}^{(d)}\mathbf{D} & \mathbf{0} & \mathbf{0} \end{bmatrix} \begin{Bmatrix} \ddot{\mathbf{q}} \\ \boldsymbol{\lambda} \\ {}^{(d)}\boldsymbol{\lambda} \end{Bmatrix} = \begin{Bmatrix} \mathbf{h} \\ \boldsymbol{\gamma} \\ \ddot{f}(t) \end{Bmatrix} \tag{14.18}$$

We can consider two cases:

1. If $n_c + n_d = n_v$, then every degree of freedom is controlled by the driver constraints and, therefore, this is a *kinematic/inverse dynamic* analysis. Solution of Eq. (14.18) provides the values of $\ddot{\mathbf{q}}$, $\boldsymbol{\lambda}$ and ${}^{(d)}\boldsymbol{\lambda}$.
2. If $n_c + n_d < n_v$, then every degree of freedom is not controlled by the driver constraints and, therefore, this is a *forward dynamic* analysis. Solution of

Eq. (14.18) provides the values of $\ddot{\mathbf{q}}$, $\boldsymbol{\lambda}$ and ${}^{(d)}\boldsymbol{\lambda}$. Note that in forward dynamics, we may or may not have driver constraints. In the discussion of Chapter 13, we did not consider driver constraints in our formulation. But in general, we may include them.

An integration process, similar to that of the forward dynamics can be stated as:

- Define initial conditions for the coordinates and velocities—that is, ${}^{0}\mathbf{q}$ and ${}^{0}\dot{\mathbf{q}}$.
- Initialize an integration array as $\mathbf{z} = \begin{Bmatrix} {}^{0}\mathbf{q} \\ {}^{0}\dot{\mathbf{q}} \end{Bmatrix}$.
- Establish time parameters for integration.
- Invoke a numerical integrator such as `ode45`. The integrator sends $\mathbf{z}$, for $t = {}^{i}t$, to a function and asks for $\dot{\mathbf{z}}$:
  - Transfer the contents of $\mathbf{z}$ to $\mathbf{q}$, and $\dot{\mathbf{q}}$.
  - Solve Eq. (14.18) for the accelerations $\ddot{\mathbf{q}}$, $\boldsymbol{\lambda}$ and ${}^{(d)}\boldsymbol{\lambda}$.
  - Transfer the contents of $\dot{\mathbf{q}}$, and $\ddot{\mathbf{q}}$ to a $\dot{\mathbf{z}}$ array as $\dot{\mathbf{z}} = \begin{Bmatrix} \dot{\mathbf{q}} \\ \ddot{\mathbf{q}} \end{Bmatrix}$.
  - Return the $\dot{\mathbf{z}}$ array to the integrator.

  This process is repeated several times for every time step.
- Recover $\ddot{\mathbf{q}}$, $\boldsymbol{\lambda}$ and ${}^{(d)}\boldsymbol{\lambda}$ if necessary.

14.5 REDUNDANT CONSTRAINTS

In all analyses that involved solution of equations containing constraints, or the Jacobian of the constraints, the assumption has been that all the constraints that are constructed in a problem are independent. However, this may not be the case in every problem. In some multibody systems inclusion of redundant links and kinematic joints may be part of the design. This could be for a more even distribution of reaction forces, lowering the amount of structural deformations, safety reasons, or not allowing a mechanism to end up in an undesirable configuration. Regardless of the actual reason for the redundancy in the design, we must be aware of the computational consequences in modeling such systems.

The following example is provided to show how redundancy may be part of a design. Consider the lift mechanism of a delivery truck shown in Figure 14.5. A hydraulic actuator lifts the container to the desired height for goods to be delivered. There are no redundancies in the design of this lift mechanism. Assuming that this system is modeled by the body coordinate formulation, for the three moving bodies we need $n_v = 9$ coordinates and for the three pin joints and two revolute-translational joints we have $n_c = 8$ constraints. This leaves the formulation with 1 DoF which is controlled by the hydraulic actuator.

Next we consider a modified design of the lift mechanism as shown in the inset of Figure 14.5. In this design, to eliminate some undesirable vibrations of the container due to slight imperfections in the joints, the container is attached to a

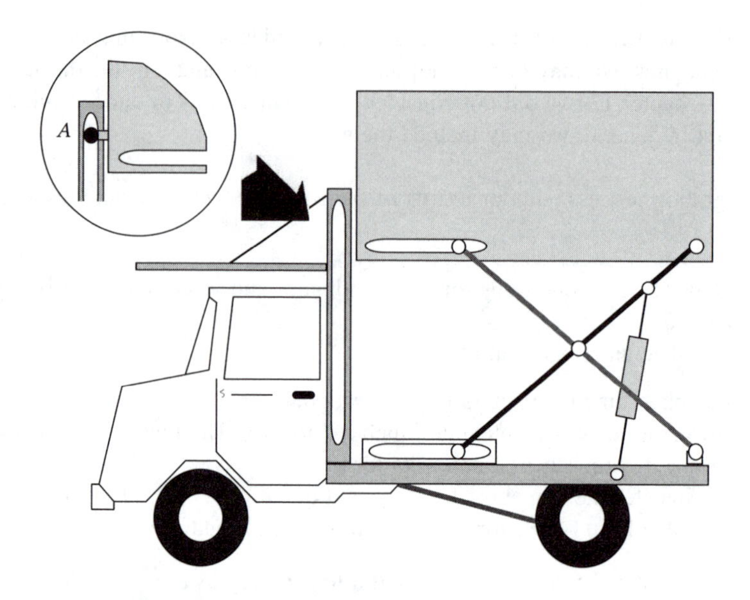

FIGURE 14.5 A lift mechanism for a delivery truck and a redundant joint (inset) that could be included in the system.

vertical support by an additional revolute-translational joint. If we model this system, we will have $n_v = 9$ and $n_c = 9$. Because the actual system still contains 1 DoF, obviously one of the nine constraints must be redundant.

When redundancy is encountered in a formulation, if we know exactly which constraint is redundant, we can manually remove that constraint from the model. If we are not aware of the redundancy, or if we are not certain which constraint should be removed, we can identify and remove the redundant constraint(s) computationally. For this purpose we can perform some form of matrix operation, such as Gaussian elimination or LU factorization with pivoting (row exchange), on the Jacobian matrix.

14.6 FRICTION

In Chapter 4 we presented two types of friction that may appear in multibody systems. In this section we provide some details in implementing these types of friction models in the dynamic analysis of such systems.

The necessary formulas to model viscous friction in rotational and sliding joints were presented in Section 4.4.7. These formulas require the relative velocity between the two contacting bodies. For a sliding joint the relative velocity can be obtained from the time derivative of a vector connecting the two bodies along the axis of the sliding joint. For a pin joint the relative angular velocity can be computed based on the angular velocities of the two bodies. If the joint coordinate formulation is used, these relative velocities could directly be one of the joint velocities. It should be obvious that modeling viscous friction is exactly the same as including a damper element in a system.

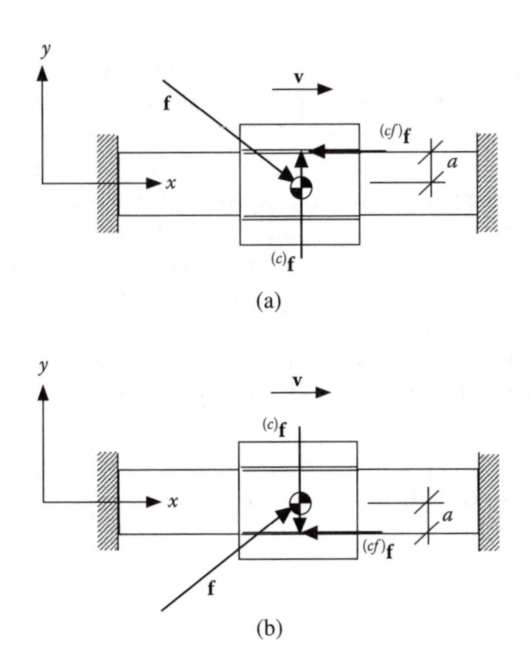

FIGURE 14.6 A slider block under two different applied loads.

Inclusion of Coulomb friction in a multibody system was briefly discussed in Section 4.5.3. Implementing this type of friction in the equations of motion is more complicated than that of the viscous friction. The difficulty arises from the fact that Coulomb friction is a function of the reaction force between the contacting surfaces. Reaction forces in the equations of motion, as well as the accelerations, are unknowns. Therefore the magnitude of the friction force cannot be determined prior to the solution of the equations of motion. Meanwhile, the equations of motion cannot be solved without knowing the exact value of the friction force! This phenomenon can be better understood through a simple example.

Consider the slider block shown in Figure 14.6(a). This block can move only in the horizontal direction and it has a velocity of $\mathbf{v} = \{0.6 \quad 0\}'$. A known force $\mathbf{f} = \{4.0 \quad -3.0\}'$ acts on the mass center of the block and a reaction force, ${}^{(c)}\mathbf{f}$, perpendicular to the x-axis acts on the contact surface of the block as shown. If we consider Coulomb friction between the contacting surfaces, the magnitude of the friction force can be expressed as ${}^{(cf)}f = \mu\, {}^{(c)}f$, where μ is the friction coefficient and ${}^{(c)}f$ is the magnitude of the reaction force ${}^{(c)}\mathbf{f}$. This friction force acts in the opposite direction of the relative velocity. If we assume $m = 1.0$, $J = 0.1$, $\mu = 0.5$, and $a = 0.2$, the equations of motion for this constrained body can be expressed as

$$\begin{bmatrix} 1.0 & 0 & 0 \\ 0 & 1.0 & 0 \\ 0 & 0 & 0.1 \end{bmatrix} \begin{Bmatrix} \ddot{x} \\ \ddot{y} \\ \ddot{\phi} \end{Bmatrix} - \begin{bmatrix} 0 & 0 \\ 1 & 0 \\ 0 & 1 \end{bmatrix} \begin{Bmatrix} \lambda_1 \\ \lambda_2 \end{Bmatrix} = \begin{Bmatrix} 4.0 \\ -3.0 \\ 0 \end{Bmatrix} + \begin{Bmatrix} -0.5\, {}^{(c)}f \\ 0 \\ 0.5\, {}^{(c)}f\, a \end{Bmatrix} \tag{14.19}$$

Because the body is not allowed to rotate or to move along the y-axis, two simple constraints provide the necessary constraints. Also note that the friction force results in a moment that must be included in the equations of motion. If we append the acceleration constraints for the two simple constraints to the equations of motion, we still have more unknowns than the number of equations—the extra unknown is ${}^{(c)}f$.

In a forward dynamic analysis where Coulomb friction is considered, there are two ways to solve the equations of motion, such as Eq. (14.19). One method is to use the value of the reaction force ${}^{(c)}f$ from the previous time step in determining the Coulomb friction force. Although this is based on an approximated value of the friction force, the process is simple to implement in a program and the results would not be too far off.

Another method is to note that ${}^{(c)}f$ and λ_1 both represent the reaction force perpendicular to the contact surface of the block. Therefore, we could express Eq. (14.19) as

$$\begin{bmatrix} 1.0 & 0 & 0 \\ 0 & 1.0 & 0 \\ 0 & 0 & 0.1 \end{bmatrix} \begin{Bmatrix} \ddot{x} \\ \ddot{y} \\ \ddot{\phi} \end{Bmatrix} - \begin{bmatrix} 0 & 0 \\ 1 & 0 \\ 0 & 1 \end{bmatrix} \begin{Bmatrix} \lambda_1 \\ \lambda_2 \end{Bmatrix} = \begin{Bmatrix} 4.0 \\ -3.0 \\ 0 \end{Bmatrix} + \begin{Bmatrix} -0.5\lambda_1 \\ 0 \\ 0.5\lambda_1 a \end{Bmatrix}$$

or

$$\begin{bmatrix} 1.0 & 0 & 0 \\ 0 & 1.0 & 0 \\ 0 & 0 & 0.1 \end{bmatrix} \begin{Bmatrix} \ddot{x} \\ \ddot{y} \\ \ddot{\phi} \end{Bmatrix} - \begin{bmatrix} -0.5 & 0 \\ 1 & 0 \\ 0.1 & 1 \end{bmatrix} \begin{Bmatrix} \lambda_1 \\ \lambda_2 \end{Bmatrix} = \begin{Bmatrix} 4.0 \\ -3.0 \\ 0 \end{Bmatrix} \quad (14.20)$$

Using the acceleration constraints $\ddot{y} = 0$ and $\ddot{\phi} = 0$, Eq. (14.20) can be solved to find $\lambda_1 = 3.0$, $\lambda_2 = 0.3$, and $\ddot{x} = 2.5$. If we inspect the results, we determine that they are correct. Without the friction force the acceleration of the block would have been $\ddot{x} = 4.0$.

Next we consider the same slider block under a different applied load $\mathbf{f} = \{4.0 \quad 3.0\}'$, as shown in Figure 14.6(b). This force causes the reaction force, but not the friction force, to change direction as shown. The equations of motion for the slider block can be expressed as

$$\begin{bmatrix} 1.0 & 0 & 0 \\ 0 & 1.0 & 0 \\ 0 & 0 & 0.1 \end{bmatrix} \begin{Bmatrix} \ddot{x} \\ \ddot{y} \\ \ddot{\phi} \end{Bmatrix} - \begin{bmatrix} 0 & 0 \\ 1 & 0 \\ 0 & 1 \end{bmatrix} \begin{Bmatrix} \lambda_1 \\ \lambda_2 \end{Bmatrix} = \begin{Bmatrix} 4.0 \\ 3.0 \\ 0 \end{Bmatrix} + \begin{Bmatrix} -0.5\,{}^{(c)}f \\ 0 \\ -0.5\,{}^{(c)}f\,a \end{Bmatrix} \quad (14.21)$$

Note that the sign of the moment caused by the friction force has been adjusted. If we substitute λ_1 for ${}^{(c)}f$, the equations of motion become

$$\begin{bmatrix} 1.0 & 0 & 0 \\ 0 & 1.0 & 0 \\ 0 & 0 & 0.1 \end{bmatrix} \begin{Bmatrix} \ddot{x} \\ \ddot{y} \\ \ddot{\phi} \end{Bmatrix} - \begin{bmatrix} -0.5 & 0 \\ 1 & 0 \\ -0.1 & 1 \end{bmatrix} \begin{Bmatrix} \lambda_1 \\ \lambda_2 \end{Bmatrix} = \begin{Bmatrix} 4.0 \\ 3.0 \\ 0 \end{Bmatrix} \quad (14.22)$$

Using $\ddot{y} = 0$ and $\ddot{\phi} = 0$, these equations can be solved to find $\lambda_1 = -3.0$, $\lambda_2 = 0.3$, and $\ddot{x} = 5.5$. If we inspect the answers, we determine that they are not correct! The negative value of λ_1 has caused the friction force to further accelerate the slider block.

To circumvent this type of problem, we can write the equations of motion for this simple system as

$$\begin{bmatrix} 1.0 & 0 & 0 \\ 0 & 1.0 & 0 \\ 0 & 0 & 0.1 \end{bmatrix} \begin{Bmatrix} \ddot{x} \\ \ddot{y} \\ \ddot{\phi} \end{Bmatrix} - \begin{bmatrix} 0 & 0 \\ 1 & 0 \\ 0 & 1 \end{bmatrix} \begin{Bmatrix} \lambda_1 \\ \lambda_2 \end{Bmatrix} = \begin{Bmatrix} f_{(x)} \\ f_{(y)} \\ 0 \end{Bmatrix} + \begin{Bmatrix} \pm 0.5 u_{(x)} \lambda_1 \\ 0 \\ 0.5 \lambda_1 a \end{Bmatrix}$$

or

$$\begin{bmatrix} 1.0 & 0 & 0 \\ 0 & 1.0 & 0 \\ 0 & 0 & 0.1 \end{bmatrix} \begin{Bmatrix} \ddot{x} \\ \ddot{y} \\ \ddot{\phi} \end{Bmatrix} - \begin{bmatrix} \pm 0.5 u_{(x)} & 0 \\ 1 & 0 \\ 0.1 & 1 \end{bmatrix} \begin{Bmatrix} \lambda_1 \\ \lambda_2 \end{Bmatrix} = \begin{Bmatrix} f_{(x)} \\ f_{(y)} \\ 0 \end{Bmatrix} \tag{14.23}$$

where $\mathbf{u} = \{u_{(x)} \quad u_{(y)}\}' = -\mathbf{v} / v$ is a unit vector in the opposite direction of the velocity vector. To decide on the sign of the term $\mp 0.5 u_{(x)}$, we consider the sign of λ_1 from the previous time step. If in the last time step $\lambda_1 > 0$, then we pick the positive sign for the term, otherwise we pick the negative sign. In situations when λ_1 is about to change sign (i.e., its value is close to zero), it would not make much difference whether we include the friction force or not because the reaction force is close to zero. Furthermore, note that the sign of the moment that is caused by the friction force is automatically taken care of by the sign of its Lagrange multiplier.

The preceding methodology for the inclusion of Coulomb friction between two bodies can be generalized and implemented in a forward dynamic analysis program. Whether only one or both contacting bodies are allowed to move, the Jacobian of the constraints and the corresponding Lagrange multipliers provide the reaction forces between the bodies. Therefore, the friction force can be expressed in terms of these Lagrange multipliers and be included in the Jacobian transpose. In this process we must assure that the friction force acts in the opposite direction of the relative velocity regardless of the sign of the Lagrange multiplier. For this purpose appropriate logic must be included in the program to determine the correct sign for the added terms in the Jacobian.

14.7 DEFORMABLE BODY

In some applications of multibody dynamics, the structural deformation of a body may not be negligible and it could influence the overall response of a system. In these applications, the deformation of such bodies must be incorporated into the corresponding multibody models. Finite element (FE) method provides a powerful tool to determine structural deformations. Finite element formulation can also be used to incorporate structural deformation in the bodies of a multibody system. However, it is outside

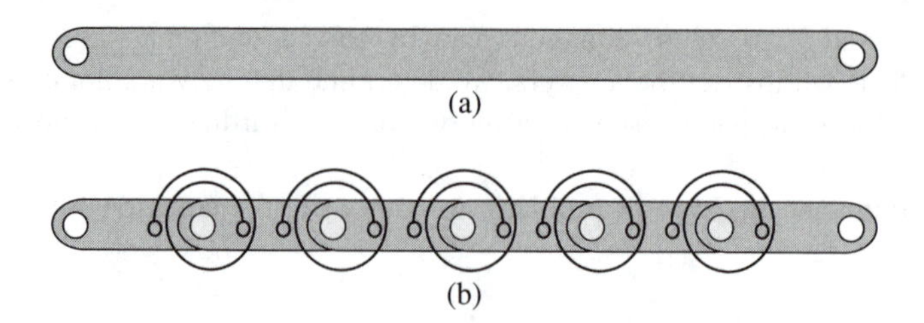

FIGURE 14.7 (a) A deformable rod and (b) its representation as an assembly of smaller rods connected by pin joints and torsional springs.

the scope of this textbook to discuss the finite element formulation within the subject of multibody dynamics. Instead, we introduce a simple approximation process that could be considered as a *crude* finite element representation for a deformable body.

The approximation method is demonstrated here for a deformable rod (a beam). Assume that the rod shown in Figure 14.7(a), with a length ℓ, exhibits bending deformation. The rod can be split into n smaller rods, each having a length ℓ/n, and connected to each other by $n-1$ pin joints as shown in Figure 14.7(b). A torsional spring is defined about the axis of each pin joint. The rotational stiffness of each spring can be determined as

$$^{(r)}k_{n-1} = \frac{EJ_A}{\ell}\frac{(n-1)(2n-1)}{2n} \tag{14.24}$$

where E is the Young's modulus and J_A is the area moment of inertia of the rod's cross-section. Values of Young's modulus, or the modulus of elasticity, for different materials can be found in engineering handbooks. The area moment of inertia for most common cross-sectional shapes can be found in most textbooks on statics and dynamics. For circular and rectangular cross sections, the area moment of inertia about the so-called *neutral axis* is shown in Figure 14.8.

It should be obvious that representing a deformable body as an assembly of several smaller bodies, connected by joints and springs, increases the size of a problem and hence the computation time. If a multibody system is modeled with the joint coordinate formulation, the increase in the problem size will not be as significant as with the body coordinate formulation. Another contributor to the

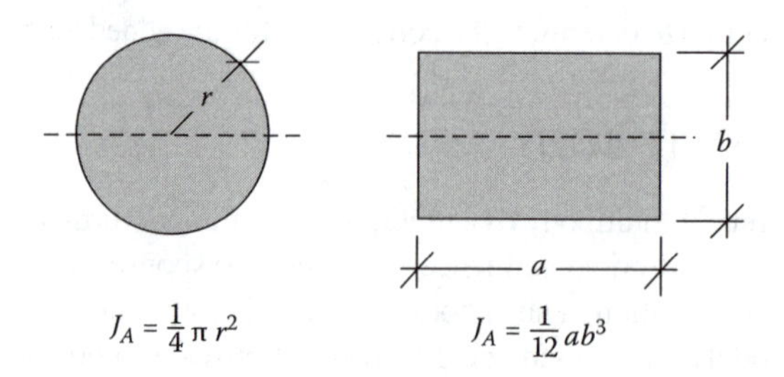

FIGURE 14.8 Area moment of inertia about the neutral axis for two common cross sections.

increase in the computation time is that realistic values for the Young's modulus yield high stiffness values for the springs. This in turn requires small integration time steps during forward dynamic analyses.

14.8 REMARKS

This chapter provided some complementary topics for modeling and analyses of multibody systems. With the methodologies learned in this and previous chapters, we can model and analyze the response of a variety of multibody systems under different scenarios. We can experiment with the learned methodologies, techniques, and formulations on some project type examples that are outlined in the following chapter.

14.9 PROBLEMS

14.1 Develop a program for static analysis of multibody systems with body coordinate formulation. For this purpose the program `BC_kinematics` can be revised by removing the driver constraints and allowing the number of coordinates to be equal to the number of kinematic constraints.

14.2 Develop a program for static analysis of multibody systems with point coordinate formulation.

14.3 Revise the program `BC_forward` to determine the static equilibrium of a system as well as the performing forward dynamics. Provide a logic that would allow the inclusion of fictitious damping if static equilibrium analysis is requested.

14.4 Repeat Problem 14.3 with the program `JC_forward`.

14.5 Include the program `BC_ic_correct` to be an integral part of the program `BC_forward`. Prior to any forward dynamic analysis, the combined programs could correct the initial conditions.

14.6 Develop a program to correct the initial conditions in the joint coordinates. Name the main M-file `JC_ic_correct`.

14.7 Include the newly developed program in Problem 14.6 with the program `JC_forward` similar to that of Problem 14.5.

14.8 Revise the program `BC_forward` to perform kinematic, inverse dynamic, and forward dynamic analyses by integration. Some of the function M-files that are used in the program `BC_kinematics` can be used in this revised program as well. The step that solves for the accelerations and Lagrange multipliers can be formulated in two ways:
   (a) Construct the coefficient matrix and the right-hand-side array in Eq. (14.18) and solve for the unknowns in one step.
   (b) Solve for the Lagrange multipliers and the accelerations in two steps.

14.9 Repeat Problem 14.8 for the joint coordinate formulation.

14.10 Develop a program for the point coordinate formulation to perform kinematic, inverse dynamic, and forward dynamic analyses by integration.

14.11 In some simulations we may want to know the amount of time it takes MATLAB to perform certain computation. For example, to find out how much time the program BC_forward spends to integrate the equations of motion, we include the following statements in BC_forward before and after invoking the integrator ode45:

```
% Start simulation time clock
    t0_clock = clock;
% Integrate
    [T, zT] = ode45(@BC_eqsmo, Tspan, z);
% Show elapsed simulation time
    elapsed_time = etime(clock,t0_clock);
    disp(['Runtime (seconds): ', num2str(elapsed_time)])
```

Include these statements in BC_forward and in other programs.

14.12 The MATLAB programs in this textbook communicate data between scripts in arrays, cells, and global statements. The same data can be organized in "structure arrays". Learn how to use struct and incorporate it in the programs. This is a very useful form of organizing and communicating data between M-files.

14.13 Computer programs should provide an attractive interface for the user to enter input data. Develop user-friendly interfaces for the programs we have developed in this textbook. Take advantage of MATLAB'S Graphical User Interface (GUI).

14.14 As an extension of Problem 14.13, develop a user-friendly interface to create plots and animations from the output of any simulation.

14.15 After completing a simulation, we may decide to restart the simulation from the point that the first simulation was ended. In other words, results from the last step of one simulation could be used as initial conditions for another simulation. Provide this capability in any of the programs, such as xx_forward, by writing the coordinates and velocities of the last step of simulation in a file similar to xx_vardata. The format of the data in the file xx_vardata could be revised if necessary.

14.16 Consider the two-body system shown. Link *1* is connected to the ground and to link *2* by pin joints. Link *2* can only slide relative to the ground. Consider the following data for the system: $\ell_1 = 1.0$, $m_1 = 1.0$, $J_1 = 1.0$, and $m_2 = 1.0$. This system in its present form with two rigid bodies has zero DoF. Now let us assume that link *1* is a deformable body made of aluminum with a Young's modulus of $E = 69$ GPa (69×10^9 Pa) and an area moment of inertia of $J_A = \frac{1}{12}10^{-8}$ (1×1 cm^2 cross-section). Split this link into several smaller links (e.g., $n = 5$), and compute the corresponding stiffness from Eq. (14.24) for each of the rotational springs. Apply a force in the *y*-direction on the middle link for a short period of time (e.g., for 0.2 seconds) to cause a deflection. Perform a forward dynamic analysis and monitor the deflection of the deformable body.

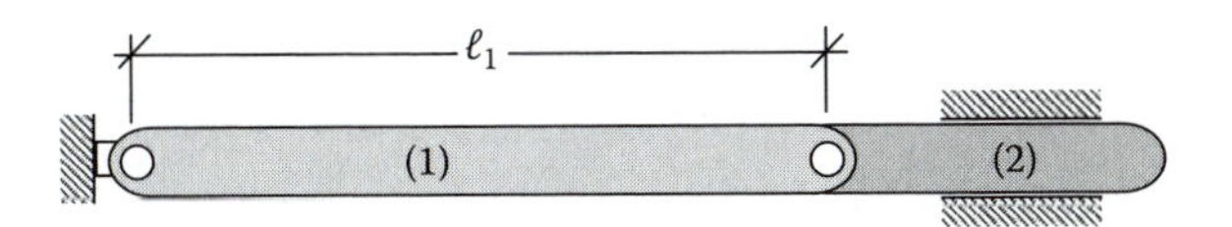

14.17 Select a mechanism and consider one of its links as deformable. For example, the coupler link of a four-bar or the connecting rod of a slider-crank can be considered as deformable. Compare the response of the rigid body model against that of its corresponding rigid-deformable body model under the same loading and initial conditions.

15 Projects

This chapter provides several examples of planar multibody systems that can be set up as simulation projects. These examples cover a variety of applications of multibody systems, and it is hoped that they help broaden the view of the reader in seeking further applications for multibody dynamics.

The projects are not listed in any particular order. For each project a short description is provided. The necessary data for modeling and simulation are listed for most examples. Some of the listed data, however, are not realistic—some have been scaled or changed completely to simplify numerical values, and some are quite arbitrary. Some examples require the reader to search and determine some missing data as it is the case in most engineering practices.

Many of the suggested multibody representations have room for improvement. The ideas and strategies that are presented for modeling and simulating certain phenomena are not unique. The reader is encouraged to be creative in developing other ideas and experimenting with different strategies. In some projects the suggested scenarios for simulation should be very clear, and in some examples the scenarios for simulation may be vague. Here again, it is left to the reader to use his/her imagination and engineering intuition to come up with interesting and useful scenarios. Sometimes a scenario which could be interesting for simulation may not necessarily be realistic, whereas in other cases, a realistic scenario may not be an exciting one to undertake. In either case, by experimentation with different scenarios, we learn more about the strength and the limitations of different methodologies and formulations.

15.1 WINDSHIELD WIPER MECHANISM

Applications of multibody dynamics in automotives have been demonstrated throughout this textbook with several examples of suspension systems. Many other automotive related examples can be found for multibody modeling and analysis. The objective of the project in this section is to analyze a windshield wiper system and to determine the power requirement for its motor-gearbox unit. As shown in Figure 15.1, the system is a six-bar mechanism where the wiper blades are attached to the two rockers. The crank is assumed to rotate one revolution per second. The lengths of the links and the coordinates of the attachment points to the frame are provided in Table 15.1. Because all of the links are slender, their inertial data are negligible. We may assume a mass density of 0.3 per unit length.

The main resistive force in this system comes from the friction between the wiper blades and the windshield. Each wiper blade is pressed against the windshield by a spring as shown in the schematic side view of Figure 15.2. We assume that each spring provides a force of ${}^{(w)}f = 10$, perpendicular to the surface of the windshield,

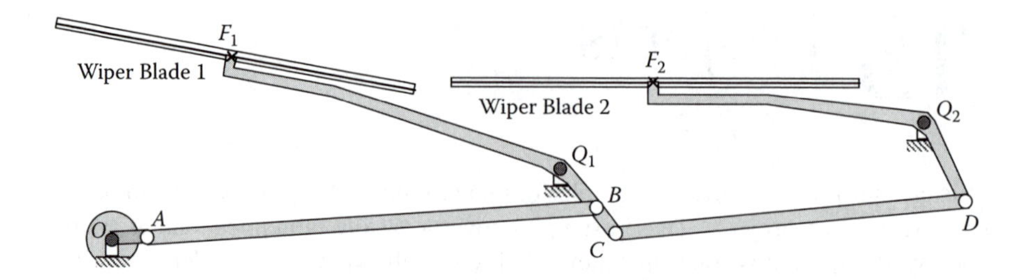

FIGURE 15.1 Schematic representation of a windshield wiper mechanism.

at F_1 and F_2. We can use the following formula to compute the friction force for each wiper arm:

$$^{(fric)}f = \frac{\omega_{arm}}{\omega_{mot}} {}^{(Cf)}d_c \, {}^{(w)}f$$

In this formula $^{(Cf)}d_c$ is the coefficient of Coulomb friction, ω_{arm} is the angular velocity of a wiper arm, and ω_{mot} is the angular velocity of the motor-gearbox combination

TABLE 15.1
Lengths, Coordinates, and Angles (SI Units)

Dimensions	**Link**	**Length**
	OA	0.052
	AB	0.648
	BC	0.041
	CQ_1	0.115
	CD	0.509
	DQ_2	0.125
	Q_1F_1	0.503
	Q_2F_2	0.390
	blade 1	0.533 (21")
	blade 2	0.584 (23")
Coordinates	**Point**	**x - y**
	O	$\{0 \quad 0\}'$
	Q_1	$\{0.649 \quad 0.103\}'$
	Q_2	$\{1.173 \quad 0.162\}'$
Angle	**Links**	**Degrees**
	CQ_1_Q_1F_1	148
	DQ_2_Q_2F_2	127
	Q_1F_1_blade 1	8
	Q_2F_2_blade 2	8
Coefficients	**Coulomb**	
	1.0–4.0	

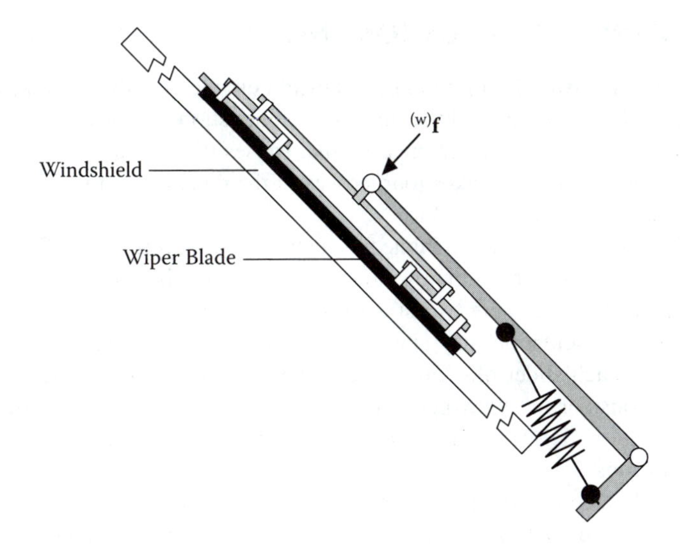

FIGURE 15.2 Side view of a wiper blade pressing against the windshield.

(2π rad/sec in our case). The higher values of the friction coefficient provided in the table can be used for the worst-case scenario when the wipers are turned on in dry conditions. The two computed friction forces must be applied to F_1 and F_2 in the opposite direction of their respective velocities. In this friction model we have made each friction force to be a function of the velocity of the corresponding wiper as well as the Coulomb friction and the normal load.*

An inverse dynamic analysis of this model will provide the torque of the motor-gearbox combination. Plotting the computed torque versus time, or versus the crank angle, for one complete revolution of the crank will reveal that the required torque is not a constant.

---

* A simple formula for computing the power requirement for windshield wiper motors can be found in: *BOSCH Automotive Handbook, 6th ed.*, Robert Bosch Gmbh, 2005. The formula is stated here (without changes in notations) as:

$$M_{An} = F_{WfN}\mu_{\max} f_S f_T L_A(\omega_{H\max} / \omega_{Mot})(1/n_{gear})(R_{AW}/R_{Ak})$$

The parameters are defined as:

F_{WfN} Downward nominal load of the wiper arm (same as our ${}^{(w)}f$)
$\mu_{\max}$ Maximum dry coefficient of friction of rubber (same as our ${}^{(cf)}d_c$)
f_S Multiplier to account for wiper-arm joint friction (usually 1.15)
f_T Tolerance factor to account for wiper-arm load tolerance (usually 1.12)
L_A Wiper arm length
$\omega_{H\max}$ Maximum angular velocity of the wiper arm
ω_{Mot} Mean angular velocity of the wiper motor crank
n_{gear} Efficiency of the gear unit (usually 0.8)
R_{AW} Electrical resistance of the rotor winding heated by nominal operation
R_{Ak} Electrical resistance of the cold rotor winding

15.2 INTERNAL COMBUSTION ENGINE

The basic mechanism of an internal combustion engine is a slider-crank where the slider block (the piston) is the input link. The combustion of the air and fuel mixture in the cylinder generates a large pressure that causes the piston to move. In an Otto four-stroke cycle engine, it takes four strokes of the piston (i.e., two full revolutions of the crank) to complete one Otto cycle.

In this exercise we model an inline four-cylinder Otto cycle engine. The schematic representation of a planar multibody model for this system is shown in Figure 15.3(a). This representation contains two slider-crank mechanisms sharing the crank link. Although this presentation may appear to be that of a two-cylinder engine, in this planar model, each slider block represents two pistons. The reason should become clear if we consider the view perpendicular to the crankshaft, as shown in Figure 15.3(b). In the model, slider 1 could represent pistons *1* and *4* and slider 2 could represent pistons *2* and *3*.

A typical gas pressure curve for two revolutions of the crank angle (one cycle) is shown in Figure 15.4. Such a curve can be discretized and accurately represented by a spline (or some other functions), or approximated by several straight lines as shown. The force associated with the pressure must be applied on pistons *1*, *3*, *4*, *2* in 180° phase lags. In our multibody model, this means that the force must be applied to each slider twice in one cycle with a 360° phase lag.

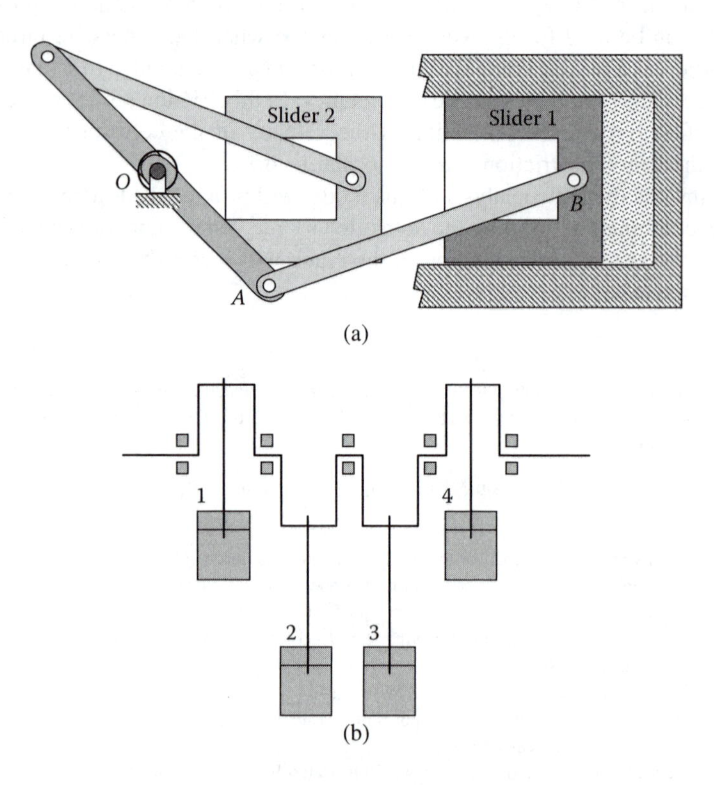

FIGURE 15.3 Schematic representation of a two-cylinder engine.

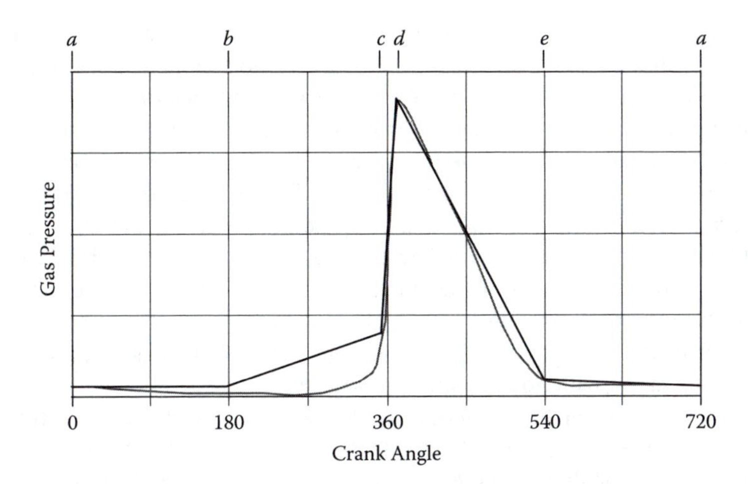

FIGURE 15.4 Gas pressure curve versus crank angle.

A set of data for the multibody model is provided in Table 15.2. If we use a realistic set of data, because we deal with large forces and high accelerations, we may encounter numerical integration problems in a forward dynamic analysis. This would require tightening the integration error tolerance or switching to a more suitable integration algorithm.

In forward dynamic simulations, we can consider Coulomb friction between the pistons and the cylinders. Regardless of the load on the engine, there is a constant reaction force (given as 100 N in the table) that acts on each piston normal to the slider axis. This force must be added to the reaction force generated by the motion (from the Lagrange multipliers) and then multiplied by the coefficient of friction to determine the friction force.

TABLE 15.2
Inertia, Length, Force, and Friction Data (SI Units)

Inertia/Length	Body	Mass	Moment of inertia	Length
	Crank *(OA)*	1.0	—	0.10
	Connecting rod	0.5	—	0.20
	Slider	0.5	—	
	Load	—	10	
Gas force	**Angle (degrees)**	**Force**		
	$a = 0$	0.3×10^4		
	$b = 180$	0.3×10^4		
	$c = 358$	1.6×10^4		
	$d = 362$	7.3×10^4		
	$e = 540$	0.5×10^4		
Friction	**Normal force**	**Coulomb**		
	100	0.4		

We can also consider a load on the crank by assuming that the engine powers a car on which an aerodynamic drag force acts. If we assume that for a gear ratio of one to one, every rotation of the crank results into one rotation of the wheel, then the speed of the car can be computed as $v = R\omega$. Here R is the radius of the wheel and ω is the angular velocity of the crank. The drag force can be calculated as ${}^{(drag)}f = \frac{1}{2} d_{drag} \rho A v^2$ where d_{drag} is the drag coefficient, ρ is the air density, and A is the cross-sectional area of the car's body perpendicular to the direction of its velocity. The resistive moment caused by the drag force is then included in the model.*

If the moment of inertia of the crank is assumed to be small, we may start a forward dynamic analysis with zero initial velocities. However, if we increase this moment of inertia by including the moment of inertia of the load, we *may* need to start the simulation with nonzero initial velocities.

Another refinement to this model could be to represent the force (pressure) versus angle curve more accurately than by several straight lines. We should be able to find piece-wise analytical expressions to represent the pressure curve.

This model could be setup with either of the three formulation methods. The joint-coordinate formulation should provide a much faster computation than the body-coordinate formulation. With the point coordinate formulation, the body inertia could be distributed to the primary points with either the approximation or the exact method. In textbooks on mechanisms and machines, we can find classical approximation formulas for analyzing the kinematics and dynamics of internal combustion engines that are similar to our point-coordinate representation.

15.3 SLED TEST AND BELTED DUMMY

Laboratory and field crash tests are performed in a variety of forms to gain better understanding of the dynamics of a crash to design and manufacture safer vehicles. Valuable data are collected from crash tests using crash dummies. In this exercise we attempt to set up a multibody model of a crash dummy and to simulate a sled crash test.** A multibody representation of the dummy and the seat/sled are shown in Figure 15.5. The simulation scenario would allow the belted dummy and the seat to move forward with a specific speed and then to come to a stop in a short duration of time.

Geometric data for individual bodies of the model are shown in Figure 15.6. It is assumed that the bodies that form the dummy are connected by pin joints. The neck is not considered as a body—it is represented as a pin joint and a torsional spring-damper.

In the model, a sliding joint must be provided between the seat and the ground. The steering column is rigidly attached to the seat, and dummy's hand is connected to the steering wheel by a pin or a bracket joint. The inertial and spring-damper data are provided in Table 15.3.

---

* Reasonable values for the parameters are: $R = 0.25$, $d_{drag} = 0.5$, $\rho = 1.2$, $A = 2$.

** The concept for this model and some of the data have been adopted from the website: www.ftss.com/pcat/fem (First Technology Safety Systems, Plymouth, MI, 48170 USA)

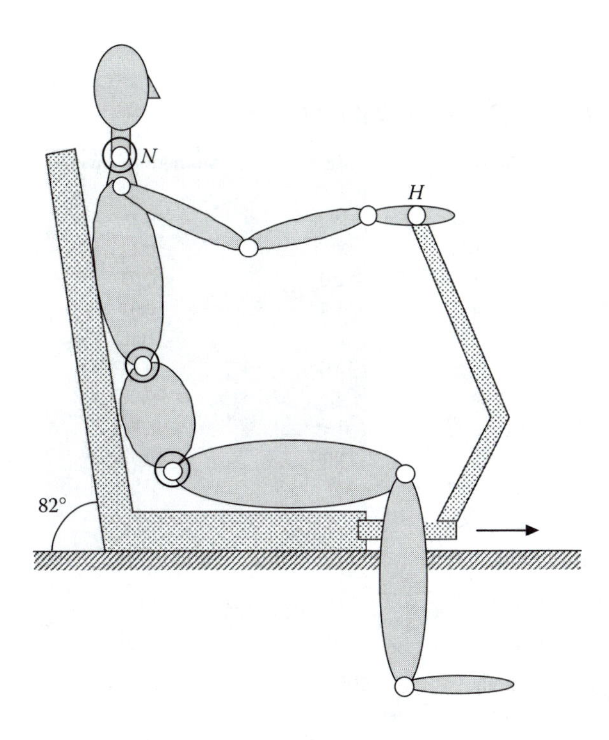

FIGURE 15.5 A multibody representation of a belted dummy.

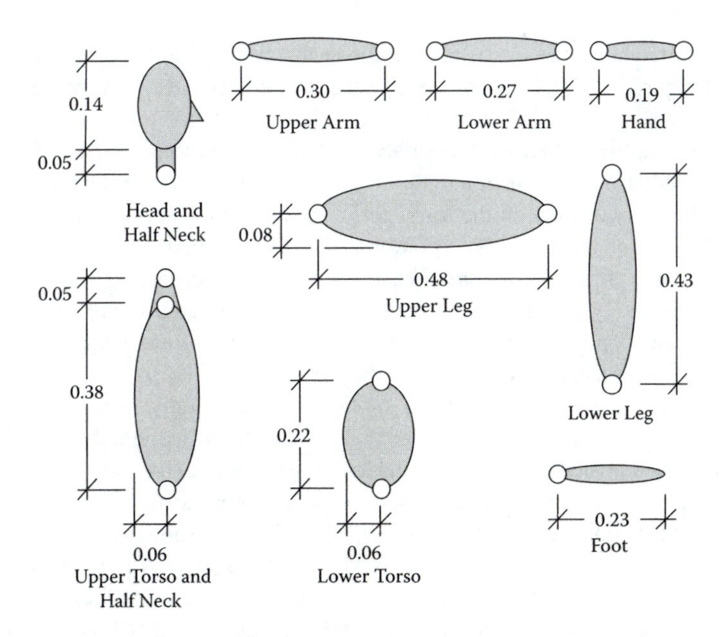

FIGURE 15.6 Individual bodies and attachment pin joints for the dummy.

TABLE 15.3
Inertia and Spring-Damper Data (SI Units)

	Body	Mass	Moment of inertia	
Inertia	**Body**	**Mass**	**Moment of inertia**	
	Head	5.25	0.04	
	Upper torso	18.0	0.35	
	Lower torso	23.0	0.13	
	Upper arm	2.0	0.02	
	Lower arm	1.7	0.01	
	Hand	0.6	0.002	
	Upper leg	6.0	0.13	
	Lower leg	4.3	0.07	
	Foot	1.4	0.007	
Dimensions	**Body**	**Point**	**ψ-coordinate**	**η-coordinate**
	Upper torso	B	0.09	0.09
	Upper torso	C	0.09	–0.09
	Seat	Q^B	–0.15	0.85
	Seat	Q^C	–0.15	–0.15
	Seat	H	0.60	0.60
Spring-damper	**Body**	**Stiffness**	**Damping**	**${}^0\ell$ or ${}^0\theta$**
	Neck (N)	34.0	1.5	
	Seatback	$k_{seat} = 50{,}000$	$d_{c,seat} = 150$	
	Seat belt	$k_{belt} = 60{,}000$	$d_{c,belt} = 200$	${}^0\ell^B = 0.35$ ${}^0\ell^C = 0.60$

The upper and lower torsos rest on the seatback. Unilateral spring-dampers, as shown in Figure 15.7(a), can provide the necessary compliance. To model the resting of the upper and lower torsos on the seatback, the distance of the mass center of each torso from the surface of the seatback should be monitored. When this distance becomes smaller than a known length, ${}^0\ell_{seat}$, the unilateral spring-damper should be activated to apply a force on the torso perpendicular to the surface of the seatback. Although it is not too difficult to compute the distance between each mass center and the seatback, the deformation of each spring-damper can be approximated, with acceptable accuracy, using a simple technique. We can assume a *long* spring-damper element that is connected between a mass center and an imaginary point fixed to the seat at some distance, as shown in Figure 15.7(b). The force of this element, regardless of any slight up or down motion of the dummy, remains practically perpendicular to the surface of the seatback. The undeformed length of such a long spring must be adjusted depending on the location of the attachment point on the seat.

The seat belt can be included in the model as another form of unilateral spring-damper element connecting points Q^B, B, C, and Q^C as shown in Figure 15.8(a). Points Q^B and Q^C are attached to and move with the seat, and points B and C are defined on the upper torso. We assume that a spring-damper is attached between Q^B and B, and another one is attached between C and Q^C. There is no need to consider a spring-damper between B and C because both points are on the same

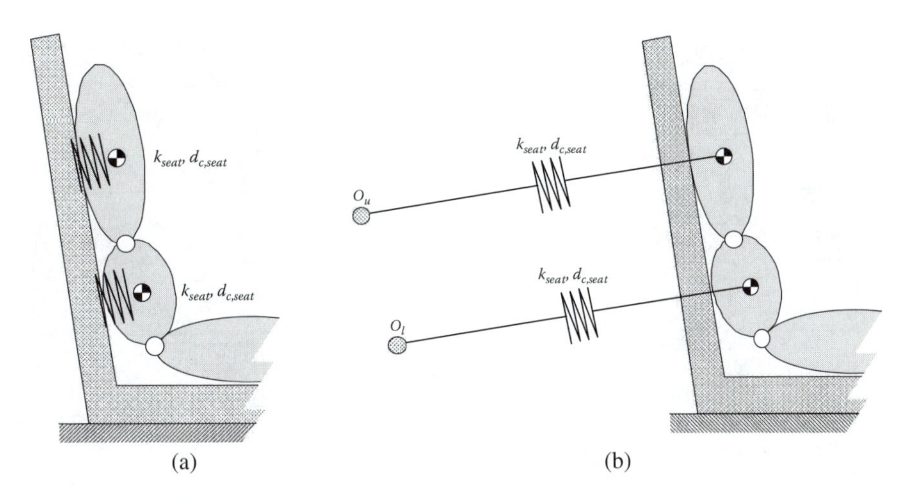

FIGURE 15.7 Modeling of the compliance of the seatback.

body and, therefore, the distance between them will not change. We monitor the lengths ℓ^B and ℓ^C and compare their sum against the corresponding undeformed sum. If $\ell^B + \ell^C > {}^0\ell^B + {}^0\ell^C$, then the spring-damper element should be activated as

$$f_{belt} = k_{belt}(\ell^B + \ell^C - {}^0\ell^B + {}^0\ell^C) + d_{c,belt}(\dot{\ell}^B + \dot{\ell}^C)$$

This force must be applied to points B and C as shown in Figure 15.8(b).

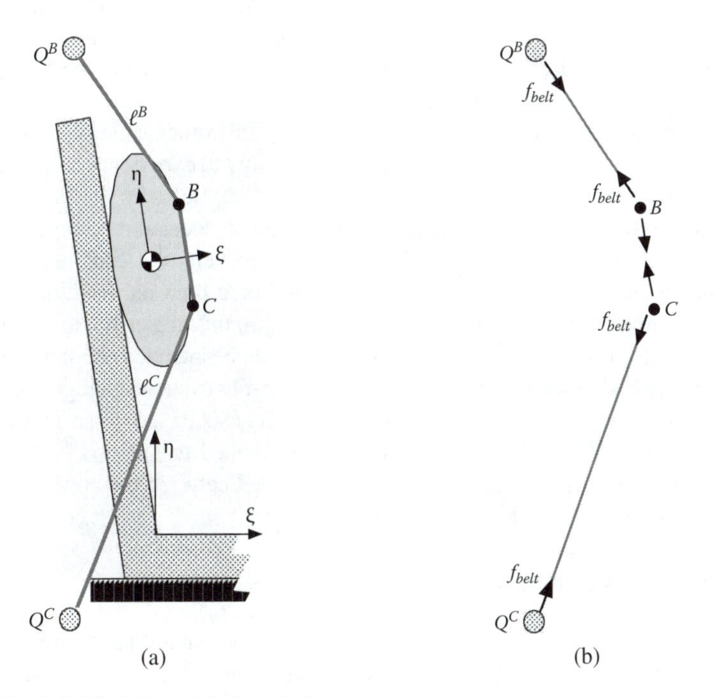

FIGURE 15.8 Modeling of the seat belt.

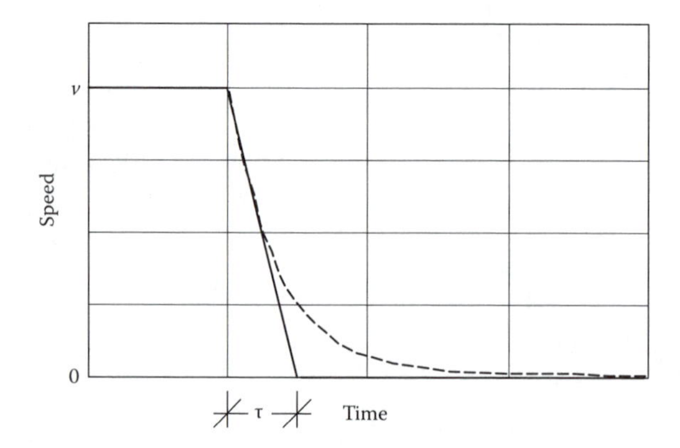

FIGURE 15.9 A predefined velocity function for the seat.

For simulating different crash scenarios, we can consider either a simple model or a full model. In the simple model, we do not include the lower leg and the foot. We rigidly fix the upper leg to the seat. In the full model, we allow the upper leg to slide relative to the seat (using a sliding joint). We can also include the lower leg and the foot.

In a forward dynamic analysis, we can specify the velocity of the seat to follow a predefined function (a driver constraint) representing a sudden stop/crash. All the bodies in the model must have the same initial velocity v. Then during a period of τ seconds, the velocity of the seat must drop to zero, as shown in solid lines in Figure 15.9. We can also cause the velocity to drop more smoothly (e.g., according to the curve shown in dashed lines in Figure 15.9). Such a curve can be constructed as $\dot{x}_{seat} = ve^{-t/\tau}$. By adjusting the values of v and τ, we can make the dummy to experience any desired deceleration. For example, an initial velocity of 10 m/sec and a *time constant* of $\tau = 0.1$ seconds causes the dummy to experience a deceleration of roughly $10g$.

We can perform several simulations with different deceleration functions. We can experiment with more realistic data for the seat belt. We can include many refinements in the model. For example, there should be a limit on the angle between the upper and lower arms. As shown in Figure 15.10(a), this angle should not become negative. For this purpose, a torsional spring-damper can be included about the pin joint between the two bodies with nonlinear spring characteristics as shown in Figure 15.10(b). A function such as ${}^{(s)}n = be^{-\theta/a}$ ($b = 100$ and $a = 5\pi/180$; i.e., 5°) can produce this nonlinear behavior. The computed torque must be applied to the two bodies in the opposite directions. Similar refinements can be included between the upper and lower torsos, the lower torso and the upper leg, and so forth.

15.4 HEAD AND NECK

In the past several decades, numerous analytical and computational models for head and neck have been developed to study whiplash or other accident related trauma. Most of these models are based on the finite-element method and a few are based

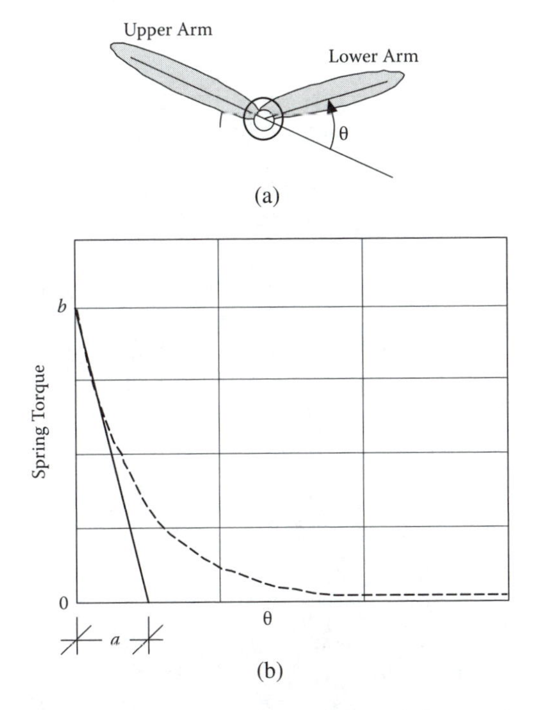

FIGURE 15.10 Torsional spring with nonlinear characteristics can be used to limit the relative rotation between two bodies.

on multibody dynamics. In this exercise a planar multibody model for the human head-neck is presented to examine the dynamics of such system under the action of arbitrary forcing conditions.*

The proposed multibody model, as shown in Figure 15.11, is composed of ten bodies (torso, seven vertebrae-disc combinations, skull, and brain), eight pin joints, one sliding joint, nine torsional and four point-to-point spring-dampers. The torso is allowed to translate horizontally with respect to the ground (sliding joint). Seven vertebrae-disc combinations attach the torso to the skull by pin joints. Torsional spring-damper elements are placed about each pin joint. The body representing the brain is connected to the skull body by spring-damper elements. Proper attachment points on the skull must be chosen to place four point-to-point spring-damper elements as shown. A torsional spring-damper element must be defined between these two bodies as well. These five force elements roughly represent the compliance between the brain and the skull. The data for the model are provided in Table 15.4.

In a forward dynamic simulation of this model, the torso could be subjected to either an external applied load or a predefined motion (a driver function) such

* This model has been adopted from:
Tsai, D-L. and Arabyan, A., "A Multibody Dynamic Model of the Human Head and Neck," Technical Report No. CAEL-91-1, Department of Aerospace and Mechanical Engineering, University of Arizona, January 1991.

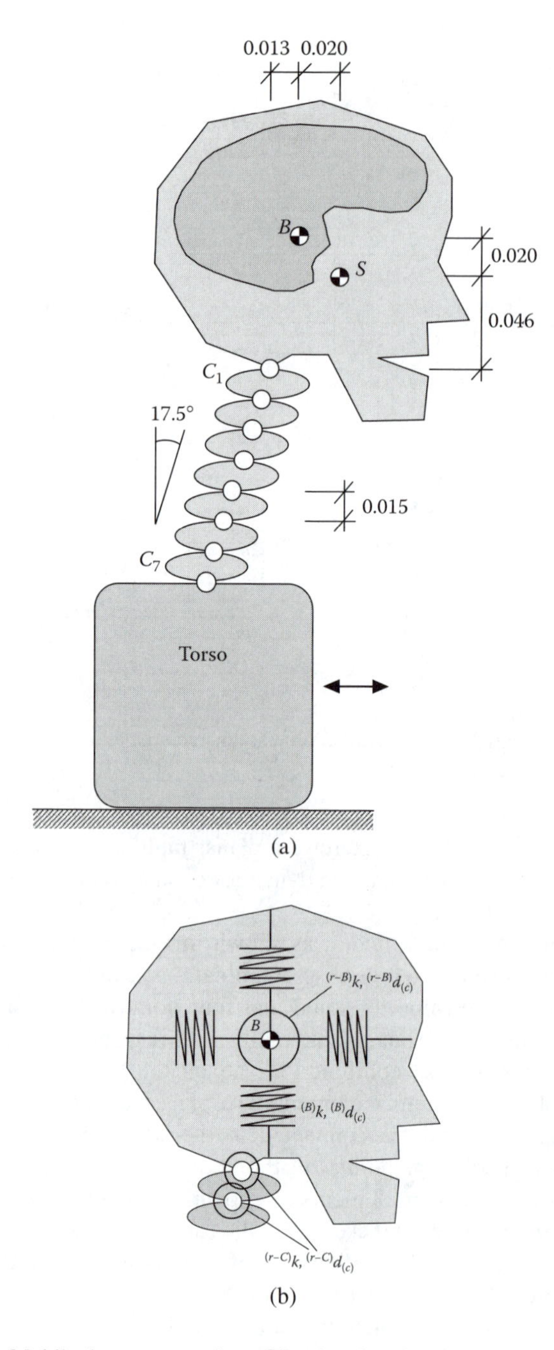

FIGURE 15.11 Multibody representation of head and neck.

TABLE 15.4
Inertia and Spring-Damper Data (SI Units)

Inertia	Body	Mass	Moment of inertia	
	skull	2.0	0.03	
	brain	1.4	0.004	
	One vertebra	0.25	0.0004	
	torso	50.0	1.2	
Spring-damper	**Bodies**	**Stiffness**	**Damping**	$^0\ell$ **or** $^0\theta$
	brain-skull	$^{(B)}k = 8{,}750$	$^{(B)}d_{(c)} = 90$	—
		$^{(r-B)}k = 23$	$^{(r-B)}d_{(c)} = 0.05$	
	skull-C_1	$^{(r-C)}k = 271$	$^{(r-C)}d_{(c)} = 0.22$	—
	two vertebrae	$^{(r-C)}k = 271$	$^{(r-C)}d_{(c)} = 0.11$	—
	C_7-torso	$^{(r-C)}k = 271$	$^{(r-C)}d_{(c)} = 0.11$	—

as a shaking function $x_{torso} = {}^0x + a\cos(\omega t)$. We can experiment with different values for the parameters a and ω. Range of values around $a = 0.1$ and $\omega = 2$ Hz are suggested. If the model data are adjusted to represent a young child, a refined version of the model could be used to study a medical trauma known as the shaken-baby syndrome.

15.5 MOUNTAIN BIKE

In this section we provide data to construct multibody models for two mountain-bikes. The two models are very similar except for the rear suspension mechanism. Schematic planar views of both bikes are shown in Figure 15.12. In either model, the frame can be connected to the front wheel by a revolute-translational joint, where the axis of the sliding joint is positioned on the frame, and the revolute joint is positioned at the center on the wheel. The front suspension spring-damper can be placed between the wheel center and a point on the frame along the axis of the sliding joint. Individual components for these models are shown in Figure 15.13. The necessary data for both models are provided in Table 15.5.

We can consider various scenarios for forward dynamic simulations with each model. One scenario could be to drop the bike from a short height and observe its response. The tire model for these simulations could be a simple unilateral spring-damper in the radial direction. Another simulation scenario could be for a bike to travel over an obstacle. For these simulations we need more sophisticated tire models.

In a simple model we ignore the tire compliance completely and consider a wheel/tire as a rolling rigid disc. For such a model we need to define appropriate constraint to keep the disc in contact with the ground and also to use the no-slip constraint from Eq. (7.49).

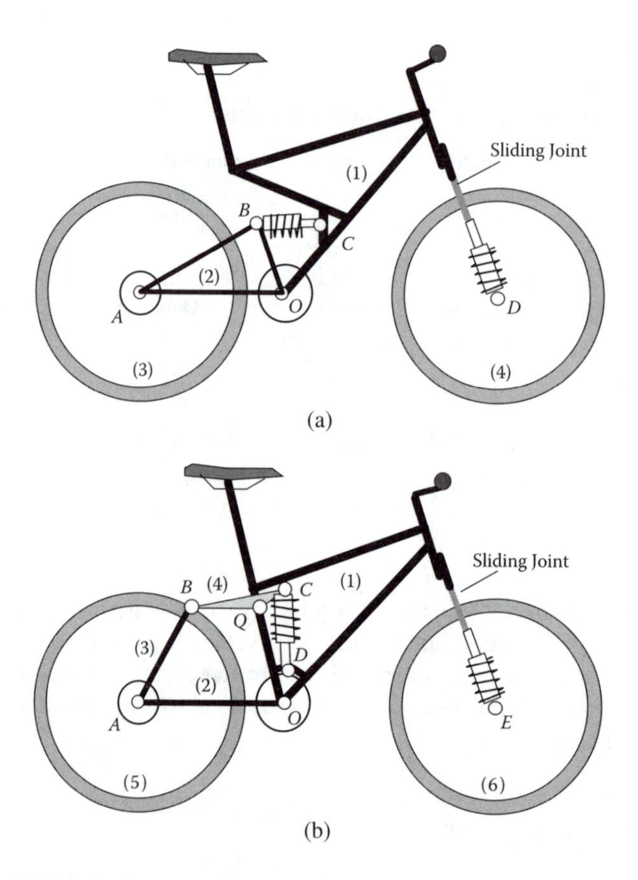

FIGURE 15.12 Multibody representation of two mountain bikes.

In a more realistic model, we consider a spring-damper in the radial direction and also a spring element in the longitudinal direction as shown in Figure 15.14(a).* The radial spring-damper force can easily be determined based on the position of the wheel center relative to the ground, the undeformed radius, and the radial tire characteristics. The longitudinal force that acts on the tire at its contact point with the ground can be determined from a force-slip curve as shown schematically in Figure 15.14(b). The force-slip characteristics are determined experimentally by tire manufacturers. The longitudinal slip, or simply *slip* in planar motion, is defined as

$$s = \begin{cases} \dfrac{v_x - v_c}{v_x} > 0 & : \text{in braking} \\ \dfrac{v_x - v_c}{v_c} < 0 & : \text{in traction} \end{cases}$$

where v_x is the magnitude of the forward velocity of the center of the wheel ($v_x = \dot{x}$ if the wheel is moving parallel to the x-axis), and $v_c = R\omega$ is the circumferential

* This type of tire model is usually developed for handling simulation of automobiles.

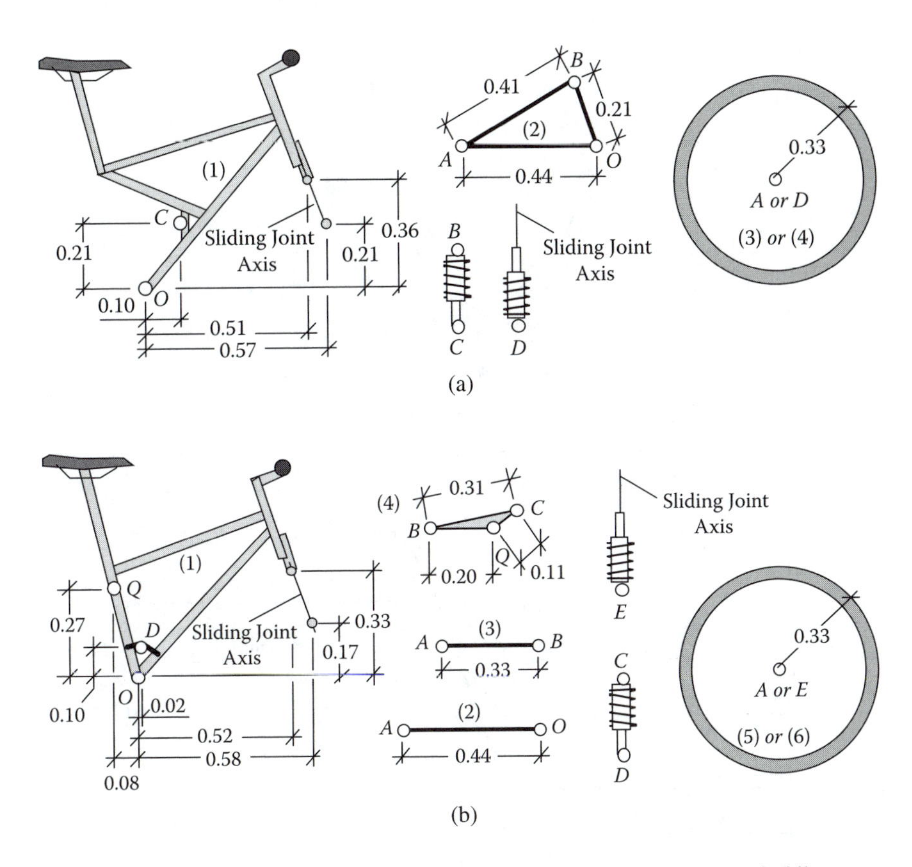

FIGURE 15.13 Individual components of multibody models for two mountain bikes.

TABLE 15.5
Inertia, Suspension, and Tire Properties (SI Units)

Inertia	**Body (Index)**	**Mass**	**Moment of Inertia**	
	(1)	3.2	—	
	bike (a): (2)	1.8	—	
	bike (b): (2)	0.5	—	
	bike (b): (3)	0.5	—	
	bike (b): (4)	0.8	—	
	each wheel	1.5	—	
Suspension	**Free Length**	**Stiffness**	**Damping**	
(front)	—	10,000	1,350	
(rear)	—	120,000	5,300	
Tire	Undeformed radius	Deformed radius at rest	Stiffness (radial)	Damping (radial)
	0.33	—	60,000	2,700

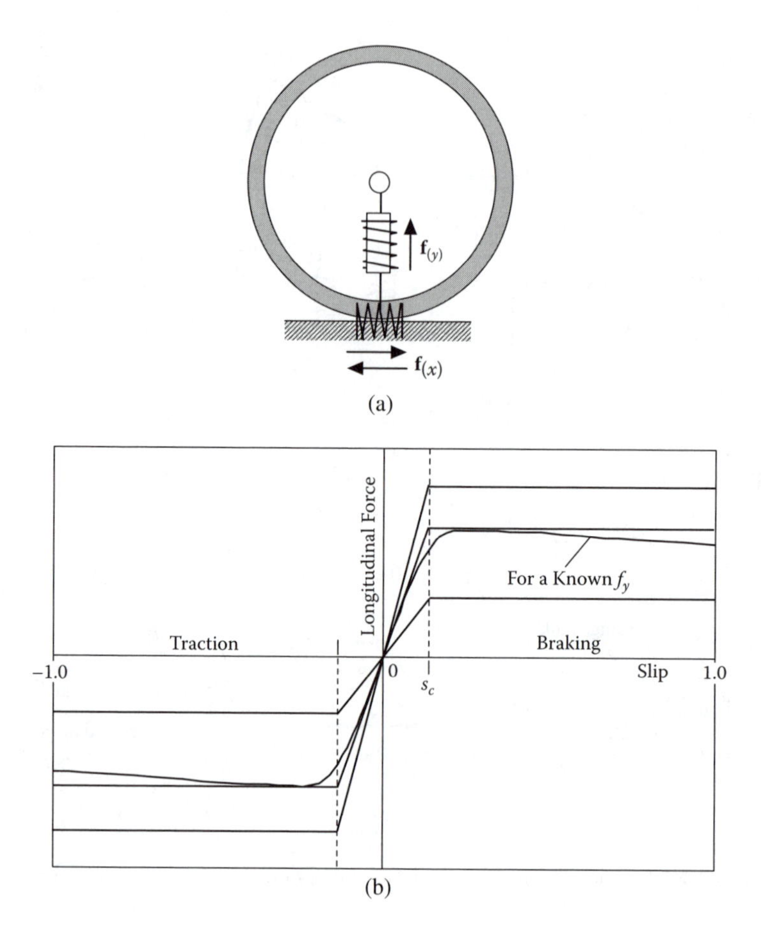

FIGURE 15.14 (a) Spring-damper elements representing radial and longitudinal tire compliance, and (b) a simplified longitudinal force-slip characteristics.

velocity of the wheel (R is the effective or deformed radius of the wheel and $\omega = \dot{\phi}$ is the angular velocity of the wheel). A typical force-slip characteristic for braking is shown as a curve in Figure 15.14(b). The associated curve for traction is symmetric with respect to the intersect of the two axes. We can use straight lines to approximate a curve: one line tangent to the curve at zero slip and one parallel to the s-axis as shown. Each curve is determined experimentally for a known load $\mathbf{f}_{(y)}$ acting on the wheel. For different values of the applied load $\mathbf{f}_{(y)}$, different curves, and consequently different straight-line approximation, can be found as depicted in the figure. The switching point from a tangent line to a constant force line occurs at a critical slip s_c regardless of the applied load $\mathbf{f}_{(y)}$. The longitudinal force can be defined in terms of two approximated straight lines as

$$f_x = \begin{cases} \dfrac{{}^{(Cf)}d_c f_y}{s_c} s & : s < s_c \\ {}^{(Cf)}d_c f_y & : s \geq s_c \end{cases}$$

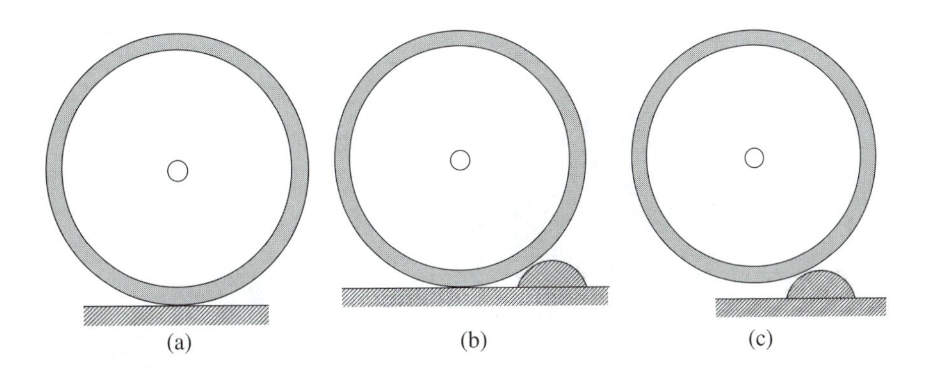

FIGURE 15.15 A wheel going over an obstacle.

This formula is stated for braking. For traction, the signs for slip and critical slip must be adjusted accordingly. The Coulomb coefficient of friction can be selected in the range of $0.9 < {}^{(Cf)}d_c < 1.3$. A value of ${}^{(Cf)}d_c = 1.0$ is acceptable for our simulation purpose. The critical slip can be set to $s_c = 0.1$ for most tires. The longitudinal force $\mathbf{f}_{(x)}$ acts on a wheel at its contact point with the ground.

Whether we model a wheel/tire as a rigid disc or model it more realistically, we must consider different possible contact situations which may arise in a simulation. Several contact scenarios of a wheel rolling over a bump are shown in Figure 15.15. In some situations the wheel may have more than one contact points.

15.6 MOTORCYCLE

This section provides data to construct a multibody model of a motorcycle to perform forward dynamic analyses. As shown in Figure 15.16, the front wheel suspension system is a four-bar mechanism in which two trailing links, attached to the frame, enable the wheel fork to move. The front suspension spring-damper is attached between the lower of the two trailing links and the frame. The rear suspension

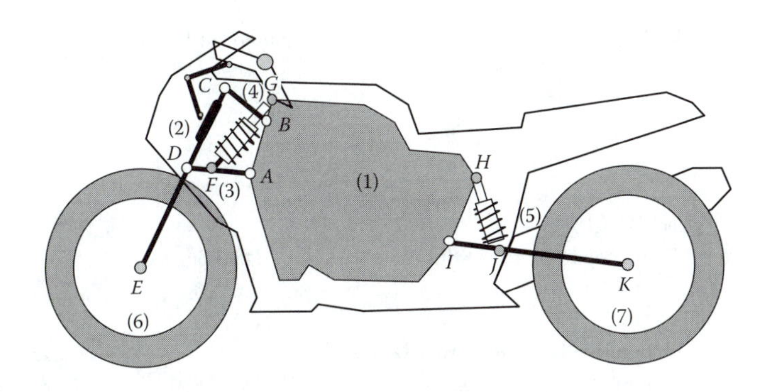

FIGURE 15.16 A multibody representation of a motorcycle.

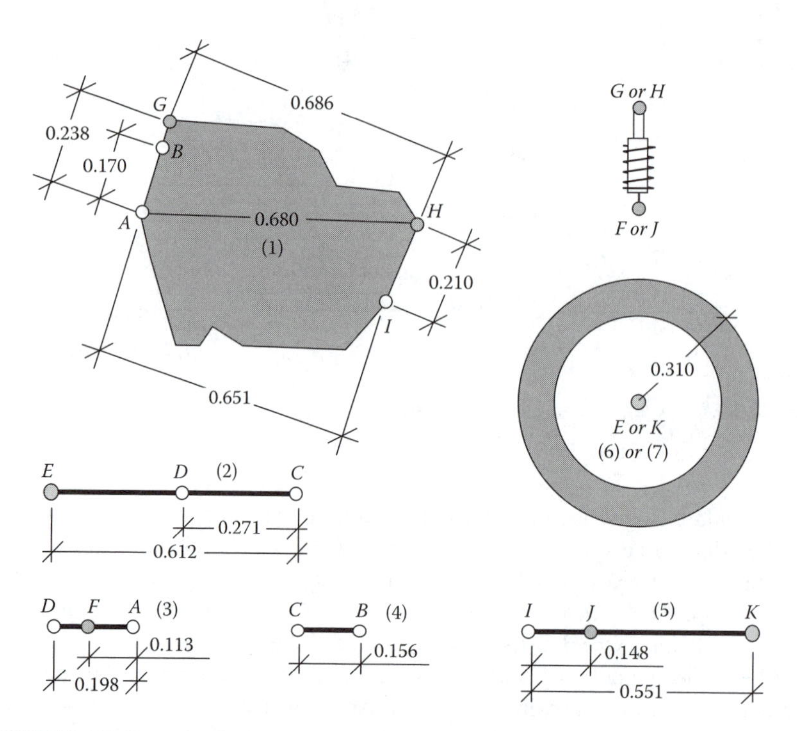

FIGURE 15.17 Components for a multibody model of the motorcycle.

spring-damper is attached between a one-arm rocker and the frame. Individual components of the model with the necessary dimensions are shown in Figure 15.17. The inertia, suspension, and tire data are provided in Table 15.6.

In the multibody model, we need to decide on the location of the mass center of the frame depending on whether we include a rider in the model or not. For

TABLE 15.6
Inertia, Suspension, and Tire Properties (SI Units)

Inertia	**Body (Index)**	**Mass**	**Moment of Inertia**	
	(1)	230	—	
	(2)	10	—	
	(3)	2	—	
	(4)	2	—	
	(5)	8	—	
	each wheel	10	—	
Suspension	**Free Length**	**Stiffness**	**Damping**	
(front)	0.29	75,000	6,700	
(rear)	0.25	120,000	8,500	
Tire	**Undeformed Radius**	**Deformed Radius at Rest**	**Stiffness (Radial)**	**Damping (Radial)**
	0.31	—	175,000	11,000

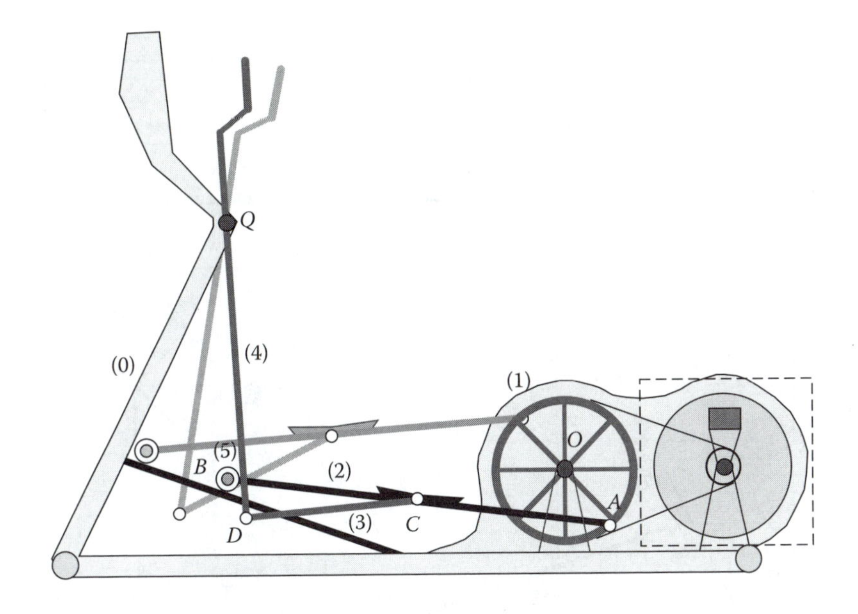

FIGURE 15.18 A multibody representation of an exercise machine.

modeling the tire-ground interaction and possible ride scenarios, we can refer to the previous section on mountain bike models.

15.7 ELLIPTICAL EXERCISE MACHINE

One of the more popular fitness equipment is the elliptical exercise machine. The machine is mainly comprised of two slider-crank mechanisms, 180° out-of-phase with one another. A schematic multibody representation of such a machine is shown in Figure 15.18. Identical left and right footrests cause a flywheel to turn when the person exercising shifts his/her weight from one leg to the other. Each footrest sits on a bar connected to the flywheel at one end and slides on an inclined surface through a roller. Two additional links provide support for each hand, which may or may not be included in the model. An adjustable resistive torque is applied to the flywheel via a system shown within a dotted box. There is no need to include a multibody representation of this braking system in the model—we can apply the resistive torque directly on the flywheel.

The individual components and the necessary dimensions for the multibody model are shown in Figure 15.19. It is left to the reader to come up with appropriate inertial values for each component. In the multibody model, we can take advantage of composite joints, such as revolute-revolute and revolute-translational, to reduce the size of the problem.

For forward dynamic analysis, the weight of a person could be shifted from one footrest to the other based on a simple strategy as shown in Figure 15.20. We can experiment in different simulations by changing: the moment of inertia of the flywheel;

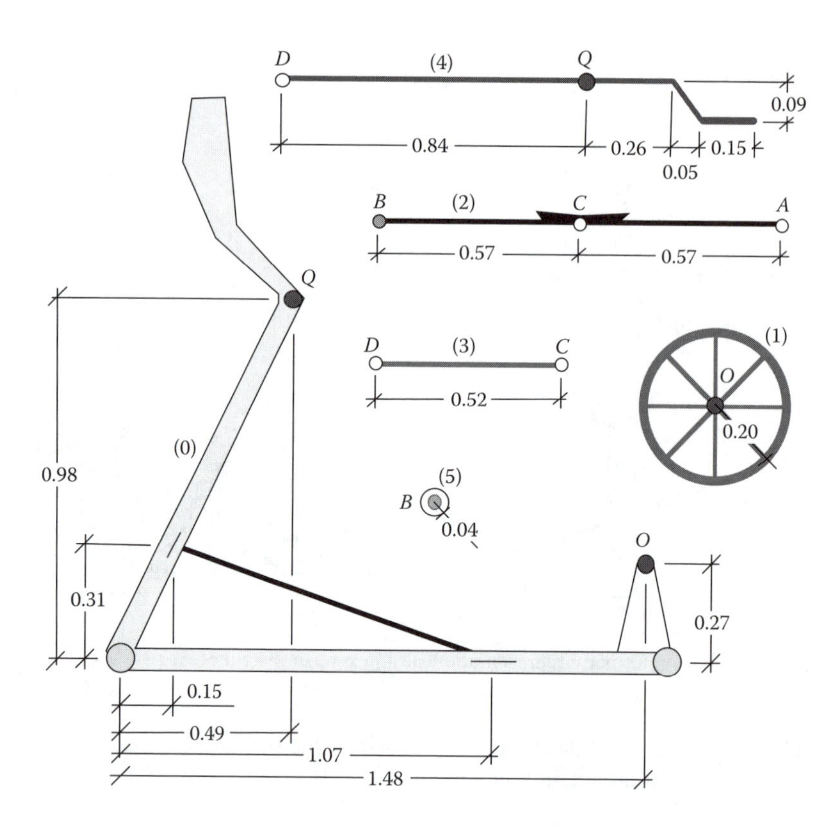

FIGURE 15.19 Components for a multibody model of the exercise machine.

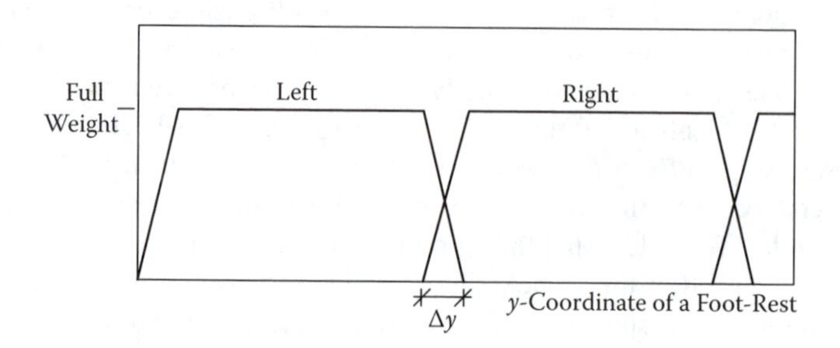

FIGURE 15.20 Shifting of the weight between the two footrests.

the value of the resistive torque; the timing of initiating the shifting of the weight; the period during which the weight is distributed on both feet; and the inclination angle of the slider.

As an extension of this exercise, a simplified model for the exercising person could be considered. For such a model, some ideas could be adopted from the *dummy-sled test* model.

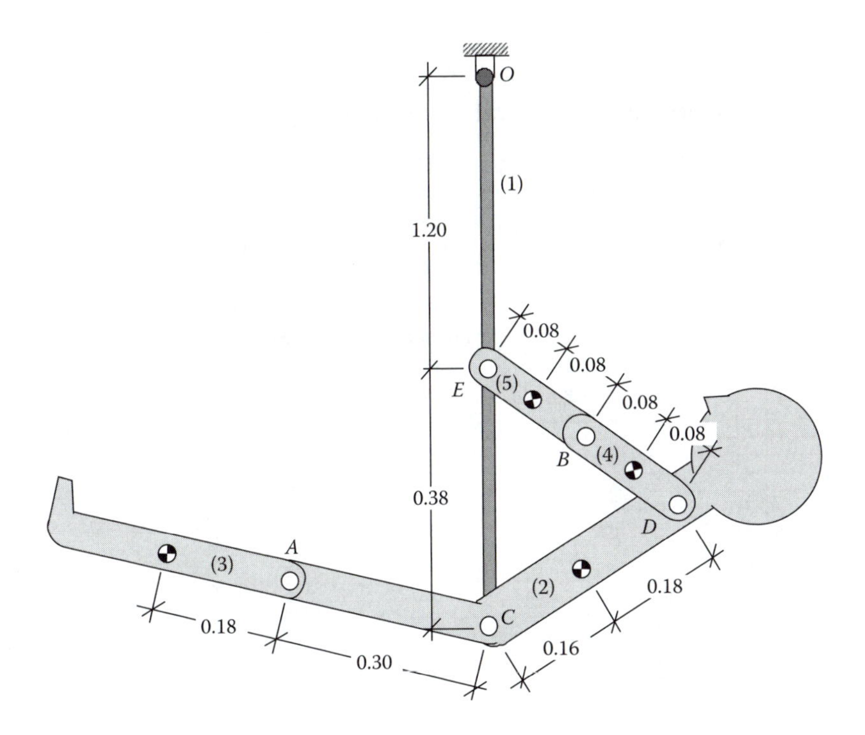

FIGURE 15.21 A multibody representation of a child on a swing.

15.8 SWING

Children are quick to figure out how to swing higher and higher, without being pushed by someone else. They intuitively learn to pump their legs and torsos at a certain points in the swing cycle. This *body-pump* changes the child's overall mass center with respect to the swing and further alters the child's moment of inertia. In this exercise we construct a multibody model of a child on a swing and include a body-pump strategy for forward dynamic analyses.

A simple multibody representation of a child on a swing is shown in Figure 15.21. The model contains five bodies and six pin joints. This represents a system with three degrees of freedom, where one is the swinging DoF and the other two can be used to control the body-pump motion of the child. A set of inertial data for the model is provided in Table 15.7.

When a child swings forward, the legs and the arms are straight to increase the moment of inertia. We will refer to this as the *straight leg-arm* configuration. When the child swings backward, the legs and the arms are tucked in. We will refer to this as the *bent leg-arm* configuration. In Figure 15.21, the child is in the straight leg-arm configuration. In our multibody model, we need to construct two driver functions to control the orientation of the legs and the arms between the two configurations.

We define two angles, θ_A and θ_B, as shown in Figure 15.22(a). These angles could be controlled and varied from a minimum value to a maximum value, and

TABLE 15.7
Inertia and Angle Data (SI Units)

Inertia	Body	Mass	Moment of Inertia
	(1)	3.0	0.50
	(2)	30.0	2.1
	(3)	2.5	0.04
	(4)	1.8	0.007
	(5)	1.4	0.006
Angles	–	**Minimum**	**Maximum**
	A	20°	175°
	B	30°	175°

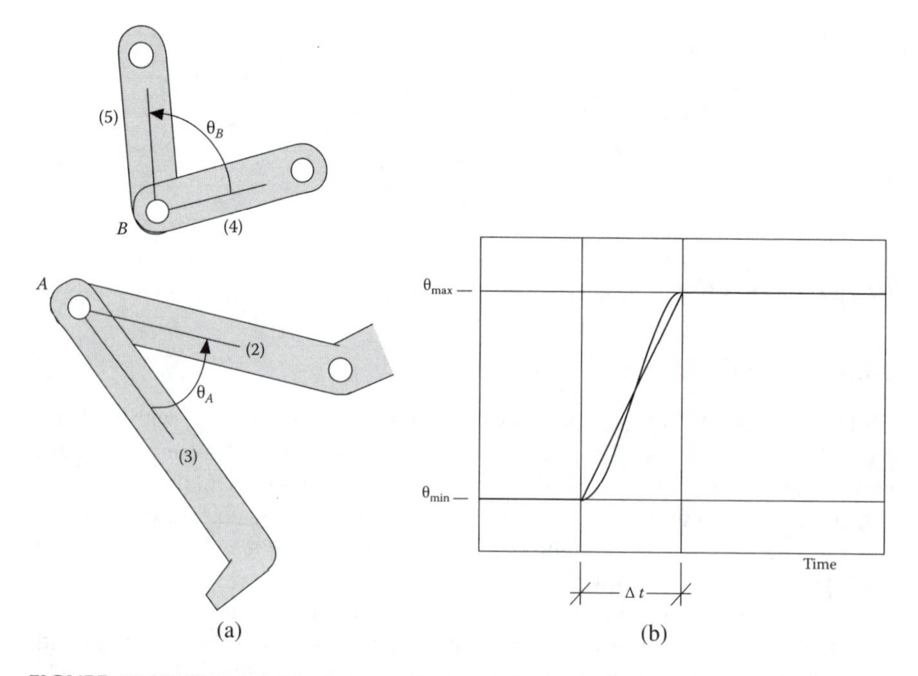

FIGURE 15.22 Changing the relative angle about the elbow or knee joints by driver functions.

vice-versa, according to the plot shown in Figure 15.22(b). The change should take place in a short period of time (e.g., $\Delta t = 0.5$ seconds). The challenging part of this project is to establish the logic to determine when to initiate the change of the angles.

One possible strategy would be to monitor the angular velocity and acceleration of the swing, body (1). When the swing is moving forward, $\dot{\phi}_1 < 0$, and it is when the child should be in the straight leg-arm configuration. When the swing is moving backward, $\dot{\phi}_1 > 0$, and the child should be in the bent leg-arm configuration. When the swing reaches the highest orientation at either end (i.e., when $\dot{\phi}_1 = 0$), it is time to initiate a change. However, the problem is that when the equations of motion are

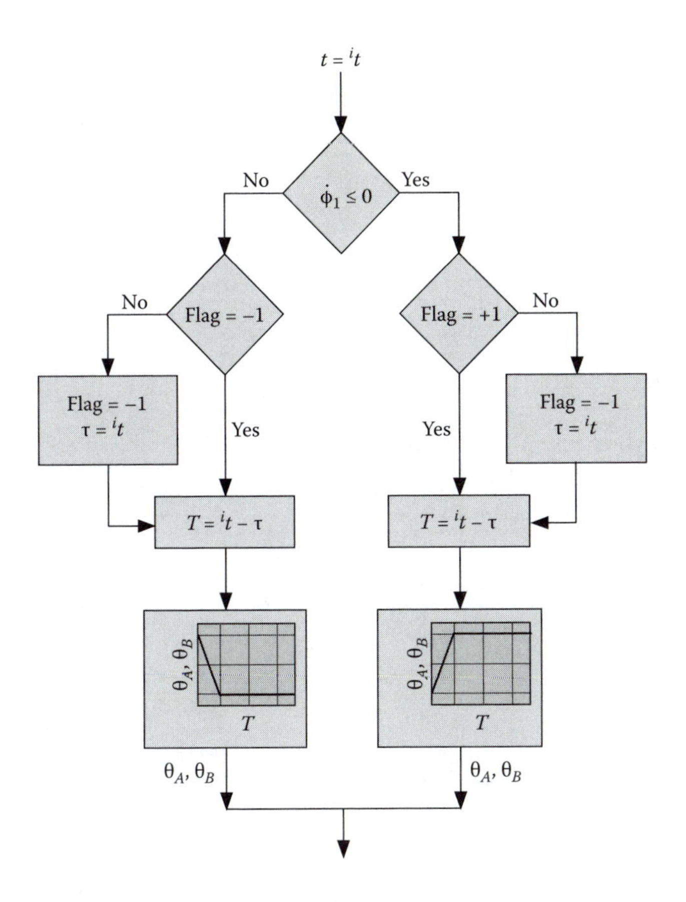

FIGURE 15.23 A flowchart for the body-pump strategy.

integrated numerically, we cannot guarantee that we will have an integration time ${}^it$ at which $\dot{\phi}_1$ is exactly zero. Therefore, to deal with this problem, we suggest the logic that is presented in the flow diagram of Figure 15.23.

In the flow diagram of Figure 15.23, we have introduced a flag that is set to +1 or –1 depending on whether the swing is going forward or backward. Assume that the swing is going forward; flag = +1 and $\dot{\phi}_1 < 0$. Then the swing reaches its limit and starts going backward; flag is still +1 but $\dot{\phi}_1 > 0$. This is the instant at which we should initiate the bending of the leg-arm driver function. We set flag = –1 and also save that instant of time as $\tau = {}^it$. For the next Δt seconds the arm and leg are bent, and after that they are kept in their bent configuration until the swing reaches the other end. Then, the reverse process will start.

We can perform a forward dynamic analysis from a starting configuration to the right of the vertical (e.g., 30°) with zero initial velocities. We should experiment with different values of Δt, and also revise the recommended strategy if necessary to make the simulation more realistic.

A Mass Center and Moment of Inertia

Location of mass center and moment of inertia for uncommon and complex shaped objects can be found experimentally or by a CAD software. If an uncommon shape can be split into several common geometric shapes, one could determine the location of the mass center and the moment of inertia by hand or by a simple computer program. In this appendix formulas for computing the moment of inertia for several commonly shaped objects are provided. Two simple processes are also presented to determine the location of mass center and the moment of inertia for bodies with uncommon geometric shapes.

A.1 COMMON GEOMETRIC SHAPES

The position of the mass center and the mass moment of inertia about an axis passing through the mass center (i.e., polar moment of inertia) for some commonly shaped planar objects are shown in Table A.1. It is assumed that the mass is distributed uniformly. For other commonly shaped objects, the reader is referred to any textbook on the fundamentals of statics and dynamics.

A.1.1 Parallel Axis Theorem

The moment of inertia of a body with respect to any axis can be determined from the parallel axis theorem, if the polar moment of inertia is known. Assume that the axis is perpendicular to the plane and passes through point O. The polar moment of inertia, J, the mass, m, and the distance between O and the mass center, ℓ, are known. The parallel axis theorem provides the moment of inertia about the axis passing through O as

$$J^O = J + m\,\ell^2 \tag{A.1}$$

Example A.1

Determine the moment of inertia of the rectangular plate shown in Figure A.1 with respect to an axis passing through the corner A. The mass of the plate is 0.8 units.

Solution

The moment of inertia of the plate can be determined from Table A.1 as

$$J = \tfrac{1}{12}(1.0^2 + 4.0^2) = 1.42$$

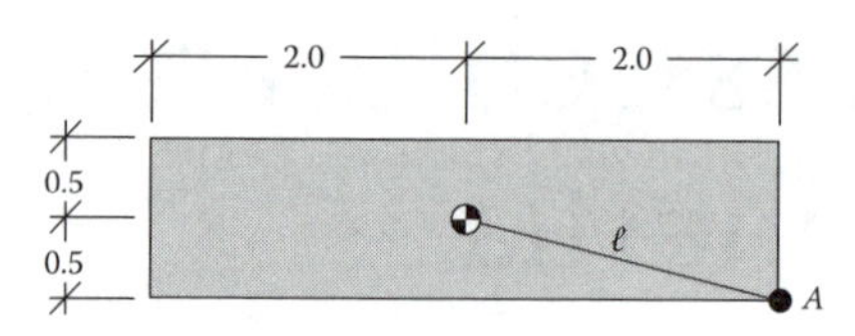

FIGURE A.1 A rectangular plate.

The distance between A and the mass center is $\ell = (0.5^2 + 2.0^2)^{\frac{1}{2}}$. The moment of inertia about A is computed as

$$J^A = 1.42 + 0.8(0.5^2 + 2.0^2) = 4.82$$

TABLE A.1
Polar Moments of Inertia for Some Commonly Shaped Planar Objects

Object	Shape and data	J
Slender Rod	$a/2$, $a/2$	$\frac{1}{12} m a^2$
Rectangular Plate	2.0, 2.0; 0.5, 0.5	$\frac{1}{12} m (a^2 + b^2)$
Ring	r	$m r^2$
Disk	r	$\frac{1}{2} m r^2$
Triangular plate	e, d, $b/2$, $b/2$, $c/2$, $c/2$, $a/2$, $a/2$	$\frac{1}{36} m (a^2 + b^2 + c^2)$ $= \frac{1}{18} m (a^2 + d^2 + e^2 - ae)$

A.2 COMPOSITE SHAPES

To determine the moment of inertia of an uncommonly shaped object, we need to split the object into several common geometric shapes. We then apply the procedures shown in the following subsections to determine the composite mass center and the composite moment of inertia.

A.2.1 Mass Center

The process of finding the mass center for a body with an uncommon geometric shape is simple. The shape is split into n (two or more) commonly shaped segments. The mass of each segment is determined as $m_i; i = 1, \cdots, n$. The mass of the composite object is computed as

$$m = m_1 + m_2 + \cdots \tag{A.2}$$

An x-y frame is defined at a convenient location. The x-y coordinates of each mass centers are measured and denoted as $\mathbf{r}_i^C; i = 1, \cdots, n$. Then, according to Eq. (4.4) the coordinates of the composite mass center is computed as

$$\mathbf{r}^C = \frac{1}{m}(m_1\mathbf{r}_1^C + m_2\mathbf{r}_2^C + \cdots) \tag{A.3}$$

As an example consider the object shown in Figure A.2(a). This object can be split into a rectangle, body (1), and a triangle, body (2), as shown in Figure A.2(b). The mass centers of these two segments are denoted as C_1 and C_2. The coordinates of these mass centers in an x-y frame are measured as $\mathbf{r}_1^C$ and $\mathbf{r}_2^C$. The masses of the two bodies are determined as m_1 and m_2. The total mass is $m = m_1 + m_2$ and the position of the composite mass center is computed as $\mathbf{r}^C = (m_1\mathbf{r}_1^C + m_2\mathbf{r}_2^C) / m$. The same composite shape can be split in a different way as shown in Figure A.2(c). Here we have a larger rectangle and a triangle that must be subtracted from the rectangle. The triangle is considered to have a *negative mass*. The total mass is determined as $m = m_1 - m_2$ and the coordinates of the mass center are computed as $\mathbf{r}^C = (m_1\mathbf{r}_1^C - m_2\mathbf{r}_2^C) / m$.

A.2.2 Moment of Inertia

The polar moments of inertia for a variety of common geometric shapes have been listed in Table A.1. The polar moment of inertia of bodies that have uncommon shapes may be determined with the aid of this table and the concept of composite shapes in a process similar to that of the mass center. The object is split into several common shapes and the mass and mass centers are identified. Then the moments of inertia of each segment, with respect to a single common axis (e.g., the axis passing through the composite mass center), are determined. The summation of these moments of inertia yields the moment of inertia of the composite body with respect to that single axis.

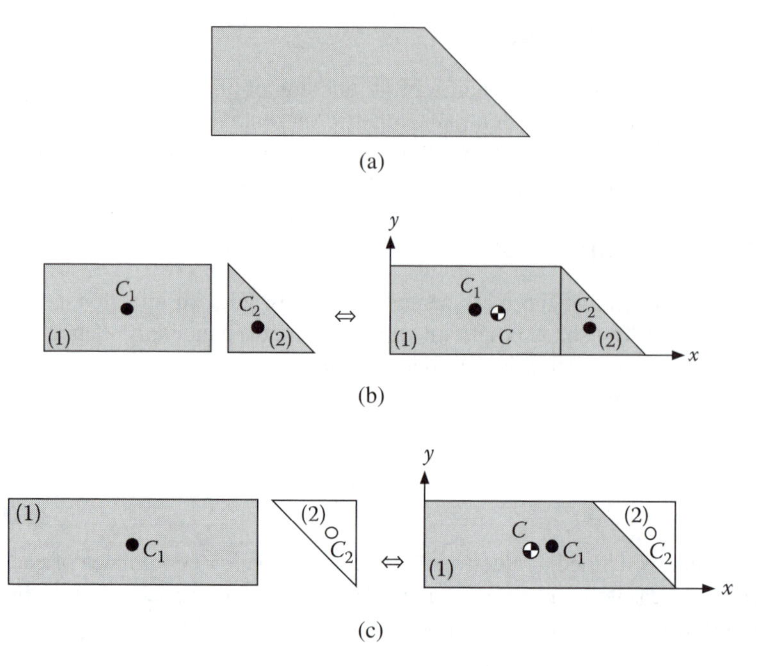

FIGURE A.2 An uncommon geometric shape and two ways to split it into two common shapes.

EXAMPLE A.2

Determine the moment of inertia of the bicycle frame, shown in Figure A.3(a), which is made of four tubular welded segments. The cross sections and materials of all four segments are identical. Assume that the mass of one unit length is ρ. Take measurements directly from the figure.

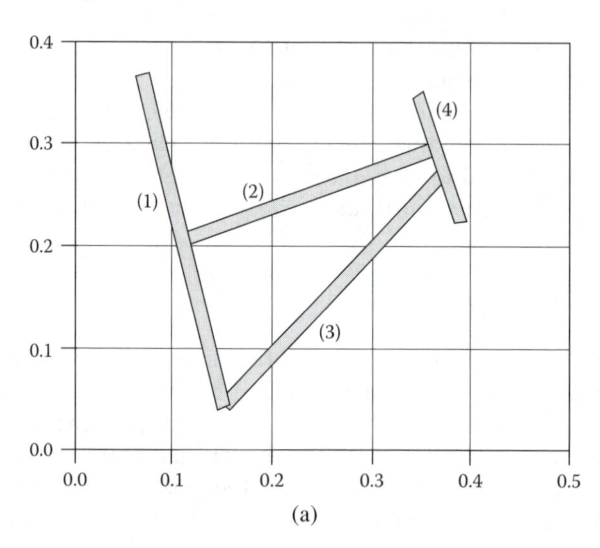

FIGURE A.3 A bicycle frame.

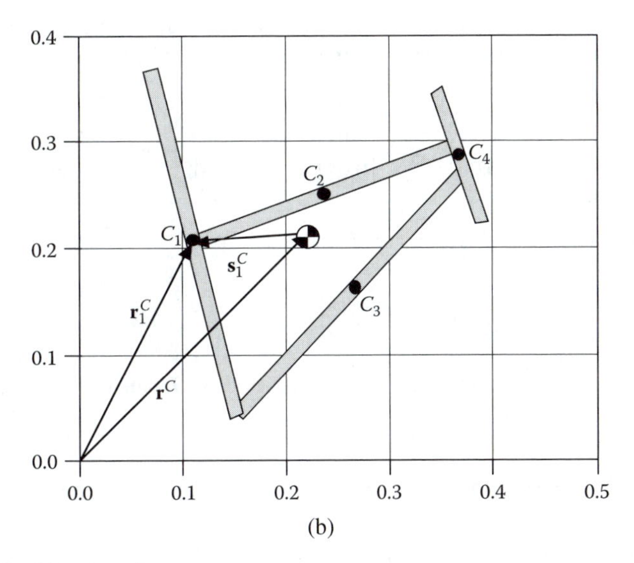

FIGURE A.3 (Continued)

Solution

The four bodies (tubes) are numbered for easy reference. The length of the bodies are measured to be

$$\ell_1 = 0.33,\ \ell_2 = 0.25,\ \ell_3 = 0.30,\ \ell_4 = 0.13$$

The coordinates of the mass center of each body is marked on Figure A.3(b) and measured to be

$$\mathbf{r}_1^C = \begin{Bmatrix} 0.11 \\ 0.21 \end{Bmatrix},\ \mathbf{r}_2^C = \begin{Bmatrix} 0.24 \\ 0.25 \end{Bmatrix},\ \mathbf{r}_3^C = \begin{Bmatrix} 0.26 \\ 0.16 \end{Bmatrix},\ \mathbf{r}_4^C = \begin{Bmatrix} 0.36 \\ 0.28 \end{Bmatrix}$$

The mass of each body is determined to be

$$m_1 = 0.33\rho,\ m_2 = 0.25\rho,\ m_3 = 0.30\rho,\ m_4 = 0.13\rho$$

The mass of the frame is found as

$$m = m_1 + m_2 + m_3 + m_4 = 1.01\rho$$

The coordinates of the mass center of the frame is computed based on Eq. (A.3):

$$\mathbf{r}^C = \frac{1}{m}(m_1\mathbf{r}_1^C + m_2\mathbf{r}_2^C + m_3\mathbf{r}_3^C + m_4\mathbf{r}_4^C) = \begin{Bmatrix} 0.22 \\ 0.21 \end{Bmatrix}$$

This mass center is marked on Figure A.3(b). The mass center of each rod is positioned from the mass center of the frame as $\mathbf{s}_i^C = \mathbf{r}_i^C - \mathbf{r}^C; i = 1, \cdots, 4$. This yields

$$\mathbf{s}_1^C = \begin{Bmatrix} -0.11 \\ -0.00 \end{Bmatrix}, \ \mathbf{s}_2^C = \begin{Bmatrix} 0.02 \\ 0.04 \end{Bmatrix}, \ \mathbf{s}_3^C = \begin{Bmatrix} 0.04 \\ -0.05 \end{Bmatrix}, \ \mathbf{s}_4^C = \begin{Bmatrix} 0.14 \\ 0.06 \end{Bmatrix}$$

The magnitudes of these vectors are found to be

$$s_1^C = 0.11, \ s_2^C = 0.04, \ s_3^C = 0.07, \ s_4^C = 0.16$$

The moment of inertia of each body (rod), based on Table A.1, is $J_i = \frac{1}{12} m_i \ell_i^2$; $i = 1, \cdots, 4$. This yields

$$J_1 = 0.0030\rho^2, \ J_2 = 0.0013\rho^2, \ J_3 = 0.0022\rho^2, \ J_4 = 0.0002\rho^2$$

The moment of inertia of each rod with respect to C is computed based on the parallel-axis formula (i.e., $J_i^C = J_i + m_i s_i^2; i = 1, \cdots, 4$) as

$$J_1^C = 0.0069\rho^2, \ J_2^C = 0.0013\rho^2, \ J_3^C = 0.0024\rho^2, \ J_4^C = 0.0004\rho^2$$

The sum of these four moments of inertia yields the moment of inertia of the frame with respect to C:

$$J^C = J_1^C + J_2^C + J_3^C + J_4^C = 0.0109\rho^2$$

Depending on the value of the length mass density ρ, the exact numerical value of J^C can be determined.

References

Baumgarte, J., “Stabilization of Constraints and Integrals of Motion,” *Concept. Methods Appl. Mech. Eng.,* Vol. 1, 1972.

Blajer, W., “Elimination of Constraint Violation and Accuracy Aspects in Numerical Simulation of Multibody Systems,” *Multibody Systems Dynamics*, Vol. 7, No. 3, 265–284, 2002.

Chua, L. O., and Lin, P. M., *Computer-Aided Analysis of Electronic Circuits: Algorithms and Computational Techniques*, Prentice-Hall, Englewood Cliffs, N.J., 1975.

Goffman, C., *Calculus of General Variables*, Harper and Row, New York, 1965.

Nikravesh, P. E., *Computer-Aided Analysis of Mechanical Systems*, Prentice-Hall, Englewood Cliffs, N.J., 1988.

Norton, R. L., *Design of Machinery: An Introduction to the Synthesis and Analysis of Mechanisms and Machines, 3rd ed.*, McGraw -Hill, 2004.

Paul, B., *Kinematics and Dynamics of Planar Machinery*, Prentice-Hall, Englewood Cliffs, N.J., 1979.

Wehage, R. A., and Haug, E. J., “Generalized Coordinate Partitioning of Dimension Reduction in Analysis of Constrained Dynamic Systems,” *ASME J. Mech. Design*, Vol. 104, pp. 247–255, January 1982.

Yoon, S., Howe, R.M., and Greenwood, D.T., “Geometric Elimination of Constraint Violations in Numerical Simulation of Lagrangian Equations,” *Transaction of the ASME, J. of Mechanical Design*, Vol. 116, No. 4, 1058–2064, 1994.

Index

D

E

F

G

H

I

J

K

L

M

N

O

P

Q

R

S

T

U

V

W

Y